Universitext

Universitext

Universitext is a series of textbooks that presents material from a wide variety of mathematical disciplines at master's level and beyond. The books, often well class-tested by their author, may have an informal, personal even experimental approach to their subject matter. Some of the most successful and established books in the series have evolved through several editions, always following the evolution of teaching curricula, to very polished texts.

Thus as research topics trickle down into graduate-level teaching, first textbooks written for new, cutting-edge courses may make their way into *Universitext.*

More information about this series at http://www.springer.com/series/223

Sergey V. Lototsky • Boris L. Rozovsky

Stochastic Partial Differential Equations

Springer

Sergey V. Lototsky
Department of Mathematics
University of Southern California
Los Angeles
California, USA

Boris L. Rozovsky
Division of Applied Mathematics
Brown University
Providence
Rhode Island, USA

ISSN 0172-5939 ISSN 2191-6675 (electronic)
Universitext
ISBN 978-3-319-58645-8 ISBN 978-3-319-58647-2 (eBook)
DOI 10.1007/978-3-319-58647-2

Library of Congress Control Number: 2017942781

Mathematics Subject Classification (2010): 60H15, 35R60

Printed on acid-free paper

This Springer imprint is published by Springer Nature
The registered company is Springer International Publishing AG
The registered company address is: Gewerbestrasse 11, 6330 Cham, Switzerland

To our families

Preface

There are textbooks and there are research monographs. Some believe that a research monograph is supposed to bore rather than batter. The following quotation is attributed to Voltaire: "*Le secret d'etre ennuyeux s'est de tout dire*"—the art of being a bore consists in telling everything. A slight variation is attributed to a French proverb, which, in English translation, says that the art of being a bore consists in telling every detail. The reader might want to stop for a moment and think about the difference between *everything* and *every detail*, as well as the two main meanings of the verb *to bore*.

Our objective in this *textbook* is to discuss as much of the SPDE-related material as possible without going too much into the details: learning is not very effective when it is boring, and so, unlike a research monograph, we believe that a textbook should *batter* rather than *bore*. While the objective of a research monograph is to document the results, the objective of a graduate-level textbook is to prepare the reader for independent research. This preparation means providing the tools necessary not only to read and understand the current paper and monographs on the subject but also to create new mathematics. In other words, our hope is that this book can help the reader not only to understand the proofs of existing theorems about SPDEs but also to formulate and prove new theorems.

Most of the book is about linear equations; a separate volume dealing with nonlinear equation is planned for the future.

Our intention was to make the presentation as self-contained as possible; a knowledgable reader coming across a definition of the standard Brownian motion or the statement of Parseval's identity should not take it personally. We also tried to make the book readable not just from the beginning but also from a random place (for possible use as a reference). As a result, there are inevitable repetitions and redundancies, as well as page references in addition to the number of the cited formula or theorem. A reader with a good memory reading the book from the beginning should not take this personally either. Some redundancy is also built into the subject index.

A subtle message is sent to the reader at the end of many proofs: a *completed* proof is somewhat more complete than a *concluded* one, and a reader might want to spend some extra time thinking about the *concluded* proofs.

Problems and exercises facilitate transition of a subject from the research level to the level of a graduate or even undergraduate course. As the subject of SPDEs is currently making this transition, the book attempts to present enough exercise material to fill enough potential exams and homework assignments. The distinction between a problem and an exercise is not always clear. One possible distinction is provided by the following example: to solve the quadratic equation $ax^2+bx+c=0$ is a problem; to solve $x^2+x+1=0$ is an exercise. In this book, the approach to problems and exercises is somewhat different. Exercises appear throughout the text and are usually directly connected to the material discussed at the particular place in the text. The question is usually to verify something, so that the reader already knows the answer and, if pressed for time, can move on. Accordingly, there are no solutions for the exercises (but there are often hints on how to proceed). We also realize that exercises can present an extra challenge for those brave enough to actually teach a class using this book.

The letter label of each exercise is supposed to convey three pieces of information at once: the level of difficulty, the degree of relevance, and the ambition of the reader. In other words, exercises labeled with a "C" are the easiest, most relevant to the core material, and are intended for all readers; in fact, many of the "C" exercises could have been labeled as examples or (easy) lemmas, and the idea is that a different label might force the reader to think a bit more about the corresponding result. Exercises labeled with an "A" are the most difficult, least relevant, and are intended for the most ambitious readers. Exercises labeled with a "B" fall somewhere in between in all three categories. Of course, the gradation represents the subjective views of the authors.

Problems are collected at the end of each chapter and usually point to the topics that were not discussed in the main text. While most problems have at least an outline of the solution at the end of the book,

- Some of the solutions are not at the level that would earn the reader a perfect score on a homework assignment,
- Even if the solution looks complete, the reader should not treat it as a final judgement but rather as a call for further thinking and investigation.

And even if not attempting to solve any problems, the reader is encouraged to look through both the problems and solutions, as some of the potentially interesting information is hidden there. When a problem (or, for that matter, an exercise) asks to *show* something, the solution assumes a derivation of the result. An alternative, which is not necessarily an easy (or efficient) way out,[1] is to *verify* that the result is true by taking it for granted and somehow confirming that it makes sense.

---

[1]For example, would the reader rather plug in the function $y(t)=e^t\cos(t)$ into $y''-2y'+2y=0$ to verify that it satisfies the equation or solve the equation using the general theory and recognize the function as a particular solution?

As far as teaching a class, both authors did so on several occasions, more or less following the order of the material in the book. It is indeed possible to cover most of the material in about 40 h of lectures, as long as not too much time is spent on the general discussion of stochastic analysis in infinite dimensions. A more detailed description, both of the structure of this book and of the corresponding class, is in Sect. 1.2.7.

To summarize, the objective of this book is not to present all the result about stochastic partial differential equations, but instead to provide the reader with the necessary tools to understand those results (by reading other sources) and maybe even to discover a few new ones.

The authors gratefully acknowledge the support of several grants during the period the book was written: ARO (DAAD19-02-1-0374) and NSF (DMS-0237724 (CAREER) and DMS-0803378) [SL]; AFOSR (5-21024 (inter), FA9550-09-1-0613) ARO (DAAD19-02-1-0374, W911NF-07-1-0044, W911NF-13-1-0012, W911-16-1-0103), NSF (DMS 0604863, DMS 1148284), ONR (N0014-03-1-0027, N0014-07-1-0044, OSD/AFOSR 9550-05-1-0613, and SD Grant 5-21024 (inter.)) [BR]. SL gratefully acknowledges hospitality of the Division of Applied Mathematics at Brown University on several occasions that were crucial to the success of the project.

Los Angeles, CA, USA — Sergey V. Lototsky
Providence, RI, USA — Boris L. Rozovsky
March 2017

Contents

Chapter 1
Introduction

1.1 Getting Started

1.1.1 Conventions and Notations

We use the same notation x for a point in the real line $\mathbb{R}$ or in a d-dimensional Euclidean space $\mathbb{R}^{\mathrm{d}}$. For $x = (x_1, \ldots, x_{\mathrm{d}}) \in \mathbb{R}^{\mathrm{d}}$, $|x| = \sqrt{x_1^2 + \ldots + x_{\mathrm{d}}^2}$; for $x, y \in \mathbb{R}^{\mathrm{d}}$, $xy = x_1y_1 + \ldots + x_{\mathrm{d}}y_{\mathrm{d}}$. Integral over the real line can be written either as $\int_{\mathbb{R}}$ or as $\int_{-\infty}^{+\infty}$. Sometimes, when there is no danger of confusion, the domain of integration, in any number of dimensions, is omitted altogether.

The space of continuous mappings from a metric space A to a metric space B is denoted by $\mathcal{C}(A; B)$. For example, given a Banach space X, $\mathcal{C}\big((0,T); X\big)$ is the collection of continuous mappings from $(0, T)$ to X. When $B = \mathbb{R}$, we write $\mathcal{C}(A)$. For a positive integer n, $\mathcal{C}^n(A)$ is the collection of functions with n continuous derivatives; for $\gamma \in (0, 1)$ and $n = 0, 1, 2\ldots$, $\mathcal{C}^{n+\gamma}(A)$ is the collection of functions with n continuous derivatives such that derivatives of order n are Hölder continuous of order γ. Similarly, $\mathcal{C}^{\infty}(A)$ is the collection of infinitely differentiable functions and $\mathcal{C}_0^{\infty}(A)$ is the collection of infinitely differentiable functions with compact support in A.

We will encounter the space $\mathcal{S}(\mathbb{R}^{\mathrm{d}})$ of smooth rapidly decreasing functions and its dual $\mathcal{S}'(\mathbb{R}^{\mathrm{d}})$, the space of generalized functions. Recall that $f \in \mathcal{S}(\mathbb{R}^{\mathrm{d}})$ if and only if $f \in \mathcal{C}^{\infty}(\mathbb{R}^{\mathrm{d}})$ and

$$\lim_{|x|\to\infty} |x|^N |D^n f(x)| = 0$$

for all non-negative integers N and all partial derivatives $D^n f$ of every order n. When necessary (for example, here), we use the convention $D^0 f = f$. For specific partial

S.V. Lototsky, B.L. Rozovsky, *Stochastic Partial Differential Equations*,
Universitext, DOI 10.1007/978-3-319-58647-2_1

derivatives, we use the standard notations

$$u_t = \frac{\partial u}{\partial t},\ u_{x_ix_j} = \frac{\partial^2 u}{\partial x_i \partial x_j};$$

also $\dot{v} = dv/dt$.

The Laplace operator is denoted by $\boldsymbol{\Delta}$:

$$\boldsymbol{\Delta} f = \sum_{i=1}^{\mathrm{d}} f_{x_ix_i}.$$

The symbol $\mathfrak{i}$ denotes the imaginary unit: $\mathfrak{i} = \sqrt{-1}$.

Notation $a_k \sim b_k$ means $\lim_{k\to\infty} a_k/b_k = c \in (0,\infty)$, and if $c = 1$, we will emphasize it by writing $a_k \simeq b_k$. Notation $a_k \asymp b_k$ means $0 < c_1 \leq a_k/b_k \leq c_2 < \infty$ for all sufficiently larger k. The same notations $\sim$, $\simeq$, and $\asymp$ can be used for functions. For example, as $x \to \infty$, we have

$$2x^2 + x \sim x^2,\ x + 5 \simeq x,\ x^2(2 + \sin x)/(1 + x) \asymp x.$$

Following a different set of conventions, $\eta \sim \mathcal{N}(m, \sigma^2)$ mens that η is a Gaussian (or normal) random variable with mean m and variance σ^2; recall that $\mathcal{N}(0, 1)$ is called a standard Gaussian (or normal) random variable.

Here are several important simplifying conventions we use in this book:

- We do not distinguish various modifications of either deterministic or random functions. Thus, in this book, all functions from the Sobolev space $H^1(\mathbb{R})$ are continuous and so are all trajectories of the standard Brownian motion.
- We will write equations driven by Wiener process either as $du = \ldots dw$ or as $\dot{u} = \ldots \dot{w}$ (or $u_t = \ldots \dot{w}$, if it is a PDE).

With apologies to the set theory experts, we often use the words "set" and "collection" interchangeably.

We fix the `stochastic basis` $\mathbb{F} = (\Omega, \mathcal{F}, \{\mathcal{F}_t\}_{t\geq 1}, \mathbb{P})$ with the `usual assumptions` ($\mathcal{F}_t$ is right-continuous: $\mathcal{F}_t = \bigcap_{\varepsilon>0} \mathcal{F}_{t+\varepsilon}$, and $\mathcal{F}_0$ contains all $\mathbb{P}$-negligible sets, that is, $\mathcal{F}_0$ contains every sub-set of Ω that is a sub-set of an element from $\mathcal{F}$ with $\mathbb{P}$-measure zero.).

1.1.2 Dealing with Noise

A stochastic ordinary differential equation defines a function of time and is driven by a noise process in time. A stochastic partial differential equation describes a function of time and space and is driven by a noise process in time and space. In what follows, we outline a construction of such space-time noise processes.

Let $\xi_k,\ k \geq 1$, be independent standard Gaussian random variables; see Krylov [118, Lemma II.2.3] for the proof that countable many standard Gaussian random variables can exist on a suitable stochastic basis. If $\{\mathfrak{m}_k(t),\ k \geq 1\}$ is an orthonormal basis in $L_2((0,T))$, then

$$w(t) = \sum_{k\geq 1}\left(\int_0^t \mathfrak{m}_k(s)ds\right)\xi_k \tag{1.1.1}$$

is a standard Brownian motion: a Gaussian process with zero mean and covariance $\mathbb{E}\big(w(t)w(s)\big) = \min(t,s)$.

Exercise 1.1.1 (C) Verify that $\mathbb{E}w(t)w(s) = \min(t,s)$. **Hint**. Note that $\int_0^t \mathfrak{m}_k(s)ds$ is the Fourier coefficient of the indicator function of the interval $[0,t]$ and use the Parseval identity connecting the inner product of two functions to their Fourier coefficients.

We now take this construction to the next level by considering a collection $\{w_k = w_k(t),\ k \geq 1\}$ of independent standard Brownian motions, $t \in [0,T]$ and an orthonormal basis $\{\mathfrak{h}_k(x),\ k \geq 1,\ x \in G\}$ in the space $L_2(G)$, with $G = (0,L)^{\mathrm{d}} = (0,L) \times \cdots \times (0,L)$, a d-dimensional hyper-cube. For $x = (x_1, x_2, \ldots, x_{\mathrm{d}})$, define

$$h_k(x) = \int_0^{x_1}\int_0^{x_2}\cdots\int_0^{x_{\mathrm{d}}} \mathfrak{h}_k(r_1,\ldots,r_{\mathrm{d}})dr_{\mathrm{d}}\ldots dr_1.$$

Then the process

$$W(t,x) = \sum_{k\geq 1} h_k(x)w_k(t) \tag{1.1.2}$$

is Gaussian,

$$\mathbb{E}W(t,x) = 0,\ \ \mathbb{E}\big(W(t,x)W(s,y)\big) = \min(t,s)\prod_{k=1}^{\mathrm{d}}\min(x_k,y_k). \tag{1.1.3}$$

Exercise 1.1.2 (C) Verify (1.1.3).

We call the process W from (1.1.2) the `Brownian sheet`.

The derivative of the standard Brownian motion, while does not exist in the usual sense, is the standard model of the Gaussian white noise $\dot{w}$. The formal term-by-term differentiation of the series in (1.1.1) suggests a representation

$$\dot{w}(t) = \sum_{k\geq 1}\mathfrak{m}_k(t)\xi_k. \tag{1.1.4}$$

While the series certainly diverges, it does define a random generalized function on $L_2((0,T))$ according to the rule

$$\dot{w}(f) = \sum_{k\geq 1} f_k\xi_k,\ f_k = \int_0^T f(t)\mathfrak{m}_k(t)dt. \tag{1.1.5}$$

Exercise 1.1.3 (C)

(a) Verify that the series in (1.1.5) converges with probability one to a Gaussian random variable with zero mean and variance $\sum_k f_k^2 = \int_0^T f^2(t)dt$.
(b) Let w be a standard Brownian motion and $f \in L_2((0,T))$ a non-random function. Show that the random variables $\xi_k = \int_0^T \mathfrak{m}_k(t)dw(t),\ k \geq 1$, are iid standard normal. Then use these random variables to define $\dot{w}$ according to (1.1.4) and verify that $\int_0^T f(s)dw(s) = \dot{w}(f)$.

Let us now take (1.1.4) to the next level and write

$$\dot{W}(t,x) = \sum_{k\geq 1} \mathfrak{h}_k(x)\dot{w}_k(t), \tag{1.1.6}$$

where $\{\mathfrak{h}_k,\ k \geq 1\}$ is an orthonormal basis in $L_2(G)$ and $G \subseteq \mathbb{R}^{\mathrm{d}}$ is an open set. We call the process $\dot{W}$ the (Gaussian) `space-time white noise`. It is a random generalized function on $L_2((0,T) \times G)$:

$$\dot{W}(f) = \sum_{k\geq 1} \int_0^T \left(\int_G f(t,x)\mathfrak{h}_k(x)dx \right) dw_k(t). \tag{1.1.7}$$

Sometimes, an alternative notation is used for $\dot{W}(f)$:

$$\dot{W}(f) = \int_0^T \int_G f(t,x)dW(t,x). \tag{1.1.8}$$

Even more confusing, $(\dot{W},f)$ and $\dot{W}_f$ can also be used to denote $\dot{W}(f)$, although we use these notations for (slightly) different purposes.

Exercise 1.1.4 (C)

(a) Verify that

$$\mathbb{E}\big(\dot{W}(f)\big)^2 = \int_0^T \int_G f^2(t,x)dxdt. \tag{1.1.9}$$

(b) Verify that

$$\dot{W}(t,x) = \sum_{k,n} \mathfrak{m}_k(t)\mathfrak{h}_n(x)\xi_{k,n},$$

where $\xi_{k,n} = \dot{W}(\mathfrak{m}_k\mathfrak{h}_n)$.

Unlike the Brownian sheet, space-time white noise $\dot{W}$ is defined on every domain $G \subseteq \mathbb{R}^{\mathrm{d}}$ and not just on hyper-cubes $(0,L)^{\mathrm{d}}$, as long as we can find an orthonormal basis $\{\mathfrak{h}_k, k \geq 1\}$ in $L_2(G)$.

Alternative description of the Gaussian white noise is as a zero-mean Gaussian process $\dot{w} = \dot{w}(t)$ such that $\mathbb{E}\dot{w}(t)\dot{w}(s) = \delta(t-s)$, where δ is the Dirac delta function. Similarly, we have

$$\mathbb{E}\dot{W}(t,x)\dot{W}(s,y) = \delta(t-s)\delta(x-y).$$

To construct noise that is white in time and colored in space, take a sequence of non-negative numbers $\{q_k,\ k \geq 1\}$ and define

$$\dot{W}^{\mathrm{Q}}(t,x) = \sum_{k\geq 1} q_k \mathfrak{h}_k(x)\dot{w}_k(t), \tag{1.1.10}$$

where $\{\mathfrak{h}_k,\ k \geq 1\}$ is an orthonormal basis in $L_2(G)$, $G \subseteq \mathbb{R}^{\mathrm{d}}$. The noise is called `finite-dimensional` if $q_k = 0$ for all sufficiently large k.

Exercise 1.1.5 (C) Verify that if

$$\sum_{k\geq 1} q_k^2 \sup_{x\in G} \mathfrak{h}_k^2(x) < \infty$$

then the function $q(x,y) = \sum_{k\geq 1} q_k^2 \mathfrak{h}_k(x)\mathfrak{h}_k(y)$ is well defined and

$$\mathbb{E}\dot{W}^{\mathrm{Q}}(t,x)\dot{W}^{\mathrm{Q}}(s,y) = \delta(t-s)q(x,y).$$

Similarly, expressions

$$\dot{W}(x) = \sum_{k\geq 1} \mathfrak{h}_k(x)\xi_k \tag{1.1.11}$$

and

$$\dot{W}^{\mathrm{Q}}(x) = \sum_{k\geq 1} q_k \mathfrak{h}_k(x)\xi_k,$$

define, respectively, the Gaussian white noise and Gaussian colored noise in space. Similar to (1.1.8), we write $\dot{W}(f)$ or $\int_G f(x)dW(x)$ to denote $\sum_{k\geq 1} f_k\xi_k$, with $f_k = \int_G f(x)\mathfrak{h}_k(x)dx$. According to (1.1.11), if $\dot{W}$ is a Gaussian white noise in space and $\mathfrak{h}_k,\ k \geq 1$ is an orthonormal basis, then $\xi_k = \dot{W}(\mathfrak{h}_k),\ k \geq 1$, are iid standard Gaussian random variables.

Throughout this book we will be mostly working with Gaussian noise $\dot{W}$ and $\dot{W}^{\mathrm{Q}}$, although two other types of noise are becoming increasingly popular in the study of SPDEs: the fractional Gaussian noise and Lévy noise.

1.1.3 A Few Useful Equalities

`Itô formula`: If $F = F(x)$ is a smooth function and $w = w(t)$ is a standard Brownian motion, then

$$F\big(w(t)\big) = F(0) + \int_0^t F'\big(w(s)\big)dw(s) + \frac{1}{2}\int_0^t F''\big(w(s)\big)ds.$$

`Itô isometry`: if w is a standard Brownian motion and f, an adapted process, then

$$\mathbb{E}\left(\int_0^T f(t)dw(t)\right)^2 = \int_0^T \mathbb{E}f^2(t)dt.$$

`Fourier transform`:

$$\widehat{f}(y) = \frac{1}{(2\pi)^{\mathrm{d}/2}}\int_{\mathbb{R}^{\mathrm{d}}} f(x)e^{-\mathrm{i}xy}dx,\ \mathrm{i} = \sqrt{-1}.$$

Recall that $\widehat{f}$ is defined on the generalized functions from $\mathcal{S}'(\mathbb{R}^{\mathrm{d}})$ by $(\widehat{f}, \varphi) = (f, \widehat{\varphi})$, $f \in \mathcal{S}'(\mathbb{R}^{\mathrm{d}})$, $\varphi \in \mathcal{S}(\mathbb{R}^{\mathrm{d}})$ (for a quick review, see Rauch [193, Sect. 2.4]).
`Inverse Fourier transform`:

$$\check{f}(y) = \frac{1}{(2\pi)^{\mathrm{d}/2}}\int_{\mathbb{R}^{\mathrm{d}}} f(x)e^{\mathrm{i}xy}dx = \widehat{f}(-y).$$

`Standard normal density is unchanged under the Fourier transform`:

$$\frac{1}{\sqrt{2\pi}}\int_{-\infty}^{\infty} e^{-x^2/2}e^{-\mathrm{i}xy}dx = \frac{1}{\sqrt{2\pi}}e^{-y^2/2} \tag{1.1.12}$$

For three different ways to establish this, see Andrews et al. [2, Exercise 1 for Chap. 6].
`Parseval's identity`: If $\{\mathfrak{h}_k,\ k \geq 1\}$ is an orthonormal basis in a Hilbert space X and $f \in X$, then

$$\sum_{k\geq 1}(f, \mathfrak{h}_k)_X^2 = \|f\|_X^2.$$

The result is essentially an infinite-dimensional Pythagorean theorem and is named after the French mathematician MARC-ANTOINE PARSEVAL DES CHÊNES (1755–1836).

Exercise 1.1.6 (C) Let $\{\mathfrak{m}_k(t),\ k \geq 1,\ t \in [0,T]$, be an orthonormal basis in $L_2((0,T))$. Show that

$$\sum_{k=1}^{\infty}\left(\int_0^t \mathfrak{m}_k(s)ds\right)^2 = t. \tag{1.1.13}$$

Hint. $\int_0^t \mathfrak{m}_k(s)ds$ is the Fourier coefficient of what function?

`Plancherel's identity` or `isometry of the Fourier transform`: if f is a smooth function with compact support in $\mathbb{R}^{\mathrm{d}}$ and

$$\widehat{f}(y) = \frac{1}{(2\pi)^{\mathrm{d}/2}}\int_{\mathbb{R}^{\mathrm{d}}} f(x)e^{-\mathrm{i}xy}dx,$$

then

$$\int_{\mathbb{R}^{\mathrm{d}}} |f(x)|^2 dx = \int_{\mathbb{R}^{\mathrm{d}}} |\widehat{f}(y)|^2 dy.$$

This result is essentially a continuum version of Parseval's identity and is named after the Swiss mathematician MICHEL PLANCHEREL (1885–1967).

`Stirling's formula` for the `Gamma function`

$$\Gamma(x+1) = \sqrt{2\pi x}\left(\frac{x}{e}\right)^x\left(1 + \frac{1}{12x} + \frac{1}{288x^2} + +O\left(\frac{1}{x^3}\right)\right), x \to +\infty,$$

named after the Scottish mathematician JAMES STIRLING (1692–1770). Recall that

$$\Gamma(x) = \int_0^{\infty} t^{x-1}e^{-t}dt,\ x > 0,$$

and

$$\Gamma(n+1) = n!,\ n = 0, 1, 2, \ldots.$$

1.1.4 A Few Useful Inequalities

Below, we summarize several inequalities that are always good to know: Burkholder-Davis-Gundy, epsilon, Gronwall, Hölder, and Jensen, and recall one particular embedding theorem for Sobolev spaces.

To state the Burkholder-Davis-Gundy inequality, recall that we are always working on the stochastic basis $\mathbb{F} = (\Omega, \mathcal{F}, \{\mathcal{F}_t\}_{t\geq 0}, \mathbb{P})$. A `square-integrable martingale` on $\mathbb{F}$ is a process $M = M(t)$ with values in $\mathbb{R}^{\mathrm{d}}$ such that $M(0) = 0$, $\mathbb{E}|M(t)|^2 < \infty$ and $\mathbb{E}\big(M(t)|\mathcal{F}_s\big) = M(s)$ for all $t \geq s \geq 0$. The `quadratic` variation of M is the continuous non-decreasing real-valued process $\langle M\rangle$ such that $|M|^2 - \langle M\rangle$ is a martingale. A stopping (or Markov) time on $\mathbb{F}$ is a non-negative random variable τ such that $\{\omega : \tau(\omega) > t\} \in \mathcal{F}_t$ for all $t \geq 0$.

Recall that $a \wedge b$ means $\min(a, b)$.

Theorem 1.1.7 (Burkholder-Davis-Gundy (BDG) Inequality) *For every $p \in (0, +\infty)$, there exist two positive real numbers c_p, C_p such that, for every continuous square-integrable martingale $M = M(t)$ with values in $\mathbb{R}^{\mathrm{d}}$ and $M(0) = 0$, and for every stopping time τ,*

$$c_p\mathbb{E}\langle M\rangle^{p/2}(\tau) \leq \mathbb{E}\sup_t |M(t \wedge \tau)|^p \leq C_p\mathbb{E}\langle M\rangle^{p/2}(\tau). \tag{1.1.14}$$

In particular, if w is a standard Brownian motion and f, an adapted process, then

$$\mathbb{E}\sup_{0<t<T}\left|\int_0^t f(s)dw(s)\right|^p \leq C_p\mathbb{E}\left(\int_0^T f^2(t)dt\right)^{p/2}; \tag{1.1.15}$$

more generally,

$$\mathbb{E}\sup_{0<t<T}\left|\sum_{k\geq 1}\int_0^t f_k(s)dw_k(s)\right|^p \leq C_p\mathbb{E}\left(\sum_{k\geq 1}\int_0^T f_k^2(t)dt\right)^{p/2}. \tag{1.1.16}$$

Proof See, for example, Krylov [118, Theorem IV.4.1].

The result is named after three American mathematicians, DONALD L. BURKHOLDER, BURGESS J. DAVIS, and RICHARD F. GUNDY, and can be traced to their joint paper [19]. A more detailed historical analysis shows that Burkholder and Gundy [18] established the result for $p \neq 1$, and Davis, who was a Ph.D. student of Burkholder, handled $p = 1$. For more inequalities of this type, see, for example, Liptser and Shiryaev [138, Sect. I.9]. We will re-visit Theorem 1.1.7 later in the context of Hilbert space-valued martingales (see Theorem 3.2.47 on page 123).
The epsilon inequality is as useful as it is simple: for every $a, b \in \mathbb{R}$ and $\varepsilon > 0$,

$$|ab| \leq \varepsilon a^2 + \varepsilon^{-1}b^2. \tag{1.1.17}$$

Indeed, $0 \leq (\sqrt{\varepsilon}|a| - |b|/\sqrt{\varepsilon})^2 = \varepsilon a^2 + \varepsilon^{-1}b^2 - 2|ab|$.

A more general epsilon inequality is as follows;

$$|ab| \leq \varepsilon|a|^p + (\varepsilon p)^{-q/p}q^{-1}|b|^q,\ \frac{1}{p} + \frac{1}{q} = 1. \tag{1.1.18}$$

Equivalently,

$$|a|^{1/p}\,|b|^{1/q} \leq \varepsilon|a| + C(\varepsilon)|b|. \tag{1.1.19}$$

This is true because $|ab| \le |a|^p/p + |b|^q/q$. See also comments about the Hölder inequality below.

The following inequality is used very often in the study of evolution equations.

Theorem 1.1.8 *If $f = f(t)$, $t \in [0, T]$ is a non-negative integrable function and C_1, C_2 are non-negative numbers so that*

$$f(t) \le C_1 \int_0^t f(s)ds + C_2,\ t \in [0, T], \tag{1.1.20}$$

then

$$f(t) \le C_2(1 + C_1 t e^{C_1 t}),\ t \in [0, T]. \tag{1.1.21}$$

In particular,

$$\sup_{0<t<T} f(t) \le C_2(1 + C_1 T e^{C_1 T}). \tag{1.1.22}$$

Proof If $F(t) = \int_0^t f(s)ds$ then $F' \le C_1 F + C_2$, $F(0) = 0$, and

$$(F(t)e^{-C_1 t})' = e^{-C_1 t}(F' - C_1 F) \le C_2 e^{-C_1 t} \le C_2,$$

so that $F(t) \le tC_2 e^{C_1 t}$. By assumption $f(t) \le C_1 F(t) + C_2$, and the result follows.

Theorem 1.1.8 can go under several names: Gronwall's lemma, Gronwall's inequality, the Gronwall-Bellman inequality, etc. Note that the right-hand side of (1.1.21) can be written in many other ways, depending on how explicitly we want to track down dependence on t. The Swedish-American mathematician THOMAS HAKON GRONWALL (Grönwall) (1877–1932) established the result under an assumption very similar to (1.1.20) (see [67]); the American mathematician RICHARD ERNEST BELLMAN (1920–1984) established a similar result independently and under a more general assumption (see [9, Lemma 1]).

Next, we give three versions of the `Hölder inequality`:

$$\mathbb{E}|\xi\eta| \le \left(\mathbb{E}|\xi|^p\right)^{1/p}\left(\mathbb{E}|\eta|^q\right)^{1/q} \quad \texttt{(probabilistic version)} \tag{1.1.23}$$

$$\int |f(x)g(x)|dx \le \left(\int |f(x)|^p dx\right)^{1/p}\left(\int |g(x)|^q dx\right)^{1/q} \quad \texttt{(integral version)} \tag{1.1.24}$$

$$\sum_k |a_k b_k| \le \left(\sum_k |a_k|^p\right)^{1/p}\left(\sum_k |b_k|^q\right)^{1/q} \quad \texttt{(discrete version)} \tag{1.1.25}$$

Everywhere $1 < p, q < \infty$ and $\frac{1}{p} + \frac{1}{q} = 1$.

Proof Note that $|ab| \leq \frac{|a|^p}{p} + \frac{|b|^q}{q}$ (using $|ab| = \exp((1/p)\ln|a|^p + (1/q)\ln|b|^q)$ and convexity of the exponential function), and that it is enough to consider the case when (for example, in the probabilistic version) $\mathbb{E}|\xi|^p = \mathbb{E}|\eta|^q = 1$.

The inequality is named after the German mathematician OTTO LUDWIG HÖLDER (1859–1937), who discovered a version of it in 1884 while studying convergence of Fourier series. The particular case $p = q = 2$ is also known as the Cauchy-Scwarz (or Cauchy-Bunyakovsky-Schwarz) inequality and has an interesting history of its own.

Exercise 1.1.9 (A)

(a) Explain why inequality $\mathbb{E}|\xi\eta| \leq \mathbb{E}|\xi|^p\,\mathbb{E}|\eta|^q$ is impossible. **Hint.** Instead of ξ, consider $c\xi$, $c > 0$.
(b) Use a scaling argument to convince yourself that, if (1.1.24) is to hold for integrals over $\mathbb{R}$, then one has to have $1/p+1/q = 1$. **Hint.** Consider $f(x) \to f(x/c)$, $g(x) \to g(x/c)$.

Next (as we go in alphabetical order), we have

Theorem 1.1.10 (Jensen's Inequality) *If $F = F(x)$ is a convex function and $\mathbb{E}|\xi| < \infty$, then*

$$F(\mathbb{E}\xi) \leq \mathbb{E}F(\xi). \tag{1.1.26}$$

Proof Out of many characterizations of a convex function, the following one works the best for the purpose of the proof: for every x_0 there exists a number $\lambda = \lambda(x_0)$ such that, for all x,

$$F(x) \geq F(x_0) + \lambda(x_0)(x - x_0) \tag{1.1.27}$$

(if F is smooth, then $\lambda(x_0) = F'(x_0)$, and (1.1.27) says that the graph of F is above the tangent to the graph at x_0). Now put $x_0 = \mathbb{E}\xi$, $x = \xi$, and take expectation on both sides.

JOHAN LUDWIG WILLIAM VALDEMAR JENSEN (1859–1925) was from Denmark and worked for the Copenhagen Telephone Company most of his life. He was doing mathematics essentially as a hobby. In his 1906 paper [95] he considers separately the cases of discrete and absolutely continuous random variables ξ, but without using any probabilistic terminology. Still, the probabilistic interpretation of the result is not only convenient, but also natural: the underlying measure always has to be normalized to one.

For more on the epsilon, Gronwall, Hölder, Jensen, and some other related inequalities, see Evans [50, Appendix B].

There are many different spaces of generalized functions beside $\mathcal{S}'(\mathbb{R}^{\mathrm{d}})$: see, for example, Triebel [218]. Some of these spaces are associated with the Russian mathematician SERGEI L'VOVICH SOBOLEV (1908–1989), who studied the topic in the 1930s. A popular family of the Sobolev spaces is

$$H_p^\gamma(\mathbb{R}^{\mathrm{d}}) = \{f \in \mathcal{S}'(\mathbb{R}^{\mathrm{d}}) : \mathcal{F}^{-1}(\widehat{f}\, w_\gamma) \in L_p(\mathbb{R}^{\mathrm{d}})\},$$

where $1 \le p < \infty$, $\gamma \in \mathbb{R}$, $w_\gamma(y) = (1 + |y|^2)^{\gamma/2}$, $\widehat{f}$ is the Fourier transform of f, $\mathcal{F}^{-1}$ is the inverse Fourier transform;

$$\|f\|^p_{H_p^\gamma(\mathbb{R}^{\mathrm{d}})} = \|\mathcal{F}^{-1}(\widehat{f}\, w_\gamma)\|^p_{L_p(\mathbb{R}^{\mathrm{d}})} = \int_{\mathbb{R}^{\mathrm{d}}} \left|\mathcal{F}^{-1}(\widehat{f}\, w_\gamma)(x)\right|^p dx.$$

In the particular case $p = 2$, we will often omit the subscript p and write $H^\gamma(\mathbb{R}^{\mathrm{d}})$ instead of $H_2^\gamma(\mathbb{R}^{\mathrm{d}})$.

Given a domain $G \subset \mathbb{R}^{\mathrm{d}}$ with smooth boundary, the Sobolev space $H_p^\gamma(G), \gamma > 0$, is the collection of functions from $H_p^\gamma(\mathbb{R}^{\mathrm{d}})$, restricted to G. For more information on the Sobolev spaces in this book, see Example 3.1.9, page 79, about $H_2^\gamma(\mathbb{R}^{\mathrm{d}})$, and Example 3.1.29, page 87, about $H_2^\gamma(G)$.

The parameter γ corresponds to the number of generalized derivatives. One type of the `Sobolev embedding theorems` is a statement asserting that existence of sufficiently many generalized derivatives implies existence of a certain number of classical derivatives. For example,

Theorem 1.1.11 *If $r > 0$ is not an integer and $f \in H_p^{r+(\mathrm{d}/p)}(\mathbb{R}^{\mathrm{d}})$, then $f \in \mathcal{C}^r(\mathbb{R}^{\mathrm{d}})$.*

Proof See Triebel [218, Theorem 2.8.1].

We will be working with $p = 2$ and write $H_2^\gamma = H^\gamma$. On the one hand, having $p = 2$ significantly simplifies the description of the spaces:

$$f \in H^\gamma(\mathbb{R}^{\mathrm{d}}) \iff \int_{\mathbb{R}^{\mathrm{d}}} |\widehat{f}(y)|^2 (1 + |y|^2)^\gamma dy < \infty.$$

On the other hand, the corresponding Sobolev embedding for $p = 2$ is far from optimal, especially for large d: one would need at least $n + (\mathrm{d}/2)$ generalized derivatives to have n classical derivatives. Still, this is good enough for our purposes.

`Stirling's formula` for the Gamma function, an alternative form:

$$\sqrt{2\pi x}\left(\frac{x}{e}\right)^x \le \Gamma(x+1) \le 4\sqrt{x}\left(\frac{x}{e}\right)^x, \ x \ge 0. \tag{1.1.28}$$

In fact, with some extra effort, the 4 on the right can be taken down to $e = 2.718\ldots$.

1.2 Some Sources of SPDEs

1.2.1 Biology

The three famous equations connected with biology are

$$v_t = v_{xx} + \sqrt{v}\,\dot{W}(t,x), \tag{1.2.1}$$

$$v_t = v_{xx} + \sqrt{v(1-v)}\,\dot{W}(t,x), \tag{1.2.2}$$

$$u_t = u_{xx} - u + g(u)\dot{W}(t,x). \tag{1.2.3}$$

Equations (1.2.1) and (1.2.2) describe time evolution of the (one-dimensional) population density in a certain continuum limit. We come back to these equations in our discussion of particle systems. The "biological" difference between the equations is that (1.2.1) models distribution of individuals (humans, animals), while (1.2.2) describes distribution of genes.

Equation (1.2.3) describes the passive nerve cylinder undergoing random stimulus along its length [222].

1.2.2 Classical Probability Theory

Here are some of the sources of SPDEs in this area: (a) limits of particle systems; (b) analysis of distributions of diffusion processes; (c) study of flows of stochastic diffeomorphisms. Below, we (briefly) discuss each of these area.

Particle Systems Consider n particles $X_1, \ldots, X_n$ evolving in (discrete or continuous) time, in (discrete or continuous) space. The number n of particles can also change in time. Define the measure $V_n(t)$ by

$$V_n(t) = \sum_{i,j=1}^{n} A_{ij}(n,t)\delta_{X_j(t)}, \tag{1.2.4}$$

where $A_{ij}(n,t)$ are weights and δ_x is the point mass at x. If $A_{ij}(n,t) = \delta_{ij}/n$, then $V_n(t)$ is the empirical (also known as occupation) measure of the system. Other dependence of A_{ij} on t and n makes it possible to consider different scaling procedures. The double summation makes it possible to consider the interactions between the particles in the definition of V_n. The possibility to have the weights A_{ij} random provides even more flexibility.

The problem is to study the limit of the measure $V_n(t)$ as $n \to \infty$. The result $V(t,dx)$ is a measure-valued process. The description of this limit is a hard problem and is impossible without a precise description of the evolution of the individual particles, their interaction, and the scaling procedure. The two famous classes of processes that are constructed according to this procedure are the Dawson-Watanabe and the Fleming-Viot processes (see Ethier and Krone [49]). If the measure $V(t,dx)$ has a density $v(t,x)$ with respect to the Lebesgue measure, then the function $v(t,x)$ is a solution of a stochastic partial differential equation. For Dawson-Watanabe and Fleming-Viot processes, the density exists only when x is one-dimensional (that is, particles moving on the line). Examples of the equations satisfied by the corresponding densities are

$$v_t = v_{xx} + \sqrt{v}\,\dot{W}(t,x) \text{ (Dawson-Watanabe)}, \tag{1.2.5}$$

$$v_t = v_{xx} + \sqrt{v(1-v)}\,\dot{W}(t,x) \text{ (Fleming-Viot)}. \tag{1.2.6}$$

The particles that in the limit produce the process with density (1.2.5) move as Brownian motions. Accordingly, the measure-valued process with density (1.2.5) is known as the super-Brownian motion. Reimers [194] discusses other examples leading to equations of the type $v_t = v_{xx} + f(v)\,\dot{W}(t,x)$.

Other particle systems on the line lead to other SPDEs. We mention two example: the long-range contact process and the long-range voter process [170]. In these models, the suitably re-scaled densities in the limit satisfy

$$u_t = u_{xx} + u(1-u) + \sqrt{|u|}\,\dot{W}(t,x) \text{ (contact process)},$$

$$u_t = u_{xx} + u(1-u) + \sqrt{|u(1-u)|}\,\dot{W}(t,x) \text{ (voter process)}.$$

In all the above examples, the weights A_{ij} are non-random. For examples with random A_{ij} see Kurtz and Xiong [127].

Distribution of Diffusion Processes Let ξ be a Gaussian random variable with zero mean and unit variance and let $X = X(t)$ be the solution of the stochastic ordinary differential equation driven by a standard Brownian motion w:

$$dX(t) = -X(t)dt + \sqrt{2}\,dw(t),\ X(0) = 0. \tag{1.2.7}$$

Since X is a Gaussian process and $\mathbb{E}X^2(t) = 1 - e^{-t}$, it follows that, as $t \to \infty$, the random variable $X(t)$ converges in distribution to ξ.

More generally, for a suitable function $V = V(x)$, $x \in \mathbb{R}^{\mathrm{d}}$, the solution $X = (X_1, \ldots, X_{\mathrm{d}})$ of the system of stochastic equations

$$dX_i(t) = -\frac{\partial V}{\partial x_i}\big(X(t)\big)dt + \sqrt{2}\,dw_i(t), \tag{1.2.8}$$

with arbitrary initial condition and independent standard Brownian motions w_i, converges in distribution to an $\mathbb{R}^{\mathrm{d}}$-valued random variable with the probability density function $p(x) = ce^{-V(x)}$, where c is a normalizing constant; see, for example, Robert and Casella [195, Sect. 7.8.5]. Note that (1.2.7) is a particular case of (1.2.8), with $\mathrm{d} = 1$ and $V(x) = x^2/2$, $x \in \mathbb{R}$. For complicated functions V, numerical solution of equation (1.2.8) as a way of generating a random variable with the probability density function $p(x) = ce^{-V(x)}$ could be a viable practical alternative to the traditional inverse transformation or acceptance-rejection methods [195, Chap. 2].

Now instead of a random variable we consider a continuous Gaussian process $\xi = \xi(t)$, $0 \leq t \leq 1$, defined by the equation

$$d\xi(t) = a\xi(t)dt + dw(t), \ \xi(0) = 0, \ a \in \mathbb{R}; \tag{1.2.9}$$

ξ is also known as the `Ornstein-Uhlenbeck process`. It is a random element with values in a suitable function space H, for example, $H = L_2((0,1))$, and generates a measure P_ξ in that space:

$$P_\xi(A) = \mathbb{P}\big(\xi(\cdot) \in A\big), \ A \subset H.$$

We just saw how distribution of a random vector can appear as a limiting distribution of the solution of a stochastic ordinary differential equation. It turns out that the distribution P_ξ of a random process can appear as a limiting distribution of the solution of a stochastic partial differential equation. Indeed, let $u = u(t,x)$ be the solution of

$$u_t(t,x) = u_{xx}(t,x) - a^2u(t,x) + \sqrt{2}\dot{W}(t,x), \ t > 0, \ x \in (0,1), \tag{1.2.10}$$

with zero initial condition $u(0,x) = 0$ and boundary conditions $u(t,0) = 0$, $u_x(t,1) - au(t,1) = 0$. Then, as $t \to \infty$, the distribution of $u(t,\cdot)$ converges to the distribution of $\xi(\cdot)$:

$$\lim_{t\to\infty} \mathbb{P}\big(u(t,\cdot) \in A\big) = \mathbb{P}\big(\xi(\cdot) \in A\big), \ A \subset L_2((0,1));$$

see Hairer et al. [71, Theorem 3.3]. Changing the boundary conditions in Eq. (1.2.10) leads to the distribution of various modifications of the process ξ. For example, boundary conditions $u(t,0) = u(t,1) = 0$ correspond to the `Ornstein-Uhlebeck bridge`, that it, the process ξ defined by (1.2.9) and conditioned on coming back to zero at time $t = 1$: $\xi(0) = \xi(1) = 0$ (see [71, Theorem 3.6]). Note that, unlike (1.2.9), Eq. (1.2.10) depends only only $|a|$. This, in particular, implies that the Ornstein-Uhlebeck bridge obtained by conditioning the stable Ornstein-Uhlebeck process $dX(t) = -X(t)dt + dw(t)$, $X(0) = 0$, on coming back to zero at time $t = 1$ has the same distribution in the space of continuous functions as the Ornstein-Uhlebeck bridge obtained by conditioning the unstable Ornstein-Uhlebeck process $dY(t) = Y(t)dt + dw(t)$, $Y(0) = 0$, on coming back to zero at time $t = 1$.

Many other Gaussian and non-Gaussian diffusion processes can be interpreted as the stationary distribution of a suitable stochastic partial differential equation. For details, see Hairer et al. [71, 72].

Flows of Stochastic Diffeomorphisms Consider a stochastic ordinary differential equation in $\mathbb{R}^{\mathrm{d}}$:

$$X_i(t) = x_i + \int_0^t b_i(s, X(s))ds + \sum_{k=1}^{m} \sigma_{ik}(s, X(s))dw_k(s), \tag{1.2.11}$$

$i = 1, \ldots, \mathrm{d}$, w_k are independent standard Brownian motions. To simplify the discussion, assume that all the functions $b_i = b_i(t, x)$, $\sigma_{ik} = \sigma_i(t, x)$ are non-random, bounded, infinitely differentiable in x, with all the derivatives also bounded. Then, for every $x \in \mathbb{R}^{\mathrm{d}}$, Eq. (1.2.11) has a unique solution.

The difference between X as a random process and X as a flow is that, as a flow (or dynamical system) the process X is studied for all initial conditions $x \in \mathbb{R}^{\mathrm{d}}$. In other words, the flow is a random function $X = X(t, x)$ of $\mathrm{d} + 1$ variables. While Eq. (1.2.11) describes $X(\cdot, x)$ for each fixed x, behavior of X as a function of x requires further investigation.

Uniqueness of solution implies that the trajectories of $X(t, x)$ do not intersect. In other words, for every $t \geq 0$, the mapping $x \mapsto X(t, x)$ is one-to-one. A natural question then is to describe the inverse mapping, which we denote by $Y(t, x)$, such that $X(t, Y(t, x)) = x$ for all t and x. It turns out that the individual components Y_ℓ, $\ell = 1, \ldots, \mathrm{d}$ of the function Y satisfy stochastic partial differential equations

$$\begin{aligned} Y_\ell(t, x) = x_\ell + \int_0^t \Bigg(& \sum_{i,j=1}^{\mathrm{d}} \sum_{k=1}^{m} \sigma_{ik} \frac{\partial}{\partial x_i} \left(\sigma_{jk} \frac{\partial Y_\ell}{\partial x_j} \right) \\ & - \frac{1}{2} \sum_{i,j=1}^{\mathrm{d}} \sum_{k \geq 1} \sigma_{ik}\sigma_{jk} \frac{\partial^2 Y_\ell}{\partial x_i \partial x_j} - \sum_{i=1}^{\mathrm{d}} b_i \frac{\partial Y_\ell}{\partial x_i} \Bigg) ds \\ & - \int_0^t \sum_{k=1}^{\ell} \sum_{i=1}^{\mathrm{d}} \sigma_{ik} \frac{\partial Y_\ell}{\partial x_i} dw_k(s). \end{aligned} \tag{1.2.12}$$

We will discuss this result below (see Theorem 4.4.18 on page 214).

1.2.3 *Economics and Finance*

The main application of SPDEs in this area is to modeling term structure of interest rates. The starting point is the Heath-Jarrow-Morton (HJM) model of bond prices. Musiela's formulations of the model replaces time of maturity with time to maturity and results in the first-order hyperbolic equation

$$u_t = u_x + a(t, x) + \sum_{k \geq 1} \sigma_k(t, x)\dot{w}_k(t), \tag{1.2.13}$$

where w_k, $k \geq 1$ are independent standard Brownian motions and the functions σ_k and a are related through the HJM no-arbitrage condition. For details, see Carmona and Tehranchi [22, Sect. 2.4]. Various modifications of (1.2.13) have been suggested, such as a parabolic version [30]:

$$u_t = u_{xx} + u_x + a(t,x) + \sum_{k\geq 1} \sigma_k(t,x)\dot{w}_k(t),$$

and a second-order hyperbolic version [203]:

$$u_{tt} + u_t = u_{xx} + u_x + a(t,x) + \sum_{k\geq 1} \sigma_k(t,x)\dot{w}_k(t).$$

1.2.4 Engineering

We discuss two applications: (a) Nonlinear filtering of diffusion processes; (b) Simulated annealing in shape optimization.

Nonlinear Filtering Although stated in purely mathematical terms, nonlinear filtering is part of any problem with incomplete information and nonlinear sensors. Nonlinear filtering was also one of the main early incentives for the development of the theory of SPDEs.

The statement of the problem is as follows. Consider two diffusion processes (X, Y) defined by

$$\begin{aligned} \dot{X}_i(t) &= b_i(t, X(t)) + \sum_{\ell=1}^{m} \sigma_{i\ell}(t, X(t))\dot{\widetilde{w}}_\ell(t) \\ \dot{Y}_k(t) &= h(t, X(t)) + \dot{w}_k(t), \end{aligned} \tag{1.2.14}$$

where $0 < t \leq T$, $i = 1, \ldots, \mathrm{d}$, $k = 1, \ldots, n$, $w_k, \widetilde{w}_\ell$ are independent standard Brownian motions, $X(0) \in \mathbb{R}^{\mathrm{d}}$ is a random variable independent of $w_k, \widetilde{w}_\ell$, $Y(0) = 0$. Thinking of X as an unobservable process and Y as observations, the `filtering problem` is to find the conditional distribution of $X(t)$ given the observations $Y(s)$, $s \leq t$. The problem is called nonlinear because Eq. (1.2.14) is nonlinear.

Define $a_{ij}(t,x) = \sum_\ell \sigma_{i\ell}(t,x)\sigma_{j\ell}(t,x)$. Under some regularity conditions on the coefficients in (1.2.14),

$$\mathbb{P}\big(X(t) \in G | Y(s), 0 \leq s \leq T\big) = \frac{\int_G u(t,x)dx}{\int_{\mathbb{R}^{\mathrm{d}}} u(t,x)dx},$$

where the function $u = u(t,x)$ satisfies

$$u_t = \sum_{i,j=1}^{\mathrm{d}} \frac{\partial^2 (a_{ij}u)}{\partial x_i \partial x_j} - \sum_{i=1}^{\mathrm{d}} \frac{\partial (b_i u)}{\partial x_i} + \sum_{k=1}^{n} h_k\, u\, \dot{Y}_k;$$

the initial condition $u(0,x)$ is the distribution density of $X(0)$. Later, we establish a much more general result (Theorem 4.4.27 on page 227).

Simulated Annealing in Shape Optimization In shape optimization problems, the objective is to find the global extremal of a functional defined on the set of shapes (curves or surfaces). In this context, simulated annealing refers to a randomization procedure intended to prevent the corresponding optimization algorithm from getting stuck at a local extremum.

While randomizing a point in a Euclidean space means adding a random number, randomizing a curve or a surface is less straightforward. Let us assume that the curve or surface $S = S(t)$ in question is a level set of a function $u = u(t,x)$:

$$S(t) = \{x : u(t,x) = c\}$$

for a fixed number c in the range of the function u. If $x = x(t)$ is a position vector of a point on $S(t)$, then

$$\frac{d}{dt}u\big(t,x(t)\big) = 0, \text{ or } u_t + \sum_i u_{x_i}\dot{x}_i(t) = 0, \text{ or}$$

or

$$u_t + \nabla u \cdot \dot{x}(t) = 0. \tag{1.2.15}$$

We will take for granted that the standard procedure to perturb $S(t)$ is to move every point of S in the direction that is normal (perpendicular) to S at that point; we also assume that S is smooth enough to have the normal direction at every point. We know from calculus that the normal direction at a point on a level curve is along the gradient of the function at that point. That is,

$$\dot{x}(t) = b\big(t,x(t)\big)\frac{\nabla u(t,x(t))}{|\nabla u(t,x(t))|},$$

with a scalar function b representing the (signed) speed. Together with (1.2.15), we get $u_t = -b|\nabla u|$, or, writing $f = -b$,

$$u_t = f(t,x)|\nabla u|. \tag{1.2.16}$$

It is now clear that a random evolution of a curve will correspond to a stochastic version of (1.2.16). This was the idea of Juan et al. [98], who argued that

- the noise should be regular in space, not to destroy the original shape;
- the resulting stochastic equation must be interpreted in the Stratonovich sense, to preserve invariance of the time evolution under smooth transformations and to ensure well-posedness of the equation.

In [98], Juan et al. suggested the following equation for random evolution of a level set:

$$u_t = f(t,x)|\nabla u| + \sum_{k=1}^{n} \sigma_k(t,x)|\nabla u| \circ \dot{w}_k(t), \tag{1.2.17}$$

where the functions σ_k are smooth and compactly supported in x and $w_k,\ k \geq 1$ are independent standard Brownian motions. The numerical experiments in [98] show improvement of the resulting randomized algorithms for shape recognition, when compared to corresponding non-randomized algorithms.

Solvability issues for fully nonlinear equations such as (1.2.17) were studied by Lions and Souganidis [134–137].

1.2.5 Physics

With many deterministic PDEs coming from physical models, it is natural that many stochastic partial differential equations have connections with physics as well.

To introduce some of the terminology, let us consider three basic evolution equations:

- The heat equation $u_t = \boldsymbol{\Delta} u + f(t,x)$;
- The Schrödinger equation $\mathrm{i}\, u_t + \boldsymbol{\Delta} u + V(t,x)u = 0$;
- The wave equation $u_{tt} = \boldsymbol{\Delta} u + f(t,x)$.

The term $f = f(t,x)$ in the heat and wave equations is usually called the (external) `force`, even though in the case of the heat equation f represents the external sources of heat. The function $V = V(t,x)$ in the Schrödinger equation is called `potential` because it is related to the potential energy of the moving particle (the Laplace operator $\boldsymbol{\Delta}$ corresponds to the kinetic energy). Just as with the force, the term potential is used for all other equations where the term $V(t,x)u$ might appear. For example, we can talk about heat equation with a potential: $u_t = u_{xx} + V(t,x)u$ even if there is no immediate connection with potential energy. Moreover, terms of the form $f(t,x,u)$ are also sometimes called a (non-linear) potential.

There are two ways to get a stochastic partial differential equation from a physical model described by a deterministic equation:

- direct randomization of the components of the equation;
- randomization at an earlier stage of the derivation of the equation.

The direct randomization approach works as follows: given a deterministic equation, we introduce randomness by considering random driving force, random potential, random initial and boundary conditions, etc. This randomness may or may not make sense from the physical point of view, but the resulting equation is usually an interesting mathematical object to study. The most direct way is to add the term $\big(f(t,x,u)+g(t,x)\big)\dot{N}(t,x)$ to the equation, with $\dot{N}$ representing the noise. The easiest choice is $f = 0$ and $g = 1$. If the positivity of the solution must be preserved, then the easiest choice is the random potential $u\dot{N}$, corresponding to $f(t,x,u) = u$ and $g = 0$.

The most popular random perturbation $\dot{N}$ is Gaussian white noise $\dot{W}$. Other options are colored-in-space Gaussian noise $\dot{W}^{Q}$, fractional Gaussian noise, and Lévy noise.

Applying this approach to the three deterministic equations (heat, Schrödinger, wave), we easily get six stochastic counterparts:

$$u_t = \boldsymbol{\Delta} u + \dot{W}(t,x),\ \mathrm{i}\,u_t + \boldsymbol{\Delta} u + u\dot{W}(t,x) = 0,$$

$$u_{tt} = \boldsymbol{\Delta} u + \dot{W}(t,x),\ u_t = \boldsymbol{\Delta} u + u\dot{W}(t,x),$$

$$\mathrm{i}\,u_t + \boldsymbol{\Delta} u = \dot{W}(t,x),\ u_{tt} = \boldsymbol{\Delta} u + u\dot{W}(t,x).$$

For yet another approach to randomizing the heat equation, by considering fluctuations in the probability representation of the solution, see [210, Sect. 1]

Here are some other popular equations coming from physical models:

1. `Reaction-diffusion` equation $u_t = \boldsymbol{\Delta} u + f(u)$. In particular,
   - `Kolmogorov-Petrovskii-Piskunov-Fisher` equation, corresponding to $f(u) = u(1-u)$; this equation can also go under the names KPPF, Kolmogorov-Petrovskii-Piskunov, and KPP.
   - `Allen-Cahn` equation, corresponding to $f(u) = F'(u)$ (that is, f is a derivative of some other function; in the original model of S.M. Allen and J.W. Cahn, $F(u) = -(u^2-1)^2/4$).
2. Equations coming from gas and fluid mechanics:
   - `Burgers equation` $u_t + uu_x = u_{xx}$;
   - `Navier-Stokes equations`

$$\frac{\partial u_i}{\partial t} + \sum_{j=1}^{\mathrm{d}} u_j \frac{\partial u_i}{\partial x_j} = \nu \boldsymbol{\Delta} u_i - \frac{\partial p}{\partial x_i},\ \ i = 1,\dots,\mathrm{d} \tag{1.2.18}$$

$$\sum_{i=1}^{\mathrm{d}} \frac{\partial u_i}{\partial x_i} = 0, \tag{1.2.19}$$

with unknowns $u_1,\dots,u_{\mathrm{d}}$ (components of the velocity field) and the pressure p;

- Euler equations: same as (1.2.18), (1.2.19), but with $\nu = 0$ in (1.2.18);
- Korteweg-de Vries (KdV) equation $u_t + uu_x + u_{xxx} = 0$;
- Camassa-Holm equation $(u - u_{xx})_t + uu_x + u_x = u_x u_{xx} + uu_{xxx}$.

3. Equations coming from quantum physics
   - Ginzburg-Landau equation $u_t = \boldsymbol{\Delta} u + u - |u|^2 u$;
   - Gross-Pitaevskii equation $\mathrm{i}\, u_t = -\boldsymbol{\Delta} u + |u|^2 u$.
4. Other equations:
   - (Generalized) porous medium equation $u_t = \boldsymbol{\Delta}\big(f(u)\big)$ or, equivalently, $u_t = \mathrm{div}\big(f'(u)\mathrm{grad}u\big)$; the original porous medium equation corresponds to $f(u) = u^\gamma$, $\gamma > 0$; the heat equation is a particular case with $f(u) = u$.
   - Kuramoto-Sivashinsky equation $u_t = -\boldsymbol{\Delta}^2 u - \boldsymbol{\Delta} u - |\nabla u|^2$, where $\boldsymbol{\Delta}^2 f = f_{x_1x_1x_1x_1} + f_{x_2x_2x_2x_2} + \dots$; $|\nabla f|^2 = f_{x_1}^2 + f_{x_2}^2 + \dots$.
   - Cahn-Hillard equation $u_t = -\boldsymbol{\Delta}^2 u + \boldsymbol{\Delta}\big(F'(u)\big)$; in the original model of J.W. Cahn and J.E. Hillard, $F(u) = -(u^2 - 1)^2/4$.

In all the above equations we scaled out most real-valued physical and mathematical constants to 1.

In the analysis of a physical model, the general philosophy is that at the **microscopic level** of individual particles (atoms, molecules, etc.) the system is inherently random, because of the quantum effects and/or because of the multiple collisions between the individual particles. The huge numbers of particles make the overall system effectively untractable, and the idea is to re-scale the system and pass to the limit.

At the **macroscopic level** [the objects we can see with a naked eye; mathematically, this is the scaling corresponding to the law of large numbers] the randomness at the microscopic level averages out and leads to the corresponding deterministic PDE (or an ODE) describing the system [an example would be a derivation of the heat equation from a random walk model].

In between the microscopic and macroscopic levels, there is the **mesoscopic level** [the objects visible in the optical microscope; mathematically, this is the scaling corresponding to the central limit theorem]. At this level, the microscopic random effects average out enough to be tractable but do not disappear completely [an example is the original Brownian motion of pollen particles suspended in liquid, as observed in 1827 by the Scottish botanist Robert Brown (1773–1858)]. Thus, a careful passage to the limit from the microscopic to mesoscopic level can lead to a stochastic differential equation, either ordinary or with partial derivatives. The Kardar-Parisi-Zhang (KPZ) equation [104] is an example:

$$u_t = \boldsymbol{\Delta} u + |\nabla u|^2 + \dot{W}(t,x).$$

A similar equation was suggested by Sandow and Trimper [202]:

$$u_t = \boldsymbol{\Delta} u + \sqrt{1 + |\nabla u|^2} + \dot{W}(t,x).$$

For more examples of this type, see Bertini and Giacomin [10].

For a detailed description of the transition from microscopic to mesoscopic to macroscopic levels, see Kotelenez [114]. For an idea how the method works at the macroscopic level, see Erdős et al. [47, 48] for a derivation of the (deterministic) Gross-Pitaevskii equation from several microscopic models.

It is possible to introduce randomness at a later stage of the derivation of the macroscopic limit. For example, re-tracing the derivation of the heat equation leads to a more general form:

$$u_t = \operatorname{div}\big(a(t,x)\operatorname{grad} u\big) + f(t,x),$$

where div and grad denote the divergence and the gradient, respectively, and (assuming inhomogeneous, non-stationary, but isotropic medium) the scalar function $a = a(t,x)$ is the thermal conductivity of the medium [for an inhomogeneous, non-stationary anizotropic medium the coefficient $a = a(t,x)$ is a tensor field]. Now, making a random leads to another interesting class of SPDEs. The straightforward randomization of a, $a(t,x) = a_0 + \varepsilon \dot{W}(t,x)$, can have some physical sense if ε is small compared to a_0, but presents major analytical problems. To keep a positive, one can try $a(t,x) = a_0 \exp\big(\varepsilon \dot{W}^Q(t,x)\big)$ or, with some extra effort, $a(t,x) = a_0 \exp\big(\varepsilon \dot{W}(t,x)\big)$.

As another example, Mikulevicius and Rozovskii [163] derive a stochastic version of the Navier-Stokes equations (1.2.18), (1.2.19) by considering random trajectories of individual particles in the flow. The resulting random perturbation in (1.2.18) is

$$\Big(\sum_{j=1}^{d} \sigma_{ij}(t,x)\frac{\partial u_i}{\partial x_j} - \frac{\partial \tilde{p}}{\partial x_i}\Big)\dot{W}(t,x).$$

1.2.6 Literature

Below is a brief survey of some of the major publications related to SPDEs. The publications, mostly books, are separated into four groups:

- [A] Classical references;
- [B] Textbooks;
- [C] Specialized;
- [D] Everything else.

This classification is certainly incomplete and subjective, and should be mostly considered an invitation to the reader to come up with something better.

Group A includes two books (Da Prato and Zabczyk [31], recently appearing as the second edition, and Rozovskii [199]), a long paper by Krylov and Rozovskii [122], and lecture notes by Walsh [223]. Most SPDE-related publications over the past 20 years cite at least one of these references.

Group B includes publications that can be (and often have been) used to teach a class on SPDEs: the books by Chow [24], Gawarecki and Mandrekar [59], Kallianpur and Xiong [102], Liu and Röckner [142], Prévôt and Röckner [191]; a collection of papers from a summer school [36], and (for now, unpublished) lecture notes by Hairer [70].

Group C is the biggest and included various research monographs and edited volumes on the subject of SPDEs; just as all other lists, it is far from complete: Blömker [12] on amplitude equations for SPDEs, a collection of six major papers on SPDEs [21], Da Prato and Zabczyk [32] on ergodicity of stochastic evolution equations, Flandoli [52] on flows for parabolic SPDEs, Grecksch and Tudor [65] on certain approximation problems for stochastic evolution equations, Holden, Øksendal, Ubøe, and Zhang on chaos solutions of SPDEs, Jentzen and Kloeden [96] on numerical methods, Khoshnevisan [109] on the stochastic heat equation, Kotelenez [114] on S(P)DEs as mesoscopic limits of microscopic systems, Liu [141] on stability of SPDEs, Métivier [158] on strongly non-linear SPDEs, Peszat and Zabczyk [185] on evolution equations driven by Lévy noise, Rozanov [198] on elliptic SPDEs, Sanz-Solé [204] on Malliavin calculus for SPDEs.

Group D includes books where SPDEs appear in at least one chapter, although not necessarily as the central topic: Carmona and Tehranchi [22] on interest rate theory, Khoshnevisan [108] on multi-parameter processes, Kunita [125] on stochastic flows, Nualart [175] [a standard general reference] on Malliavin calculus, Omatu and Seinfeld [182] on distributed-parameter systems, as well as books on optimal non-linear filtering, such as Bain and Crisan [5], Kallianpur [100], and Xiong [230].

1.2.7 The Structure of This Book

Even the most general discussion of stochastic partial differential equations is not possible without the knowledge of different types of noise that can appear in the equation. The same knowledge is necessary for a meaningful discussion of basic examples of SPDEs. This discussion happens in the beginning of every SPDE-related book, including ours. It can be the subject of first few lectures of the corresponding class.

In Chap. 2 we survey the basic ideas in the theory of stochastic processes and deterministic partial differential equations, including the Kolmogorov criterion for continuity of a random field and different notions of solution of a differential equation. Then we present a collection of stochastic partial differential equations, both linear and non-linear, that admit a closed-form solution. From the authors' experience, examples of solvable SPDEs are always well received by the students.

Chapter 3 provides further background in functional analysis and infinite-dimensional stochastic analysis that is necessary for the development of general theory of SPDEs. Most of the chapter can probably be omitted in a typical class and left to the students for independent reading.

Chapter 4 presents the core material of the book. All three main types of equations (elliptic, hyperbolic, and parabolic) are discussed in some detail, together with the standard material (method of characteristic, change of variables, Zakai equation) related to parabolic equations.

Chapter 5 reflects the research interests of the authors and presents an alternative approach to SPDEs using a Fourier-type series expansion in the space of random variables. This approach makes it possible to greatly enlarge the class of admissible equations and leads to a new type of numerical methods for such equations.

Chapter 6 is another example of a topic of special interest to the authors. From the point of view of statistical inference, a stochastic parabolic equation can lead to a model with independent but not identically distributed observations, which, in turn, produces new asymptotic behavior of the corresponding estimators and some interesting mathematics.

A typical course using this book might consist of five to eight 50-min lectures on Chaps. 1 and 2, 20–25 lectures on the material from Chap. 4, with digressions to Chap. 3 as necessary. The rest of the time can be used to cover topics of particular interest to the instructor, including some of the material from Chaps. 5 and 6, as well as students' presentations.

Chapter 2
Basic Ideas

2.1 Some Useful Facts

2.1.1 Continuity of Random Functions

Given a probability space $(\Omega, \mathcal{F}, \mathbb{P})$ and two measurable spaces, $(A, \mathcal{A})$ and $(B, \mathcal{B})$, a `random function` X is a (measurable) mapping from $(A\times\Omega, \mathcal{A}\times\mathcal{F})$ to $(B, \mathcal{Y})$. In the traditional terminology, the `random process` corresponds to $A, B \subset \mathbb{R}$; a `random field` corresponds to $A = \mathbb{R}^{\mathrm{d}}$, $B = \mathbb{R}$. A `sample path`, or `sample trajectory` of X is the function $X(\cdot, \omega)$ for fixed $\omega \in \Omega$. A `modification` of X is a random function $\bar{X}$ such that, for every $a \in A$, $\mathbb{P}\big(X(a) = \bar{X}(a)\big) = 1$; note that this, in general, DOES NOT mean that $\mathbb{P}\big(X(a) = \bar{X}(a) \text{ for all } a \in A\big) = 1$ (although it does if both X and $\bar{X}$ are continuous).

Many results about random functions state that, under certain conditions, X has a modification with some good properties; sometimes, a different notation is used for the modification, sometimes not. Keeping in mind that one could consider the particular modification of X from the very beginning, we will not distinguish between different modifications of a random function.

When both A and B are metric spaces, it is natural to study the continuity properties of the sample trajectories. In the case $A = \mathbb{R}^{\mathrm{d}}$, the basic result is commonly known as `Kolmogorov's continuity criterion`, named after the Russian mathematician ANDREY NIKOLAEVICH KOLMOGOROV (1903–1987).

We start with the easiest version, when X is a random process and $A = [0, T]$.

S.V. Lototsky, B.L. Rozovsky, *Stochastic Partial Differential Equations*,
Universitext, DOI 10.1007/978-3-319-58647-2_2

Theorem 2.1.1 (Kolmogorov's Continuity Criterion, Part I) *Let T be a positive real number and X, a real-valued random process on $[0, T]$. If there exists a positive number $C > 0$, a positive number $p > 1$, and a positive number $q \geq p$ such that, for all $t, s \in [0, T]$,*

$$\mathbb{E}|X(t) - X(s)|^q \leq C|t - s|^p, \tag{2.1.1}$$

then (there is a modification of X such that) the sample trajectories are Hölder continuous of every order less than $(p - 1)/q$. More precisely, if (2.1.1) holds, then, for every $\gamma < (p-1)/q$, there exists a random variable η such that $\mathbb{P}(0 < \eta < \infty) = 1$ and, for all $t, s \in [0, T]$,

$$|X(t) - X(s)| \leq \eta\, |t - s|^\gamma \text{ (with probability one)}.$$

Moreover, if $\mathbb{E}|X(t_0)|^q < \infty$ for some $t_0 \in [0, T]$, then

$$\mathbb{E} \sup_{t\in[0,T]} |X(t)|^q < \infty.$$

Proof This is a particular case of the two more general results, presented below. For a direct proof, see Karatzas and Shreve [103, Theorem 2.2.8]. Also, remember that we do not distinguish between the different modifications of X, and will not mention the modifications in the future.

Recall that a process X is called Gaussian if $\sum_{k=1}^{N} a_k X(t_k)$ has normal distribution for all $a_k \in \mathbb{R}$, $t_k \in [0, T]$ and $N \geq 1$. For such processes, it is enough to consider (2.1.1) with $p = 2$.

Corollary 2.1.2 *Assume that the process X is zero-mean ($\mathbb{E}X(t) = 0$, $t \in [0, T]$) and Gaussian. Also assume that there exists an $\varepsilon \in (0, 2)$ with the following property: for every $\delta \in (0, \varepsilon)$ there exists a number C, depending on δ such that, for all $s, t \in [0, T]$,*

$$\mathbb{E}|X(t) - X(s)|^2 \leq C|t - s|^{\varepsilon-\delta}. \tag{2.1.2}$$

Then the trajectories of X are Hölder continuous of every order less than $\varepsilon/2$.

Proof If ξ is a zero-mean Gaussian random variable, then, for every positive integer n, direct computations show that

$$\mathbb{E}\xi^{2n} = \left(\mathbb{E}\xi^2\right)^n \prod_{k=1}^{n}(2k - 1). \tag{2.1.3}$$

Thus, (2.1.1) holds with $q = 2n$ and $p = n(\varepsilon - \delta)$. Since n can be arbitrarily large and δ arbitrarily close to 0, the result follows.

In the setting of Corollary 2.1.2, we say that X is `almost` $C^{\varepsilon/2}((0, T))$.

Exercise 2.1.3 (A)

(a) Verify (2.1.3).
(b) Verify that (2.1.1) cannot hold with $q < p$ (unless X does not depend on t).
(c) In Corollary 2.1.2, we assume that $\mathbb{E}X(t) = 0$; in Theorem 2.1.1, assumption $\mathbb{E}X(t) = 0$ is not necessary. Why?
(d) State and prove a version of Corollary 2.1.2 without assuming $\mathbb{E}X(t) = 0$.

As an example of Corollary 2.1.2, the reader can investigate the fractional Brownian motion.

Exercise 2.1.4 (B) A `fractional Brownian motion` with `Hurst parameter` $H \in (0, 1)$ is a zero-mean Gaussian process $w^H = w^H(t)$ with covariance

$$\mathbb{E}\big(w^H(t)w^H(s)\big) = \frac{1}{2}\Big(t^{2H} + s^{2H} - |t-s|^{2H}\Big). \tag{2.1.4}$$

(a) Show that $w^{1/2} = w$ (that is, fractional Brownian motion with Hurst parameter $H = 1/2$ is the standard Brownian motion).
(b) Show that the trajectories of w^H are Hölder continuous of every order less than H. **Hint.** $\mathbb{E}|w^H(t) - w^H(s)|^2 = |t-s|^{2H}$.

Theorem 2.1.1 continues to hold when X takes values in a Banach space.

Theorem 2.1.5 (Kolmogorov's Continuity Criterion, Part II) *Let T be a positive real number, V, a Banach space with norm $\|\cdot\|_V$, and X, a measurable mapping from $[0,T] \times \Omega$ to V. If there exists a positive number $C > 0$, a positive number $p > 1$, and positive number $q \geq p$ such that, for all $t, s \in [0,T]$,*

$$\mathbb{E}\|X(t) - X(s)\|_V^q \leq C|t-s|^p,$$

then the sample trajectories $X = X(t)$ are Hölder continuous of every order less than $(p-1)/q$. More precisely, for every $\gamma < (p-1)/q$, there exists a random variable η such that, for all $t, s \in [0,T]$,

$$\|X(t) - X(s)\|_V \leq \eta\,|t-s|^\gamma.$$

Moreover, if $\mathbb{E}\|X(t_0)\|_V^q < \infty$ for some $t_0 \in [0,T]$, then

$$\mathbb{E}\sup_{t\in[0,T]}\|X(t)\|_V^q < \infty.$$

Proof See Kunita [125, Theorem 1.4.1].

Corollary 2.1.6 *Let X be a zero-mean Gaussian process with values in V, that is,* $\sum_{k=1}^{N} \ell_k(X(t_k))$ *is a Gaussian random variable for all* $N \geq 1$, $t_k \in [0,T]$, *and bounded linear functionals* ℓ_k *on V. Assume that there exists an* $\varepsilon \in (0,2)$ *with the following property: for every* $\delta \in (0,\varepsilon)$ *there exists a number C, depending on* δ *such that, for all* $s,t \in [0,T]$,

$$\mathbb{E}\|X(t)-X(s)\|_V^2 \leq C|t-s|^{\varepsilon-\delta}. \tag{2.1.5}$$

Then the trajectories of X are Hölder continuous of every order less than $\varepsilon/2$.

Finally, we state the result for real-valued random fields.

Theorem 2.1.7 (Kolmogorov's Continuity Criterion, Part III) *Let* $L > 0$ *be non-random,* $G = (0,L)^{\mathrm{d}}$, *a hype-cube in* $\mathbb{R}^{\mathrm{d}}$, *and X, a measurable mapping from* $G \times \Omega$ *to* $\mathbb{R}$. *If there exists a positive number* $C > 0$, *a positive number* $p > \mathrm{d}$, *and positive number* $q \geq p$ *such that, for all* $x,y \in G$,

$$\mathbb{E}|X(x)-X(y)|^q \leq C\left(\sum_{i=1}^{\mathrm{d}} |x_i - y_i|^2\right)^{p/2},$$

then the sample trajectories $X = X(t)$ *are Hölder continuous of every order less than* $(p-d)/q$. *More precisely, for every* $\gamma < (p-d)/q$, *there exists a random variable* η *such that, for all* $x,y \in [0,L]^{\mathrm{d}}$,

$$|X(x)-X(y)| \leq \eta \left(\sum_{i=1}^{\mathrm{d}} |x_i - y_i|^2\right)^{\gamma/2}.$$

Moreover, if $\mathbb{E}|X(x^*)|^p < \infty$ *for some* $x^* \in G$, *then*

$$\mathbb{E}\sup_{x\in G}|X(x)|^p < \infty.$$

Proof See Walsh [223, Corollary 1.2]. Note that, while X can be defined in an arbitrary region in $\mathbb{R}^{\mathrm{d}}$, the result about Hölder continuity is stated in the hyper-cube G. There is no loss of generality here, as long as the domain of X is a bounded open set with a sufficiently smooth boundary.

As in the previous two cases, we have a special version when X is a zero-mean Gaussian field; notice that the result in this case does not depend on the dimension d of G.

Corollary 2.1.8 *Assume that* $\mathbb{E}X(x) = 0$, $x \in G$ *and* $\sum_{k=1}^{N} a_k X(x_k)$ *is Gaussian for all* $a_k \in \mathbb{R}$, $x_k \in G$ *and* $N \geq 1$. *Suppose there exists an* $\varepsilon \in (0,2)$ *with the following property: for every* $\delta \in (0,\varepsilon)$ *there exists a number C, depending on* δ *such that, for all* $x,y \in G$,

$$\mathbb{E}|X(x)-X(y)|^2 \leq C|x-y|^{\varepsilon-\delta}. \tag{2.1.6}$$

Then the trajectories of X are Hölder continuous of every order less than $\varepsilon/2$.

In the setting of Corollary 2.1.8, we say that X is `almost` $\mathcal{C}^{\varepsilon/2}(G)$.

In the next two exercises, the reader is invited to apply Corollary 2.1.8 to two particular Gaussian fields: the Brownian sheet and the Lévy Brownian motion.

Exercise 2.1.9 (C) Consider the Brownian sheet $W = W(t,x)$, $t \in [0,1]$, $x \in [0,\pi]$:

$$W(t,x) = \sqrt{\frac{2}{\pi}} \sum_{k\geq 1} \left(\int_0^x \sin(ky)dy \right) w_k(t). \tag{2.1.7}$$

Verify that the trajectories of W are Hölder continuous of order less than $1/2$ in both t and x.

Exercise 2.1.10 (B) The `Lévy Brownian motion` on $\mathbb{R}^{\mathrm{d}}$, $\mathrm{d} \geq 2$, is, by definition, a zero-mean Gaussian field $L = L(x)$, $x \in \mathbb{R}^{\mathrm{d}}$, such that $L(0) = 0$ and $\mathbb{E}(L(x) - L(y))^2 = |x-y|$. This random field was introduced in the 1940s by the French mathematician PAUL PIERRE LÉVY (1886–1971) as an extension of the standard Brownian motion to a multi-dimensional time parameter [132, Chap. VIII]. Verify that $\mathbb{E}\big(L(x)L(y)\big) = (|x| + |y| - |x-y|)/2$ and that the trajectories of L are Hölder continuous of every order less than $1/2$.

Theorems 2.1.1, 2.1.5, and 2.1.7 are enough for our purposes, but the story about Kolmogorov's criterion does not end here. In particular, the result of Kunita [125, Theorem 1.4.1] provides even more general version of the continuity criterion: for the Banach space-valued random functions $X = X(\omega, x)$, $x \in \mathbb{R}^{\mathrm{d}}$; the result of Walsh [223, Corollary 1.2] provides a more precise bound on $|X(x) - X(y)|$ in the setting of Theorem 2.1.7.

2.1.2 Connection Between the Itô and Stratonovich Integrals

Recall that if $w = w(t)$ is a standard Brownian motion (or Wiener process), then

$$\int_0^T w(s)dw(s) = \frac{w^2(T) - T}{2} \qquad \text{(Itô)} \tag{2.1.8}$$

$$\int_0^T w(s) \circ dw(s) = \frac{w^2(T)}{2} \qquad \text{(Stratonovich).} \tag{2.1.9}$$

Even though most of the time in this book we will be using the Itô integral, in what follows we briefly review the Stratonovich integral (also known as Fisk-Stratonovich integral).

Let $T > 0$ be fixed and non-random, and let A, B, P, Q be adapted processes such that

$$\mathbb{E}\int_0^T \big(|A(t)| + |P(t)| + B^2(t) + Q^2(t)\big)\,dt < \infty.$$

Consider two processes X and Y defined for $0 \le t \le T$ by

$$\begin{aligned} X(t) &= X_0 + \int_0^t A(s)ds + \int_0^t B(s)dw(s),\\ Y(t) &= Y_0 + \int_0^t P(s)dt + \int_0^t Q(s)dw(s), \end{aligned} \tag{2.1.10}$$

and let

$$\langle X, Y\rangle(t) = \int_0^t B(s)Q(s)ds.$$

Also, recall the notation

$$a \wedge b = \min(a, b).$$

Given a non-random partition $0 = t_1 < t_2 < \ldots < t_N = T$ of the interval $[0, T]$, let $\Delta_N = \max_k \big(t_{k+1} - t_k\big)$. Recall that the `Itô integral` $\int_0^t X(s)dY(s)$ is defined by

$$\int_0^t X(s)dY(s) = \lim_{\Delta_N\to 0}\sum_k X(t_k)\big(y(t_{k+1}\wedge t) - Y(t_k \wedge t)\big),$$

where the limit is in probability [192, Theorem II.21]. This integral is named after the Japanese mathematician Kiyosi Itô (1915–2008), who introduced it in 1944 [89].

It is possible to show that each of the following can be taken as the definition of the `Stratonovich integral` $\int_0^t X(s)\circ dY(s)$, $0 \le t \le T$:

$$\int_0^t X(s)\circ dY(s) = \lim_{\Delta_N\to 0}\sum_k \frac{X(t_{k+1}) + X(t_k)}{2}\big(y(t_{k+1}\wedge t) - Y(t_k\wedge t)\big);$$

$$\int_0^t X(s)\circ dY(s) = \lim_{\Delta_N\to 0}\sum_k X\left(\frac{t_{k+1} + t_k}{2}\right)\big(y(t_{k+1}\wedge t) - Y(t_k\wedge t)\big);$$

$$\int_0^t X(s)\circ dY(s) = \int_0^t X(s)dY(s) + \frac{1}{2}\langle X, Y\rangle(t), \tag{2.1.11}$$

where the limits are in probability. For details, see Protter [192, Theorems V.26 and V.30]. The integral is named after the Soviet scientist RUSLAN LEONT'EVICH STRATONOVICH (1930–1997), who described it in 1964 [211]. It was also discovered independently by D.L. FISK, but only appeared as a part of his Ph.D. dissertation, written in 1964 under HERMAN RUBIN at the Department of Statistics, Michigan State University. For more details about the history of stochastic calculus, see the paper by Jarrow and Protter [94].

Note that $\int_0^t X(s) \circ dY(s) = \int_0^t X(s)dY(s)$ when $\langle X, Y\rangle = 0$, which is the case, for example, if either $B = 0$ or $Q = 0$, that is, either X or Y has no martingale component.

The Stratonovich integral appears naturally in the following situation. Suppose that the trajectories of the Brownian motion are approximated by piece-wise continuously-differentiable functions $w_n = w_n(t)$ so that

$$\lim_{n\to\infty} \sup_{0<t<T} |w_n(t) - w(t)| = 0$$

with probability one. Consider the ordinary differential equation

$$\frac{dX_n(t)}{dt} = b\big(t, X_n(t)\big) + \sigma\big(t, X_n(t)\big)\frac{dw_n(t)}{dt},\ X_n(0) = x.$$

By a result of Wong and Zakai [229], $\lim_{n\to\infty} \sup_{0<t<T} |X_n(t) - X(t)| = 0$ with probability one, where X is the solution of the stochastic ordinary differential equation in the Stratonovich form:

$$dX(t) = b\big(t, X(t)\big)dt + \sigma\big(t, X(t)\big) \circ dw(t),\ X(0) = x.$$

The connection between the stochastic differential equations in the Itô and Stratonovich forms is as follows.

Theorem 2.1.11 *If $X = X(t)$ satisfies*

$$dX(t) = b\big(t, X(t)\big)dt + \sigma\big(t, X(t)\big) \circ dw(t),$$

and the function $\sigma = \sigma(t, x)$ is continuously differentiable in x for all t, then X satisfies

$$dX(t) = b\big(t, X(t)\big)dt + \frac{1}{2}\sigma\big(t, X(t)\big)\sigma_x\big(t, X(t)\big)dt + \sigma\big(t, X(t)\big)dw(t).$$

Exercise 2.1.12 (B) Use (2.1.11) to prove Theorem 2.1.11. **Hint.** At least when σ_{xx} exists, we have, by the Itô formula,

$$d\sigma(t, X(t)) = A(t)dt + \sigma_x(t, X(t))\sigma(t, X(t))dw(t),$$

and so $\big\langle\sigma(\cdot, X(\cdot)), w\big\rangle(t) = \int_0^t \sigma_x(s, X(s))\sigma(s, X(s))ds$.

2.1.3 Random Change of Variables in Random Functions

As a motivation, we consider the following example. Let $v = v(t,x)$, $t > 0$, $x \in \mathbb{R}$, be a smooth solution of the heat equation

$$v_t(t,x) = \frac{1}{2} v_{xx}(t,x); \tag{2.1.12}$$

as usual, $v_t = \partial v/\partial t,\ v_{xx} = \partial^2 v/\partial x^2$.

Let $w = w(t)$ be a standard Brownian motion and define the function $u(t,x) = v(t, x + w(t))$. By the Itô formula, for every $x \in \mathbb{R}$,

$$du(t,x) = v_t\big(t, x + w(t)\big)dt + v_x\big(t, x + w(t)\big)dw(t) + \frac{1}{2} v_{xx}\big(t, x + w(t)\big)dt.$$

By (2.1.12),

$$v_t(t, x + w(t)) = \frac{1}{2} v_{xx}\big(t, x + w(t)\big) = \frac{1}{2} u_{xx}(t,x),$$

and we conclude that the function $u = u(t,x)$ satisfies the stochastic equation

$$du(t,x) = u_{xx}(t,x)dt + u_x(t,x)dw(t). \tag{2.1.13}$$

Let us now try to go in the opposite direction, and start with the function $u = u(t,x)$, a smooth solution of (2.1.13). Next, define the function $v = v(t,x)$ by $v(t,x) = u(t, x - w(t))$. By construction, the function v must satisfy (2.1.12). On the other hand, if we apply the usual Itô formula to the function $u(t, x - w(t))$, we get

$$dv = u_t dt - u_x dw + \frac{1}{2} u_{xx} dt, \text{ or } v_t = \frac{3}{2} v_{xx} \quad (!?!) \tag{2.1.14}$$

The apparent difference between (2.1.12) and (2.1.14) suggests that the Itô formula must be modified when the function is itself random: there could be some interaction between the Wiener process in the argument of the function u and the (same) Wiener process in the differential of the function. The corresponding modification is known as the `generalized Itô formula` or `Itô-Wentzell formula`.

Theorem 2.1.13 (Itô-Wentzell Formula, I) *If $dY(t) = b(t)dt + \sigma(t)dw(t)$ and $dF(t,x) = J(t,x)dt + H(t,x)dw(t)$, then, assuming all the necessary smoothness of F and H,*

$$\begin{aligned} dF\big(t, Y(t)\big) &= J\big(t, Y(t)\big)dt + H\big(t, Y(t)\big)dw(t) \\ &\quad + F_x\big(t, Y(t)\big)\Big(b(t)dt + \sigma(t)dw(t)\Big) + \frac{1}{2}\sigma^2(t)F_{xx}\big(t, Y(t)\big)dt \\ &\quad + H_x\big(t, Y(t)\big)\sigma(t)dt. \end{aligned} \tag{2.1.15}$$

Proof See Kunita [125, Theorem 3.3.1] or Rozovskii [199, Theorem 1.4.9]. The reader can easily recover the proof from the following observation. With $\Delta Y = Y(t+\Delta t) - Y(t)$, we have

$$
\begin{aligned}
F\big(t+\Delta t, Y(t) + \Delta Y\big) - F\big(t, Y(t)\big) &\\
&= \Big(F\big(t+\Delta t, Y(t)\big) - F\big(t, Y(t)\big)\Big)\\
&\quad + F\big(t+\Delta t, Y(t)+\Delta Y\big) - F\big(t+\Delta t, Y(t)\big);\\
F\big(t+\Delta t, Y(t) + \Delta Y\big) - F\big(t+\Delta t, Y(t)\big) &\\
&\approx F_x\big(t+\Delta t, Y(t)\big)\Delta Y + \frac{1}{2}F_{xx}\big(t+\Delta t, Y(t)\big)(\Delta Y)^2\\
F_x\big(t+\Delta t, Y(t)\big)\Delta Y = F_x\big(t, Y(t)\big)\Delta Y + \Delta F_x \Delta Y;&
\end{aligned}
$$

remember that $\Delta F \sim \Delta F_x \sim \Delta F_{xx} \sim \Delta Y \sim \sqrt{\Delta t}$. In the limit $\Delta t \to dt$, the term $\Delta F_x \Delta Y$ becomes $H_x\big(t, Y(t)\big)\sigma(t)dt$. That is, we have to distinguish between $F_x\big(t, Y(t)\big)\Delta Y$ and $F_x\big(t+\Delta t, Y(t)\big)\Delta Y$.

It is possible to write (2.1.15) in a more compact form. For two processes $X(t) = X_0 + \int_0^t A(s)ds + \int_0^t B(s)dw(s)$ and $Y(t) = Y_0 + \int_0^t P(s)ds + \int_0^t Q(s)dw(s)$ (see (2.1.10) on page 30), we define

$$
\langle X, Y\rangle(t) = \int_0^t B(s)Q(s)ds,
$$

the cross variation of the martingale components of X and Y. Then

$$
\begin{aligned}
J\big(t, Y(t)\big)dt + H\big(t, Y(t)\big)dw(t) &= dF(t,x)\Big|_{x=Y(t)},\\
F_x\big(t, Y(t)\big)\Big(b(t)dt + \sigma(t)dw(t)\Big) + \frac{1}{2}\sigma^2(t)F_{xx}\big(t, Y(t)\big)dt &\\
&= F_x\big(t, Y(t)\big)dY(t) + \frac{1}{2}F_{xx}\big(t, Y(t)\big)d\langle Y\rangle(t),\\
H_x\big(t, Y(t)\big)\sigma(t)dt &= d\langle F_x, Y\rangle(t),
\end{aligned}
$$

and (2.1.15) becomes

$$
\begin{aligned}
dF\big(t, Y(t)\big) &= dF(t,x)\Big|_{x=Y(t)}\\
&\quad + F_x\big(t, Y(t)\big)dY(t) + \frac{1}{2}F_{xx}\big(t, Y(t)\big)d\langle Y\rangle(t) \qquad (2.1.16)\\
&\quad + d\langle F_x, Y\rangle(t);
\end{aligned}
$$

the last term, $d\langle F_x, Y\rangle(t)$, represents the interaction between the randomness in F and Y.

Example 2.1.14 If $du(t,x) = \frac{1}{2}u_{xx}dt + u_x dw(t)$, then $du_x = \frac{1}{2}u_{xxx}dt + u_{xx}dw$, $d\langle u_x, w\rangle(t) = u_{xx}dt$, and, for $v(t,x) = u(t, x - w(t))$, (2.1.16) becomes $dv = du - u_x dw(t) + \frac{1}{2}u_{xx}dt - u_{xx}dt = \frac{1}{2}v_{xx}dt$, which now agrees with (2.1.12).

Exercise 2.1.15 (C) Let $du = au_{xx}dt + \sigma u_x dw(t)$ and, for $b \in \mathbb{R}$, define $v(t,x) = u(t, x + bw(t))$. Find the equation satisfied by v and find b for which the equation is deterministic. What is the relation between a and b to ensure that the equation for v is well-posed? **Hint.** $b = -\sigma$ makes the equation deterministic; the equation is well-posed if $a \geq b^2/2$.

A multi-dimensional version of (2.1.15) also exists.

Theorem 2.1.16 (Itô-Wentzel Formula, II) *If $dY(t) = b(t)dt + \sigma(t)dW(t)$ and $dF(t,x) = J(t,x)dt + H(t,x)dW(t)$, where $x \in \mathbb{R}^{\mathrm{d}}$, $Y \in \mathbb{R}^{\mathrm{d}}$, $W \in \mathbb{R}^m$ (m-dimensional Wiener process with independent components), $\sigma \in \mathbb{R}^{\mathrm{d}\times m}$, $H \in \mathbb{R}^m$, $F \in \mathbb{R}$, then, assuming all the necessary smoothness of F and H and writing $D_i = \partial/\partial x_i$,*

$$dF\big(t, Y(t)\big) = dF(t,x)\Big|_{x=Y(t)}$$
$$+ \sum_{i=1}^{\mathrm{d}} D_i F\big(t, Y(t)\big) dY_i(t) + \frac{1}{2}\sum_{i,j=1}^{\mathrm{d}}\sum_{k=1}^{m} D_i D_j F\big(t, Y(t)\big)\sigma_{ik}(t)\sigma_{jk}(t)dt$$
$$+ \sum_{i=1}^{\mathrm{d}}\sum_{k=1}^{m} D_i H_k\big(t, Y(t)\big)\sigma_{ik}(t)dt.$$

Proof See Kunita [125, Theorem 3.3.1] or Rozovskii [199, Theorem 1.4.9].

When the Stratonovich integral is used throughout, the rules of the classical calculus apply: for example, in the one-dimensional setting, if $dY(t) = b(t)dt + \sigma(t) \circ dw(t)$ and $dF(t,x) = J(t,x)dt + H(t,x) \circ dw(t)$, then

$$dF\big(t, Y(t)\big) = J\big(t, Y(t)\big)dt + H\big(t, Y(t)\big) \circ dw(t) + F_x\big(t, Y(t)\big) \circ dY(t); \qquad (2.1.17)$$

see Kunita [125, Theorem 3.3.2].

The original Itô formula goes back to the 1951 paper by Itô [91]. Theorem 2.1.13 can be traced to a 1965 note [220] by ALEXANDER DMITRIEVICH WENTZELL.[1] In fact, the note is so short that it does not appear as a separate paper in the journal, but as a part of the collection "Summary Of Papers Presented At The Meetings Of The Probability And Statistics Section Of The Moscow Mathematical Society."

[1]Ventzell and Ventzel' are two of the several other possible spellings of the name.

2.1.4 Problems

Problem 2.1.1 Let w_k^H, $k \geq 1$, be a collection of independent fractional Brownian motions with the same Hurst parameter $H \in (0,1)$ (see Exercise 2.1.4 on page 27). For $t \geq 0$, $x \in (0,\pi)$, define

$$B^H(t,x) = \sqrt{\frac{2}{\pi}} \sum_{k\geq 1} \frac{1-\cos(kx)}{k} w_k^H(t).$$

What are the continuity properties of B^H in t and x?

Problem 2.1.2 Similar to the previous problem, define

$$B^{\bar{H}}(t,x) = \sqrt{\frac{2}{\pi}} \sum_{k\geq 1} \frac{1-\cos(kx)}{k} w_k^{H_k}(t),$$

where the fractional Brownian motions $w_k^{H_k}$ are independent but have different Hurst parameters H_k. What are the continuity properties of $B^{\bar{H}}$ in t and x?

Problem 2.1.3 There are at least two objects that can be considered an extension of the standard Brownian motion to a multi-parameter case:

1. the Brownian sheet $W = W(x)$, $x \in \mathbb{R}^{\mathrm{d}}$, a zero-mean Gaussian field with

$$\mathbb{E}W(x)W(y) = \prod_{i=1}^{\mathrm{d}} \min(x_i, y_i);$$

2. the Levy Brownian motion $L = L(x)$, $x \in \mathbb{R}^{\mathrm{d}}$, a zero-mean Gaussian field with

$$\mathbb{E}L(x)L(y) = \frac{1}{2}\left(|x| + |y| - |x-y|\right).$$

Indeed, if $\mathrm{d} = 1$ and $x, y \geq 0$, then $\mathbb{E}W(x)W(y) = \mathbb{E}L(x)L(y) = \min(x,y)$, which is the covariance of the standard Brownian motion.

(a) Compare and contrast W and L when $\mathrm{d} \geq 2$;
(b) Think of another possible generalization by constructing a covariance function $C = C(x,y)$ on $\mathbb{R}^{\mathrm{d}\times\mathrm{d}}$ such that $C(x,y) = \min(x,y)$ when $\mathrm{d} = 1$ and $x, y \geq 0$.

2.2 Classification of SPDEs

As the name suggests, a stochastic partial differential equation (SPDE) is an equation combining the features of equations with partial derivatives and stochastic differential equations. In the most general sense, an SPDE is a partial differential equation in which at least one of the following is random: coefficients, initial conditions, boundary conditions, the region in which the equation is considered, including the terminal time, and the driving force (free term).

There are three main types of problems for SPDEs: forward problems, inverse problems, and synthesis problems.

The `forward problem` is the study of the equation given the `input data`: coefficients, initial conditions, boundary conditions, the region in which the equation is considered (including the terminal time), and the driving force (free term). The first question is `well-posedness` of the equation: existence, uniqueness, and continuous dependence on the input data. Sometimes we can do more and show that the equation establishes a homeomorphism (one-to-one and onto mapping, continuous in both directions) between suitably chosen spaces of the solution and the input data. Other questions include Markov properties of the solution, properties of the solution as a random variable (existence and regularity of the density and various statistical moments), asymptotic problems (small parameter: large deviations, averaging, and homogenization;large time behavior: existence of attractors and invariant measures, ergodicity, stability), fine properties of the sample paths (local times, level sets, potential theory, small ball probabilities),and methods of computing the solution numerically.

The `inverse problems` consist in determining (some of) the input data from the observations of the solution of the equation. The solution of an inverse problem is only possible for equations that are well-posed. In the stochastic setting, inverse problems are usually solved using the methods of statistical inference.

Examples of the `synthesis problems` are optimal filtering and optimal control of processes governed by SPDEs.

The emphasis of this book is on well-posedness of various equations: existence, uniqueness, and regularity of the solution.

2.2.1 SPDEs as Stochastic Equations

Recall that a process $X = X(t),\ t \geq 0$, on a stochastic basis $(\Omega, \mathcal{F}, \{\mathcal{F}_t\}_{t\geq 0}, \mathbb{P})$ is called `adapted` if $X(t)$ is $\mathcal{F}_t$-measurable for every $t \geq 0$. Note that this definition implicitly assumes that the process is indexed by a scalar parameter, which is interpreted as time.

A `stochastic differential equation` (SDE) describes an adapted stochastic process with values in a finite-dimensional Euclidean space, and has a finite-dimensional initial condition.This interpretation of an SDE excludes delayed equations (in which the initial condition is a function), backward equations (in which a forward-adapted process is determined by the terminal rather than initial value), anticipating equations (in which the solution does not have to be adapted), equations describing multi-parameter processes, and ordinary differential equations in which the independent variable has no clear direction (such as two-point boundary value problems).

A stochastic ordinary differential equation (SODE) is a particular example of an SDE. More generally, an SDE can allow dependence of the coefficients on all past history of the solution. Beside a formal generalization of the theory, this dependence on the past history can be a useful alternative representation of the equation. For example, the second-order equation $\ddot{X}(t) + a^2X(t) = \dot{w}(t)$ (with $\dot{}$ denoting the derivative with respect to time and w, the standard Brownian motion) can be interpreted as a system of two first-order SODEs: $dX = Ydt$, $dY = -a^2Xdt + dw(t)$. Alternatively, the process $Y(t) = \dot{X}(t)$ solves an SDE (but not an SODE)

$$dY(t) = -a^2\left(\int_0^t Y(s)ds\right) dt + dw(t);$$

this interpretation can be useful in the study of the measure generated by the process Y in the space of continuous functions.

Stochastic partial differential equations do not need to involve time (for example, stochastic elliptic equations, including two-point boundary value problems on an interval). In `evolution equations`, the time is present and is assumed to go forward; the solution is adapted and determined by the initial conditions. We will not discuss either backward or delayed equations.

A general form of the deterministic partial differential equation is

$$F(u, Du, D^2u, \ldots) = 0$$

where F is some functional relation (which can include all the independent variables) and D^ku is a k-th order partial derivative of the unknown function u. Accordingly, a general stochastic partial differential equation is

$$F(u, Du, D^2u, \ldots) = G(u, Du, D^2u, \ldots) \cdot N, \tag{2.2.1}$$

where F and G can now be random (that is, depend on $\omega \in \Omega$), N is the noise term, and $\cdot$ represents the suitable operation of stochastic integration. Sometimes, the left-hand side of (2.2.1) is called the `deterministic part` of the equation and the right-hand side, the `stochastic part`. In stochastic evolution equations

$$u_t = F(u, Du, D^2u, \ldots) + G(u, Du, D^2u, \ldots) \cdot N, \tag{2.2.2}$$

with D^ku now representing only spacial derivatives, the terms deterministic part and stochastic part usually refer to F and $G \cdot N$, respectively.

As stochastic equations, SPDEs can be classified according to the following features:

1. the **type** of the noise entering the equation;
2. the **manner** the noise enters the equation;
3. the type of the stochastic integral.

Some of the popular types of noise are Gaussian white, Gaussian fractional, Poisson, and α-stable. In this book, we will mostly consider Gaussian white noise.

The noise can enter the equation in two main ways: **additively** (that is, when the function G in (2.2.1) and (2.2.2) does not depend on u or any of $D^k u$), or **multiplicatively** (otherwise). Accordingly, we talk about equations with additive or multiplicative noise. In the special case, when both F and G are linear, equations with multiplicative noise are called `bi-linear`. For example, $du(t,x) = u\,u_{xx}(t,x)dt + e^{-x^2}dw(t)$ is an equation with additive noise, and $du(t,x) = u_{xx}(t,x)dt + u_x(t,x)dw(t)$ is a bi-linear equation.

The main types of stochastic integrals that appear in SPDEs are

1. Itô;
2. Stratonovich;
3. Skorokhod;
4. Wick-Itô-Skorokhod.

The only consistent extension of the Lebesgue-Stieltjes integral to the stochastic setting is the integral with respect to a semi-martingale; the result is known as the `Bichteler-Dellacherie theorem`, see Protter [192, Sect. III.7]. In other words, the only process with unbounded variation that can be an integrator is a continuous martingale. For such integrators, there are essentially two types of stochastic integrals: Itô and Stratonovich. The Skorokhod integral is the extension of the Itô integral to anticipating (non-adapted) setting.

For objects that are not semi-martingales (for example, fractional Brownian motion w^H with $H \neq 1/2$, or random functions with no time parameter), there are different possible definitions of the stochastic integral.

2.2.2 *SPDEs as Partial Differential Equations*

As partial differential equations, SPDEs can be classified according to the following features:

1. the order of the equation;
2. the type of the nonlinearity in the equation;
3. the type of the initial and boundary conditions;
4. elliptic/hyperbolic/parabolic.

The order of an SPDE is the highest order of the partial derivative appearing in the equation. This order can depend on the type of the stochastic integral used in the equation. For example, consider the first-order Stratonovich equation

$$du(t,x) = u_x(t,x) \circ dw(t). \tag{2.2.3}$$

If the solution u is a smooth function of x, then u also satisfies the Itô equation

$$du(t,x) = \frac{1}{2} u_{xx} dt + u_x dw(t) \tag{2.2.4}$$

which is second-order.

Exercise 2.2.1 (C) Verify that if u_{xx} exists and is continuous, then Eqs. (2.2.3) and (2.2.4) are equivalent. **Hint.** Use (2.1.11) on page 30. Note that $du_x = u_{xx} \circ dw(t)$.

Classification by the type of nonlinearity is identical to the deterministic setting (see Evans [50, Sect. 1.1]). Consider the equation

$$F(u, Du, D^2u, \ldots) = G(u, Du, D^2u, \ldots) \cdot N. \tag{2.2.5}$$

Equation (2.2.5) is called `linear`, if the operators

$$u \mapsto F(u, Du, D^2u, \ldots) \text{ and } u \mapsto G(u, Du, D^2u, \ldots) \tag{2.2.6}$$

are linear. For example, equations $du = u_{xx}dt + dw(t)$ and $du = u_{xx}dt + u_{xx}dw(t)$ are linear.

Equation (2.2.5) is called `semilinear` if the operators (2.2.6) are linear in the highest-order derivative, with the coefficients independent of u or any of the derivatives of u, for example,

$$du = (u_{xx} - u^3)dt + uu_x dw(t)$$

(this equation is second-order, and the operator $u \mapsto G(u, u_x, u_{xx})$ does not depend on u_{xx} at all).

Equation (2.2.5) is called `quasilinear` if the operators (2.2.6) are linear in the highest-order derivative, but the coefficients are allowed to depend on u and its lower-order derivatives, for example,

$$du = (u_x u_{xx} - u^3)dt + uu_x dw(t).$$

Equation (2.2.5) is called `fully nonlinear` if the dependence on the highest-order derivative is nonlinear, for example,

$$du = u_x^2 \circ dw(t).$$

For a more detailed discussion of this classification and numerous references, see Kotelenez [114, Sect. 13.1].

The only difference from the deterministic setting is the use of the term `bilinear equation` as an alternative name for a linear equation with multiplicative noise. For example, equation $du = u_{xx}dt + u_x dw(t)$ is bilinear. One reason for this special terminology is that bilinear equations are much more difficult to study than linear equations with additive noise. This difficulty manifests in several ways, such as

1. The solution of a bilinear equation is not a Gaussian process, even if the noise and the initial/boundary conditions are.
2. The fundamental solution of a bilinear equation is random, which complicates the use of the variation of parameters formula (see Problem 2.2.2 on page 50).
3. The stochastic part can influence the deterministic part. For example, initial-value problem for equation $du = au_{xx}dt + \sigma u_x dw(t)$ is well-posed in $L_2(\Omega \times \mathbb{R})$ if and only if $\sigma^2 \leq 2a$ (see Exercise 2.2.2).

Exercise 2.2.2 (C) Verify that the equation $du = au_{xx}dt + \sigma u_x dw(t)$ is well-posed in $L_2(\Omega \times \mathbb{R})$ if and only if $\sigma^2 \leq 2a$. **Hint.** If u is a classical solution, then its Fourier transform $\widehat{u}(t, y)$ satisfies

$$d\widehat{u}(t, y) = -ay^2\widehat{u}(t, y)dt + \mathrm{i}\sigma\widehat{u}(t, y)dw(t),$$

or

$$\widehat{u}(t, y) = \widehat{u}(0, y) \exp\Big(-(2a - \sigma^2)y^2t/2 + \mathrm{i}\sigma w(t)\Big).$$

The terminology concerning initial and boundary value problems for SPDEs is the same as for deterministic equations; the only difference is that initial and boundary conditions can now be random. The initial value problem is also known as the Cauchy problem. Two main types of boundary value problems are the Dirichlet problem, when the value of the solution is specified on the boundary of the domain, and the Neumann problem, when the value of the normal derivative of the solution (derivative in the direction of the outer normal vector to the boundary) is specified.

Stochastic elliptic/hyperbolic/parabolic equations with additive noise inherit the type of the corresponding deterministic equation. For example, $\boldsymbol{\Delta} u(x) = \dot{W}(x)$ is elliptic, $u_{tt} = u_{xx}dt + \dot{W}(t, x)$ is hyperbolic, and $du = (u_{xx} + uu_x)dt + dW(t, x)$ is parabolic. Equations with multiplicative noise are more difficult to classify because of possible interaction between the deterministic and stochastic parts. Section 4.1.2 (page 146) provides more information about the classification of deterministic equations into elliptic/hyperbolic/parabolic type.

A useful alterative representation of partial differential equations, both deterministic and stochastic, is provided by abstract equations, in which the spatial variable is ignored and the solution, free terms, and initial and boundary conditions are considered as elements of suitable function spaces. As a result, it is no longer necessary to consider partial derivatives. Among the advantages of this representation are a more compact form of the equation, flexibility in choosing the function spaces, and availability of tools from functional analysis. One of the disadvantages is absence of point-wise multiplication in many abstract spaces and the resulting difficulty in the study of some bilinear and nonlinear equations.

For example, the abstract formulation of the Dirichlet problem for the elliptic equation $\boldsymbol{\Delta} u(x) = f(x)$, $x \in G$, $u|_{\partial G} = 0$, is $\mathrm{A}u = f$, where A is the Laplace operator with zero boundary conditions. One can then study this equation in the scales of Sobolev or Hölder spaces, such as $u \in H_0^1(G)$, $f \in H^{-1}(G)$ or $u \in \mathcal{C}^{2+\beta}(G), f \in \mathcal{C}^{\beta}(G)$.

When the time variable is present, the result is an `abstract evolution equation`. The equation is often written in the integral form. For example, the initial-boundary value problem for the heat equation

$$u_t(t,x) = \boldsymbol{\Delta} u(t,x) + f(t,x),\ t > 0, x \in G;\ u(0,x) = h(x),\ u|_{\partial G} = 0,$$

becomes

$$u(t) = h + \int_0^t \mathrm{A}u(s)ds + \int_0^t f(s)ds,\ t \geq 0,$$

where A is the Laplace operator with zero boundary conditions. This representation is especially convenient for stochastic equations, when the irregular nature of the noise prevents the solution from having classical derivative in time. For example, equation

$$du(t,x) = \boldsymbol{\Delta} u(t,x)dt + dw(t)$$

becomes

$$u(t) = u(0) + \int_0^t \mathrm{A}u(s)ds + w(t).$$

A similar interpretation of the stochastic equation with multiplicative space-time white noise,

$$du(t,x) = u_{xx}dt + u(t,x)dW(t,x)$$

is possible, but is much more complicated (see Krylov [120, Sect. 8]).

2.2.3 *Various Notions of a Solution*

In the probabilistic sense, a solution of an SPDE can be either strong (constructed on a given probability space) or weak (the probability space is constructed as part of the solution). Solution of a `martingale problem` is often a synonym of a probabilistically weak solution. This is completely analogous to the stochastic ordinary differential equations.

For reasons that might not be clear at this point, we also separate `square-integrable solutions` from other possible solutions. A square-integrable solution has, in some sense, a finite second moment; in the probabilistic sense, the solution can be either weak or strong. As with many other classifications in this book, this distinction is not widely accepted, partly because hardly anybody considers solutions that are not square-integrable. We talk more about square-integrable solutions in Sect. 4.1.1 (see page 143 below).

In the PDE sense the solution of an SPDE can be

1. classical;
2. generalized:
   - closed-form;
   - mild;
   - variational;
     - strong;
     - weak;
     - measure-valued;
     - chaos;
   - viscosity;

A `classical solution` of an SPDE is a continuous function satisfying the equation and the initial and boundary conditions point-wise on the same set of probability one. If the equation has the time variable and can be formulated as an abstract evolution equation, then the goal is to satisfy the integral form of this evolution equation. For example, the function $u(t,x) = x + w(t)$ is a classical solution of the equation $du(t,x) = u_x(t,x)dw(t),\ u(0,x) = x$, which in the integral form becomes

$$u(t,x) = x + \int_0^t u_x(s,x)dw(s).$$

Two factors can prevent the equation from having a classical solution:

1. structure of the equation
2. insufficient regularity of the input data (even the heat equation $u_t = u_{xx}$ might fail to have a classical solution if the initial condition is discontinuous).

In particular, if the noise is not regular in space, the equation is unlikely to have a classical solution.

Every solution that is not classical is usually called generalized. A generalized solution extends certain properties of a classical solution without requiring existence of partial derivatives. The idea is to take one particular property satisfied by the classical solution and then say that every function having this property is a generalized solution.

A `closed-form solution` is a (more-or-less explicit) formula expressing the solution in terms of the input data. For example, a closed-form solution of the initial-value problem $u_t + u_x = 0$, $u(0,x) = f(x)$ is $u(t,x) = f(x-t)$. The solution does not have to be classical: in this example, it is classical if and only if f is continuously differentiable; for functions that are not continuously differentiable, we define the solution to be $f(x-t)$. In general, the meaning of an explicit formula is not always clear. As a general guideline, we will be thinking of a quadrature solution, that is, a formula involving (at most) integrals of the input data. With this definition, every linear equation $y'(t) = a(t)y(t)$ has a closed-form solution $y(t) = y(0)\exp(\int_0^t a(s)ds)$, but the second-order linear equation $y''(t)+(2+\sin(t))y(t) = 0$ does not have a closed-form solution.

Closed-form solutions are available only for a very limited number of equations and every partial differential equation admitting a closed-form solution deserves special attention. A direct analysis of a closed form solution is not always the best route to studying the properties of the solution. For example, the closed-form solution of the heat equation $u_t = u_{xx}$, $u(0,x) = f(x)$, $x \in \mathbb{R}$, is

$$u(t,x) = \frac{1}{\sqrt{4\pi t}} \int_{\mathbb{R}} e^{-(x-y)^2/(4t)} f(y)dy,$$

and it is not at all obvious from this expression that

$$\int_{\mathbb{R}} u^2(t,x)dx \le \int_{\mathbb{R}} f^2(x)dx$$

or equivalently, $\|u(t,\cdot)\|^2_{L_2(\mathbb{R})} \le \|f\|^2_{L_2(\mathbb{R})}$. On the other hand, multiplying the equation by u and integrating by parts shows right away that the function $\|u(t,\cdot)\|^2_{L_2(\mathbb{R})}$ is non-increasing in time. For stochastic equations, working with closed-form solutions can be even more difficult, as the corresponding formulas can involve anticipating stochastic integrals (see Problem 2.2.2 on page 50).

A `mild solution` is an extension of the closed-form solution and is mostly used for equations that can be considered perturbations of a linear equation with a closed-form solution. For example, consider the heat equation

$$v_t(t,x) = v_{xx}(t,x) + f(t,x),\ t > 0,\ x \in \mathbb{R}.$$

The closed-form solution of this equation is

$$\begin{aligned} v(t,x) &= \frac{1}{\sqrt{4\pi t}} \int_{\mathbb{R}} e^{-(x-y)^2/(4t)} v(0,y)dy \\ &+ \int_0^t \frac{1}{\sqrt{4\pi(t-s)}} \left(\int_{\mathbb{R}} e^{-(x-y)^2/(4(t-s))} f(s,y)dy \right) ds; \end{aligned} \tag{2.2.7}$$

see Evans [50, Theorem 2.3.2].

Consider now a nonlinear equation

$$du(t,x) = \big(u_{xx} - u^3(t,x)\big)dt + \sin\big(u(t,x)\big)dw(t).$$

A mild solution of this equation is, by definition, the solution of the integral equation

$$\begin{aligned} u(t,x) = &\frac{1}{\sqrt{4\pi t}}\int_{\mathbb{R}} e^{-(x-y)^2/(4t)}u(0,y)dy \\ &-\int_0^t \frac{1}{\sqrt{4\pi(t-s)}}\left(\int_{\mathbb{R}} e^{-(x-y)^2/(4(t-s))}u^3(s,y)dy\right)ds \\ &+\int_0^t \frac{1}{\sqrt{4\pi(t-s)}}\left(\int_{\mathbb{R}} e^{-(x-y)^2/(4(t-s))}\sin\big(u(s,y)\big)dy\right)dw(s). \end{aligned} \tag{2.2.8}$$

It is convenient to introduce the notation

$$\big(\Phi h\big)(t,x) = \frac{1}{\sqrt{4\pi t}}\int_{\mathbb{R}} e^{-(x-y)^2/(4\pi t)}h(y)dy,$$

and then to re-write Eq. (2.2.8) in a more compact form:

$$\begin{aligned} u(t,x) = &\big(\Phi u(0,\cdot)\big)(t,x) - \int_0^t \big(\Phi u^3(s,\cdot)\big)(t-s,x)ds \\ &+\int_0^t \big(\Phi\sin(u(s,\cdot))\big)(t-s,x)dw(s). \end{aligned} \tag{2.2.9}$$

The mild solution is especially useful in the study of equations that can be formulated as abstract evolution equations. Let A be a linear operator and let $\Phi = \Phi(t)$ be the semi-group generated by A Alternatively, we can say that Φ is the `fundamental solution` of the homogeneous linear equation

$$u(t) = u_0 + \int_0^t \mathrm{A}u(s)ds.$$

Recall that, in the finite-dimensional case, if A is represented by a matrix A, then $\Phi(t) = e^{tA}$, the matrix exponential. For more information in the general case, see page 156 below.

With the semi-group notation, a closed-form solution of the inhomogeneous linear equation

$$u(t) = u_p + \int_0^t \mathrm{A}u(s)ds + \int_0^t f(s)ds$$

is

$$u(t) = \Phi(t)u_0 + \int_0^t \Phi(t-s)f(s)ds,$$

as long as all expressions on the right-hand side are well-defined. By definition, the mild solution of the stochastic equation

$$u(t) = u_0 + \int_0^t \mathrm{A}u(s)ds + \int_0^t F(s, u(s))ds + \int_0^t G(s, u(s))dw(s)$$

is the solution of the integral equation

$$u(t) = \Phi(t)u_0 + \int_0^t \Phi(t-s)F(s, u(s))ds + \int_0^t \Phi(t-s)G(s, u(s))dw(s).$$

For a survey of the subject, see [210].

Working with the mild solution, one should keep in mind that

- Just as a closed-form solution, a mild solution is not necessarily classical;
- While a generalization to time-dependent non-random operators $\mathrm{A} = \mathrm{A}(t)$ is relatively straightforward (the corresponding semi-group becomes a two-parameter process $\Phi(t, s)$, $0 \le s \le t$), it is much harder to work with mild solutions when the operator A is both time-dependent and random (such as a linear partial differential operator with random and adapted coefficients), because, for fixed t, the process $\Phi(t, s)$ is not adapted as a function of s (see Problem 2.2.2 on page 50).
- For nonlinear equations, it is often a difficult problem to show that a mild solution is the same as a variational solution.

The idea of the mild solution helps to reduce many stochastic equation with additive noise to random equations. For example, consider the equation

$$u(t) = \int_0^t \mathrm{A}u(s)ds + \int_0^t F(s, u(s))ds + N(t), \tag{2.2.10}$$

where A is a non-random linear operator and N is the noise term; the exact nature of the noise is not important. Under very general assumptions on N, one can define the **stochastic convolution**

$$N_A(t) = \int_0^t \Phi(t-s)dN(s),$$

where Φ is the semi-group of A; since Φ is non-random, there is usually no need to develop special theory of stochastic integration to compute N_A. The mild solution of (2.2.10) is the solution of the integral equation

$$u(t) = \Phi(t)u_0 + \int_0^t \Phi(t-s)F\big(s, u(s)\big)ds + N_A(t).$$

Define the new process $v(t) = u(t) - N_A(t)$. Then $v(0) = u(0)$ and

$$v(t) = \Phi(t)v_0 + \int_0^t \Phi(t-s)F\big(s, v(s) + N_A(s)\big)ds,$$

that is, v is a mild solution of the random evolution equation

$$v(t) = v(0) + \int_0^t \mathrm{A}v(s)ds + \int_0^t \tilde{F}\big(s, v(s)\big)ds, \tag{2.2.11}$$

where $\tilde{F}(t,u) = F\big(t, u + N_A(t)\big)$. Problem 2.2.4 invites the reader to think about an alternative substitution $\bar{v}(t) = u(t) - N(t)$.

In the theory of deterministic PDEs, the idea of both the variational and the viscosity solutions is to avoid computing partial derivatives of the unknown function by using smooth test functions. This is best illustrated on an example. Consider the heat equation on the whole line

$$u_t(t,x) = u_{xx}(t,x),\ 0 < t \le T;\ u(0,x) = f(x). \tag{2.2.12}$$

For simplicity, assume that f is a continuous function with compact support in $\mathbb{R}$.

Here are five different variational solutions:

[W1] `Strong variational solution` is a function u from $L_1\big([0,T]; H^2(\mathbb{R})\big)$ such that

$$u(t,x) = f(x) + \int_0^t u_{xx}(s,x)ds$$

for almost all $t \in [0,T]$, $x \in \mathbb{R}$;

[W2] `Weak variational solution I` is a function u from $L_1\big([0,T]; H^1(\mathbb{R})\big)$ such that, for every smooth function $v = v(x)$ with a compact support in $\mathbb{R}$, the equality

$$\big(u(t,\cdot), v\big)_{L_2(\mathbb{R})} = (f, v)_{L_2(\mathbb{R})} - \int_0^t \big(u_x(s,\cdot), v_x\big)_{L_2(\mathbb{R})}ds$$

holds for almost all $t \in [0,T]$ (as usual, $(f,g)_{L_2(\mathbb{R})} = \int_{\mathbb{R}} f(x)g(x)dx$);

[W3] `Weak variational solution II` is a function u from $L_1\big([0,T]; L_2(\mathbb{R})\big)$ such that, for every smooth function $v = v(x)$ with a compact support in

$\mathbb{R}$, the equality

$$\big(u(t,\cdot),v\big)_{L_2(\mathbb{R})} = (f,v)_{L_2(\mathbb{R})} + \int_0^t \big(u(s,\cdot),v_{xx}\big)_{L_2(\mathbb{R})}ds \qquad (2.2.13)$$

holds for almost all $t \in [0,T]$;

[W4] `Weak variational solution III` is a function u that is locally integrable on every compact subset of $[0,T]\times\mathbb{R}$ and such that, for every smooth function v with compact support in $[0,T)\times\mathbb{R}$,

$$\int_{\mathbb{R}} v(0,x)f(x)dx + \int_0^T\int_{\mathbb{R}} u(t,x)\Big(v_t(t,x)+v_{xx}(t,x)\Big)dxdt = 0.$$

[W5] (`Measure-valued solution`) is a collection $\mu_t = \mu_t(dx)$ of sigma-finite signed measures on $(\mathbb{R},\mathcal{B}(\mathbb{R}))$ such that, for every smooth compactly supported on $\mathbb{R}$ function $\varphi = \varphi(x)$,

$$\mu_t[\varphi] = \int_{\mathbb{R}} \varphi(x)f(x)dx + \int_0^t \mu_s[\varphi_{xx}]ds$$

where, for a bounded measurable function g, $\mu_t[g] = \int_{\mathbb{R}} g(x)\mu_t(dx)$.

Furthermore,

[W6] (`Viscosity solution`) of Eq. (2.2.12) is a continuous on $[0,T]\times\mathbb{R}$ function with $u(0,x) = f(x)$ and with the following two properties:

- if $v = v(t,x)$ is a smooth function and $u - v$ has a local maximum at a point (t_0,x_0), then $v_t(t_0,x_0) \le v_{xx}(t_0,x_0)$;
- if $v = v(t,x)$ is a smooth function and $u - v$ has a local minimum at a point (t_0,x_0), then $v_t(t_0,x_0) \ge v_{xx}(t_0,x_0)$.

The main difference between the five variational solutions is the a priori condition on the function u. There is certain similarity among [W1]–[W3]: in all three cases, the underlying equation is required to hold in a suitable function space (which is $L^2(\mathbb{R})$ for [W1], $H^{-1}(\mathbb{R})$ for [W2], and $H^{-2}(\mathbb{R})$ for [W3]; the larger the space, the weaker the notion of the solution. The reason for calling [W1] strong is that the solution u is required to belong to the natural domain of the operator $\boldsymbol{\Delta}$.

Exercise 2.2.3 **(C)** Show that if μ_t is a measure-valued solution of (2.2.12) and, for every $t \in [0,T]$, the measure μ_t has a density $u = u(t,x)$ with respect to the Lebesgue measure such that $u \in \mathcal{C}\big([0,T];L_2(\mathbb{R})\big)$, then u is a weak variational solution [W3].

The classical solution satisfies each of the definitions (see Exercise 2.2.4), and the corresponding generalization relies on moving the derivatives from the solution u to a `test function` v (in [W1], test functions are used to define the derivatives in the space $H^2(\mathbb{R})$). For [W1]–[W5], the basic idea is integration by parts; for [W6], the basic idea is the second partials test for the local min/max of a function. Unlike [W1]–[W4] and [W6], the initial condition in [W5] can be a measure, such as a point mass.

Definitions [W1]–[W6] extend to more general equations. In particular, [W1]–[W4] can be used for a large class of linear, semi-linear, and quasi-linear equations, as well as abstract evolution equations (when time is the only independent variable and the unknown function is an element of an infinite-dimensional Banach space); the viscosity solution [W6] is the only option for fully nonlinear equations.

When there is no classical solution, there is no obvious reason to claim that, for example, [W1] and [W3] define the same object. In fact, comparing different variational solutions can be difficult, especially for nonlinear equations.

Exercise 2.2.4 (C) Verify that a classical solution of the heat equation $u_t = u_{xx}$ satisfies [W1]–[W6]. **Hint.** For [W6], note that $v_t - u_t = 0$ at every critical point, and $v_{xx} - u_{xx} < 0$ at the local maximum; for a classical solution, $u_t = u_{xx}$ at all points.

Now consider an abstract evolution equation

$$u(t) = u_0 + \int_0^t \mathrm{A}u(s)ds + \int_0^t g(s)ds, 0 \leq t \leq T, \tag{2.2.14}$$

where A is a bounded linear operator from $L_1((0,T);H)$ to $L_1((0,T);X)$ (A can depend on time), $u_0 \in U$, $g \in L_1((0,T);X)$, and H, X, U are Banach spaces such that $U \subseteq X$, $H \subseteq X$. In the interest of generality, we allow Banach spaces rather than Hilbert space and require only L_1 integrability in time rather than stronger L_2 integrability.

For this equation, there is essentially one definition of a `variational solution`:

[WA] $w(H,X)$ `variational solution` of an abstract evolution equation is a process u with values in $L_1\big((0,T);H\big)$ such that equality (2.2.14) holds in $L_1((0,T);X)$.

There are often several ways to set up a concrete equation in the abstract form (2.2.14), and the flexibility is usually in the choice of the spaces H, X, U. Depending on the set-up, [WA] can correspond to one of the variational solutions [W1]–[W5]. Let us illustrate this correspondence for Eq. (2.2.12), in which $\mathrm{A}(s)u = u_{xx}$ and $F = 0$. If $H = H^2(\mathbb{R})$, $U = X = L_2(\mathbb{R})$, then [WA] corresponds to [W1]. If $H = H^1(\mathbb{R})$, $U = L_2(\mathbb{R})$, $X = H^{-1}(\mathbb{R})$, then [WA] corresponds to [W2]. The reader is encouraged to think how to get [W3]–[W5] out of [WA]. The viscosity solution [W6] cannot be defined for (2.2.14).

Depending on the type of the equation, each of the definitions [W1]–[W5], [WA] can be formulated for an SPDE.

In an SPDE, the presence of yet another variable, the elementary outcome ω, leads to an even larger selection of test functions. Of special interest is the situation when the test functions are random and do not depend on t and x. For example, if $h = h(t)$ is a smooth, compactly supported function on $(0, 1)$, then the collection of the random variables

$$\mathcal{E}_h = \exp\left(\int_0^1 h(t)dw(t) - \frac{1}{2}\int_0^1 h^2(t)dt\right),\ h \in C_0^\infty((0,1)),$$

is everywhere dense in the space of square-integrable random variables that are measurable with respect to the sigma-algebra of $w(t)$, $t \in [0,1]$, (see, for example, Øksendal [181, Lemma 4.3.2]). Thus, a square-integrable random process $u = u(t,x)$ is uniquely determined by the collection of deterministic functions

$$u_h(t,x) = \mathbb{E}\big(u(t,x)\mathcal{E}_h\big),\ h \in C_0^\infty((0,1)).$$

In the case of an SPDE, every function u_h will satisfy a deterministic PDE, which can often be solved under more general conditions than the original SPDE. This is the idea behind the `chaos solution` for SPDEs. While a chaos solution is a particular case of a variational solution, it has no obvious analogy for deterministic equations.

As an illustration, consider the stochastic equation

$$du(t,x) = au_{xx}(t,x)dt + \sigma u_x(t,x)dw(t),\ 0 < t < 1,\ x \in \mathbb{R}, \tag{2.2.15}$$

where w is a standard Brownian motion, a, σ are non-random constants, $a > 0$, and the initial condition $u(0,x)$ is non-random. If $u = u(t,x)$ is a classical solution of this equation, then u_h is a classical solution of

$$\frac{\partial u_h(t,x)}{\partial t} = a\frac{\partial^2 u_h(t,x)}{\partial x^2} + \sigma h(t)\frac{\partial u_h(t,x)}{\partial x},\ 0 < t < 1,\ x \in \mathbb{R}, \tag{2.2.16}$$

where the initial condition $u_h(0,x) = u(0,x)$ is the same for all functions h.

Exercise 2.2.5 (C) Verify (2.2.16). **Hint.** Define $\mathcal{E}_h(t)$ by

$$d\mathcal{E}_h(t) = h(t)\mathcal{E}_h(t)dw(t),\ \mathcal{E}(0) = 1,$$

and note that, since $\mathcal{E}_h(t)$ is a martingale and u_h is adapted,

$$u_h(t,x) = \mathbb{E}\big(u(t,x)\mathcal{E}_h(t)\big).$$

Then use Itô formula.

We saw in Exercise 2.2.2 that Eq. (2.2.15) is well-posed in $L_2(\Omega\times\mathbb{R})$ if and only if $2a \geq \sigma^2$. On the other hand, as long as $a > 0$, Eq. (2.2.16) can have a classical, or any other type of solution available for deterministic PDEs, without any restrictions on σ. The resulting collection of functions $u_h = u_h(t,x)$, $h \in C_0^\infty((0,1))$, is called the `chaos solution` of Eq. (2.2.15) and can be used to establish well-posedness of the equation in a suitable space for all $\sigma \in \mathbb{R}$. Depending on the type of the solution used to find u_h, one can have classical chaos solutions, mild chaos solutions, strong chaos solutions, viscosity chaos solutions, etc.

To summarize, the first step in the analysis of an SPDE is to prove that the equation is well-posed. The successful completion of this step essentially depends on two choices: (a) the type of the solution, and (b) the function spaces for the input data and the solution. We just saw that there are plenty of choices for the type of solution. Later, we will see that there are also many choices for the function spaces.

2.2.4 Problems

Problem 2.2.1

(a) Let $a = a(t)$ be a non-random continuous function and define

$$\Psi(t,s) = e^{\int_s^t a(r)dr},\ 0 \le s \le t.$$

Not that $\Psi(s,s) = 1$ and $\partial\Psi(t,s)/\partial t = a(t)\Psi(t,s),\ t \ge s$. Given a continuous function f, verify that the function $y(t) = \int_0^t \Psi(t,s)f(s)ds$ satisfies $y'(t) = a(t)y(t) + f(t),\ y(0) = 0$.

(b) Let $a = a(t)$ be a continuous function such that $1 \le a(t) \le 2$ for all t. Repeat part (a) for the heat equation $u_t = a(t)u_{xx} + f(t,x), t > 0,\ x \in \mathbb{R}$, with $u(0,x) = 0$, by finding a family of linear operators $\Phi = \Phi(t,s)$ such that $u(t,x) = \int_0^t \Big(\Phi(t,s)f(s,\cdot)\Big)(x)ds$.

Problem 2.2.2

(a) Consider the stochastic ordinary differential equation

$$dX(t) = X(t)dw(t),\ X(0) = 1. \tag{2.2.17}$$

Verify that $X(t) = e^{w(t)-(t/2)}$; the process X is known as the `geometric Brownian motion`. Note that (2.2.17) is a bilinear equation.

(b) Let $f = f(t)$ be a non-random continuous function and consider the equation

$$dY(t) = Y(t)dw(t) + f(t)dt,\ Y(0) = 0. \tag{2.2.18}$$

From the point of view of general theory of linear differential equations, (2.2.18) is the inhomogeneous version of (2.2.17). With $X = X(t)$ from (2.2.17), verify that

$$Y(t) = \int_0^t \frac{X(t)}{X(s)} f(s)ds. \tag{2.2.19}$$

In other words, $\Phi(t,s) = X(t)/X(s),\ 0 \le s \le t$, is the `fundamental solution` of (2.2.19). Note that, for fixed $T > 0$, the process $g(t) = \Phi(T,t)$, $0 \le t \le T$, is not $\mathcal{F}_t^w$-adapted.

(c) What complications arise if we replace the term $f(t)dt$ on the right-hand side of (2.2.18) with the term $f(t)dw(t)$ and still try to write

$$Y(t) = \int_0^t \frac{X(t)}{X(s)} f(s)dw(s)?$$

(d) Do parts (a)–(c) starting with the equation $dX(t) = a(t)X(t)dw(t)$, $X(0) = 1$, where $a = a(t)$ is an adapted process (precise conditions on a is a part of the problem).

Problem 2.2.3

(a) Define a classical solution for the abstract evolution equation

$$u(t) = u_0 + \int_0^t \mathrm{A}u(s)ds.$$

What changes if we consider the same equation in the differential form $\dot{u} = \mathrm{A}u$?

(b) Let A be a partial differential operator. Define the measure-valued solution for the evolution equation $u_t(t,x) = \mathrm{A}u(t,x)$ using the notion of a signed sigma-finite measure. Then consider three concrete operators on $L_2(\mathbb{R})$:

$$\mathbf{A}_1 f(t,x) = a(t,x)\frac{\partial^2}{\partial x^2} f(t,x);$$

$$\mathbf{A}_2 f(t,x) = \frac{\partial}{\partial x}\left(a(t,x)\frac{\partial}{\partial x} f(t,x)\right);$$

$$\mathbf{A}_3 f(t,x) = \frac{\partial^2}{\partial x^2}\left(a(t,x)f(t,x)\right).$$

In each case, find the minimal conditions on the function a that are necessary to define the measure-valued solution.

Problem 2.2.4 In Eq. (2.2.10) on page 45, make the substitution $\bar{v}(t) = u(t) - N(t)$ and compare the result with (2.2.11). Which substitution results in a better equation and why?

Problem 2.2.5 Consider the stochastic heat equation with multiplicative space-time white noise:

$$du(t,x) = u_{xx}dt + u(t,x)\dot{W}(t,x),\ t > 0,\ x \in (0,1),$$

with zero Neumann boundary conditions. (a) Can this equation have a classical solution? (b) Define the variational solutions [W1]–[W4] for this equation (see page 46). Which ones can actually exist? What about the measure-valued solution?

Problem 2.2.6 Compare and contrast the four variational solutions [W1]–[W4] and think of further possible modifications, especially in the case of stochastic equations.

2.3 Closed-Form Solutions

While a closed-form solution is not always the best way to study the equation, any mathematical problem with a closed-form solution is somewhat of a gem and deserves special attention. Accordingly, in this section we discuss several linear and nonlinear SPDEs that admit a closed-form solution. To simplify the presentation, we will usually work in one space variable.

2.3.1 *Heat Equation*

Let a be a positive real number. Recall that the solution of the heat equation

$$u_t(t,x) = au_{xx}(t,x),\ t > 0,\ x \in \mathbb{R}, \tag{2.3.1}$$

is

$$u^{\text{heat}}(t,x) = \frac{1}{\sqrt{4a\pi t}} \int_{\mathbb{R}} e^{-|x-y|^2/(4at)} u(0,y)dy. \tag{2.3.2}$$

If $u(0,\cdot) \in L_p(\mathbb{R})$ for some $1 \leq p \leq +\infty$, then $u(t,x)$ is a smooth function of (t,x) for $t > 0$.

An equivalent way to write (2.3.2) is

$$u^{\text{heat}}(t,x) = \mathbb{E}u\big(0, x + \sqrt{2a}\,B(t)\big), \tag{2.3.3}$$

where $B = B(t)$ is a standard one-dimensional Brownian motion (Wiener process); formula (2.3.3) is known as a probabilistic representation or the `Feynmann-Kac formula` for the heat equation.

Note that, for $a = 1/2$, the solution of $u_t = au_{xx}$ is

$$u(t,x) = \mathbb{E}u\big(0, x + B(t)\big).$$

Exercise 2.3.1 (C) Let $c = c(t)$ be a locally integrable function. Show that the function

$$v(t,x) = u^{\text{heat}}(t,x)e^{\int_0^t c(s)ds} \tag{2.3.4}$$

satisfies $v_t = v_{xx} + c(t)v$. Note that $v(t,x)$ is not necessary a smooth function of t for $t > 0$: the regularity of v in time now depends on the regularity of the function c.

Next, consider the equation

$$du(t,x) = au_{xx}(t,x)dt + \sigma u(t,x)dw(t),\ t>0,\ x\in\mathbb{R}, \tag{2.3.5}$$

where a, σ are real numbers, $a > 0$. By analogy with (2.3.4), let us look for a solution in the form $u(t,x) = u^{\text{heat}}(t,x)X(t)$ for some process $X(t)$ with $X(0) = 1$. Then $du = u_t^{\text{heat}}X(t)dt + u^{\text{heat}}dX(t) = au_{xx}dt + udX(t)/X(t)$. Therefore, $dX(t) = \sigma X(t)dw(t)$ or $X(t) = e^{\sigma w(t)-(\sigma^2 t/2)}$. As a result,

$$u(t,x) = u^{\text{heat}}(t,x)e^{\sigma w(t)-(\sigma^2 t/2)}. \tag{2.3.6}$$

Exercise 2.3.2 (C) Establish (2.3.6) using the Fourier transform.

Formula (2.3.6) suggests that, similar to the deterministic heat equation, Eq. (2.3.5) is natural to call parabolic: if $u(0,\cdot) \in L_p(\mathbb{R})$, $1 \le p \le +\infty$, then the solution $u(t,x)$ is a smooth function of x, $t > 0$, and, as a function of t, has the same regularity as the Wiener process w (Hölder continuous of any order less than $1/2$).

If B is a Wiener process independent of w then (2.3.3) and (2.3.6) imply

$$u(t,x) = e^{\sigma w(t)-(\sigma^2 t/2)}\,\mathbb{E}u\big(0, x+\sqrt{2a}\,B(t)\big). \tag{2.3.7}$$

Exercise 2.3.3 (B) Find representations of the solution of the equation $du(t,x) = au_{xx}(t,x)dt + h(t)u(t,x)dw(t)$ similar to (2.3.6) and (2.3.7). What conditions on the process h do you need? Can h be random? **Hint.** See (2.3.8) below.

To summarize, the equation

$$du = (au_{xx} + c(t)u)dt + h(t)udw(t)$$

has the closed-form solution

$$u(t,x) = u^{\text{heat}}(t,x)\exp\left(\int_0^t c(s)ds + \int_0^t h(s)dw(s) - \frac{1}{2}\int_0^t h^2(s)ds\right), \tag{2.3.8}$$

where u^{heat} is defined in (2.3.2).

Next, consider the equation

$$du(t,x) = au_{xx}(t,x)dt + \sigma u_x(t,x)dw(t),\ t>0,\ x\in\mathbb{R}, \tag{2.3.9}$$

a, σ are real numbers, $a > 0$. To find a representation of the solution, take the Fourier transform in the x variable,

$$\widehat{u}(t,y) = \frac{1}{\sqrt{2\pi}}\int_{\mathbb{R}} e^{-\mathrm{i}xy}u(t,x)dx,\ \mathrm{i} = \sqrt{-1}.$$

Then (2.3.9) implies that, for each y, $\widehat{u}(t,y)$ is a geometric Brownian motion

$$d\widehat{u}(t,y) = -ay^2\widehat{u}(t,y)dt + \mathrm{i}\sigma\, y\widehat{u}(t,y)dw(t)$$

or

$$\widehat{u}(t,y) = \widehat{u}(0,y)e^{-(a-(\sigma^2/2))y^2t+\mathrm{i}\sigma yw(t)}.$$

If $c = a - (\sigma^2/2) > 0$, then we can invert the Fourier transform to find

$$u(t,x) = v(t,x+\sigma w(t)),\ v(t,x) = \frac{1}{\sqrt{4c\pi t}}\int_{\mathbb{R}} e^{-|x-y|^2/(4ct)}u(0,y)dy. \tag{2.3.10}$$

Note by comparison that there are no restriction on σ to solve (2.3.5).

Alternatively, we can write (2.3.10) as

$$u(t,x) = \mathbb{E}\left(u\big(0,x+\sqrt{2a-\sigma^2}\widetilde{w}(t)+\sigma w(t)\big)|\mathcal{F}_t^w\right), \tag{2.3.11}$$

where $\widetilde{w}$ is a Wiener process independent of w and $\mathcal{F}_t^w$ is the sigma-algebra generated by w up to time t.

If $2a = \sigma^2$, then $u(t,x) = u_0(x+\sigma w(t))$, and, for $t > 0$, the solution, as a function of x, is as smooth as the initial condition. In this case, Eq. (2.3.9) is called `degenerate parabolic`.

If $2a < \sigma^2$, then we cannot invert the Fourier transform unless $\widehat{u}(0,y)$ is very special, for example, has compact support. In this case, the initial value problem for (2.3.9) is ill-posed in $L_q(\Omega;L_p((0,T);\mathbb{R}))$ for every $p,q \geq 1$.

Note that Eq. (2.3.9) in the Stratonovich form is

$$du = (a-(\sigma^2/2))u_{xx}dt + \sigma u_x \circ dw(t). \tag{2.3.12}$$

In particular, if $2a = \sigma^2$, we get a first-order equation $du = \sigma u_x \circ dw(t)$ and a full analogy with the deterministic `transport equation` $v_t = h(t)v_x$, whose solution is $v(t,x) = v\Big(0,x+\int_0^t h(s)ds\Big)$.

Exercise 2.3.4 (C)

(a) Verify (2.3.10)–(2.3.12).
(b) Write (2.3.7) as a conditional expectation similar to (2.3.11).

We now summarize some of the above observations as follows:

- The equation

$$du = \big((a+\frac{\sigma^2}{2})u_{xx} + b(t)u_x + c(t)u\big)dt + \big(\sigma u_x + h(t)u\big)dw(t), x \in \mathbb{R},$$

has closed-form solution

$$u(t,x) = u^{\text{heat}}\left(t, x + \int_0^t b(s)ds + \sigma w(t)\right)$$
$$\times \exp\left(\int_0^t c(s)ds + \int_0^t h(s)dw(s) - \frac{1}{2}\int_0^t h^2(s)ds\right), \tag{2.3.13}$$

where u^{heat} is defined in (2.3.2).

- In some sense, the first derivative in the stochastic part of an SPDE can be as powerful as the second derivative in the deterministic part.
- Probabilistic representation of solutions of SPDEs involves an additional Wiener process and conditional expectation.

To finish with parabolic equations on the whole line, the reader can think about the following:

Exercise 2.3.5 (A) Can you extend formula (2.3.13) to time-dependent a and σ? What conditions on the function $a = a(t)$ will you need? How much of the above will work for equations in $\mathbb{R}^{\mathrm{d}}$, $\mathrm{d} > 1$? **Hint.** Use the Fourier transform.

Next, consider the heat equation driven by additive space-time white noise on the bounded interval $(0,\pi)$: $u_t = u_{xx} + \dot{W}(t,x)$. Equivalently, if $\mathfrak{h}_k$, $k \geq 1$, is an orthonormal basis in $L_2((0,\pi))$, and w_k, $k \geq 1$, are independent standard Brownian motions, then

$$du = u_{xx}dt + \sum_{k=1}^{\infty} \mathfrak{h}_k(x)dw_k(t),\ t > 0,\ x \in (0,\pi). \tag{2.3.14}$$

We assume zero boundary conditions $u(t,0) = u(t,\pi) = 0$, and, for simplicity, zero initial condition: $u(0,x) = 0$. Given zero boundary condition, it is natural to take

$$\mathfrak{h}_k(x) = \sqrt{2/\pi}\,\sin(kx),\ k \geq 1.$$

We look for the solution in the form

$$u(t,x) = \sum_{k=1}^{\infty} u_k(t)\mathfrak{h}_k(x). \tag{2.3.15}$$

Then

$$du_k(t) = -k^2 u_k(t)dt + dw_k(t),\ u_k(t) = \int_0^t e^{-k^2(t-s)}dw_k(s), \tag{2.3.16}$$

and

$$u(t,x) = \sum_{k=1}^{\infty} \mathfrak{h}_k(x)\int_0^t e^{-k^2(t-s)}dw_k(s). \tag{2.3.17}$$

Note that

$$\mathbb{E}u^2(t,x) = \sum_{k=1}^{\infty} \mathfrak{h}_k^2(x) \int_0^t e^{-2k^2(t-s)}ds = \sum_{k=1}^{\infty} \mathfrak{h}_k^2(x)\, \frac{1-e^{-2k^2t}}{2k^2}, \tag{2.3.18}$$

that is, the series in (2.3.17) converges with probability one for each $(t,x) \in (0,\infty)\times[0,\pi]$. Moreover, by the Burkholder-Davis-Gundy inequality (1.1.16) on page 8,

$$\begin{aligned} \mathbb{E}|u(t,x_1) - u(t,x_2)|^p &\le C(p,\delta)|x_1 - x_2|^{(1-2\delta)(p/2)}, \\ \mathbb{E}|u(t_1,x) - u(t_2,x)|^p &\le C(p,\delta)|t_1 - t_2|^{(1/2-\delta)(p/2)} \end{aligned} \tag{2.3.19}$$

for every $p \ge 1$ and $\delta \in (0, 1/2)$. By the Kolmogorov criterion we conclude that the solution $u = u(t,x)$ is Hölder continuous of any order less than $1/4$ as a function of t, and is Hölder continuous of any order less than $1/2$ as a function of x. The details are left to the reader:

Exercise 2.3.6 (C) Verify (2.3.19) and confirm the stated Hölder continuity of the solution. **Hint.** Note that, for every $\beta \in (0,1)$, we have $|\mathfrak{h}_k(x_1) - \mathfrak{h}_k(x_2)|^{2\beta} \le 2k^{2\beta}|x_1 - x_2|^{2\beta}$ and, as long as $t_2 > t_1$, we also have $|1 - e^{-k^2(t_2-t_1)}| \le k^{2\beta}|t_1 - t_2|^{\beta}$. Then take $\beta = 1/2 - \delta$.

Note that the coefficients u_k defined by (2.3.16) are independent Gaussian processes. In the limit $t \to +\infty$, each u_k converges in distribution to a normal random variable with zero mean and variance

$$\lim_{t\to\infty} \int_0^t e^{-2k^2(t-s)}ds = \frac{1}{2k^2}.$$

Therefore, as $t \to +\infty$, $u(t,x)$ from (2.3.17) converges in distribution to

$$v(x) = \sum_{k=1}^{\infty} \frac{\xi_k}{\sqrt{2}\,k}\, \mathfrak{h}_k(x), \tag{2.3.20}$$

where $\xi_k,\ k \ge 1$, are independent standard normal random variables.

Exercise 2.3.7 (C) Consider the equation $du = u_{xx}dt + dw(t)$ on the interval $(0,\pi)$ with zero initial condition, where w is a standard Brownian motion. You have a choice of three possible boundary conditions: zero Dirichlet, zero Neumann, and periodic.

(a) Which choice of the three boundary conditions is the least convenient to study the equation?
(b) With your choice of boundary conditions, find the solution and determine its regularity in time and space. Is there a limit, as $t \to \infty$, similar to (2.3.20)? (Note that $u(t,x) = w(t)$ CAN be a solution.)

Analysis of equation $du = \boldsymbol{\Delta} u dt + dW(t,x)$, $x \in \mathbb{R}^{\mathrm{d}}$, is Problem 2.3.5.

2.3.1.1 Further Directions

Equations with multiplicative space-time white noise $du = (u_{xx} + f(u, u_x))dt + g(u)dW(t,x)$ usually do not have closed-form solutions, but have been studied extensively. In particular, under suitable assumptions on f and g, the solution has the same regularity in time and space as in the additive case; see the lecture notes by Walsh [223, Chap. 3] for equations on an interval, and the paper by Krylov [120, Sect. 8.3] for equations on the whole line.

2.3.2 Wave Equation

Consider the stochastic wave equation in $\mathbb{R}$ with zero initial conditions, driven by the space-time white noise:

$$u_{tt}(t,x) = u_{xx}(t,x) + \dot{W}(t,x),\ t > 0,\ x \in \mathbb{R},\ u(0,x) = u_t(0,x) = 0. \tag{2.3.21}$$

The function $u = u(t,x)$ is the displacement of the string at time t at point x. Define $\bar{t} = (t-x)/\sqrt{2}$, $\bar{x} = (t+x)/\sqrt{2}$, $\bar{u}(\bar{t},\bar{x}) = u(t,x)$, $\dot{\widetilde{W}}(\bar{t},\bar{x}) = \dot{W}(t,x)$. Then (2.3.21) becomes

$$2\frac{\partial^2 \bar{u}}{\partial \bar{t}\partial \bar{x}} = \dot{\widetilde{W}}(\bar{t},\bar{x}),$$

and so

$$u(t,x) = \frac{1}{2}\widetilde{W}\left(\frac{t-x}{\sqrt{2}}, \frac{t+x}{\sqrt{2}}\right) = \dot{W}(G(t,x)) = (\dot{W}, \mathbf{1}_{G(t,x)}), \tag{2.3.22}$$

where $G(t,x)$ is the domain of dependence of the point (t,x) (the equilateral right triangle ABC on Fig. 2.1) and $\mathbf{1}_{G(t,x)}$ is the indicator function of the set $G(t,x)$ (see also (1.1.7) on page 4). It is known (see, for example, Evans [50, Page 81, Example (i)]) that if $f = f(t,x)$ is a smooth function, then the solution of the one-dimensional

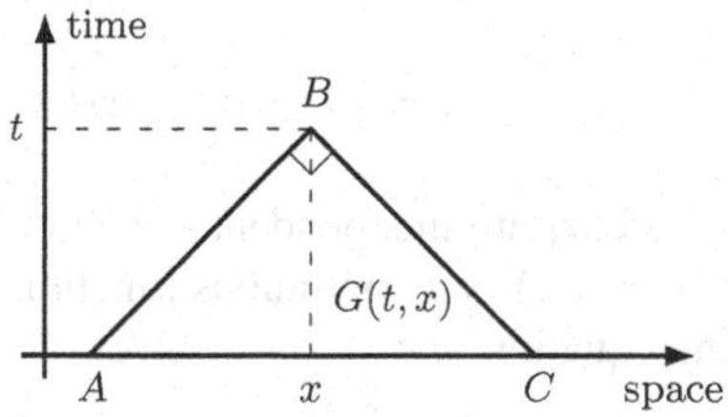

Fig. 2.1 Solving wave equation

wave equation on the line, $u_{tt} = u_{xx} + f(t,x)$ with zero initial conditions, is

$$u(t,x) = \int_{G(t,x)} f(s,y)\, dsdy. \tag{2.3.23}$$

Formula (2.3.23) can be used to get an alternative derivation of (2.3.22).

Two more examples of the stochastic wave equation on the line are left as an exercise.

Exercise 2.3.8 (C)

(a) Show that the solution of the equation $u_{tt} = u_{xx} + \dot{w}(t)$, $x \in \mathbb{R}$, where $\dot{w}(t)$ is the time-only white noise, is $u(t,x) = \int_0^t (t-s)dw(s)$.
(b) Use the Fourier transform to solve the equation $u_{tt} = u_{xx} + u\dot{w}$, $t > 0$, $x \in \mathbb{R}$, where $\dot{w}$ is white noise in time. At which point do you have to stop? (See the remark below.)

Remark 2.3.9 In the case of the heat equation, our ability to consider equations with multiplicative noise was guaranteed by solvability in closed form of the first-order differential equation $y'(t) = a(t)y(t)$ for any locally integrable function a. The corresponding second-order equation $y''(t) = a(t)y(t)$ does not have a closed-form solution for general functions a. As a result, we are unable to find closed-form solutions for the stochastic wave equation with multiplicative noise.

Let us now consider the wave equation on an interval; for simplicity, we assume zero initial and boundary conditions:

$$u_{tt} = u_{xx} + \dot{W}(t,x),\ t > 0,\ x \in (0,\pi), \tag{2.3.24}$$

where $u(0,x) = u_t(0,x) = u(t,0) = u(t,\pi) = 0$, and $\dot{W}(t,x)$ is space-time white noise. The physical interpretation of (2.3.24) could be the sound produced by a one-string guitar left out in a sandstorm. Similar to (2.3.15) on page 55 we look for the solution as a Fourier series

$$u(t,x) = \sqrt{\frac{2}{\pi}} \sum_{k=1}^{\infty} u_k(t) \sin(kx).$$

Substitution into (2.3.24) results in

$$u_k''(t) = -k^2 u_k(t) + \dot{w}_k(t),\ t > 0,\ u_k(0) = u_k'(0) = 0, \tag{2.3.25}$$

where w_k are independent standard Brownian motions. To solve (2.3.25) recall that if $f = f(t)$ is a continuous function and b, c are real numbers, then the solution of the equation

$$y''(t) + by'(t) + cy(t) = f(t),\ t > 0 \tag{2.3.26}$$

with zero initial conditions $y(0) = y'(0) = 0$ is given by

$$y(t) = \int_0^t g(t-s)f(s)ds, \tag{2.3.27}$$

where $g = g(t)$ satisfies $g''(t) + bg'(t) + cg(t) = 0$, $g(0) = 0$, $g'(0) = 1$. Therefore, it is natural to expect that

$$u_k(t) = \frac{1}{k}\int_0^t \sin(k(t-s))dw_k(s), \tag{2.3.28}$$

and direct computations show that this function u_k indeed satisfies (2.3.25).

Exercise 2.3.10 (B)

(a) Verify that (2.3.27) indeed solves (2.3.26).
(b) Using integration by parts, verify that (2.3.28) satisfies (2.3.25), that is, $u_k(t) = \int_0^t v_k(s)ds$, and $v_k(t) = -k^2\int_0^t u_k(s)ds + w_k(t)$.

As a result, the solution of (2.3.24) is

$$u(t,x) = \sqrt{\frac{2}{\pi}}\sum_{k=1}^{\infty}\frac{1}{k}\left(\int_0^t \sin(k(t-s))dw_k(s)\right)\sin(kx). \tag{2.3.29}$$

Note that, for each t and x, $u(t,x)$ is a Gaussian random variable with zero mean and variance

$$\begin{aligned}\mathbb{E}u^2(t,x) &= \frac{2}{\pi}\sum_{k=1}^{\infty}\frac{1}{k^2}\left(\int_0^t \sin^2(k(t-s))ds\right)\sin^2(kx)\\ &= \frac{2}{\pi}\sum_{k=1}^{\infty}\frac{1}{k^2}\left(\frac{t}{2}-\frac{\sin(2kt)}{4k}\right)\sin^2(kx).\end{aligned} \tag{2.3.30}$$

Exercise 2.3.11 (A) Using estimates similar to (2.3.19) on page 56 (see also Exercise 2.3.6), verify that, as a function of x, $u(t,\cdot)$ is Hölder continuous of order less that $1/2$. What can you say about regularity of u as a function of t?

Note that $\mathbb{E}u^2(t,x)$ does not have a limit as $t \to \infty$. As a result, unlike the heat equation (2.3.14) on page 55, the solution of the stochastic wave equation (2.3.24) does not have a limiting distribution. One possibility to have a wave equation with a limiting distribution is to introduce damping; see Problem 2.3.2(c) on page 68.

2.3.3 Poisson Equation

Consider the equation

$$-u''(x) = \sin x + \dot{W}(x),\ x \in (0,\pi),\ u(0) = u(\pi) = 0. \tag{2.3.31}$$

We look for the solution in the form of a Fourier series

$$u(x) = \sum_{k=1}^{\infty} u_k \mathfrak{h}_k(x),$$

where the Fourier coefficients u_k are random variables and

$$\mathfrak{h}_k(x) = \sqrt{2/\pi}\,\sin(kx),\ k \ge 1. \tag{2.3.32}$$

Writing $\dot{W}(x) = \sum_{k=1}^{\infty} \xi_k \mathfrak{h}_k(x)$, where $\xi_k,\ k \ge 1$, are independent standard Gaussian random variables, we conclude that $u_k = \xi_k/k^2,\ k \ge 2$, and

$$u(x) = \sin x + \sum_{k=1}^{\infty} \frac{\xi_k}{k^2}\,\mathfrak{h}_k(x). \tag{2.3.33}$$

The series in (2.3.33) converges uniformly in x for each $\omega \in \Omega$, and also converges in $L_2\big(\Omega; L_2((0,\pi))\big)$. Also, $\mathbb{E}u(x) = \sin x$ and $v(x) = \sin x$ is the solution of the unperturbed equation $-v''(x) = \sin x$.

Exercise 2.3.12 (C) Derive the following representation for the function u:

$$u(x) = \sin x + x\int_0^{\pi} W(y)dy - \int_0^x W(y)dy. \tag{2.3.34}$$

Suggestion: start by deriving the formula

$$u(x) = x\int_0^{\pi} (\pi - y)f(y)dy - \int_0^x (x-y)f(y)dy$$

for the solution of $u_{xx}(x) = f(x)$ on $(0,\pi)$ with zero boundary conditions and a continuous function f.

In the following exercise, you have an option of using either (2.3.33) or (2.3.34). Make sure to use the representation that is more convenient for the given question.

Exercise 2.3.13 (C)

(a) Compute $\mathbb{E}\int_0^{\pi} u^2(x)dx$. **Hint**: $\sum_{k\ge1} k^{-4} = \pi^4/90$.
(b) Verify that the function $u = u(x)$ is continuously differentiable for $x \in (0,\pi)$ and the first derivative u' is Hölder continuous of any order less than $1/2$.

Next, consider

$$-u''(x) = \sin x + \xi\, u(x),\ x \in (0, \pi),\ u(0) = u(\pi) = 0, \tag{2.3.35}$$

where ξ is a standard normal random variable. Direct computations show that

$$u(x) = \frac{\sin x}{1 - \xi}, \tag{2.3.36}$$

and, while $u(x)$ is well-defined with probability one, $\mathbb{E}u(x)$ does not exist in the usual sense. The principal-value trick (see Problem 2.3.13 below) makes it possible to define $\mathbb{E}u(x)$ in the generalized sense, but the result is not zero.

To summarize, a simple multiplicative perturbation of an elliptic equation does not preserve the mean value of the solution.

Let us modify Eq. (2.3.35) to preserve the mean valuc of thc solution. To this end, recall the `Hermite polynomials` $\mathrm{H}_n(x)$, best understood via the equality

$$e^{xt-(t^2/2)} = \sum_{n=0}^{\infty} \frac{\mathrm{H}_n(x)}{n!}\, t^n. \tag{2.3.37}$$

In particular, $\mathrm{H}_0(x) = 1$, $\mathrm{H}_1(x) = x$, $\mathrm{H}_2(x) = x^2 - 1$. The reason for considering the Hermite polynomials is the following: if ξ is a standard normal random variable, then, by direct computation,

$$\mathbb{E}\mathrm{H}_n(\xi) = 0,\ n \geq 1. \tag{2.3.38}$$

Exercise 2.3.14 (C) Verify that

$$\mathbb{E}\big(\mathrm{H}_n(\xi)\mathrm{H}_k(\xi)\big) = \begin{cases} n! & \text{if } n = k; \\ 0 & \text{if } n \neq k. \end{cases} \tag{2.3.39}$$

Hint. Use (2.3.37) to write $\Upsilon(t, s) = e^{s\xi-(s^2/2)}e^{t\xi-(t^2/2)}$ as a power series in (t, s). Take expected value on both sides and note that $\mathbb{E}\Upsilon(t, s) = e^{((t+s)^2-t^2-s^2)/2} = e^{ts}$.

Let us modify Eq. (2.3.35) in such a way that

$$u(x) = \sin x \sum_{n=0}^{\infty} u_n(x)\mathrm{H}_n(\xi),$$

with suitable functions $u_n(x)$, is a solution of the modified equation. Even if the series diverges, the equality $\mathbb{E}\mathrm{H}_n(\xi) = 0$ for $n \geq 1$ allows us to claim that, in some sense, $\mathbb{E}u(x) = \sin x$.

The only natural place for a modification of the equation is the definition of the product $\xi\, u$: instead of the usual multiplication, we will use a different operation

$\xi \diamond u$, defined so that

$$\mathrm{H}_n(\xi) \diamond \mathrm{H}_k(\xi) = \mathrm{H}_{n+k}(\xi),\ n, k \geq 0. \tag{2.3.40}$$

By linearity, this definition extends to a larger class of random variables, because the collection $\{\mathrm{H}_n(\xi),\ n \geq 0\}$ is an orthogonal basis in the space of square-integrable random variables that are functions of ξ. Then direct computations show that the formal series

$$u(x) = \sin(x) \sum_{n=0}^{\infty} \mathrm{H}_n(\xi) \tag{2.3.41}$$

satisfies

$$-u_{xx}(x) = \sin x + \xi \diamond u(x),\ x \in (0, \pi),\ u(0) = u(\pi) = 0. \tag{2.3.42}$$

Exercise 2.3.15 (C) Keeping in mind that $\mathrm{H}_1(x) = x$, verify that (2.3.41) satisfies (2.3.42).

Note that (2.3.41) is a formal series. The series diverges with probability one, and, similar to (2.3.36), we have $\mathbb{E}|u(x)|^p = +\infty$ for all $x \in (0, \pi)$ and $p \geq 1$. Still, because of (2.3.38), it is reasonable to expect that, in some generalized sense, we have $\mathbb{E}u(x) = \sin x$. The following exercise provides more evidence that the operation $\diamond$ preserves the mean structure of the equation.

Exercise 2.3.16 (B)

(a) Show that the function $u(t, x) = e^{-t} \sin x$ satisfies the heat equation $u_t = u_{xx},\ u(0, x) = \sin x$.
(b) Let ξ be a standard normal random variable. Show that the function $u(t, x) = \big(e^{-t} \sin x\big) e^{\xi t}$ belongs to $L_2\big(\Omega \times (0, T) \times (0, \pi)\big)$, satisfies the stochastic heat equation $u_t = u_{xx} + \xi u,\ u(0, x) = \sin x$, and $\mathbb{E}u(t, x) = e^{-t+(t^2/2)} \sin x$.
(c) Let ξ be a standard normal random variable. Show that the function

$$u(t, x) = e^{-t} \sin x \sum_{n=0}^{\infty} \frac{t^n \mathrm{H}_n(\xi)}{n!} = \big(e^{-t} \sin x\big) e^{\xi t - (t^2/2)}$$

belongs to $L_2\big(\Omega; L_2\big((0, T); L_2((0, \pi))\big)\big)$, satisfies the stochastic heat equation $u_t = u_{xx} + \xi \diamond u,\ u(0, x) = \sin x$, and $\mathbb{E}u(t, x) = e^{-t} \sin x$.

We now summarize some of the above observations as follows:

- The solution of an elliptic equation with multiplicative noise is unlikely to belong to $L_2(\Omega)$.
- To preserve the mean structure of an equation with multiplicative spatial noise, the operation $\diamond$ from (2.3.40) is the right choice.

The operation $\diamond$ is known as the `Wick product`, after the Italian theoretical physicist GIAN-CARLO WICK (1909–1992) who introduced it in 1950 in connection with a problem from quantum field theory (see [225]). The paper [75] provides one of the first mathematical discussions of $\diamond$.

2.3.4 *Nonlinear Equations*

Let us start by reviewing the reduction of certain semi-linear parabolic equations to the heat equation; for details, see Evans [50, Sect. 4.4.1]. In what follows, a and b are real numbers and $a > 0$.

If $u = u(t,x)$ satisfies $u_t = au_{xx} - bu_x^2$, $x \in \mathbb{R}$, and $v(t,x) = e^{-bu(t,x)/a}$, then, by direct computation, $v_t = av_{xx}$, and

$$u(t,x) = u^*(t,x) = -\frac{a}{b}\ln\left(\frac{1}{\sqrt{4\pi at}}\int_{\mathbb{R}} e^{-(x-y)^2/(4at)}e^{-bu(0,y)/a}dy\right). \tag{2.3.43}$$

Similarly, if $u_t = au_{xx} - uu_x$, then $u(t,x) = \widetilde{u}_x(t,x)$, where $\widetilde{u}_t = a\widetilde{u}_{xx} - (\widetilde{u}_x^2)/2$, and

$$u(t,x) = u^{**}(t,x) = -2a\frac{\partial}{\partial x}\ln\left(\int_{\mathbb{R}} e^{-(x-y)^2/(4at)}e^{-\widetilde{u}(0,y)/(2a)}dy\right), \tag{2.3.44}$$

where $\widetilde{u}(0,x)$ is an anti-derivative of $u(0,x)$.

Exercise 2.3.17 (B)

(a) Verify that (2.3.44) follows from (2.3.43).
(b) Write (2.3.43) and (2.3.44) in the probabilistic form similar to (2.3.3) on page 52.
(c) Show that u^{**} does not depend on the choice of an anti-derivative of $u(0,x)$.

The Austrian-American mathematician EBERHARD FREDERICH FERDINAND HOPF (1902–1983), of the Wiener-Hopf equations fame, and the American mathematician JULIAN DAVID COLE (1925–1999) came up with formula (2.3.44) independently and at about the same time [29, 79]. The `Hopf-Cole (or Cole-Hopf) transformation` can mean either one of the formulas (2.3.43), (2.3.44), or the corresponding change of the unknown function.

The Hopf-Cole transformation also works for "inhomogeneous" equations.

Exercise 2.3.18 (C)

(a) Verify that if

$$u_t = au_{xx} - bu_x^2 + g(t,x),$$

then $v = e^{-bu/a}$ satisfies

$$v_t = av_{xx} - \frac{b}{a}vg$$

(b) Verify that if

$$u_t = au_{xx} - uu_x + G_x(t,x), \tag{2.3.45}$$

then $u = -2av_x/v$, where v satisfies

$$v_t = av_{xx} - \frac{1}{2a}vG.$$

As usual, $G_x = \partial G/\partial x$.

We will now investigate the Hopf-Cole transformation for the corresponding equations with additive noise. To begin, consider

$$du = (au_{xx} - bu_x^2)dt + g(t,x)dw(t),\ t > 0,\ x \in \mathbb{R}, \tag{2.3.46}$$

where w is a standard Brownian motion and g is a known function. Substitution $v = e^{-bu/a}$ results in $dv = -(bv/a)du + (1/2)(bg/a)^2 v dt$ (Itô formula) and the following equation for v:

$$dv = \left(av_{xx} + \frac{b^2g^2}{2a^2}\,v\right)dt - \frac{b}{a}\,v\,g\,dw(t). \tag{2.3.47}$$

Given the amount of calculations, all the details are left to the reader.

Exercise 2.3.19 (C)

(a) Verify (2.3.47).
(b) Show that, in the special case $g(t,x) = 1$, the solution of (2.3.46) is $u(t,x) = u^*(t,x) + w(t)$, where u^* is given by (2.3.43). **Hint.** The corresponding equation for v is solvable by (2.3.13) on page 55.

Next, consider

$$du = (au_{xx} - uu_x)dt + dw(t),\ t > 0,\ x \in \mathbb{R}. \tag{2.3.48}$$

This time, $u(t,x) = -2av_x(t,x)/v$, and

$$dv = \left(av_{xx} + \frac{x^2}{8a^2}\,v\right)dt - \frac{1}{2a}\,v\,x\,dw(t). \tag{2.3.49}$$

Once again, all the details are left to the reader.

Exercise 2.3.20 (B)

(a) Verify (2.3.49).
(b) Show that the solution of (2.3.48) is

$$u(t,x) = u^{**}\left(t, x - \int_0^t w(s)ds\right) + w(t), \tag{2.3.50}$$

where u^{**} is given by (2.3.44). **Hint.** One more substitution

$$v(t,x) = \exp(-xw(t)/(2a))V(t,x)$$

leads to

$$V_t = V_{xx} - w(t)V_x + \frac{w^2(t)}{4a}V,$$

which can be solved by (2.3.13) on page 55.

Next, we investigate some non-linear equations with multiplicative time-only noise. We start with a generalization of the result from Exercise 2.3.1 on page 52 to nonlinear equations. Consider the equation

$$v_t = F(v, v_x, v_{xx}, \ldots). \tag{2.3.51}$$

Definition 2.3.21

(a) The function F is called `homogeneous of degree` $\gamma \geq 1$ if, for every $\lambda > 0$,

$$F(\lambda x, \lambda y, \lambda z, \ldots) = \lambda^\gamma F(x, y, z, \ldots). \tag{2.3.52}$$

(b) Equation (2.3.51) is called `homogeneous of degree` $\gamma \geq 1$ if the function F is homogeneous of degree γ.

Let $f = f(t)$ be a locally integrable function and define

$$g_\gamma(t) = \int_0^t e^{(\gamma-1)\int_0^s f(r)dr}ds. \tag{2.3.53}$$

Next, consider the following equation:

$$u_t = F(u, u_x, u_{xx}, \ldots) + uf(t). \tag{2.3.54}$$

Proposition 2.3.22 *Assume that F is homogenous of degree γ. A function $v = v(t,x)$ is a solution of* (2.3.51) *if and only if $u(t,x) = v(g_\gamma(t), x)e^{\int_0^t f(s)ds}$ is a solution of* (2.3.54).

Proof Assume that v satisfies (2.3.51). To simplify the notations, denote $e^{\int_0^t f(s)ds}$ by $\widetilde{f}(t)$. We have $u_x = v_x(g_\gamma(t),x)\widetilde{f}(t)$, $u_{xx} = v_{xx}(g_\gamma(t),x)\widetilde{f}(t)$, etc., and, since $g'_\gamma(t) = e^{(\gamma-1)\int_0^t f(s)ds} = \widetilde{f}^{\gamma-1}(t)$, $u_t = v_t\widetilde{f}^\gamma(t) + f(t)u$. Therefore, by (2.3.52), $u_t = F(\widetilde{f}(t)v, \widetilde{f}(t)v_x, \ldots) + f(t)u$, that is, u satisfies (2.3.54). The proof in other direction is left as an exercise.

Exercise 2.3.23 (B) Complete the proof of the proposition, that is, show that if u solves (2.3.54), then v solves (2.3.51).

Consider next the stochastic Itô equation for the random field $u = u(t,x)$, $t > 0$, $x \in \mathbb{R}$:

$$du = F(u, u_x, u_{xx}, \ldots)dt + u\left(g(t)dt + f(t)dw(t)\right), \tag{2.3.55}$$

where $f = f(t)$ is a locally square-integrable deterministic function, g is a locally integrable deterministic function, and w is a standard Brownian motion. Define the functions

$$\begin{aligned} h(t) &= \exp\left(\int_0^t g(s)ds + \int_0^t f(s)dw(s) - \frac{1}{2}\int_0^t f^2(s)ds\right), \\ H_\gamma(t) &= \int_0^t h^{\gamma-1}(s)ds. \end{aligned} \tag{2.3.56}$$

Proposition 2.3.24 *Assume that the function F is homogeneous of degree γ. A function $v = v(t,x)$ is a solution of* (2.3.51) *if and only if $u(t,x) = v(H_\gamma(t),x)h(t)$ is a solution of* (2.3.55).

Exercise 2.3.25 (C)

(a) Prove Proposition 2.3.24.
(b) Find a solution of the equation $du(t,x) + u(t,x)u_x(t,x)dt = u(t,x)dw(t)$, $t > 0$, $x \in \mathbb{R}$, where w is a standard Brownian motion. **Hint.** For the equation $v_t + vv_x = 0$ use separation of variables $v(t,x) = F(t)G(x)$ to find $v(t,x) = x/t$. Then put $f = 1, g = 0$ in (2.3.56) to find $u(t,x) = xh(t)/H_2(t)$.

In the special case of equation $u_t = \boldsymbol{\Delta}(u^\gamma) + au$, $a = const.$, the result of Proposition 2.3.24 appears in [68]. For the general case, see [146].

As an application of Proposition 2.3.24, consider the `porous medium equation`

$$v_t(t,x) = \left(v^\gamma(t,x)\right)_{xx},\ t > 0,\ x \in \mathbb{R};\ \gamma > 1. \tag{2.3.57}$$

This equation is homogeneous of degree γ. The function

$$v^{\mathrm{BT}}(t,x) = \frac{1}{t^\alpha}\left(\max\left(b - \frac{\gamma-1}{2\gamma}\beta\frac{|x|^2}{t^{2\beta}},\ 0\right)\right)^{1/(\gamma-1)}, \tag{2.3.58}$$

where $b > 0$ and $\alpha = \beta = 1/(\gamma + 1)$, is known as `Barenblatt's solution` of (2.3.57), named after GRIGORY ISAAKOVICH BARENBLATT, who introduced and studied it in [7]. For a derivation of (2.3.58), see Evans [50, Sect. 4.2.2]; for even more information on the porous medium equation, see Aronson [3] and Vázquez [219]. Notice that the function v^{BT} has compact support in x for every $t > 0$, and the boundary points x^* of the support (the `interface`) satisfy

$$|x^*| = ct^{\beta} \tag{2.3.59}$$

for some number $c = c(b, \gamma)$. Now consider the stochastic porous medium equation

$$du(t,x) = (u^{\gamma}(t,x))_{xx}dt + u(t,x)f(t)dw(t),\ t > 0,\ x \in \mathbb{R}, \tag{2.3.60}$$

where f is a locally square-integrable deterministic function and w is a standard Brownian motion. Notice that it is important to use multiplicative noise to preserve positivity (and hence the potential physical interpretation) of the solution.

Let $h = h(t)$ and $H_{\gamma} = H_{\gamma}(t)$ be defined by (2.3.56). By Proposition 2.3.24,

$$u^{\mathrm{BT}}(t,x) = v^{\mathrm{BT}}(H_{\gamma}(t), x)h(t) \tag{2.3.61}$$

is a solution of (2.3.60), where $v^{\mathrm{BT}}(t,x)$ is given by (2.3.58). This solution has compact support for all $t > 0$, and the interface points x^* satisfy

$$|x^*| = c\,(H_{\gamma}(t))^{\beta}. \tag{2.3.62}$$

If $\int_0^{\infty} f^2(t)dt < \infty$, then $\int_0^t f(s)dw(s)$, being a zero-mean square-integrable Gaussian martingale, converges with probability one, as $t \to \infty$, to a Gaussian random variable. Therefore, for large t, we have

$$|x^*| \approx \eta\, t^{\beta} \tag{2.3.63}$$

for some random variable η.

For a more detailed analysis of (2.3.60), including physical interpretations, see [146].

Exercise 2.3.26 (A) Show that if $\int_0^{\infty} f^2(t)dt < \infty$, then (2.3.63) holds. Find the distribution of the random variable η.

2.3.5 Problems

The following problems provide more examples of equations with closed-form solutions. Problem 2.3.1 extends (2.3.9) on page 53 to several dimensions of x and many Brownian motions. Problem 2.3.2 is purely computational and reinforces several ideas about the heat and wave equations. Problem 2.3.3 is another illustration that the condition $2a \geq \sigma^2$ is necessary to have a square-integrable solution for the heat equation (2.3.9) on page 53. Problem 2.3.6 is an extension of the Poisson equation (2.3.31) on page 60. Problem 2.3.7 is a stochastic Poisson equation on the whole line. Problem 2.3.8 provides examples of deterministic equations with stochastic initial or boundary conditions. Problem 2.3.9 describes several representations of the Brownian bridge, one of them using a stochastic heat equation. Problems 2.3.10–2.3.12 are about Hermite polynomials. Problem 2.3.13 shows how one can define the expected value of the random variable (2.3.36) using the principal value method. Problem 2.3.14 leads deeper into the subject of chaos solution for stochastic elliptic equations. Problem 2.3.15 is about Eq. (2.3.48) with a more general random perturbation. Problem 2.3.16 is an extension of the Hopf-Cole transformation to a multi-dimensional version of (2.3.45).

Problem 2.3.1 Let $w_k, k \geq 1$, be independent standard Brownian motions, and let $\sigma_{ik},\ i = 1, \ldots, d,\ k \geq 1$, be real numbers with the property $\sum_{k=1}^{\infty} \sigma_{ik}\sigma_{jk} < \infty$ for all i, j. Let $a_{ij},\ i, j = 1, \ldots, \mathrm{d}$ be real numbers. For $t > 0, x \in \mathbb{R}^{\mathrm{d}}$ consider the equation

$$du(t, x) = \sum_{i,j=1}^{\mathrm{d}} a_{ij} D_i D_j u(t, x) dt + \sum_{i=1}^{\mathrm{d}} \sum_{k=1}^{\infty} \sigma_{ik} D_i u(t, x) dw_k(t). \tag{2.3.64}$$

(a) Find a diffusion process $X = X(t)$ with $X(0) = 0$ such that the function $v(t, x) = u(t, x - X(t))$ satisfies a deterministic equation.
(b) Under what condition on the numbers a_{ij}, σ_{ik} will the initial value problem for the function v be well-posed?
(c) Assume that, instead of $x \in \mathbb{R}^{\mathrm{d}}$, we consider Eq. (2.3.64) in a smooth bounded domain $G \in \mathbb{R}^{\mathrm{d}}$ with zero Dirichlet boundary conditions. Let $X = X(t)$ and $v = v(t, x)$ be from part (a). Find the equation satisfied by v, including the new domain and the boundary conditions.

Problem 2.3.2 For each of the following equations, find a representation of the solution $u = u(t, x)$, compute $\mathbb{E}u^2(t, x)$, establish regularity of u in t, x, and determine whether the solution has a limiting distribution as $t \to +\infty$.

(a)

$$du = \frac{1}{2}\boldsymbol{\Delta} u dt + \exp\left(-\frac{|x|^2}{f(t)}\right) dw(t),\ t > 0,\ x \in \mathbb{R}^{\mathrm{d}}, \tag{2.3.65}$$

where $\boldsymbol{\Delta}$ is the Laplace operator, $u(0, x) = 0$, w is a standard Brownian motion, and $f = f(t)$ is a positive, continuous on $(0, +\infty)$, deterministic function.

(b)

$$du = u_{xx} dt + u dw(t),\ t > 0,\ x \in (0, \pi), \tag{2.3.66}$$

where $u(0,x) = x(\pi - x)$, $u(t,0) = u(t,\pi) = 0$, and w is a standard Brownian motion.

(c)

$$u_{tt} + bu_t = u_{xx} + f(t)\dot{W}(t,x),\ t > 0,\ x \in (0,\pi), \tag{2.3.67}$$

where $u(0,x) = u_t(0,x) = u(t,0) = u(t,\pi) = 0$, b is a real number, f is a continuous deterministic function, and $\dot{W}(t,x)$ is space-time white noise.

Problem 2.3.3 Let w be a standard Brownian motion. Show that the initial-boundary value problem

$$du = u_{xx}dt + 2u_x dw(t),\ t > 0,\ x \in (0,\pi), \tag{2.3.68}$$

with periodic boundary conditions, is ill-posed in $L_2(\Omega; L_2((0,T) \times (0,\pi)))$ for every $T > 0$.

Problem 2.3.4 Let G be the unit disk. Consider the boundary value problem

$$u_t = \boldsymbol{\Delta} u + \dot{W},\ t > 0,\ x \in G,\ u|_{t=0} = u|_{\partial G} = 0,$$

where $\dot{W}$ is space-time white noise. Show that $\mathbb{E}\|u(t,\cdot)\|^2_{L_2(G)} = +\infty$ for all $t > 0$.

Problem 2.3.5 Consider the initial value problem

$$u_t = \boldsymbol{\Delta} u + \dot{W},\ t > 0,\ x \in \mathbb{R}^{\mathrm{d}},\ u|_{t=0} = 0,$$

where $\dot{W}$ is space-time white noise. Verify that if $\mathrm{d} > 1$, then $\mathbb{E}u^2(t,x) = +\infty$ for all $t > 0$ and $x \in \mathbb{R}^{\mathrm{d}}$.

Problem 2.3.6 Let $\dot{W}$ be the space white noise in the d-dimensional cube $G = (0,\pi)^{\mathrm{d}}$. Consider the Poisson equation

$$\boldsymbol{\Delta} u = \dot{W},\ u|_{\partial G} = 0. \tag{2.3.69}$$

(a) For $\mathrm{d} = 2$, find the solution of the equation in the form

$$u(x,y) = \frac{2}{\pi}\sum_{n,k=1}^{\infty} u_{nk} \sin(nx)\sin(ky)$$

with suitable random variables u_{nk} and compute $\iint_G \mathbb{E}u^2(x,y)dxdy$.

(b) For what values of d is $\mathbb{E}\|u\|^2_{L_2(G)}$ finite?

Problem 2.3.7 Consider the equation

$$u(x) - u_{xx}(x) = \dot{W}(x),\ x \in \mathbb{R}.$$

(a) Find the solution for which $\sup_{x\in\mathbb{R}} \mathbb{E}u^2(x) < \infty$ and verify that such a solution is unique.
(b) Verify that u is continuously differentiable and u' is locally Hölder continuous of any order less that $1/2$.
(c) For what values of γ is the expression $\mathbb{E}\|u\|^2_{H_2^\gamma(\mathbb{R})}$ finite?

Problem 2.3.8

(a) Consider the heat equation $u_t = u_{xx}$ on the interval $x \in (0, \pi)$ with boundary conditions $u_x(t, 0) = u_x(t, \pi) = 0$ and initial condition $u(0, x) = \dot{W}(x)$. Find the solution and determine its regularity in (t, x).
(b) Consider the wave equation $u_{tt} = u_{xx}$ on the interval $(0, \pi)$ with zero boundary conditions and with initial conditions $u(0, x) = 0$, $u_t(0, x) = \dot{W}(x)$. Find the solution and determine its regularity in (t, x).
(c) Consider the Laplace equation $\boldsymbol{\Delta} u = 0$ in the unit disk $G = \{(x, y) : x^2 + y^2 < 1\}$ with the boundary condition $u|_{\partial G} = \dot{W}$. Find the solution and determine its regularity in (x, y). **Suggestion**: switch to polar coordinates.

Problem 2.3.9 A `Brownian bridge` $B = B(t),\ t \in (0, T)$ is a zero-mean Gaussian process with the covariance function

$$\mathbb{E}\big(B(t)B(s)\big) = \min(t, s) - \frac{ts}{T}. \tag{2.3.70}$$

(a) Verify that, with probability one, $B(0) = B(T) = 0$.
(b) Verify that each of the following represents the Brownian bridge:

1. $B(t) = w(t) - \frac{tw(T)}{T}$, where w is a standard Brownian motion;
2. $B(t) = (T - t)\int_0^t (T - s)^{-1} dw(s)$, that is, $B(t)$ is the solution of

$$dB(t) = -\frac{B(t)}{T - t}dt + dw(t),\ B(0) = 0;$$

3. $B(t) = \frac{\sqrt{2T}}{\pi}\sum_{k\geq 1} \xi_k \frac{\sin(\pi kt/T)}{k}$, where $\xi_k,\ k \geq 1$ are iid standard random variables;
4. $B(\cdot)$ is the limit in distribution, as $t \to \infty$, of $u(t, \cdot)$, where $u = u(t, x)$ is the solution of the stochastic parabolic equation

$$u_t(t, x) = u_{xx}(t, x) + \sqrt{2}\,\dot{W}(t, x), t > 0,\ x \in (0, T),$$

with zero initial and boundary conditions: $u(0, x) = u(t, 0) = u(t, T) = 0$; $\dot{W}(t, x)$ is the space-time white noise.

5. $B(\cdot)$ is the solution of the stochastic elliptic equation $\sqrt{-\Delta}\, u(x) = \dot{W}(x)$, $x \in (0, T)$, with zero boundary conditions $u(0) = u(T) = 0$.

Comment on pros and cons of using each representation for numerical simulation of the Brownian bridge.

Problem 2.3.10 Let $\mathrm{H}_n = \mathrm{H}_n(x)$ be the Hermite polynomial defined in (2.3.37) on page 61.

Define the operator D and δ by

$$Df(x) = f'(x), \quad \delta f(x) = -f'(x) + xf(x).$$

(a) Verify that $D\delta f - \delta Df = f$,

$$\int_{-\infty}^{+\infty} \big(Df(x)\big) g(x) e^{-x^2/2} dx = \int_{-\infty}^{+\infty} f(x) \big(\delta g(x)\big) e^{-x^2/2} dx,$$

and then verify the following properties of the Hermite polynomials:

$$\delta \mathrm{H}_n = \mathrm{H}_{n+1}, \; D\mathrm{H}_n = n\mathrm{H}_{n-1}, \; (D + \delta)H_n = x\mathrm{H}_n, \delta D\mathrm{H}_n = n\mathrm{H}_n.$$

In physics, in particular, quantum mechanics, the operators D and δ are examples of the `annihilation` and `creation` operators, respectively.

(b) Show that

$$\mathrm{H}_n(x) = (-1)^n e^{x^2/2} \frac{d^n}{dx^n} e^{-x^2/2}.$$

(c) Let (ξ, η) be a two-dimensional Gaussian vector such that both ξ and η are standard normal random variables. Show that, for $n \neq m$,

$$\mathbb{E}\Big(\mathrm{H}_n(\xi)\mathrm{H}_m(\eta)\Big) = 0.$$

Note that ξ and η do not have to be independent.

Problem 2.3.11 Verify that $\{\mathrm{H}_n(x)e^{-x^2/4}, \; n \geq 0\}$ is a complete system in $L_2(\mathbb{R})$. Equivalently, $\{\mathrm{H}_n(\xi), \; n \geq 1\}$ is a complete system in the space of square-integrable integrable functions of a standard Gaussian random variable ξ.

Here is a possible outline. Let $f = f(x)$ be such that $\int_{-\infty}^{+\infty} f^2(x)e^{-x^2/2}dx < \infty$ and $\int_{-\infty}^{+\infty} f(x)\mathrm{H}_n(x)e^{-x^2/2}dx = 0$ for all n. We need to show that $f = 0$. For a complex number z, define $F(z) = \int f(x)e^{zx-(z^2/2)}e^{-x^2/2}dx$. Argue that F is analytic everywhere and $F(z) = 0$ for all z. Then note that $F(-iy)e^{-y^2/2}$ is the Fourier transform of $f(x)e^{-x^2/2}$ and invert the Fourier transform.

Problem 2.3.12 Sometimes, it is convenient to consider a two-parameter family of Hermite polynomials $\mathrm{H}_n(s,x)$, $n = 0, 1, \ldots$, $s > 0$, $x \in \mathbb{R}$ defined by

$$e^{tx-(t^2 s/2)} = \sum_{n=0}^{\infty} \frac{\mathrm{H}_n(s,x)}{n!}$$

Verify that the following identities:

$$\mathrm{H}_n(s,x) = s^{n/2}\mathrm{H}_n(x/\sqrt{s}) \ [\mathrm{H}_n(x) = \mathrm{H}_n(1,x)],$$

$$\mathrm{H}_n(s,x) = \frac{(-s)^n}{n!}\exp\left(\frac{x^2}{2s}\right)\frac{\partial^n}{\partial x^n}\exp\left(-\frac{x^2}{2s}\right), \quad n = 0, 1, \ldots$$

$$\mathrm{H}_n(s,x) = x\mathrm{H}_{n-1}(s,x) - (n-1)s\mathrm{H}_{n-2}(s,x),\ n = 2, 3, \ldots,$$

$$\sum_{k=0}^{n} \frac{\mathrm{H}_{n-k}(s,x)}{(n-k)!}\frac{\mathrm{H}_k(r,y)}{k!} = \frac{\mathrm{H}_n(s+r, x+y)}{n!},$$

and verify that, for fixed $s > 0$, the collection of functions $\{\mathrm{H}_n(s,x)e^{-x^2/(4s)},\ n \geq 0,\ x \in \mathbb{R}\}$ is an orthogonal basis in $L_2(\mathbb{R})$.

Problem 2.3.13 Take $\varphi \in \mathcal{S}(\mathbb{R})$ and define the action $\langle\varphi, f\rangle$ of $f(x) = 1/(1-x)$ on φ as

$$\langle\varphi, f\rangle = \lim_{\varepsilon\to 0}\int_{\varepsilon<|y|<1/\varepsilon} \frac{\varphi(1-y)}{y}dy.$$

Show that this operation defines a distribution on the set $\mathcal{S}(\mathbb{R})$. Convince yourself that $\langle f, \varphi_0\rangle$, where φ_0 is the standard normal density, can be taken as a generalized expected value of the random variable $1/(1-\xi)$. Show that $\langle f, \varphi_0\rangle \neq 0$.

Problem 2.3.14 Consider the equation $-u_{yy} = \sin y + u \diamond \xi$, $y \in (0,\pi)$, with zero boundary conditions (see (2.3.42) on page 62). Let x be a real number and define

$$\mathcal{E}_x = e^{x\xi-(x^2/2)}.$$

(a) Assume that $u(y) = \sum_{k\geq 0} u_k(y)\mathrm{H}_k(\xi)$ and suppose that $\mathbb{E}u^2(y) < \infty$ for all $y \in (0,\pi)$. Define the function $\breve{u}(y;x) = \mathbb{E}\big(u(y)\mathcal{E}_x\big)$. Show that $\breve{u}(y;x)$ satisfies

$$-\frac{\partial^2\breve{u}(y;x)}{\partial y^2} = \sin y + x\breve{u}(y;x),\ \breve{u}(0;x) = \breve{u}(\pi;x) = 0; \tag{2.3.71}$$

and therefore

$$\breve{u}(y;x) = \frac{\sin y}{1-x}. \tag{2.3.72}$$

(b) We know from (2.3.41) on page 62 that $u(y) = \sum_{k=0}^{\infty} u_k(y)\mathrm{H}_k(\xi)$ with $u_k(y) = \sin y$, and so $\mathbb{E}u^2(y) = \infty$ for all $y \in (0, \pi)$. On the other hand, the function $\breve{u}(y; x)$ is well-defined for all $x \in (-1, 1)$, and

$$u_k(y) = \frac{1}{k!} \frac{\partial^k \breve{u}(y; x)}{\partial x^k}\bigg|_{x=0} = \sin y. \tag{2.3.73}$$

Why does (2.3.73) hold even though u is not square-integrable?

Problem 2.3.15 Investigate existence of closed-form solutions for equation

$$du = (au_{xx} - uu_x)dt + G_x(t, x)dw(t),\ t > 0,\ x \in \mathbb{R}, \tag{2.3.74}$$

where $G = G(t, x)$ is a known non-random function, and $G_x = \partial G/\partial x$. (On Page 64 we discussed a particular case $G(t, x) = x$.)

Problem 2.3.16 The d-dimensional version of equation

$$u_t + uu_x = au_{xx} + G_x(t, x)$$

is

$$\frac{\partial u_i}{\partial t} + \sum_{j=1}^{\mathrm{d}} u_j \frac{\partial u_i}{\partial x_j} = a \sum_{j=1}^{\mathrm{d}} \frac{\partial^2 u_i}{\partial x_j^2} + \frac{\partial G(t, x)}{\partial x_i} \tag{2.3.75}$$

or, with $U = (u_1, \ldots, u_{\mathrm{d}})$,

$$U_t + (U \cdot \nabla)U = a\boldsymbol{\Delta} U + \nabla G.$$

Derive the Hopf-Cole transformation for (2.3.75). An even more ambitious reader is then invited to look at the multi-dimensional version of (2.3.74).

Chapter 3
Stochastic Analysis in Infinite Dimensions

This chapter contains somewhat abstract but necessary, material on functional analysis and stochastic calculus. To save time, one can move on to the following chapters and come back as necessary.

The starting point in the study of stochastic ordinary differential equations is the standard Brownian motion (also known as Wiener process). While this process can also be used as a driving force in equations with partial derivatives, one of the distinctive features of stochastic partial differential equations is the possibility to have spacial structure in the noise. This spacial structure can be introduced in several ways, and the objective of this chapter is to describe the main constructions of the space-dependent noise and the corresponding stochastic integral. As always, we fix the stochastic basis $(\Omega, \mathcal{F}, \{\mathcal{F}_t\}_{t\geq 0}, \mathbb{P})$ with the usual assumptions ($\mathbb{P}$-completeness of $\mathcal{F}_0$ and right-continuity of $\mathcal{F}_t$). All processes are assumed $\mathcal{F}_t$-adapted, and we do not distinguish between different modifications of a process.

3.1 An Overview of Functional Analysis

A possible smooth transition from ODEs to PDEs (either deterministic or stochastic) is to replace finite-dimensional Euclidean space with three infinite-dimensional Hilbert spaces, and to replace matrices with operators acting between those Hilbert spaces. The good news is that this transition from finite to infinite dimensions requires only finitely many new definitions and constructions. The bad news is that, depending on reader's background in functional analysis, this number can be large. One of the objectives of this section is to provide all the necessary background to a reader who might not have systematic knowledge of functional analysis and related topics.

S.V. Lototsky, B.L. Rozovsky, *Stochastic Partial Differential Equations*,
Universitext, DOI 10.1007/978-3-319-58647-2_3

The following objects frequently appear in the study of stochastic partial differential equations:

1. Normal triple of Hilbert spaces;
2. Hilbert scale;
3. Hilbert space tensor product;
4. Nuclear (trace class) operator;
5. Hilbert-Schmidt operator;
6. Reproducing kernel Hilbert space.

It is difficult to find all this in a single text book on functional analysis, and the objective of the following two sections is to present a self-contained summary.

3.1.1 Spaces

A `measure space` $(S, \mathcal{S}, \mu)$ is a set S, the sigma-algebra $\mathcal{S}$ of sub-sets of S, and a measure μ on $\mathcal{S}$. Recall that $\mathcal{S}$ contains S and is closed under the operations of taking complements and countable unions and intersections. Also recall that the measure μ is a countably additive function $\mu : \mathcal{S} \to \mathbb{R}$. The measure space is called `positive` if the measure μ is positive, that is, $\mu(A) \geq 0$ for all $A \in \mathcal{S}$. The measure space is called `finite` if the measure μ is finite: $|\mu|[S] = \int_S |\mu|(ds) < \infty$.

A `topological space` V is a set with a collection of sub-sets, called `topology`, that contains the set itself, the empty set, and is closed under the operations of taking arbitrary unions and finite intersections; every element of the topology is, by definition, an `open set`. The availability of open sets makes it possible to define

1. `closed set` as a complement of an open set;
2. `closure of a set` as the intersection of all closed sets containing it;
3. `compact set`, if every covering of the set by open sets has a finite sub-covering;
4. `neighborhood of a point`, as an open set containing the point;
5. `convergent sequence of points` (every neighborhood of the limit contains all but finitely many element of the sequence);
6. `Borel sigma-algebra` $\mathcal{B}(V)$, as the smallest sigma-algebra containing all open sets in V;
7. `continuous mapping` between two topological spaces (pre-image of an open set is open).

A `dense subset` of the topological space V is a set whose closure is V. A `separable space` is a topological space possessing a countable dense subset. A measure μ on a topological space is called `regular` if, for every Borel set A and every $\varepsilon > 0$, there exist an open set A_o and a closed set A_c such that $A_c \subset A \subset A_o$ and $\mu(A_o) - \mu(A_c) < \varepsilon$.

A `linear topological space` is a linear space and a topological space such that the operations of addition and multiplication by a scalar are continuous. Unless mentioned otherwise, all scalars are assumed to be real numbers, and thus all linear spaces are real.

In a `metric space`, we always consider the topology generated by open balls. A particular case of a metric space is a `normed space`; unlike a general metric space, the definition of the norm requires the underlying space to be linear. In a metric space, we have the notion of a Cauchy sequence; the space is called `complete` if every Cauchy sequence has a limit. If the space is not complete, we take the `completion` (equivalently, `closure`) of the space with respect to the corresponding metric, by considering all equivalence classes of all Cauchy sequences.

A subset A of a complete metric space X is compact if and only if every sequence of points in A has a converging sub-sequence (in the metric of X) and the limit is an element of A. In particular, compact sets in a complete metric space are closed and bounded. In the finite-dimensional case, a set is compact *if and only* if it is closed and bounded.

A norm in a normed space X is denoted by $\|\cdot\|_X$. Two norms $\|\cdot\|_{X,1}$ and $\|\cdot\|_{X,2}$ in X are called `equivalent` if there exist two positive numbers c_1, c_2 such that

$$c_1\|x\|_{X,1} \le \|x\|_{X,2} \le c_2\|x\|_{X,1}$$

for all $x \in X$.

One useful fact about a complete metric space is the `contraction mapping theorem`: a contracting mapping of a complete metric space to itself has a unique fixed point. Below is a precise statement, with a proof, for a normed space.

Definition 3.1.1 Let X be a normed space. A mapping $F : X \to X$ is called a `contraction` if there exists a number $\alpha \in (0, 1)$ such that $\|F(x) - F(y)\|_X \le \alpha\|x - y\|_X$ for all $x, y \in X$.

Theorem 3.1.2 (Contraction Mapping Theorem) *If X is complete and $F : X \to X$ is a contraction, then F has a unique fixed point: there exists a unique $x^* \in X$ such that $F(x^*) = x^*$.*

Proof Uniqueness is immediate: if x^* and y^* are two fixed points, then $\|x^*-y^*\|_X = \|F(x^*) - F(y^*)\|_X \le \alpha\|x^* - y^*\|$, meaning $\|x^* - y^*\|_X = 0$. To prove existence, do the `fixed point iterations`: take $x_0 \in X$ and define $x_n = F(x_{n-1})$, $n \ge 1$. Then $x^* = x_0 + \sum_{n\ge1}(x_n - x_{n-1})$ (that is, $x^* = \lim_n x_n$). The series converges because $\|x_n - x_{n-1}\|_X \le \alpha^{n-1}\|x_1 - x_0\|_X$, and, by completeness of X, the sum is in X.

A `Polish metric space` is a complete separable metric space; a `Banach space` is a complete normed space; a `Hilbert space` is a Banach space in which the norm is generated by an inner product. An inner product in a Hilbert space H is denoted by $(\cdot,\cdot)_H$; if H is a real Hilbert space (as we usually assume), then

$$(x,y)_H = (y,x)_H.$$

Two *topological spaces* are called `homeomorphic` if there exists a `bijective` (one-to-one and onto) mapping between them that is continuous in both directions. Two *metric spaces* are called `isomorphic` (or `isometrically isomorphic`) if there exists a bijective mapping between them that is metric-preserving; such a mapping is necessarily continuous in both directions. Two *Hilbert spaces* X, Y are called `isomorphic` if there exists a *linear* bijective mapping A between them that preserves the inner product:

$$(\mathrm{A}x, \mathrm{A}y)_Y = (x,y)_X \text{ for all } x,y \in X. \tag{3.1.1}$$

Exercise 3.1.3 (C) Show that (3.1.1) is equivalent to

$$\|\mathrm{A}x\|_Y^2 = \|x\|_X^2 \text{ for all } x \in X. \tag{3.1.2}$$

Hint. Use the parallelogram identity:

$$(x,y)_H = \frac{1}{4}\left(\|x+y\|_H^2 - \|x-y\|_H^2\right). \tag{3.1.3}$$

A Banach space Y is `continuously embedded` into a Banach space X if Y is a subset of X and there exists a number $c > 0$ such that $\|y\|_X \le c\|y\|_Y$ for all $y \in Y$.

Exercise 3.1.4 (C) Assume that a Banach space Y is a sub-set of a Banach space X and define the `embedding operator` $\imath : Y \to X$ by $\imath(y) = y$. Show that Y is continuously embedded into X if and only if $\imath$ is continuous.

Definition 3.1.5 A `normal triple` of Hilbert spaces is an ordered collection (V, H, V') of three Hilbert spaces with the following properties:

1. V is a dense sub-set of H and is continuously embedded into H,
2. H is dense subset of V' and is continuously embedded into V',
3. The inequality

$$|(v,h)_H| \le \|v\|_V\, \|h\|_{V'} \tag{3.1.4}$$

holds for all $v \in V$, $h \in H$,

Exercise 3.1.6 (C) For smooth functions f, g with a compact support in $\mathbb{R}^{\mathrm{d}}$ define

$$(f,g)_{n,r} = \sum_{0\le k\le n} \int_{\mathbb{R}^{\mathrm{d}}} (1+|x|^2)^r D^k f(x)\, D^k g(x) dx,\ \|f\|_{n,r}^2 = (f,f)_{n,r}. \tag{3.1.5}$$

where $n \in \{0, 1, 2, \dots\}$, $r \in \mathbb{R}$, and the sum is taken over all partial derivatives of order at most n; $D^0 f = f$. For example, for d $= 2$, $n = 2$, we have

$$\|f\|_{2,r}^2 = \iint_{\mathbb{R}^2} (1 + x_1^2 + x_2^2)^r (f^2 + f_{x_1}^2 + f_{x_2}^2 + f_{x_1x_1}^2 + f_{x_1x_2}^2 + f_{x_2x_2}^2) dx_1 dx_2.$$

Define the `weighted Sobolev spaces` $W^{n,r,\mathrm{d}}$ as the closure of $\mathcal{C}_0^\infty(\mathbb{R}^\mathrm{d})$ in the norm $\|\cdot\|_{n,r}$.

(a) Show that if $f \in \mathcal{C}^\infty(\mathbb{R}^\mathrm{d})$ and $|D^k f(x)| \leq C(1 + |x|^q)$ for all $x \in \mathbb{R}^\mathrm{d}$ and $k = 0, \dots, n$, then $f \in W^{n,r,\mathrm{d}}$ for all $r < -(\mathrm{d} + 2q)/2$. (b) Show that $(W^{n+2,r,\mathrm{d}}, W^{n+1,r,\mathrm{d}}, W^{n,r,\mathrm{d}})$ is a normal triple of Hilbert spaces. (c) Use the Sobolev embedding theorem to argue that if $n > N + (\mathrm{d}/2)$, then $W^{n,r,\mathrm{d}}$ is continuous embedded into $\mathcal{C}^N(\mathbb{R}^\mathrm{d})$.

If (V, H, V') is a normal triple of Hilbert spaces $v \in V$, $y \in V'$, $h_n \in H$ and $\lim_n \|y - h_n\|_{V'} = 0$, define

$$[y, v] = \lim_n (v, h_n)_H. \tag{3.1.6}$$

Exercise 3.1.7 (C)

(a) Verify that if (3.1.4) is replaced with

$$|(v, h)_H| \leq c\|v\|_V \, \|h\|_{V'}$$

with some c independent of v and h, then an equivalent norm can be defined in either V or V' so that (3.1.4) holds.

(b) Verify that the limit on the right-hand side of (3.1.6) exists and does not depend on the particular sequence $\{h_n,\ n \geq 1\}$ converging to y.

There is no coincidence that the notation of the third space in the normal triple coincides with the commonly used notation for the dual of the first space: see Exercise 3.1.22 on page 85 below.

Definition 3.1.8 A collection $\{H^r,\ r \in \mathbb{R}\}$ of Hilbert spaces is called a `Hilbert scale` if

1. For every $\gamma < \nu$, the space H^ν is densely and continuously embedded in H^γ;
2. For every $\alpha < \beta < \gamma$ and $x \in H^\gamma$,

$$\|x\|_{H^\beta} \leq \|x\|_{H^\alpha}^{(\gamma-\beta)/(\gamma-\alpha)} \|x\|_{H^\gamma}^{(\beta-\alpha)/(\gamma-\alpha)}. \tag{3.1.7}$$

Example 3.1.9 (Sobolev Spaces on $\mathbb{R}^\mathrm{d}$) Define $H^r = H^r(\mathbb{R}^\mathrm{d})$, $r \in \mathbb{R}$, as the completion of $\mathcal{C}_0^\infty(\mathbb{R}^\mathrm{d})$ (the set of smooth function with compact support in $\mathbb{R}^\mathrm{d}$) in the norm $\|\cdot\|_r$, where

$$\|f\|_r^2 = \int_{\mathbb{R}^\mathrm{d}} (1 + |y|^2)^r \, |\widehat{f}(y)|^2 dy < \infty, \ r \in \mathbb{R}, \tag{3.1.8}$$

and

$$\widehat{f}(y) = \frac{1}{(2\pi)^{d/2}} \int_{\mathbb{R}^d} e^{-ixy} f(x)dx \tag{3.1.9}$$

is the `Fourier transform` of f. Then $\{H^r,\ r \in \mathbb{R}\}$ is a Hilbert scale and, for every $\gamma \in \mathbb{R}$ and $r > 0$,

$$(H^{\gamma+r}, H^{\gamma}, H^{\gamma-r}) \tag{3.1.10}$$

is a normal triple with

$$[f, g]_\gamma = \int_{\mathbb{R}^d} (1 + |y|^2)^\gamma \widehat{f}(y)\overline{\widehat{g}(y)}dy,$$

where $\overline{\widehat{g}(y)}$ is the complex conjugate of $\widehat{g}(y)$.

If we define the operator $\Lambda = \sqrt{1 - \boldsymbol{\Delta}}$ by

$$\widehat{\Lambda f}(y) = (1 + |y|^2)^{1/2}\widehat{f}(y), \tag{3.1.11}$$

then we get, for all $\gamma, r \in \mathbb{R}$ and $f, g \in \mathcal{C}_0^\infty(\mathbb{R}^d)$,

$$(f, g)_\gamma = (\Lambda^r f, \Lambda^r g)_{\gamma-r}, \quad [f, g]_\gamma = (\Lambda^m f, \Lambda^{-m} g)_\gamma. \tag{3.1.12}$$

This concludes Example 3.1.9.

Exercise 3.1.10 (C)

(a) Using the L_2-isometry of the Fourier transform ($\|f\|_0^2 = \|\widehat{f}\|_0^2$) verify that $\|f\|_1^2 = \|f\|_0^2 + \sum_{n=1}^{d} \|f_{x_n}\|_0^2$, at least for $f \in \mathcal{C}_0^\infty(\mathbb{R}^d)$. **Hint.** Recall the the Fourier transform of f_{x_n} is $\mathrm{i}y_n\widehat{f}$.

(b) Verify that the norms (3.1.8) satisfy (3.1.7).

(c) Verify that, for an positive integer n, $\|f^{(n)}\|_0^2 + \|f\|_0^2$ is an equivalent norm in $H^n(\mathbb{R})$, where $f^{(n)}$ is n-th derivative of f. **Hint.** Use isometry of the Fourier transform to conclude that the norm $\|f\|_n$ in $H^n(\mathbb{R})$ is equivalent to $\|f\|_0 + \|f'\|_0 + \ldots + \|f^{(n)}\|_0$. Then use (3.1.7) with $\alpha = 0$, $\gamma = n$, and $\beta = 1, \ldots, n-1$, followed by the epsilon inequality (1.1.19) on page 8, to estimate every $\|f^{(m)}\|_0$ in terms of $\|f\|_0$ and $\|f^{(n)}\|_0$.

(d) Verify that if $n = 2k$ is a positive even integer and $\boldsymbol{\Delta}$ is the Laplace operator, then the norm $\|f\|_n$ in $H^n(\mathbb{R}^d)$ is equivalent to $\|f\|_0 + \|\boldsymbol{\Delta}^k f\|_0$.

The following result establishes a connection between normal triples and Hilbert scales.

Theorem 3.1.11

(a) If $\{H^r,\ r \in \mathbb{R}\}$ is a Hilbert scale, $\gamma \in \mathbb{R}$ and $m > 0$, then $(H^{\gamma+m}, H^\gamma, H^{\gamma-m})$ is a normal triple.

(b) If (V, H, V') is a normal triple, then there is a unique Hilbert scale $\{H^r,\ r \in \mathbb{R}\}$ such that $H^1 = V$, $H^0 = H$, $H^{-1} = V'$.

Proof See Kreĭn et al. [117, Sect. IV.1.10].

Yet another result shows that the analogue of the operator $(1 - \boldsymbol{\Delta})^{1/2}$ exists in a much more general situation.

Theorem 3.1.12 *Let H and X be two Hilbert spaces such that H is densely and continuously embedded into X. Then there exist a Hilbert scale $\{H^r,\ r \in \mathbb{R}\}$ and a positive-definite self-adjoint operator Λ with the following properties:*

1. $H^1 = H$, $H^0 = X$;
2. *for every $r > 0$, the domain of Λ^r is dense in X and H^r is the closure of the domain of Λ^r in the norm $\|\Lambda^r \cdot\|_X$;*
3. *for every $r \leq 0$, the domain of Λ^r contains X and H^r is the closure of X in the norm $\|\Lambda^r \cdot\|_X$.*
4. *The scale $\{H^r,\ r \in \mathbb{R}\}$ is uniquely determined by the spaces H and X.*

Proof The operator Λ is defined as follows: $\Lambda = \mathrm{A}^{-1/2}$, where $\mathrm{A} : X \to H$ is the bounded linear operator such that, for every $x \in H\, y \in X$, $(x, y)_X = (x, \mathrm{A}y)_H$. Existence and uniqueness of A follows from the Riesz representation theorem. The rest of the proof is direct verification of the corresponding definitions; see Kreĭn et al. [117, Theorem 4.1.12].

Exercise 3.1.13 (B) Verify existence of the operator A in the proof of Theorem 3.1.12 and show that the operator is self-adjoint and positive-definite.

Definition 3.1.14 In the setting of Theorem 3.1.12, we say that the operator Λ generates the scale $\{H^r,\ r \in \mathbb{R}\}$, or, equivalently, the scale is generated by the operator.

Remark 3.1.15 Uniqueness of the scale in Theorem 3.1.12 is in the sense that any other scale with the same properties will have equivalent norms in every space H^r. Many such equivalent scales can be generated if instead of the operator Λ we use, for example, $a + b\Lambda$ or $\sqrt{a + b\Lambda^2}$, $a, b > 0$.
To summarize, every Hilbert scale is generated by a positive-definite self-adjoint operator, and every positive-definite self-adjoint operator on a Hilbert space generates a Hilbert scale.

Next, we review the tensor product of Hilbert spaces. If $f = f(t)$ and $g = g(s)$ are real-valued square-integrable functions on $(0, 1)$, we can define a square-integrable function $h = h(t, s)$ on $(0, 1) \times (0, 1)$ by $h(t, s) = f(s)g(s)$. In general Hilbert spaces, such an operation of point-wise multiplication is not defined, but can often be replaced with the tensor product.

Given two linear spaces X, Y, the (algebraic) tensor product $X \otimes Y$ can be defined in one of three equivalent ways:

- As the linear space such that every linear mapping from the Cartesian product $X \times Y$ to a linear space V factors through $X \otimes Y$ via a commutative diagram. In other words, tensor product turns bi-linear maps into linear maps.
- As a factor space of $X \times Y$ modulo certain equivalence relations (see Weidmann [224, Sect. 3.4]);

- as the collection of objects

$$\Big\{\sum_{i=1}^{N} c_i x_i \otimes y_i,\ \ c_i \in \mathbb{R},\ x_i \in X, y_i \in Y, N \geq 1\Big\}. \tag{3.1.13}$$

In (3.1.13) we have elementary building blocks, *denoted* by $x \otimes y$, $x \in X$, $y \in Y$, and the *algebraic* tensor product $X \otimes Y$ of the spaces X and Y consists of all *finite* linear combinations of these elementary building blocks.

For two Hilbert spaces X, Y, define

$$(x_1 \otimes y_1, x_2 \otimes y_2)_{X\otimes Y} = (x_1, x_2)_X\,(y_1, y_2)_Y, \tag{3.1.14}$$

and then extend the operation $(\cdot,\cdot)_{X\otimes Y}$ by linearity to $X \otimes Y$:

$$\begin{aligned}&(ax_1 \otimes y_1 + bx_2 \otimes y_2, x_3 \otimes y_3)_{X\otimes Y}\\ &\qquad = a(x_1 \otimes y_1, x_3 \otimes y_3)_{X\otimes Y} + b(x_2 \otimes y_2, x_3 \otimes y_3)_{X\otimes Y}.\end{aligned}$$

The operation $(\cdot,\cdot)_{X\otimes Y}$ defines an inner product on the space $X \otimes Y$, and the corresponding norm $\|\cdot\|_{X\otimes Y}$ satisfies

$$\|x \otimes y\|_{X\otimes Y} = \|x\|_X\,\|y\|_Y.$$

The closure (equivalently, completion) of the algebraic tensor product $X \otimes Y$ of two Hilbert spaces X, Y with respect to the norm $\|\cdot\|_{X\otimes Y}$ is called the `Hilbert space tensor product` of X and Y and again denoted by $X \otimes Y$. The completion is usually necessary: the algebraic tensor product of two Hilbert spaces is complete if and only if at least one of the spaces is finite-dimensional (Weidmann [224, Exercise 3.12]).

While we use the same notation $X \otimes Y$ for two different constructions, there will be no ambiguity, as we will always consider the Hilbert space tensor product if X and Y are Hilbert spaces.

Exercise 3.1.16 (C)

(a) Verify that if X is a separable Hilbert space with an orthonormal basis $\{\mathfrak{m}_k,\ k \geq 1\}$ and Y is a separable Hilbert space with an orthonormal basis $\{\mathfrak{u}_k,\ k \geq 1\}$, then $X \otimes Y$ is a separable Hilbert space with an orthonormal basis $\{\mathfrak{m}_i \otimes \mathfrak{u}_j,\ i,j \geq 1\}$.

(b) Let X be a separable Hilbert space and let $Y = L_2(S)$ be the space of real-valued square-integrable functions on a measure space $(S, \mathcal{S}, \mu)$. In this situation, we do have the possibility to multiply point-wise the elements of X and the element of Y. If $\{\mathfrak{m}_k,\ k \geq 1\}$ is an orthonormal basis in X and $\{\mathfrak{u}_k,\ k \geq 1\}$ is an orthonormal basis in Y, then show that the orthonormal basis in $X \otimes Y$ is $\{\mathfrak{u}_i\mathfrak{m}_j,\ i,j \geq 1\}$.

(c) Let $X = Y = L_2((0,1))$. Verify that the mapping

$$(f \otimes g) \mapsto f(t)\, g(s)$$

defines an isomorphism between $L_2((0,1)) \otimes L_2((0,1))$ and $L_2((0,1) \times (0,1))$.

(d) Let $(S, \mathcal{S}, \mu)$ and $(R, \mathcal{R}, \nu)$ be measure spaces. Show that $L_2(S; L_2(R))$ (the space of square-integrable functions on S with values in the space of square-integrable functions on R) is isomorphic to $L_2(S \times R)$.

Hint. Show that both are isomorphic to $L_2(S) \otimes L_2(R)$.

3.1.2 Linear Operators

Let X, Y be real Banach spaces with norms $\|\cdot\|_X$, $\|\cdot\|_Y$. Recall that a mapping $\mathrm{A} : X \to Y$ is called a `linear operator` if $\mathrm{A}(ax + by) = a\mathrm{A}x + b\mathrm{A}y$ for all $a, b \in \mathbb{R}$ and $x, y \in X$; if $Y = \mathbb{R}$, then A is called a `linear functional` on X.

Definition 3.1.17 A linear operator $\mathrm{A} : X \to Y$ is called

- `bounded`, if A is defined on all of X and there exists a number c such that $\|\mathrm{A}x\|_Y \le c\|x\|_X$ for all $x \in X$;
- `closed`, if the set $\{(x, \mathrm{A}x),\ x$ in the domain of $\mathrm{A}\}$ is closed in the Cartesian product space $X \times Y$;
- `compact` (or completely continuous), if the image of every sequence that is bounded in the norm of X has a sub-sequence that converges in the norm of Y.
- `nuclear`, if there exist a sequence $\{\ell_n,\ n \ge 1\}$ of bounded linear operators from X to $\mathbb{R}$, a sequence $\{y_n,\ n \ge 1\}$ of elements from Y, and a sequence $\{c_n,\ n \ge 1\}$ of real numbers, such that
  - $\sup\limits_n \sup\limits_{x:\|x\|_X \le 1} |\ell_n(x)| < \infty$, $\sup_n \|y_n\|_Y < \infty$, $\sum_{n\ge1} |c_n| < \infty$;
  - $\lim\limits_{N\to\infty} \|\mathrm{A}x - \sum\limits_{n=1}^{N} c_n \ell_n(x) y_n\|_Y = 0$, $x \in X$, that is,

$$\mathrm{A}x = \sum_{n\ge1} c_n \ell_n(x) y_n. \tag{3.1.15}$$

Note that a bounded linear operator $\mathrm{A} : X \to Y$ is continuous and closed. The interesting situation is when $\mathrm{A} : X \to X$ is an unbounded closed operator.

Theorem 3.1.18 *Let X be a Banach space and let $\mathrm{A} : X \to X$ be an unbounded linear operator such that the domain $H \subset X$ of A is a Hilbert space and $\mathrm{A} : H \to X$ is a bounded operator: $\|\mathrm{A}x\|_X \le C_0\|x\|_H$.*

If there exists a positive number C such that, for all $x \in H$, $\|x\|_H \le C\big(\|\mathrm{A}x\|_X + \|x\|_X\big)$, then $\mathrm{A} : X \to X$ is a closed (unbounded) operator.

Proof Let us point out that A is trivially a closed operator as a mapping from H to X. We have to show that A is closed as a mapping from X to X.

By definition, A is closed as an operator from X to X if the set $(x, \mathrm{A}x), x \in H$, is closed in $X \times X$. In other words, we have to show that, for every $\{x_n\} \subset X$ such that $x_n \in H$, $\|x_n - x\|_X \to 0$, and $\|\mathrm{A}x_n - y\|_X \to 0$, it follows that $x \in H$ and $\mathrm{A}x = y$. We have

$$\|x_n - x_m\|_H \leq C\big(\|\mathrm{A}x_n - \mathrm{A}x_m\|_X + \|x_n - x_m\|_X\big),$$

and therefore $\{x_n\}$ is a Cauchy sequence in H (because by assumption both $\{\mathrm{A}x_n\}$ and $\{x_n\}$ are Cauchy sequences in X). By completeness of H, $\lim_n x_n$ exists in H. Therefore, $\lim_n x_n = x \in H$. By continuity of A on H, $\lim_n \mathrm{A}x_n = \mathrm{A}x$, that is $\mathrm{A}x = y$.

Next, we discuss some properties of compact operators.

Theorem 3.1.19

(a) A linear operator with a finite-dimensional range is compact.

(b) If A_n, $n \geq 1$, *are compact linear operators from X to Y and* $\mathrm{A} : X \to Y$ *is a bounded linear operator such that* $\lim_{n\to\infty} \sup_{x:\|x\|_X \leq 1} \|\mathrm{A}_n x - \mathrm{A}x\|_Y = 0$, *then* A *is compact.*

Proof

(a) This follows from compactness of closed bounded sets in a finite-dimensional Euclidean space.

(b) See, for example, Dunford and Schwartz [42, Lemma VI.5.3] or Yosida [231, Sect. X.2].

Exercise 3.1.20 (B) Verify that a nuclear operator is compact. **Hint.** Note that $\mathrm{A}_N : x \mapsto \sum_{n=1}^{N} c_n \ell_n(x) y_n$ is an operator with at most N-dimensional range, and

$$\|(\mathrm{A}_N - \mathrm{A})x\|_Y \leq C\|x\|_X \sum_{n=N+1}^{\infty} |c_n|.$$

Recall that the (strong) `dual space` of a Banach Space X is the collection X' of bounded linear functionals on X, equipped with the norm $\|\ell\|_{X'} = \sup_{x:\|x\|_X=1} |\ell(x)|$; X' is a Banach space (see, for example, Yosida [231, Theorem IV.7.1]). A sequence $\{x_n,\ n \geq 1\}$ is said to converge `weakly` to $x \in X$ if $\lim_n \ell(x_n) = \ell(x)$ for every $\ell \in X'$.

When X is a Hilbert space, then the following `Riesz representation theorem` allows us to identify X with X'.

Theorem 3.1.21 (Riesz Representation Theorem) *Let X be a Hilbert space.*

(a) If $\ell \in X'$, *then there exists a unique* $y_\ell \in X$ *such that* $\|\ell\|_{X'} = \|y_\ell\|_X$ *and* $\ell(x) = (x, y_\ell)_X$ *for all* $x \in X$.

(b) If $y \in X$, *then the mapping* $x \mapsto (x, y)_X$ *defines a bounded linear functional* ℓ_y *on X and* $\|\ell_y\|_{X'} = \|y\|_X$.

Proof See, for example, Yosida [231, Theorem III.6]. Keep in mind that we assume that all the spaces are real.

The Riesz representation theorem (which is a common name for any result providing a characterization of the dual of a Banach space, not just for Hilbert spaces) is named after the Hungarian mathematician FRIGYES (FREDERIC) RIESZ (1880–1956), who, around 1910, established the result in the setting of L_p spaces, first for $p = 2$, and then for $p \neq 2$. Frigyes Riesz is not to be confused with his younger brother Marcel Riesz (1886–1969), also a mathematician after whom the Riesz kernel is named.

Exercise 3.1.22 (C)

(a) Use the Riesz representation theorem to show that if H is a (real) Hilbert space, then H is isomorphic to H'. **Hint.** $\mathrm{J}_0 : x \mapsto (x, \cdot)_H$ is the required isomorphism.
(b) Let (V, H, X) be a normal triple of Hilbert spaces. Verify that X is isomorphic to V'. Is there a contradiction with part (a)? **Hint.** Show that the required isomorphism $\mathrm{J} : X \to V'$ is defined by

$$(\mathrm{J}x)(v) = [x, v];$$

see (3.1.6) on page 79. There is no contradiction: according to part (a), V is isomorphic to V' relative to the inner product in V; in the normal triple (V, H, X), X is isomorphic to V' relative to the inner product in H.

The following generalization of Theorem (3.1.21) helps in the study of elliptic equations.

Theorem 3.1.23 (Lax-Milgram Theorem) *Let H be a (real) Hilbert space, and let $\mathrm{B} : H \times H \to \mathbb{R}$ be a mapping which is*

- ***Bi-linear**: for $a, b \in \mathbb{R}$, $x, y, h \in H$, $\mathrm{B}(ax+bh, y) = a\mathrm{B}(x, y)+b\mathrm{B}(h, y)$, $\mathrm{B}(x, ay+bh) = a\mathrm{B}(x, y) + b\mathrm{B}(x, h)$;*
- ***Bounded**: There exists $K > 0$ such that, for all $x, y \in H$, $|\mathrm{B}(x, y)| \leq K\|x\|_H \|y\|_H$;*
- ***Coercive (or strongly positive)**: There exists a $c > 0$ such that, for all $x \in H$, $\mathrm{B}(x, x) \geq c\|x\|_H^2$.*

Then, for every bounded linear operator ℓ on H there exists a unique $u \in H$ such that, for every $y \in H$, $\mathrm{B}(u, y) = \ell(y)$.

Alternatively, for every bounded linear operator ℓ on H there exists a unique $v \in H$ such that, for every $x \in H$, $\mathrm{B}(x, v) = \ell(x)$.

Proof The idea is to fix the first argument of B and to consider the bounded linear functional $y \mapsto \mathrm{B}(u, y)$. By the Riesz representation theorem, $\mathrm{B}(u, y) = (x_u, y)$, and then boundedness of B implies that the correspondence $u \mapsto x_u$ defines a bounded linear operator $\mathrm{A} : H \to H$ so that $\mathrm{B}(u, x) = (\mathrm{A}u, x)$. Also by Theorem 3.1.21, $\ell(y) = (h, y)_H$ for some $h \in H$. Coercivity of B implies that the operator A has a bounded inverse, and then $u = \mathrm{A}^{-1}h$. For details, see Evans [50, Theorem 6.2.1].

To prove the alternative statement of the theorem, we fix the second argument of B and repeat the argument.

The Riesz representation theorem is a particular case corresponding to $\mathrm{B}(x, y) = (x, y)_H$.

This concludes the proof of Theorem 3.1.23.
Theorem 3.1.23 is named after the Hungarian-American mathematician PETER DAVID LAX and the American mathematician ARTHUR NORTON MILGRAM (1912–1961), and can be traced to their joint paper [129].

Exercise 3.1.24 (A) Let B be from Theorem 3.1.23. Show that

$$|\mathrm{B}(u, v)\,\mathrm{B}(v, u)| \leq \mathrm{B}(u, u)\mathrm{B}(v, v).$$

Hint. Note that $\mathrm{B}(u + \lambda v, u + \lambda v) \geq 0$, expand and optimize in λ.
To conclude our discussion of the dual spaces, recall that a Banach space X is called `reflexive` if $(X')' = X$. In particular, every Hilbert space is reflexive (by Theorem 3.1.21), and so is every $L_p(G)$ for $1 < p < \infty$. A useful property of a reflexive space is `weak sequential compactness`: if $\{x_n,\ n \geq 1\}$ is a sequence of elements in X and $\sup_n \|x_n\|_X < \infty$, then there exists an $x \in X$ and a subsequence $\{x_{n'}\}$ such that $\lim_{n'} \ell(x_{n'}) = \ell(x)$ for all $\ell \in X'$; see, for example, Yosida [231, Theorem V.2.1]. In particular, if H is a Hilbert space and $\sup_n \|h_n\|_H < \infty$, then there exists an $h \in H$ such that, along a subsequence $h_{n'}$, $\lim_{n'}(h_{n'}, x)_H = (h, x)_H$ for every $x \in H$.

Next, we summarize some facts about adjoint and self-adjoint operators.

Theorem 3.1.25 *Let* $\mathrm{A} : X \to Y$ *be a linear operator and let* X, Y *be Hilbert spaces with inner products* $(\cdot, \cdot)_X$ *and* $(\cdot, \cdot)_Y$. *If the domain of* A *is dense in* X, *then there exists a unique linear operator* $\mathrm{A}^* : Y \to X$, *called the* `adjoint operator` *of* A, *defined by* $(x, \mathrm{A}^*y)_X = (\mathrm{A}x, y)_Y$ *for all* x *in the domain of* X. *If* A *is defined on all of* X *and is bounded, then* A^* *is defined on all of* Y *and is a bounded linear operator from* Y *to* X.

Proof See Yosida [231, Sects. VII.1, VII.2].

Definition 3.1.26 Let X be a Hilbert space. A linear operator $\mathrm{A} : X \to X$ is called

- `self-adjoint` if $\mathrm{A} = \mathrm{A}^*$;
- `non-negative`, if $(\mathrm{A}x, x)_X \geq 0$ for all x in the domain of A.

Exercise 3.1.27 (C) Let X, Y be Hilbert spaces and let $\mathrm{A} : X \to Y$ be a closed linear operator with a dense domain. Verify that the operators $\mathrm{A}^*\mathrm{A}$ and $\mathrm{A}\mathrm{A}^*$ are non-negative and self-adjoint.

The following theorem summarizes some useful properties of bounded self-adjoint operators.

Theorem 3.1.28 *Let* $\mathrm{A} : X \to X$ *be a bounded linear operator on a separable Hilbert space* X.

(a) If A *is non-negative and self-adjoint, then there exists a unique non-negative self-adjoint operator* B : $X \to X$, *such that* BB = A. *This operator is denoted by* $\sqrt{\mathrm{A}}$,

(b) If A *is compact and self-adjoint, then* X *has an orthonormal basis consisting of the eigenfunctions of* A.

Proof

(a) One can either use spectral representation theorem (e.g. Yosida [231, Theorem XI.6.1]) or representation of C^*-algebras (e.g. Murphy [171, Theorem 2.2.1]).
(b) See, for example, Murphy [171, Theorem 2.4.4].

Example 3.1.29 (Hilbert Scale Generated by a Self-adjoint Operator) Let H be a separable Hilbert space and $\bar{\mathrm{A}} : H \to H$, a positive-definite, compact, self-adjoint operator. Choose the eigenfunctions $\mathfrak{H} = \{\mathfrak{h}_k,\ k \geq 1\}$ of $\bar{\mathrm{A}}$ to form an orthonormal basis in H and denote by $\{a_k,\ k \geq 1\}$ the corresponding eigenvalues of $\bar{\mathrm{A}} : \bar{\mathrm{A}}\mathfrak{h}_k = a_k\mathfrak{h}_k$. Since $\bar{\mathrm{A}}$ is compact, we have $\lim_{k\to 0} a_k = 0$. With no loss of generality, assume that $a_1 \geq a_2 \geq a_3 \geq \ldots$, and define the numbers $\lambda_k = 1/a_k$ so that $\lambda_1 \leq \lambda_2 \leq \lambda_3 \leq \ldots$ and $\lim_{k\to\infty} \lambda_k = +\infty$. It turns out that the numbers λ_k and the corresponding operator $\Lambda = \big(\bar{\mathrm{A}}\big)^{-1}$ defined by

$$\Lambda\mathfrak{h}_k = \lambda_k\mathfrak{h}_k,\ k \geq 1,$$

are more convenient for our purposes than the numbers a_k and the operator $\bar{\mathrm{A}}$.

Let H_F be the collection of all finite linear combinations of the elements of $\mathfrak{H}$. Then, for $r \in \mathbb{R}$, define H^r as the closure of H_F in the norm $\|\cdot\|_r$, where, for $f = \sum_k f_k\mathfrak{h}_k$,

$$\|f\|_r^2 = \sum_k \lambda_k^{2r} f_k^2. \tag{3.1.16}$$

Then $\{H^r, r \in \mathbb{R}\}$ is a Hilbert scale, and, for every $m > 0$ and $\gamma \in \mathbb{R}$, the duality between $H^{\gamma+m}$ and $H^{\gamma-m}$ relative to the inner product in H^γ is

$$[f,g]_\gamma = \sum_k \lambda_k^{2\gamma} f_k g_k.$$

Note that, similar to (3.1.12) on page 80,

$$(f,g)_\gamma = (\Lambda^r f, \Lambda^r g)_{\gamma-r},\ [f,g]_\gamma = (\Lambda^m f, \Lambda^{-m} g)_\gamma,\ \ \gamma, r \in \mathbb{R}. \tag{3.1.17}$$

A special example is $H = L_2(G)$, where G is a smooth closed manifold or a smooth bounded domain, and $\Lambda = (-\boldsymbol{\Delta})^{1/2}$ on G (with suitable boundary conditions if G is a domain); see Shubin [208, Sect. I.7]. In this case, the spaces $H^\gamma = H^\gamma(G)$ are known as the `Sobolev spaces` on G.

This concludes Example 3.1.29.

Exercise 3.1.30 (C) Verify that the norms (3.1.16) satisfy (3.1.7) on page 79.

Next, we recall the definition of the `dual operator`.

Theorem 3.1.31 *Let* $\mathrm{A} : X \to Y$ *be a bounded linear operator from a Banach space* X *to a Banach space* Y*. If* A *is defined on all of* X*, then there exists a unique bounded linear operator* $\mathrm{A}' : Y' \to X'$*, called the* `dual operator` *of* A*, such that,* A' *is defined on all of* Y' *and, for every* $\ell \in Y'$ *and every* $x \in X$*,*

$$(\mathrm{A}'\ell)(x) = \ell(\mathrm{A}x).$$

Proof See, for example, Yosida [231, Theorem VII.1.2].

Exercise 3.1.32 (B) Let X, Y be Hilbert spaces and let $\mathrm{J}_X : X \to X'$, $\mathrm{J}_Y : Y \to Y'$ be the isomorphisms defined by the Riesz representation theorem. Show that

$$\mathrm{A}^* = \mathrm{J}_X^{-1}\mathrm{A}'\mathrm{J}_Y \tag{3.1.18}$$

The dual of a Hilbert space can be identified with various Hilbert spaces, and, as the following example illustrates, the computations of A^* and A' depend on this identification.

Example 3.1.33 Let H^γ, $\gamma \in \mathbb{R}$, be the Sobolev spaces on the interval $(0, 1)$ with zero boundary conditions: $H^\gamma = \Lambda^\gamma L_2((0, 1))$, where $\Lambda = \sqrt{-\boldsymbol{\Delta}}$ and $\boldsymbol{\Delta}$ is the Laplace operator with zero boundary conditions, defined on smooth compactly supported functions by $\boldsymbol{\Delta}u(x) = u''(x) = d^2u(x)/dx^2$. Denote by $(\cdot, \cdot)_\gamma$ the inner product in H^γ, so that

$$(f, g)_\gamma = (\Lambda^\gamma f, \Lambda^\gamma g)_0.$$

Consider the operator $\mathrm{A} : f(x) \to a\boldsymbol{\Delta}f$, where $a = a(x)$ is a smooth compactly supported function: $a \in \mathcal{C}_0^\infty$. It is a bounded linear operator from $H^{\gamma+1}$ to $H^{\gamma-1}$ for every $\gamma \in \mathbb{R}$.

The `formal adjoint` $\mathrm{A}^\top$ of A, defined by $f \to \boldsymbol{\Delta}(af)$, is also bounded linear operator from $H^{\gamma+1}$ to $H^{\gamma-1}$ for every $\gamma \in \mathbb{R}$. For a smooth compactly supported function u, we have

$$\mathrm{A}^\top u(x) = \frac{d^2(a(x)u(x))}{dx^2},$$

so that $(\mathrm{A}u, v)_0 = (u, \mathrm{A}^\top v)_0$.

First, consider the normal triple (H^1, H^0, H^{-1}) and identify $(H^1)'$ with H^{-1}; denote by $[f, g]_0$, $f \in H^1$, $g \in H^{-1}$, the duality between H^1 and H^{-1} relative to the inner product in H^0. Let us show that, for the operator A acting from H^1 to H^{-1},

$$\mathrm{A}^* = \Lambda^{-2}\mathrm{A}^\top\Lambda^{-2} : H^{-1} \to H^1, \qquad \mathrm{A}' = \mathrm{A}^\top : H^1 \to H^{-1}. \tag{3.1.19}$$

In particular,

$$\mathrm{A}^* = \Lambda^{-2}\mathrm{A}'\Lambda^{-2}, \tag{3.1.20}$$

which is consistent with (3.1.18), because Λ^2 is an isomorphism from H^1 to H^{-1} and in this setting, $H^{-1} = (H^1)'$.

To verify (3.1.19), take $u, v \in \mathcal{C}_0^\infty((0,1))$ and note that

$$\begin{aligned}(\mathrm{A}u, v)_{-1} &= (\Lambda^{-1}\mathrm{A}u, \Lambda^{-1}v)_0 = (\mathrm{A}u, \Lambda^{-2}v)_0 = (u, \mathrm{A}^\top\Lambda^{-2}v)_0 \\ &= (u, \Lambda^{-2}\mathrm{A}^\top\Lambda^{-2}v)_1 \Rightarrow \mathrm{A}^* = \Lambda^{-2}\mathrm{A}^\top\Lambda^{-2}.\end{aligned}$$

Similarly,

$$\begin{aligned}[\mathrm{A}u, v]_0 &= (\Lambda^{-1}\mathrm{A}u, \Lambda v)_0 = (\mathrm{A}u, v)_0 = (u, \mathrm{A}^\top v)_0 = (\mathrm{A}^\top v, u)_0 \\ &= (\Lambda^{-1}\mathrm{A}^\top v, \Lambda u)_0 = [\mathrm{A}^\top v, u]_0 \Rightarrow \mathrm{A}' = \mathrm{A}^\top.\end{aligned}$$

Next, consider the normal triple (H^2, H^1, H^0) and identify $(H^2)'$ with H^0. Denote by $[f,g]_1, f \in H^0, g \in H^2$, the duality between H^2 and H^0 relative to the inner product in H^1. It follows that $[f,g]_1 = (\Lambda^{-1}f, \Lambda^1 g)_1$. Let us show that, for the operator A acting from H^2 to H^0, we have

$$\mathrm{A}^* = \Lambda^{-4}\mathrm{A}^\top : H^0 \to H^2; \qquad \mathrm{A}' = \Lambda^{-2}\mathrm{A}^\top\Lambda^2 : H^2 \to H^0. \tag{3.1.21}$$

Indeed, for $u, v \in \mathcal{C}_0^\infty((0,1))$,

$$(\mathrm{A}u, v)_0 = (u, \mathrm{A}^\top v)_0 = (\Lambda^{-2}u, \Lambda^{-2}\mathrm{A}^\top v)_2 = (u, \Lambda^{-4}\mathrm{A}^\top v)_2.$$

Similarly,

$$\begin{aligned}\mathrm{A}u, vt_1 &= (\Lambda^{-1}\mathrm{A}u, \Lambda v)_1 = (\mathrm{A}u, \Lambda^2 v)_0 = (u, \mathrm{A}^\top\Lambda^2 v)_0 = (\Lambda^2 u, \Lambda^{-2}\mathrm{A}^\top\Lambda^2 v)_0 \\ &= (\Lambda^{-2}\mathrm{A}^\top\Lambda^2 v, \Lambda^2 u)_0 = (\Lambda^{-3}\mathrm{A}^\top\Lambda^2 v, \Lambda u)_1 = [\Lambda^{-2}\mathrm{A}^\top\Lambda^2 v, u]_1.\end{aligned}$$

Note that (3.1.20) still holds, which is again consistent with (3.1.18), because Λ^2 is an isomorphism from H^2 to H^0 and, in our setting, $H^0 = (H^2)'$.

This concludes Example 3.1.33.

Exercise 3.1.34 (B)

(a) In the setting of Example 3.1.33, show that, for the operator A acting from $H^{\gamma+1}$ to $H^{\gamma-1}$, $\gamma \in \mathbb{R}$, we have

$$\begin{aligned}\mathrm{A}^* &= \Lambda^{-2\gamma-2}\mathrm{A}^\top\Lambda^{2\gamma-2} : H^{\gamma-1} \to H^{\gamma+1}, \\ \mathrm{A}' &= \Lambda^{-2\gamma}\mathrm{A}^\top\Lambda^{2\gamma} : H^{\gamma+1} \to H^{\gamma-1},\end{aligned} \tag{3.1.22}$$

and verify that the result is consistent with (3.1.18).

(b) Show that (3.1.19) continues to hold if the function a is Lipschitz continuous.
(c) Show that (3.1.21) continues to hold if the function a is bounded and measurable.

Exercise 3.1.35 (C) In the setting of Example 3.1.29 show that, for the operator Λ acting from H^r to H^{r-1}, the adjoint operator Λ^* satisfies $\Lambda^* = \Lambda^{-1}$.

Hint $(f, g)_{r-1} = (\Lambda^{-1}f, \Lambda^{-1}g)_r$.

The following is a useful result about the dual and adjoint of an inverse operator.

Theorem 3.1.36 *Let X, Y be Banach spaces and $\mathrm{A} : X \to Y$, a bounded linear bijection. Then $(\mathrm{A}^{-1})' = (\mathrm{A}')^{-1}$. If, in addition, X and Y are Hilbert space, then also $(\mathrm{A}^{-1})^* = (\mathrm{A}^*)^{-1}$.*

Proof Equality $(\mathrm{A}^{-1})' = (\mathrm{A}')^{-1}$ follows from a theorem of Phillips (Yosida [231, Theorem VIII.6.1]). Then equality $(\mathrm{A}^{-1})^* = (\mathrm{A}^*)^{-1}$ follows from the result of Exercise 3.1.32.

The next result is the polar decomposition of bounded linear operators. A real number x can be written as $x = |x|\mathrm{sgn}(x)$; a complex number z can be written as $z = |z| \exp\{\mathrm{i} \arg(z)\}$. The polar decomposition an analogue of these representations for operators.

Theorem 3.1.37 (Polar Decomposition of Bounded Operators) *Let $\mathrm{A} : X \to Y$ be a bounded linear operator and let X, Y be Hilbert spaces.*

(a) There exists a unique bounded linear operator $\mathrm{S} : X \to X$ and a unique bounded linear operator $\mathrm{U} : X \to Y$ with the following properties:

- *S is non-negative and self-adjoint;*
- *$\mathrm{U}^*\mathrm{U}$ is an orthogonal projection on the orthogonal complement of the kernel of S;*
- *$\mathrm{A} = \mathrm{US}$.*

(b) If, in addition, A is compact, then there exists a collection $\{x_k,\ k \geq 1\}$ of elements in X, a collection $\{y_k,\ k \geq 1\}$ of elements in Y, and a sequence of real numbers $\{\lambda_k,\ k \geq\}$ such that

- *the collection $\{x_k,\ k \geq 1\}$ is orthonormal in X: $\|x_k\|_X = 1,\ (x_k, x_m)_X = 0,\ k \neq m$;*
- *the collection $\{y_k,\ k \geq 1\}$ is orthonormal in Y;*
- *$\lambda_k > 0$, $\lim_{k\to\infty} \lambda_k = 0$;*
- *$\mathrm{A}x = \sum_{k\geq 1} \lambda_k (x, x_k)_X\, y_k$.*

Proof

(a) Define $\mathrm{S} = \sqrt{\mathrm{A}^*\mathrm{A}}$ and then

$$\mathrm{U}x = 0,\ x \perp \mathrm{S}(X);\ \mathrm{U}(\mathrm{S}x) = \mathrm{A}x.$$

For more details, see, for example Shubin [208, Proposition A.3.4] or Murphy [171, Theorem 2.3.4].

(b) If A is compact, then so is the corresponding operator S (because $\mathrm{S} = \mathrm{U}^*\mathrm{A}$). By Theorem 3.1.28(b), we can take x_k and λ_k such that $\mathrm{S}x_k = \lambda_k x_k$; then $y_k = \mathrm{U}x_k$.

Remark 3.1.38 The operator S in the polar decomposition $\mathrm{A} = \mathrm{US}$ is often denoted by $|\mathrm{A}|$. The operator U is an example of `partial isometry`. For a bounded linear operator $\mathrm{U} : X \to Y$ each of the following can be taken as the definition of partial isometry: (a) $\mathrm{U} = \mathrm{UU^*U}$; (b) $\mathrm{UU^*}$ is an orthogonal projection; (c) $\mathrm{U^*U}$ is an orthogonal projection; (d) $\|\mathrm{U}x\|_Y = \|x\|_X$ for all x in the orthogonal complement of the kernel of U. See Murphy [171, Theorem 2.3.3].

Definition 3.1.39 Let X, Y be Hilbert spaces, and let $\mathrm{A} : X \to Y$ be a bounded linear operator. The operator A is called

- `Hilbert-Schmidt`, if $\sum_{k\geq 1} \|\mathrm{A}x_k\|_Y^2 < \infty$ for every orthonormal collection $\{x_k,\ k \geq 1\}$ in X.
- `trace-class`, if $\sum_{k\geq 1}(\sqrt{\mathrm{A^*A}}\, x_k, x_k)_X < \infty$ for every orthonormal collection $\{x_k,\ k \geq 1\}$ in X.

Exercise 3.1.40 (C) Let X, Y be separable Hilbert spaces, and let $\mathrm{A} : X \to Y$ be a Hilbert-Schmidt operator.

(a) Show that A is compact. **Hint.** Use Theorem 3.1.19.

(b) Let $\mathfrak{M} = \{\mathfrak{m}_k,\ k \geq 1\}$ be an orthonormal basis in X. Define $\|\mathrm{A}\|_2$ by

$$\|\mathrm{A}\|_2 = \left(\sum_{k\geq 1} \|\mathrm{A}\mathfrak{m}_k\|_Y^2\right)^{1/2}. \tag{3.1.23}$$

Show that $\|\mathrm{A}\|_2$ does not depend on the choice of $\mathfrak{M}$ and defines a norm in the space of Hilbert-Schmidt operators from X to Y.

Hint. Take an orthonormal basis $\{\mathfrak{h}_k,\ k \geq 1\}$ in Y and show that $\sum_k \|\mathrm{A}\mathfrak{m}_k\|_Y^2 = \sum_k \|\mathrm{A}^*\mathfrak{h}_k\|_X^2$. Alternatively, use Theorem 3.1.37(b) to show that $\sum_k \|\mathrm{A}\mathfrak{m}_k\|_Y^2 = \sum_k \lambda_k^2$.

Theorem 3.1.41 *Let X, Y be separable Hilbert spaces, and let $\mathrm{A} : X \to Y$ be a bounded linear operator.*

(a) The following conditions are equivalent:

1. *A is Hilbert-Schmidt;*
2. *A^* is Hilbert-Schmidt;*
3. *A is compact and $\sum_{k\geq 1} \lambda_k^2 < \infty$, where $\lambda_k,\ k \geq 1$, are eigenvalues of $\sqrt{\mathrm{A^*A}}$.*

(b) The following conditions are equivalent:

1. *A is nuclear;*
2. *A^* is nuclear;*

3. A *is trace-class;*
4. $\sum_{k\geq 1}(\mathrm{A}\mathfrak{m}_k, \mathfrak{h}_k)_Y < \infty$ *for every orthonormal bases* $\{\mathfrak{m}_k,\ k \geq 1\}$ *in* X *and* $\{\mathfrak{h}_k,\ k \geq 1\}$ *in* Y*;*
5. A *is compact and* $\sum_{k\geq 1}\lambda_k < \infty$, *where* λ_k, $k \geq 1$, *are eigenvalues of* $\sqrt{\mathrm{A}^*\mathrm{A}}$*;*
6. $\sqrt{\mathrm{S}}$ *is Hilbert-Schmidt, where* $\mathrm{S} = \sqrt{\mathrm{A}^*\mathrm{A}}$*;*
7. $\sum_{k\geq 1}\|\mathrm{A}\mathfrak{m}_k\|_Y < \infty$ *for at least one orthonormal basis* $\{\mathfrak{m}_k, k \geq 1\}$ *in* X*;*
8. $\mathrm{A} = \mathrm{BC}$, *where* B *and* C *are Hilbert-Schmidt;*

Proof Most of the arguments are rather straightforward and are left to the reader. The details can be found in Murphy [171, Sect. 2.4] and Gelfand and Vilenkin [61, Sect. I.2].

Exercise 3.1.42 (C) True or false: if A is nuclear, then $\sum_{k\geq 1}\|\mathrm{A}\mathfrak{m}_k\|_Y < \infty$ for *every* orthonormal basis in X.

Hint. False: let both X and Y be the collection of square-summable sequences), let $\{\mathfrak{m}_k,\ k \geq 1\}$ be the standard unit basis in X, and let A be the orthogonal projection on the vector $f = \sum_{k\geq 1}\mathfrak{m}_k/k$.

Definition 3.1.43

(a) If X, Y are separable Hilbert spaces and $\mathrm{A} : X \to Y$ is a nuclear operator, then the `trace norm` $\|\mathrm{A}\|_1$ of A is defined by

$$\|\mathrm{A}\|_1 = \sum_{k\geq 1}\lambda_k, \tag{3.1.24}$$

where $\lambda_k, k \geq 1$, are the eigenvalues of $\sqrt{\mathrm{A}^*\mathrm{A}}$.

(b) Let X be a separable Hilbert space with an orthonormal basis $\mathfrak{M} = \{\mathfrak{m}_k,\ k \geq 1\}$, let Y be a separable Hilbert space with an orthonormal basis $\mathfrak{H} = \{\mathfrak{h}_k,\ k \geq 1\}$, and let $\mathrm{A} : X \to Y$ be a nuclear operator. The `matrix trace` of A relative to $\mathfrak{M}$ and $\mathfrak{H}$ is $\sum_{k\geq 1}(\mathrm{A}\mathfrak{m}_k, \mathfrak{h}_k)_Y$.

Exercise 3.1.44 (C) Let X be a separable Hilbert space with an orthonormal basis $\mathfrak{M} = \{\mathfrak{m}_k,\ k \geq 1\}$ and let $\mathrm{A} : X \to X$ be a nuclear operator. Define the `trace` $\mathrm{tr}(\mathrm{A})$ of A by

$$\mathrm{tr}(\mathrm{A}) = \sum_{k\geq 1}(\mathrm{A}\mathfrak{m}_k, \mathfrak{m}_k)_X. \tag{3.1.25}$$

(a) Show that $\mathrm{tr}(\mathrm{A})$ does not depend on $\mathfrak{M}$ and $|\mathrm{tr}(\mathrm{A})| \leq \|\mathrm{A}\|_1$. **Hint.** Use Theorem 3.1.37(b).
(b) Show that if A is non-negative and self-adjoint, then $\mathrm{tr}(\mathrm{A}) = \|\mathrm{A}\|_1$.
(c) Show that, for two linear operators $\mathrm{A} : X \to X$ and $\mathrm{B} : X \to X$,

$$\mathrm{tr}(\mathrm{AB}) = \mathrm{tr}(\mathrm{BA}) \tag{3.1.26}$$

if (i) A and B are both Hilbert-Schmidt operators, or if (ii) one of the operators is trace-class and the other is bounded. **Hint.** See Murphy [171, Theorem 2.4.14].

(d) Show that if $X = \mathbb{R}^d$, then tr(A) is the sum of the diagonal elements of the matrix representation of A in some orthonormal basis. In particular tr(A) does not depend on the choice of the orthonormal basis in $\mathbb{R}^d$ and (3.1.26) holds for every two $d \times d$ square matrices.

This concludes Exercise 3.1.44.

Example 3.1.45 (Integral Operators) Let $(S, \mathcal{S}, \mu)$ be a positive measure space and and let $K = K(s,t)$ be a measurable real-valued function on $S \times S$ such that

$$\int_S \int_S K^2(s,t)\mu(ds)\mu(dt) < \infty. \tag{3.1.27}$$

Define the operator $\mathrm{K} : L_2(S,\mu) \to L_2(S,\mu)$ by

$$(\mathrm{K}f)(s) = \int_S K(s,t)f(t)\mu(dt). \tag{3.1.28}$$

(a) The operator K is Hilbert-Schmidt and

$$\|\mathrm{K}\|_2^2 = \int_S \int_S K^2(s,t)\mu(ds)\mu(dt)$$

(see Yosida [231, Example X.2.2]); the function K is called the `kernel` of K. The converse is also true: every Hilbert-Schmidt operator on $L_2(S,\mu)$ can be written in the form (3.1.28) with the kernel satisfying (3.1.27) (see Dunford and Schwartz [43, Problem XI.8.44]).

(b) The operator K is nuclear if there exist functions K_1, K_2 satisfying (3.1.27) and such that

$$K(s,t) = \int_S K_1(s,r)K_2(r,t)\mu(dr).$$

For such an operator,

$$\mathrm{tr(K)} = \int_S K(s,s)\mu(ds); \tag{3.1.29}$$

see Dunford and Schwartz [43, Problem XI.8.49(c)].

(c) If $K(s,t) = K(t,s)$, then the operator K is self-adjoint. In that case, if λ_i, $i \geq 1$, are the eigenvalues of K and φ_i, $i \geq 1$, are the corresponding eigenfunctions:

$$\mathrm{K}\varphi_i = \lambda_i \varphi_i,$$

then

$$K(s,t) = \sum_{i\geq 1} \lambda_i \varphi_i(s)\varphi_i(t), \tag{3.1.30}$$

and the series converges in $L_2(S \times S)$; see Dunford and Schwartz [43, Problem XI.8.56].

(d) Assume that $K(s,t) = K(t,s)$ and the corresponding operator K is non-negative. If, in addition, S is a compact topological space, $\mu(S) < \infty$, the measure μ is regular, and the function K is continuous, then the series (3.1.30) converges uniformly; see Dunford and Schwartz [43, Problem XI.8.58]. This result is known as `Mercer's theorem`, after the British mathematician JAMES MERCER (1883–1932), who published it in 1909.

This concludes Example 3.1.45.

We finish this section with a brief discussion of the `reproducing kernel Hilbert space`.

Definition 3.1.46 A Hilbert space H is called a `reproducing kernel Hilbert space` if

1. the elements of H are real-valued functions on a set S;
2. for every $s \in S$, the operation of point-wise evaluation $f \mapsto f(s)$ is a continuous mapping from H to $\mathbb{R}$.

Exercise 3.1.47 (B)

(a) Show that $L_2(\mathbb{R})$, as a set of function on $\mathbb{R}$, is not a reproducing kernel Hilbert space. **Hint.** Take f the indicator function of the set $[0, 1]$ and g, the indicator function of the set $[\delta, 1]$, $\delta > 0$, and look at $f(0) - g(0)$.

(b) Show that the Sobolev space $H^1(\mathbb{R})$ (see Example 3.1.9 on page 79) is a reproducing kernel Hilbert space. **Hint.** For a smooth compactly supported f, $f(x) = (2\pi)^{-1/2} \int_{\mathbb{R}} e^{\mathrm{i}xy}\widehat{f}(y)dy$. By the Cauchy-Schwarz inequality, $|f(x)-g(x)| \leq (1/\sqrt{2})\|f-g\|_1$.

To understand the origin of the name, *reproducing kernel* Hilbert space, we need one more definition.

Definition 3.1.48 Let S be a set. A real-valued function K defined on $S\times S$ is called a `positive-definite kernel` on S if $K(t,s) = K(s,t)$, $s,t \in S$, and, for every integer $N \geq 1$, every collection of points $s_1, \ldots, s_N$ from S, and every collection of real numbers $y_1, \ldots, y_N$, the following inequality holds:

$$\sum_{i,j=1}^{N} K(s_i, s_j) y_i y_j \geq 0. \tag{3.1.31}$$

In other words, every matrix of the form $\big(K(s_i, s_j),\ i,j = 1, \ldots, N\big)$ is symmetric and non-negative definite.

Exercise 3.1.49 (B)

(a) Verify that a positive-definite kernel K has the following properties: (i) $K(t,t) \geq 0$; (ii) $|K(s,t)|^2 \leq K(s,s)\,K(t,t)$.
(b) Verify that a real continuous symmetric function K on $[0,1]\times[0,1]$ is a positive-definite kernel on $[0,1]$ if and only if

$$\int_0^1\int_0^1 K(s,t)f(s)f(t)dsdt \geq 0$$

for every $f \in L_2((0,1))$.

The following theorem connects the reproducing kernel Hilbert spaces and the positive-definite kernels.

Theorem 3.1.50 *Let S be a set.*

(1) For every reproducing kernel Hilbert space H of functions on S, there exists a unique positive-definite kernel K_H on S, called the `reproducing kernel` *of H, such that, for every $s \in S$ and $f \in H$,*

$$f(s) = \big(f, K_H(s,\cdot)\big)_H$$

(note that, for every fixed s, $K_H(s,\cdot)$ is a function on S and therefore can be an element of H).

(2) For every positive-definite kernel K on S, there exists a reproducing kernel Hilbert space H of real-valued functions on S, such that K is the reproducing kernel of H: $K_H = K$.

Proof For the main idea, see parts (a) and (b) of the following exercise. For details, see Aronszajn [4, Sect. I.2].

Exercise 3.1.51 (C)

(a) Derive existence and uniqueness of K_H from the Riesz representation theorem. **Hint.** The continuity of $f \mapsto f(s)$ implies $f(s) = (f, g_s)_H$ for some $g_s \in H$; then $K_H(s,t) = g_s(t)$.
(b) Show that, given a kernel K, the corresponding reproducing kernel Hilbert space H can be constructed as the closure of the set of finite sums of the form $\sum_k a_k K(x_k,\cdot)$. The closure in with respect to the norm generated by the inner product

$$\left(\sum_k a_k K(x_k,\cdot), \sum_n b_n K(y_n,\cdot)\right)_H = \sum_{k,n} a_k b_n K(x_k, y_n).$$

(c) Let H be a reproducing kernel Hilbert space. Show that

$$\big(K(s,\cdot), K(t,\cdot)\big)_H = K(s,t).$$

(d) Let H be a reproducing kernel Hilbert space with an orthonormal basis $\{\mathfrak{m}_k,\ k \geq 1\}$. Show that

$$K_H(s,t) = \sum_{k\geq 1} \mathfrak{m}_k(s)\mathfrak{m}_k(t).$$

3.1.3 Problems

Together with the conclusion of Problem 3.1.1, the reader should keep in mind that while every two separable Hilbert spaces are isomorphic, they can be very different. Problem 3.1.2 shows how, given an arbitrary separable Hilbert space H, one can construct a Hilbert space $\tilde{H}$ such that the inclusion operator $\mathfrak{j} : H \to \tilde{H}$ is Hilbert-Schmidt; this construction is commonly used in the study of random elements with values in Hilbert spaces. Problem 3.1.3 provides two useful technical results: (a) a way to prove strong convergence in a Hilbert space, and (b) an analogue of the fundamental theorem of calculus in a normal triple of Hilbert spaces. Problem 3.1.4 introduces a class of Hilbert spaces involving time, useful in the study of evolution equations. Problem 3.1.5 summarizes some facts about linear operators that were not mentioned in the text. Problem 3.1.6 discusses linear operators in Hilbert space tensor products. Problems 3.1.7 and 3.1.8 illustrate the importance of the underlying set S in the study of reproducing kernel Hilbert spaces. Problems 3.1.9 and 3.1.10 present examples of constructing a reproducing kernel Hilbert space given the kernel K. Problem 3.1.11 demonstrates a connection between the Hilbert space tensor product and the Hilbert-Schmidt operators.

Problem 3.1.1 Show that every two separable Hilbert spaces are isomorphic.

Problem 3.1.2 Let H be a separable Hilbert space with an orthonormal basis $\{\mathfrak{m}_k,\ k \geq 1\}$ and let $\{q_k,\ k \geq 1\}$ be sequence of positive real numbers such that $\sum_{k\geq 1} q_k^2 < \infty$. Define the space $\tilde{H}$ as the closure of H with respect to the norm

$$\|f\|_{\tilde{H}} = \left(\sum_{k\geq 1} q_k^2 (f, \mathfrak{m}_k)_H^2\right)^{1/2}.$$

Show that the inclusion $\mathfrak{j} : H \to \tilde{H}$ is a Hilbert-Schmidt operator and $\mathrm{tr}(\mathfrak{j}\mathfrak{j}^*) = \sum_{k\geq 1} q_k^2 = \mathrm{tr}(\mathfrak{j} * \mathfrak{j})$.

Problem 3.1.3

(a) Let H be a Hilbert space and $h, h_1, h_2, \ldots \in H$. Show that $\lim_n \|h - h_n\|_H = 0$ (strong convergence) if and only if $\lim_n (h_n, x)_H = (h, x)_H$ for every x from a dense subset of H (weak convergence on a dense subset) and $\lim_n \|h_n\|_H = \|h\|_H$ (convergence of norms).

(b) Let (V, H, V') be a normal triple of Hilbert spaces and let $u = u(t)$ be an element of $L_2((0,T); V)$ such that, for all $t \in [0,T]$,

$$u(t) = u_0 + \int_0^t f(s)ds \tag{3.1.32}$$

for some $u_0 \in H$ and $f \in L_2((0,T); V')$ (equality (3.1.32) is in V'). Show that $u \in \mathcal{C}((0,T); H)$ and

$$\sup_{0<t<T} \|u\|_H^2(t) \leq \left(\|u_0\|_H^2 + \int_0^T \|u\|_V^2(s)ds + \int_0^T \|f\|_{V'}^2(s)ds \right) \tag{3.1.33}$$

A Comment Part (b) of the problem can be considered a `fundamental theorem of calculus` in a normal triple (if $V = H = V'$, then the theorem becomes the claim that an absolutely continuous function is continuous).

Problem 3.1.4 Show that the collection Y of functions from $L_2((0,T); V)$ with representation (3.1.32) is a Hilbert space with norm

$$\|u\|^2 = \|u_0\|_H^2 + \int_0^T \|u(s)\|_V^2 ds + \int_0^T \|g(s)\|_{V'}^2 ds.$$

Problem 3.1.5 Let X, Y be separable Hilbert spaces. Introduce the following notations:

- $\mathcal{F}(X,Y)$, the collection of linear operators from X to Y with a finite-dimensional range (that is, $\mathrm{A} \in \mathcal{F}(X,Y)$ if and only if A is linear and $\mathrm{A}(X)$ is finite-dimensional).
- $\mathcal{K}(X,Y)$, the collection of compact operators from X to Y;
- $\mathcal{L}_0(X,Y)$, the collection of bounded linear operators from X to Y;
- $\mathcal{L}_1(X,Y)$, the collection of nuclear operators from X to Y;
- $\mathcal{L}_2(X,Y)$, the collection of Hilbert-Schmidt operators from X to Y.

For $\mathrm{A} \in \mathcal{F}(X,Y)$, denote by λ_n, $n \geq 1$, the eigenvalues of the operator $\sqrt{\mathrm{A}^*\mathrm{A}}$ and define

$$\|\mathrm{A}\|_0 = \max_n \lambda_n, \ \|\mathrm{A}\|_1 = \sum_n \lambda_n, \ \|\mathrm{A}\|_2 = \left(\sum_n \lambda_n^2 \right)^{1/2}. \tag{3.1.34}$$

(a) Show that each $\|\cdot\|_i$, $i = 0, 1, 2$, is a norm, and

$$\|\mathrm{A}\|_i = \|\mathrm{A}^*\|_i, \tag{3.1.35}$$

$$\|\mathrm{A}\|_0 \le \|\mathrm{A}\|_2 \le \|\mathrm{A}\|_1, \tag{3.1.36}$$

$$\|\mathrm{A}\|_0 = \sup_{x:\|x\|_X \le 1} \|\mathrm{A}x\|_Y, \tag{3.1.37}$$

$$\|\mathrm{A}\|_2^2 = \|\mathrm{A}^*\mathrm{A}\|_1 = \mathrm{tr}(\mathrm{A}^*\mathrm{A}) = \|\mathrm{A}\mathrm{A}^*\|_1 = \mathrm{tr}(\mathrm{A}\mathrm{A}^*). \tag{3.1.38}$$

(b) Show that $\mathcal{K}(X, Y)$ is the closure of $\mathcal{F}(X, Y)$ in the norm $\|\cdot\|_0$, and, for $j = 1, 2$, $\mathcal{L}_j(X, Y)$ is the closure of $\mathcal{F}(X, Y)$ in the norm $\|\cdot\|_j$.
(c) Show that the spaces $\mathcal{L}_i(X, Y)$ with the corresponding norms $\|\cdot\|_i$, $i = 0, 1, 2$, are separable Banach space, $\mathcal{K}(X, Y)$ is a closed subspace of $\mathcal{L}_0(X, Y)$, and $\mathcal{L}_2(X, Y)$ is a separable Hilbert space with inner product $(\mathrm{A}, \mathrm{B})_2 = \mathrm{tr}(\mathrm{B}^*\mathrm{A})$, where tr is defined in (3.1.25). Also, show that

$$\mathcal{F}(X, Y) \subset \mathcal{L}_1(X, Y) \subset \mathcal{L}_2(X, Y) \subset \mathcal{K}(X, Y) \subset \mathcal{L}_0(X, Y).$$

(d) Let $\mathrm{A} \in \mathcal{L}_j(X, Y)$, $\mathrm{B} \in \mathcal{L}_0(Y, Y)$. Show that, for $j = 1, 2$, $\|\mathrm{BA}\|_j \le \|\mathrm{B}\|_0 \, \|\mathrm{A}\|_j$.
(e) Let $\mathrm{A} \in \mathcal{L}_2(X, Y)$, $\mathrm{B} \in \mathcal{L}_2(Y, Y)$. Show that $\|\mathrm{BA}\|_1 \le \|\mathrm{B}\|_2 \, \|\mathrm{A}\|_2$.
(f) Let $\mathrm{A} \in \mathcal{K}(X, Y)$, $\mathrm{B} \in \mathcal{L}_0(Y, Y)$. Show that $\mathrm{BA} \in \mathcal{K}(X, Y)$.

Problem 3.1.6 Let X, Y be separable Hilbert spaces and let $\mathrm{A} : X \to X, \mathrm{B} : Y \to Y$ be bounded linear operators. Consider the Hilbert space tensor product $X \otimes Y$ and consider the mapping defined by

$$x \otimes y \mapsto (\mathrm{A}x) \otimes (\mathrm{B}y). \tag{3.1.39}$$

(a) Show that the mapping (3.1.39) extends to a bounded linear operator $\mathrm{A} \otimes \mathrm{B} : X \otimes Y \mapsto X \otimes Y$ and, using the notations from the previous problem,

$$\|\mathrm{A} \otimes \mathrm{B}\|_0 = \|\mathrm{A}\|_0 \, \|\mathrm{B}\|_0.$$

(b) What can you say about the operator $\mathrm{A} \otimes \mathrm{B}$ if both operators A and B are (i) compact (ii) Hilbert-Schmidt (iii) nuclear?

Problem 3.1.7 Let H be a Hilbert space and let us identify H with its dual H'. Then every element h of H becomes a real-valued function on H defined by $h(x) = (h, x)_H, x \in H$. Show that H is a reproducing kernel Hilbert space and find the kernel K_H. Is there a contradiction with the conclusion of Exercise 3.1.47(a) on page 94 if we take $H = L_2(\mathbb{R})$?

Problem 3.1.8 Let (V, H, V') be a normal triple. Identifying V with the dual of V' relative to the inner product in H, we can think of V as a collection of real-valued functions on V'.

(a) Show that V is a reproducing kernel Hilbert space and find the reproducing kernel of V.
(b) Now take $V = L_2(\mathbb{R})$, $H = H^{-1}(\mathbb{R})$, $V' = H^{-2}(\mathbb{R})$; see Example 3.1.9 on page 79 for the definition of the Sobolev spaces $H^r(\mathbb{R})$. Is there a contradiction between the result of part (a) and the conclusion of Exercise 3.1.47(a) on page 94?

Problem 3.1.9 Let A be a non-negative self-adjoint operator on a separable Hilbert space X. For $x, y \in X$, define $K(x, y) = (\mathrm{A}x, y)_X$. Show that the reproducing kernel Hilbert space corresponding to K is $\sqrt{\mathrm{A}}(X)$.

Problem 3.1.10 Let $K(s, t) = \min(s, t)$, $s, t \in [0, 1]$. Verify that K is positive-definite kernel on $[0, 1]$ and find the corresponding reproducing kernel Hilbert space.

Problem 3.1.11 Let X and Y be separable Hilbert spaces. Show that, for $x, x_1 \in X, y \in Y$,

$$x \otimes y : \; x_1 \mapsto y(x, x_1)_X$$

establishes an isomorphism between the tensor product Hilbert space $X \otimes Y$ the space $L_2(X, Y)$ of the Hilbert-Schmidt operators from X to Y.

3.2 Random Processes and Fields

The objective of this section is to review various constructions of the noise process that can be used in the study of SPDEs: a real-valued multi-parameter random processes, a random elements with values in a linear topological space, and the canonical processes on a linear topological space with a suitable probability measure.

3.2.1 Fields (No Time Variable)

One way to define a real-valued random variable ξ is by a measurable mapping from an abstract probability space $(\Omega, \mathcal{F}, \mathbb{P})$ to the real line; the function $F(x) = \mathbb{P}(\xi \leq x)$ is called the cumulative distribution function of ξ. Alternatively, one can start with a given cumulative distribution function $F = F(x)$, define a probability measure on $(\mathbb{R}, \mathcal{B}(\mathbb{R}))$ by $\mathbb{P}((a, b]) = F(b) - F(a)$ and then define the random variable ξ on $(\mathbb{R}, \mathcal{B}(\mathbb{R}), \mathbb{P})$ by $\xi(x) = x$. By construction, the cumulative distribution function of ξ is F. The same two approaches extend to random elements with values in linear topological spaces.

Consider a measurable space $(V, \mathcal{B}(V))$, where V is a linear topological space and $\mathcal{B}(V)$ is the Borel sigma-algebra on V. A V-valued random element X defines a probability measure on μ on $(V, \mathcal{B}(V))$ as follows:

$$\mu(A) = \mathbb{P}(X \in A),\ A \in \mathcal{B}(V).$$

This measure is called the `probability distribution` of X. Conversely, if μ is a probability measure on $(V, \mathcal{B}(V))$, then $X(v) = v$, $v \in V$, defines a canonical V-valued random element on the probability space $(V, \mathcal{B}(V), \mu)$ such that the probability distribution of X is μ. In the special case when the elements of V are functions of time, the canonical random element is usually called the `canonical process`. In particular, the `Wiener measure` is the probability measure on the space $V = \{f \in \mathcal{C}([0, T]),\ f(0) = 0\}$ with the sup norm, such that the canonical process under this measure is the standard Brownian motion.

In what follows, we will be working with Gaussian measures and Gaussian random elements. For a comprehensive treatment of Gaussian measures on linear topological spaces see Bogachev [13]. Below, we present a short summary of the main ideas.

Recall that a probability measure μ on $(\mathbb{R}, \mathcal{B}(\mathbb{R}))$ is called `Gaussian` if there exists a real number m and a non-negative number σ such that, for all $t \in \mathbb{R}$,

$$\int_{\mathbb{R}} e^{\mathrm{i}tx} \mu(dx) = e^{\mathrm{i}mt - \frac{1}{2}t^2\sigma^2},\ \ \mathrm{i} = \sqrt{-1}.$$

The numbers m and σ are called the mean and the variance of μ, respectively. If $\sigma = 0$, then μ is the point mass at m. If $\sigma > 0$, then μ has the density with respect to the Lebesgue measure:

$$\mu((a, b)) = \frac{1}{\sqrt{2\pi\sigma^2}} \int_a^b e^{-\frac{(x-m)^2}{2\sigma^2}}\, dx.$$

The measure μ is called `centered` if $m = 0$. A (real-valued) random variable is called Gaussian if its probability distribution is a Gaussian measure. A `standard Gaussian random variable` has zero mean and unit variance.

Definition 3.2.1 A probability measure μ on $(V, \mathcal{B}(V))$ is called `Gaussian` if, for every linear functional ℓ on V, the probability measure $\ell^*\mu$ on $(\mathbb{R}, \mathcal{B}(\mathbb{R}))$, defined by $(\ell^*\mu)((a, b)) = \mu\{v \in V : \ell(v) \in (a, b)\}$ is Gaussian. The measure μ is called `centered` if the measure $\ell^*\mu$ is centered for every ℓ.

Exercise 3.2.2 (C) Let X be a Banach space and let X' be its dual. Show that μ is a centered Gaussian measure on $(X, \mathcal{B}(X))$ if and only if every element of X' is a zero-mean Gaussian random variable on the probability space $(X, \mathcal{B}(X), \mu)$.

Exercise 3.2.3 (B)

(a) Verify that, in the case $V = \mathbb{R}^{\mathrm{d}}$, μ is Gaussian if and only if

$$\int_{\mathbb{R}^{\mathrm{d}}} e^{\mathrm{i}y^{\top}x}\mu(dx) = e^{\mathrm{i}m^{\top}y-\frac{1}{2}y^{\top}Ry} \tag{3.2.1}$$

for some d-dimensional vector m and a d $\times$ d-dimensional symmetric non-negative definite matrix R; $y^{\top}$ means the transpose of the column-vector y. **Hint.** $R_{ij} = \int_{\mathbb{R}^{\mathrm{d}}} x_i x_j \mu(dx)$.

(b) Denote by V' the set of linear functionals on V. Verify that a centered Gaussian measure μ on V defines a positive-definite kernel C on V' by

$$C_{\mu}(\ell, h) = \int_V \ell(v)h(v)\,\mu(dv). \tag{3.2.2}$$

For example, let $V = \{f \in \mathcal{C}([0,1]),\ f(0) = 0\}$ with the sup norm. For every $s \in [0,1]$, the point mass at s (Dirac delta-function) δ_s is a continuous linear functional on V. If μ is the Wiener measure on V, then, since the canonical process on $(V, \mathcal{B}(V), \mu)$ is the standard Brownian motion, we find

$$C_{\mu}(\delta_s, \delta_t) = \int_V f(s)f(t)\,\mu(df) = \mathbb{E}\,(w(s)w(t)) = \min(t, s). \tag{3.2.3}$$

Computation of the general expression for C_{μ} in this example is the subject of Problem 3.2.5 on page 126.

Definition 3.2.4 The reproducing kernel Hilbert space H_{μ} with the reproducing kernel C_{μ} is called the reproducing kernel Hilbert space of the Gaussian measure μ. If X is a V-valued Gaussian random element with distribution μ, then H_{μ} is also called the reproducing kernel Hilbert space of X.

There are two ways to look at the above definition. Recall that one of the starting points in the construction of the reproducing kernel Hilbert space is a collection of functions on a set S. In the setting of Definition 3.2.4, we take $S = V'$, and then, for fixed $\ell \in V'$, the mapping

$$h \mapsto C_{\mu}(\ell, h) = \int_V \ell(v)\,h(v)\,\mu(dv) := K(\ell, h). \tag{3.2.4}$$

becomes a real-valued function on V'. Following the procedure outlined in Exercise 3.1.51(b) on page 95, the space H_{μ} becomes the closure of the set of finite linear combinations of the type

$$\sum_k a_k \int_V h_k(v)\,v\,\mu(dv)$$

and is, in particular, a sub-space of V. For an example illustrating this construction, see Problem 3.2.6 on page 126 below.

Alternatively, one can look at (3.2.4) as a definition of an operator $\ell \mapsto C_\mu(\ell, \cdot)$ from V' to $(V')'$, which could explain why C_μ is called the covariance operator of the Gaussian measure μ. If $V = H$ is a Hilbert space, identified with its dual, then C_μ in this interpretation becomes an operator from H to itself: for every $h \in H$, $C_\mu(h, \cdot)$ is the unique element h_μ of H such that, according to the Riesz representation theorem, $C_\mu(h, x) = (h_\mu, x)_H$.

Theorem 3.2.5 *Let H be a separable Hilbert space, identified with its dual: $H' = H$.*

(a) If μ is a Gaussian measure on H, then its covariance operator is nuclear.
(b) Conversely, if $\mathrm{K} : H \to H$ is a self-adjoint non-negative nuclear operator, then there exists a Gaussian measure μ on H such that $C_\mu(f, g) = (\mathrm{K}f, g)_H$ for all $f, g \in H$.

Proof

(a) Take an orthonormal basis $\mathfrak{h}_k,\ k \geq 1$ in H. Then

$$\sum_k C_\mu(\mathfrak{h}_k, \mathfrak{h}_k) = \sum_k \int_H (\mathfrak{h}_k, x)_H^2 \mu(dx) = \int_H \|x\|_H^2 \mu(dx) < \infty.$$

The last inequality follows from a result due to Fernique: for every Gaussian measure μ on a Hilbert space H, there exists a positive number a such that $\int_H e^{a\|x\|_H^2} \mu(dx) < \infty$; see Bogachev [13, Theorem 2.8.5].

(b) Let $\mathrm{K}\mathfrak{m}_k = \lambda_k \mathfrak{m}_k$ and assume that $\{\mathfrak{m}_k,\ k \geq 1\}$ is an orthonormal basis in H. Given iid standard Gaussian random variables ξ_k, we get the measure μ as the probability distribution of the H-valued random element $X = \sum_{k\geq 1} \sqrt{\lambda_k} \xi_k \mathfrak{m}_k$. The operator K can also be called the covariance operator of μ.

Given a Gaussian measure on a linear topological space V, we immediately get a corresponding canonical Gaussian random element with values in V. We will now discuss other ways of defining Gaussian random elements.

Definition 3.2.6 A Gaussian field on $G \subseteq \mathbb{R}^{\mathrm{d}}$ is a collection of random variables $X = X(P),\ P \in G$, such that, for every $\{P_1, \ldots, P_n\} \subset G$, the random variables $X(P_i),\ i = 1, \ldots, n$, form a Gaussian vector.
In what follows, we consider zero-mean fields: $\mathbb{E}X(P) = 0$ for all $P \in G$. A zero-mean Gaussian field is characterized by its covariance function

$$q(P, Q) = \mathbb{E}\big(X(P)X(Q)\big);$$

the field is called homogeneous if there exists a function $\varphi = \varphi(t),\ t \geq 0$ such that $q(P, Q) = \varphi(|P - Q|)$, where $|P - Q|$ is the Euclidean distance between the points P and Q. We write P, Q instead of x, y to stress that the objects are not tied to any particular coordinate system in $\mathbb{R}^{\mathrm{d}}$.

As an example and a motivation for the discussion to follow, let $\boldsymbol{\ell}$ be the Lebesgue measure $\mathbb{R}^{\mathrm{d}}$ and let $\boldsymbol{W}$ be a random set function on $\{A : A \in \mathcal{B}(\mathbb{R}^{\mathrm{d}}),\ \boldsymbol{\ell}(A) < \infty\}$. We assume that $\boldsymbol{W}$ has the following properties:

1. $\boldsymbol{W}(A)$ is a Gaussian random variable with mean zero and variance $\boldsymbol{\ell}(A)$;
2. if $A \bigcap B = \emptyset$, then $\boldsymbol{W}(A \bigcup B) = \boldsymbol{W}(A) + \boldsymbol{W}(B)$ and the random variables $\boldsymbol{W}(A)$, $\boldsymbol{W}(B)$ are independent.

To construct $\boldsymbol{W}$ take a collection $\{\xi_k,\ k \geq 1\}$ of independent standard Gaussian random variables and an orthonormal basis $\{\mathfrak{m}_k,\ k \geq 1\}$ in $L_2(\mathbb{R}^{\mathrm{d}})$. Then

$$\boldsymbol{W}(A) = \sum_{k \geq 1} \xi_k \int_A \mathfrak{m}_k(x)dx. \tag{3.2.5}$$

Indeed, if $\mathbf{1}_A$ is the `indicator function` of the set $A \subset \mathbb{R}^{\mathrm{d}}$ ($\mathbf{1}_A(x) = 1$ if $x \in A$ and $\mathbf{1}_A(x) = 0$ if $x \notin A$), then, by Parseval's identity,

$$\mathbb{E}\big(\boldsymbol{W}(A)\boldsymbol{W}(B)\big) = \int_{\mathbb{R}^{\mathrm{d}}} \mathbf{1}_A(x)\mathbf{1}_B(x)dx. \tag{3.2.6}$$

For $r > 0$ and $P \in \mathbb{R}^{\mathrm{d}}$, denote by $B_r(P)$ the ball with center at P and radius r. Then, for every fixed $r > 0$, the function

$$W_r(P) = \frac{\boldsymbol{W}(B_r(P))}{\boldsymbol{\ell}(B_r(P))} \tag{3.2.7}$$

is a zero-mean homogeneous Gaussian field on $\mathbb{R}^{\mathrm{d}}$, and $\mathbb{E}\big(W_r(P)W_r(Q)\big) = 0$ if $|P - Q| > r$. Once again, note that, up to this point, we have not been tied to any particular coordinate system in $\mathbb{R}^{\mathrm{d}}$.

Let us now fix a Cartesian coordinate system in $\mathbb{R}^{\mathrm{d}}$ and, given a point $x = (x_1, \ldots, x_{\mathrm{d}})$ with $x_i > 0$ for all $i = 1, \ldots, \mathrm{d}$, consider a rectangular box A_x defined by

$$A_x = [0, x_1] \times [0, x_2] \times \cdots \times [0, x_{\mathrm{d}}],$$

Then the random field

$$W(x) = \boldsymbol{W}(A_x) \tag{3.2.8}$$

is called `Brownian sheet`; cf. Walsh [223, Chap. 1].

Exercise 3.2.7 (B)

(a) Show that $\lim_{r\to 0} W_r(P)$ does not exist in distribution for any P. **Hint.** Consider the characteristic function of $W_r(P)$.
(b) Show that, for the Brownian sheet (3.2.8),

$$\mathbb{E}\big(W(x)W(y)\big) = \prod_{i=1}^{\mathrm{d}} \min(x_i, y_i). \tag{3.2.9}$$

Hint. Use (3.2.6).

Looking at (3.2.5), we realize that both $\lim_{r\to 0} W_r(x)$ and $\partial^{\mathrm{d}} W(x)/\partial x_1 \ldots \partial x_{\mathrm{d}}$, if existed, would have to be equal to a divergent series $\sum_{k\geq 1} \xi_k h_k(x)$. On the other hand, this divergent series can be interpreted as a generalized function $\dot{W}$, acting on the functions from the Schwartz space $\mathcal{S}(\mathbb{R}^{\mathrm{d}})$ of rapidly decreasing functions according to the rule

$$\dot{W}(f) = \sum_{k\geq 1} \xi_k f_k, \text{ where } f_k = \int_{\mathbb{R}^{\mathrm{d}}} f(x) h_k(x)dx. \tag{3.2.10}$$

Recall that $\mathcal{S}(\mathbb{R}^{\mathrm{d}})$ is the collection of infinitely differentiable functions on $\mathbb{R}^{\mathrm{d}}$ such that $\sup_{x\in\mathbb{R}^{\mathrm{d}}}(1+|x|^2)^m|D^n f(x)| < \infty$ for all positive integer m, n and all partial derivatives D^n of f of order n; $|x|^2 = x_1^2 + \cdots + x_{\mathrm{d}}^2$. Denote by $\mathbb{R}^{\mathrm{d}}_+$ the set $\{x \in \mathbb{R}^{\mathrm{d}} : x_1 > 0, \ldots, x_{\mathrm{d}} > 0\}$. Direct computations show that

$$\dot{W}(f) \sim \mathcal{N}\Big(0, \|f\|^2_{L_2(\mathbb{R}^{\mathrm{d}})}\Big), \tag{3.2.11}$$

$$\mathbb{E}\big(\dot{W}(f)\dot{W}(g)\big) = (f, g)_{L_2(\mathbb{R}^{\mathrm{d}})}, \tag{3.2.12}$$

$$\dot{W}(f) = (-1)^{\mathrm{d}} \int_{\mathbb{R}^{\mathrm{d}}_+} W(x) \frac{\partial^{\mathrm{d}} f(x)}{\partial x_1 \cdots \partial x_{\mathrm{d}}} dx,\ f \in \mathcal{C}_0^\infty(\mathbb{R}^{\mathrm{d}}_+). \tag{3.2.13}$$

Equality (3.2.13) shows that $\dot{W}$ is a generalized derivative of W. Equality (3.2.12) shows that we can extend $\dot{W}$ from $\mathcal{S}(\mathbb{R}^{\mathrm{d}})$ to $L_2(\mathbb{R}^{\mathrm{d}})$. We call this extension the `Gaussian white noise` on $L_2(\mathbb{R}^{\mathrm{d}})$ and continue to denote it by $\dot{W}$.

Exercise 3.2.8 (C) Verify (3.2.11)–(3.2.13).
Sometimes, we use an alternative notation for $\dot{W}(f)$:

$$\dot{W}(f) = \int_{\mathbb{R}^{\mathrm{d}}} f(x)dW(x).$$

The following definition combines the ideas from Gelfand and Vilenkin [61, Sects. III.1.2 and III.5.1] and from Métivier and Pellaumail [159, Sect. 15.1].

Definition 3.2.9 A `generalized random field` $\mathfrak{X}$ over a linear topological space V is a collection of random variables $\{\mathfrak{X}(v),\ v \in V\}$ with the properties

1. $\mathfrak{X}(au + bv) = a\mathfrak{X}(u) + b\mathfrak{X}(v)$, $a, b \in \mathbb{R}$, $u, v \in V$;
2. if $\lim_{n\to\infty} v_n = v$ in the topology of V, then $\lim_{n\to\infty} \mathfrak{X}(v_n) = \mathfrak{X}(v)$ in probability.

In other words, $\mathfrak{X}$ is a continuous linear mapping from V to the space of random variables. Alternative names for such an object are a `cylindrical random element` and `generalized random element`.

For a generalized Gaussian random field over a Hilbert space, an alternative definition is often used (e.g. Nualart [175, Sect. 1.1.1]).

Definition 3.2.10 A zero-mean `generalized Gaussian field` $\mathfrak{B}$ over a Hilbert space H is a collection of Gaussian random variables $\{\mathfrak{B}(f),\ f \in H\}$ with the properties

1. $\mathbb{E}\mathfrak{B}(f) = 0$ for all $f \in H$;
2. There exists a bounded, linear, self-adjoint, non-negative operator K on H (called the `covariance operator` of $\mathfrak{B}$) such that

$$\mathbb{E}\big(\mathfrak{B}(f)\mathfrak{B}(g)\big) = (\mathrm{K}f, g)_H$$

for all $f, g \in H$, where $(\cdot,\cdot)_H$ is the inner product in H.

In the special case when K is the identity operator, alternative names for $\mathfrak{B}$ are `Gaussian white noise` on (or over) H and `isonormal Gaussian process` on (or over) H.

If $\mathfrak{B}$ is a Gaussian white noise on H, then $\mathbb{E}\big(\mathfrak{B}(f)\mathfrak{B}(g)\big) = (f, g)_H$. In the particular case $H = L_2(G)$, $G \subseteq \mathbb{R}^{\mathrm{d}}$, we usually write $\dot{W} = \dot{W}(x)$ to denote Gaussian white noise on H and also use an alternative notation for $\dot{W}(f)$:

$$\dot{W}(f) = \int_G f(x)dW(x).$$

In fact, as the following exercise shows, $\mathfrak{B}$ generalizes the construction of $\dot{W}$ from (3.2.10) to an abstract Hilbert space.

Exercise 3.2.11 (C) Let H be a separable Hilbert space with an orthonormal basis $\{\mathfrak{m}_k,\ k \geq 1\}$.

(a) Let $\mathfrak{B}$ be a Gaussian white noise on H. Verify that

$$\{\mathfrak{B}(\mathfrak{m}_k),\ k \geq 1\}$$

is a collection of iid standard Gaussian random variables.

(b) Let $\{\xi_k,\ k \geq 1\}$ be a collection of iid standard Gaussian random variables, and, for $f \in H$, define

$$\mathfrak{B}(f) = \sum_{k\geq 1} (f, \mathfrak{m}_k)_H\, \xi_k. \tag{3.2.14}$$

Verify that $\mathfrak{B}$ is a Gaussian white noise on H.

The next exercise establishes a connection between Definitions 3.2.9 and 3.2.10.

Exercise 3.2.12 (B)

(a) Verify that a generalized Gaussian field is a generalized random field in the sense of Definition 3.2.9. **Hint**. Verify by direct computation that $\mathbb{E}\Big(\mathfrak{B}(af+bg) - a\mathfrak{B}(f) - b\mathfrak{B}(g)\Big)^2 = 0$ and $\mathbb{E}(\mathfrak{B}(f) - \mathfrak{B}(f_n))^2 \leq C\|f-f_n\|_H^2$.
(b) Verify that if $\mathfrak{X}$ is a generalized random field over a Hilbert space and every random variable $\mathfrak{X}(v)$ is Gaussian, then $\mathfrak{X}$ is a generalized Gaussian field in the sense of Definition 3.2.10 and the correlation operator K is uniquely defined.
Hint. Use the Riesz representation theorem.

Definition 3.2.13 A generalized field $\mathfrak{X}$ over a Hilbert space H is called `regular` if there exists an H-valued random element X such that $X \in L_2(\Omega; H)$ (that is, $\mathbb{E}\|X\|_H^2 < \infty$) and $\mathfrak{X}(f) = (X, f)_H$ for all $f \in H$.

Exercise 3.2.14 Let $G \subseteq \mathbb{R}^{\mathrm{d}}$, $H = L_2(G)$, and consider a regular zero-mean Gaussian random field $\mathfrak{X}(f) = (X, f)_H$. Show that the covariance operator K of $\mathfrak{X}$ is an integral operator on H with kernel $K(x,y) = \mathbb{E}\big(X(x)X(y)\big)$, that is,

$$\mathrm{K}f(x) = \int_G K(x,y)f(y)dy.$$

Hint. Exchange integration and expectation.

There is a close connection between regular Gaussian fields and nuclear operators.

Theorem 3.2.15 *A generalized Gaussian field* $\mathfrak{X}$ *over a separable Hilbert space* H *is regular if and only if the covariance operator* K *of* $\mathfrak{X}$ *is nuclear.*

Proof

(a) Assume that $\mathfrak{X}(f) = (X, f)_H$ and let $\{\mathfrak{h}_k,\ k \geq 1\}$ be an orthonormal basis in H. Then

$$\sum_{k\geq 1} (\mathrm{K}\mathfrak{h}_k, \mathfrak{h}_k)_H = \sum_{k\geq 1} \mathbb{E}(X, \mathfrak{h}_k)^2 = \mathbb{E}\|X\|_H^2 < \infty,$$

which, by Theorem 3.1.41(b), implies that K is nuclear.
(b) Assume that K is nuclear and let $\{\mathfrak{m}_k,\ k \geq 1\}$ be the orthonormal basis in H consisting of the eigenfunctions of K; such a basis exists because K is

compact: see Exercise 3.1.20 and Theorem 3.1.28(b). Denote by $\lambda_k,\ k \geq 1$, the eigenvalues of K and define $\xi_k = \lambda_k^{-1/2}\mathfrak{X}(\mathfrak{m}_k)$ so that $\{\xi_k,\ k \geq 1\}$ are iid standard Gaussian random variables. Then $\mathfrak{X}(f) = (X,f)_H$, where

$$X = \sum_{k\geq 1} \sqrt{\lambda_k}\xi_k\mathfrak{m}_k.$$

Indeed, $\mathbb{E}\|X\|_H^2 = \sum_{k\geq 1}\lambda_k = \|\mathrm{K}\|_1 < \infty$ (see (3.1.24)) and, with $f_k = (f,\mathfrak{m}_k)_H$,

$$\mathbb{E}\Big((X,f)_H(X,g)_H\Big) = \sum_k \lambda_k f_k g_k = (\mathrm{K}f,g)_H.$$

This completes the proof of Theorem 3.2.15.

Recall that there is a one-to-one correspondence between a regular Gaussian field X and Gaussian measure μ. The corresponding reproducing kernel Hilbert space H_μ of μ is often called the `reproducing kernel Hilbert space of` X and is denoted by H_X. Below is the precise construction.

Exercise 3.2.16 (C) Assume that $\mathfrak{X}(f) = (X,f)_H$ is a regular Gaussian field with the covariance operator K and define the Gaussian measure μ on H by $\mu(A) = \mathbb{P}(X \in A)$, $A \in \mathcal{B}(H)$. Let C_μ be the covariance operator of the measure μ.

(a) Show that $C_\mu(f,g) = (\mathrm{K}f,g)_H$. **Hint.** It is obvious.
(b) Conclude that

$$H_\mu = \sqrt{K}(H)$$

Hint. Use the result of Problem 3.1.9 on page 99.

Let us have another look at the white noise. If $\mathfrak{B}$ is a Gaussian white noise on a separable Hilbert space H, then, by Definition 3.2.10, the covariance operator of $\mathfrak{B}$ is identity:

$$\mathbb{E}\Big(\mathfrak{B}(f)\mathfrak{B}(g)\Big) = (f,g)_H, \tag{3.2.15}$$

and therefore, by Theorem 3.2.15, there is no H-valued random element B such that $\mathfrak{B}(f) = (B,f)_H$. On the other hand, if $\{\mathfrak{m}_k,\ k \geq 1\}$ is an orthonormal basis in H, then (3.2.14) suggests a representation

$$\mathfrak{B} = \sum_{k\geq 1} \mathfrak{B}(\mathfrak{m}_k)\,\mathfrak{m}_k; \tag{3.2.16}$$

we call (3.2.16) the `chaos expansion` of $\mathfrak{B}$. The following exercise justifies the representation (3.2.16).

Exercise 3.2.17 (B)

(a) Let H be a separable Hilbert space with an orthonormal basis $\{\mathfrak{m}_k,\ k \geq 1\}$, and let $\tilde{H}$ be a Hilbert space such that $H \subset \tilde{H}$ and the inclusion operator $\mathrm{j} : H \to \tilde{H}$ is Hilbert-Schmidt. Show that (3.2.16) defines a regular generalized Gaussian field over $\tilde{H}$ with the covariance operator $\mathrm{K} = \mathrm{jj}^*$. For a systematic way to construct the space $\tilde{H}$, see Problem 3.1.2 on page 96.
(b) Show that every generalized Gaussian random field becomes regular when considered on a suitably chosen extension of the original space.

Next, we will see how Theorem 3.2.15 works when $H = L_2(G)$.

Example 3.2.18 Consider a zero-mean Gaussian field $W = W(x),\ x \in G \subseteq \mathbb{R}^{\mathrm{d}}$ (in the sense of Definition 3.2.6), with the covariance function $q(x, y) = \mathbb{E}\big(W(x)W(y)\big)$. Then $\mathbb{E}W^2(x) = q(x, x)$, and, under the assumption

$$\int_G q(x, x)dx < \infty, \tag{3.2.17}$$

$W \in L_2(\Omega; L_2(G))$ so that W defines a regular generalized Gaussian field over $L_2(G)$ by

$$W(f) = \int_G W(x)f(x)dx.$$

Indeed, since $\mathbb{E}\big(W(f)W(g)\big) = \iint_{G\times G} q(x, y)f(x)f(y)dxdy$, the covariance operator Q of this field is an integral operator with kernel q: $\mathrm{Q}f(x) = \int_G q(x, y)f(y)dy$. Condition (3.2.17) ensures that the operator Q is nuclear: see (3.1.29) on page 93; note also that, by the Cauchy-Schwarz inequality, $q^2(x, y) \leq q(x, x)q(y, y)$. If $\mathfrak{M} = \{\mathfrak{m}_k,\ k \geq 1\}$ is an orthonormal basis in $L_2(\mathbb{R}^{\mathrm{d}})$ such that $\mathrm{Q}\mathfrak{m}_k = \lambda_k \mathfrak{m}_k,\ k \geq 1$, then the Fourier series expansion of W in the basis $\mathfrak{M}$ is

$$W(x) = \sum_{k\geq 1} \xi_k \mathfrak{m}_k(x),\ \xi_k = \int_G W(x)\mathfrak{m}_k(x)dx, \tag{3.2.18}$$

Note that the zero-mean Gaussian random variables ξ_k have variance λ_k and are independent for different k.

Representation (3.2.18) is known as the `Karhunen-Loève expansion` of W, after the Finnish mathematician Kari Karhunen (1915–1992) and the French–American mathematician Michel Loève (1907–1979), who established the result independently in the mid 1940's [105, 106, 143, 144].

If the function $q(x, x)$ is Lebesgue-integrable on every compact subset of G, then we can define a generalized random field $\dot{W}$ over the space $C_0^\infty(G)$ of smooth functions with compact support in G:

$$\dot{W}(f) = (-1)^{\mathrm{d}} \int_G W(x)D_x^{(\mathrm{d})}f(x)dx,\ D_x^{(\mathrm{d})} = \frac{\partial^{\mathrm{d}}}{\partial x_1 \ldots \partial x_{\mathrm{d}}}. \tag{3.2.19}$$

Since $\mathbb{E}\big(\dot{W}(f)\dot{W}(g)\big) = \iint_{G\times G} q(x,y)\big(D_x^{(\mathrm{d})}f(x)\big)\big(D_y^{(\mathrm{d})}g(y)\big)dxdy$, we see that the covariance operator of $\dot{W}$ is a generalized function on $\mathcal{C}_0^\infty(G\times G)$ and is given by $D_x^{(\mathrm{d})}D_y^{(\mathrm{d})}q(x,y)$. If there exists a $C>0$ such that, for all $f,g\in\mathcal{C}_0^\infty(G)$,

$$\left|\iint_{G\times G} q(x,y)\big(D_x^{(\mathrm{d})}f(x)\big)\big(D_y^{(\mathrm{d})}g(y)\big)dxdy\right| \le C\|f\|_{L_2(G)}\|g\|_{L_2(G)},$$

then $\dot{W}$ extends to a zero-mean generalized Gaussian field over $L_2(G)$.

This concludes Example 3.2.18.

Next we define the action of a linear operator on a generalized random field.

Definition 3.2.19 Let H, Y be Hilbert spaces, $\mathrm{A}: H\to Y$, a bounded linear operator, and $\mathfrak{B}$, a generalized Gaussian field over H. Then $\mathrm{A}\mathfrak{B}$ is a generalized Gaussian field over Y defined by $(\mathrm{A}\mathfrak{B})(f) = \mathfrak{B}(\mathrm{A}^*f)$.

Exercise 3.2.20 (C) Verify that if K is the covariance operator of $\mathfrak{B}$, then AKA^* is the covariance operator of $\mathrm{A}\mathfrak{B}$.

Example 3.2.21 Just as the dual of Hilbert space can be identified with different spaces, a generalized field can be considered over different spaces. For example, let G be a smooth bounded domain and consider the Sobolev spaces $H^r(G)$ on G (see Example 3.1.29, page 87). Let $\dot{W}$ be a Gaussian white noise over $H = L_2(G) = H^0(G)$ and let $\mathrm{A} = \Lambda = \sqrt{-\boldsymbol{\Delta}}$ with zero boundary conditions. Consider the field $\mathfrak{X} = \Lambda\dot{W}$. By Definition 3.2.19, $\mathfrak{X}$ is a Gaussian white noise over $Y = H^{-1}(G)$. Indeed, in this setting $\Lambda^* = \Lambda^{-1}$, because $(\Lambda f, g)_{-1} = (f, \Lambda^* g)_0$, $f\in H^0(G)$, $g\in H^{-1}(G)$, while the definition of the inner product in $H^{-1}(G)$ implies

$$(\Lambda f, g)_{-1} = (\Lambda^{-1}\Lambda f, \Lambda^{-1}g)_0 = (f, \Lambda^{-1}g)_0.$$

Therefore for $f,h\in H^{-1}$,

$$\mathbb{E}\Big(\mathfrak{X}(f)\mathfrak{X}(h)\Big) = (\Lambda^{-1}f, \Lambda^{-1}h)_0 = (f,h)_{-1}.$$

On the other hand, let $\{\mathfrak{m}_k,\ k\ge 1\}$ be the orthonormal basis in H consisting of the eigenfunctions of Λ, and let λ_k, $k\ge 1$, be the corresponding eigenvalues. Then

$$\dot{W} = \sum_{k\ge 1}\mathfrak{m}_k\xi_k,\ \ \mathfrak{X} = \sum_{k\ge 1}\lambda_k\mathfrak{m}_k\xi_k;$$

note that $\{\lambda_k\mathfrak{m}_k,\ k\ge 1\}$ is an orthonormal basis in $H^{-1}(G)$. Given $f = \sum_k f_k\mathfrak{m}_k\in H^1(G)$, we can define

$$\mathfrak{X}(f) = \sum_{k\ge 1} f_k\lambda_k\xi_k.$$

Then, for $f, g \in H^1(G)$,

$$\mathbb{E}\Big(\mathfrak{X}(f)\mathfrak{X}(g)\Big) = \sum_{k\geq 1} \lambda_k^2 f_k g_k = (f, g)_1,$$

that is, $\mathfrak{X}$ can also be considered as a Gaussian white noise over $H^1(G)$. Thus, $\mathfrak{X}$ is a Gaussian white noise over two different spaces.

This concludes Example 3.2.21.

We conclude this section with a discussion of the `Markov property` for the random fields. The definition of the Markov property for random processes, namely, that the past and future are independent given the present, essentially relies on the linear ordering of index set of the process, the real line. Since a random field is indexed by a set without a linear ordering, there is no clear analogue of the past and future and thus no natural way to define the Markov property.

There are three main versions of the Markov property for random fields: the `sharp` Markov property, the `germ` Markov property, and the `global` Markov property. As in the case of processes, all definitions rely on *conditions independence*, which we will now review.

Let $\mathcal{G}_i$, $i = 1, 2, 3$, be three sigma algebras of events on the the probability space $(\Omega, \mathcal{F}, \mathbb{P})$. Recall that $\mathcal{G}_1$ and $\mathcal{G}_2$ are called `conditionally independent` given $\mathcal{G}_3$ if, for every $A \in \mathcal{G}_1$ and $B \in \mathcal{G}_2$,

$$\mathbb{E}\big(\mathbf{1}_A\, \mathbf{1}_B|\mathcal{G}_3\big) = \mathbb{E}\big(\mathbf{1}_A|\mathcal{G}_3\big)\, \mathbb{E}\big(\mathbf{1}_B|\mathcal{G}_3\big)$$

In this case, $\mathcal{G}_3$ is called the `splitting sigma-algebra` or the `splitting field` for $\mathcal{G}_1$ and $\mathcal{G}_2$.

If $\mathcal{G}_3$ is the trivial sigma algebra, then conditional independence becomes the usual independence. Dependent events can become conditionally independent given a non-trivial sigma-algebra. For example, values of a Markov process at two different moments are usually dependent, but they become independent given the value at an intermediate moment.

As another example, consider events A, B, C such that $C \subseteq A \cap B$. By the Bayes formula, A and B are conditionally independent given C (all conditional probabilities are equal to one).

Exercise 3.2.22 (C)

(a) Give an example of independent events that become dependent after a conditioning. **Hint.** Roll a die twice and condition on the sum.

(b) Show that if $\mathcal{G}_1$ and $\mathcal{G}_2$ are conditionally independent given $\mathcal{G}_3$, then $\sigma(\mathcal{G}_1 \cup \mathcal{G}_3)$ and $\sigma(\mathcal{G}_2 \cup \mathcal{G}_3)$ are conditionally independent given $\mathcal{G}_3$. **Hint.** Consider the events of the type $A_1 \cap C_1$ and $B_1 \cap C_2$, where $A_1 \in \mathcal{G}_1$, $B_1 \in \mathcal{G}_2$, and $C_1, C_2 \in \mathcal{G}_3$. Taking such events helps because, for example, $\mathbf{1}_{A_1 \cap C_1} = \mathbf{1}_{A_1}\mathbf{1}_{C_1}$ and $\mathbf{1}_{C_1}$ is $\mathcal{G}_3$-measurable.

(c) Show that $\mathcal{G}_1$ and $\mathcal{G}_2$ are conditionally independent given $\mathcal{G}_3$ if and only if, for every $A \in \mathcal{G}_1$,

$$\mathbb{E}\big(\mathbf{1}_A|\sigma(\mathcal{G}_2 \cup \mathcal{G}_3)\big) = \mathbb{E}(\mathbf{1}_A|\mathcal{G}_3).$$

We are now ready to define the first version of the Markov property for (regular) random fields. Recall that, for two sets A, B, notation $A \setminus B$ means the part of A that is not in B.

Definition 3.2.23 (Sharp Markov Property) A random field $X = X(x), x \in S$ on a metric space S has the `sharp Markov property` relative to a bounded open set $B \subset S$ (with boundary ∂B and closure $\overline{B}$) if the sigma-algebras

$$\sigma\big(X(x),\ x \in B\big) \text{ and } \sigma\big(X(x),\ x \in S \setminus \overline{B}\big)$$

are conditionally independent given the sigma-algebra $\sigma\big(X(x),\ x \in \partial B\big)$.
In other words, a random field has the sharp Markov property relative to a set if the values of the field inside and outside of the set are independent given the values of the field on the boundary of the set. We take S a metric space rather than a more general topological space because we want to have an easy notion of a bounded set.

Exercise 3.2.24 (C) Verify that $\sigma\big(X(x),\ x \in S \setminus \overline{B}\big)$ can be replaced with $\sigma\big(X(x),\ x \in S \setminus B\big)$.

Hint. See Exercise 3.2.22(b).

Definition 3.2.23, fist suggested by Lévy [131], is both the most natural and the most restrictive extension of the Markov property from one-parameter processes to random fields. The definition is natural because the values of a field inside and outside of a set are clear analogues of the past and the future of a process. The definition is restrictive because the boundary sigma-algebra $\sigma\big(X(x),\ x \in \partial B\big)$ is usually not big enough to ensure the conditional independence required by the definition. As a result, to have the sharp Markov property, one has to consider either very special random fields or special sets. Here are some example:

- If a homogeneous Gaussian field on $\mathbb{R}^\mathrm{d}$, $\mathrm{d} > 1$, has the sharp Markov property with respect to all bounded open set with sufficiently regular boundary, then the field is degenerate (essentially a single random variable; for details, see Wong [227, Theorem 1]).
- The Brownian sheet on $\mathbb{R}^2$ has the sharp Markov property relative to a finite union of rectangles with the sides parallel to the axes [200], but not relative to a triangle with vertices at $(0,0)$, $(1,0)$, $(0,1)$ (Walsh [223, Sect. 1]).
- The Brownian sheet in $\mathbb{R}^2$ has the sharp Markov property relative to bounded open sets with sufficiently irregular ("thick") boundary, such as a fractal curve [35].
- Certain two-parameter pure jump processes have the sharp Markov property relative to every bounded open set in $\mathbb{R}^2$ [34].

Exercise 3.2.25 (C) Verify that the Brownian sheet on $\mathbb{R}^2$ has the sharp Markov property relative to the square $(0, 1) \times (0, 1)$.
An extension of Definition 3.2.23 to generalized fields is possible in two directions:

1. by considering fields that, although generalized, still allow the definition of some form of a boundary value (Walsh [223, Chap. 9], Wong [227, Sect. 4]);
2. by considering the values of the field on the functions supported in an open neighborhood of the boundary; this consideration leads to *the germ Markov property*.

Let $\mathfrak{X}$ be a generalized random field over a linear topological space V, and assume that V is a collection of functions on a metric space S. Given a (measurable) set $C \subset S$, define the `germ sigma algebra` (or the `germ field`) $\mathcal{B}_+(C)$ of $\mathfrak{X}$ on C by

$$\mathcal{B}_+(C) = \bigcap_{A \supseteq \overline{C}} \sigma\big(\mathfrak{X}(f),\ f \text{ supported in } A\big),$$

where the intersection is over all open sets A containing the closure of C.

Definition 3.2.26 (Germ Markov Property) A generalized random field $\mathfrak{X}$ over functions on S has the `germ Markov property` relative to a bounded open set B (with boundary ∂B and closure $\overline{B}$) if the sigma algebras

$$\sigma\big(\mathfrak{X}(f),\ f \text{ supported in } B\big) \text{ and } \sigma\big(\mathfrak{X}(f),\ f \text{ supported in } S \setminus \overline{B}\big)$$

are conditionally independent given $\mathcal{B}_+(\partial B)$.

Exercise 3.2.27 (C) Verify that $\mathfrak{X}$ has the `germ Markov property` relative to a bounded open set B if and only if the sigma algebras $\mathcal{B}_+(B)$ and $\mathcal{B}_+(S \setminus B)$ are conditionally independent given $\mathcal{B}_+(\partial B)$.
Definition 3.2.26 is usually attributed to McKean [157]. Nualart [174] showed that the Brownian sheet in $\mathbb{R}^2$ has the germ Markov property relative to every bounded open set.

For a *regular* zero-mean Gaussian field $X = X(x), x \in G \subseteq \mathbb{R}^{\mathrm{d}}$, the germ Markov property is equivalent to the reproducing kernel Hilbert space H_X corresponding to the kernel $K(x, y) = \mathbb{E}\big(X(x)X(y)\big)$ being in a certain sense local:

Theorem 3.2.28 *Let $X = X(x),\ x \in G \subseteq \mathbb{R}^{\mathrm{d}}$ be a* continuous *zero-mean Gaussian random field. The field X has a germ Markov property relative to bounded open sub-sets of G if and only if the reproducing kernel Hilbert space H_X of X has the following properties:*

1. *if v_1, v_2 from H_X have disjoint supports, then $(v_1, v_2)_{H_X} = 0$.*
2. *if $v = v_1 + v_2 \in H_X$ and v_1, v_2 have disjoint supports, then $v_1 \in H_X$ and $v_2 \in H_X$.*

Proof See Künsch [126, Theorem 5.1] or Pitt [188, Theorem 3.3].

A slight modification of the germ Markov property is the global Markov property.

Definition 3.2.29 (Global Markov Property) A generalized random field $\mathfrak{X}$ over functions on S has the `global Markov property` if, for every two open sets A, B with $A \cup B = S$, the sigma algebras

$$\sigma\big(\mathfrak{X}(f), f \text{ supported in } A\big) \text{ and } \sigma\big(\mathfrak{X}(f), f \text{ supported in } B\big)$$

are conditionally independent given $\sigma\big(\mathfrak{X}(f), f \text{ supported in } A \cap B\big)$.

Intuitively, by taking a sequence of sets A and B so that their intersection $A \cap B$ becomes smaller and smaller, one could get the germ Markov property from the global Markov property. This is indeed true: the global Markov property implies the germ Markov property relative to bounded open sets [92, Theorem 1.19]. Moreover, for regular fields, the germ and global Markov properties are equivalent [92, Remark 1.25].

Exercise 3.2.30 (C) Verify that the Gaussian white noise $\dot{W}$ on $\mathbb{R}^{\mathrm{d}}$ has the global Markov property.

For another discussion of various Markov properties for random fields see Balan and Ivanoff [6] and Iwata [92].

Let us summarize the main facts about zero-mean Gaussian fields on $\mathbb{R}^{\mathrm{d}}$:

1. A regular field W on $\mathbb{R}^{\mathrm{d}}$ is a mapping from $\mathbb{R}^{\mathrm{d}}$ to the space of zero-mean Gaussian random variables, and is characterized by the covariance function $q(x,y) = \mathbb{E}\big(W(x)W(y)\big)$.
2. A generalized field $\mathfrak{B}$ over $L_2(\mathbb{R}^{\mathrm{d}})$ is a linear mapping from $L_2(\mathbb{R}^{\mathrm{d}})$ to the space of zero-mean Gaussian random variables and is characterized by the covariance operator Q: $\mathbb{E}\big(\mathfrak{B}(f)\mathfrak{B}(g)\big) = (\mathrm{Q}f, g)_{L_2(\mathbb{R}^{\mathrm{d}})}$.
3. A regular field W on $\mathbb{R}^{\mathrm{d}}$ defines a generalized field $\dot{W}$ over $\mathcal{S}(\mathbb{R}^{\mathrm{d}})$ by
$$\dot{W}(f) = (-1)^{\mathrm{d}} \int_{\mathbb{R}^{\mathrm{d}}} W(x)\Big(D_x^{(\mathrm{d})} f(x)\Big)dx,$$
where $D^{(\mathrm{d})} = \partial^n/\partial x_1 \cdots \partial x_{\mathrm{d}}$. The kernel of the corresponding covariance operator of $\dot{W}$ is $D_x^{(\mathrm{d})} D_y^{(\mathrm{d})} q(x,y)$, and usually must be interpreted as a generalized function.
4. If the covariance operator of a generalized field $\mathfrak{B}$ is nuclear, then there exists a Gaussian field $B = B(x), x \in \mathbb{R}^{\mathrm{d}}$ such that
$$\mathfrak{B}(f) = \int_{\mathbb{R}^{\mathrm{d}}} B(x) f(x) dx.$$

3.2.2 Processes

Let V be a Banach space.

Definition 3.2.31

(a) A V-valued stochastic (or random) process $X = X(t),\ t \in [0, T]$, is a collection of V-valued random elements $X(t),\ t \in [0, T]$. The sample trajectory of X is the function $X(t, \omega), t \in [0, T]$ for fixed $\omega \in \Omega$. Given a property of a V-valued function of time (continuity, differentiability, etc.), the process is said to have this property if every sample trajectory has this property.
(b) A V-valued stochastic process X is called Gaussian if, for every $n \geq 1, t_1, \ldots, t_n \in [0, T]$ and every collection $(\ell_1, \ldots, \ell_n)$ of bounded linear functionals on V, the random variable $\ell_1(X(t_1)) + \ldots + \ell_n(X(t_n))$ is Gaussian.
(c) A generalized random process, also known as cylindrical process, is a collection of generalized random elements, indexed by points in $[0, T]$. If $X = X(t)$ is a cylindrical process over V, then we write $X_f(t)$ to denote the value of the generalized random element $X(t)$ on $f \in V$. A cylindrical process X is called Gaussian if $\{X_{f_i}(t_j), i = 1, \ldots, N,\ j = 1, \ldots, M\}$ is a Gaussian system for every finite collection of $f_i \in V$ and $t_j \in [0, T]$.

Recall that a real-valued Gaussian process $X = X(t),\ t \geq 0$, is completely specified by the mean-value function $m(t) = \mathbb{E}X(t)$ and the covariance function $r(t, s) = \mathbb{E}\big((X(t) - m(t))(X(s) - m(s)\big)$.

Definition 3.2.32 The standard Brownian motion, also known as the Wiener process $w = w(t), t \in [0, T]$, is a real-valued Gaussian process such that $\mathbb{E}w(t) = 0$ and $\mathbb{E}\big(w(t)w(s)\big) = \min(t, s)$.
Below are some properties of the standard Brownian motion:

1. $w(0) = 0$ with probability one;
2. w has independent increments: if $t_1 < t_2 \leq t_3 < t_4$, then the random variables $w(t_2) - w(t_1)$ and $w(t_4) - w(t_3)$ are independent;
3. with probability one, the trajectories of w are Hölder continuous of any order less than $1/2$ and are nowhere differentiable;

Exercise 3.2.33 (C) Verify the above properties of the standard Brownian motion.
Next, we outline the connections of the standard Brownian motion $w = w(t)$ with other objects in probability and stochastic analysis.

1. **Gaussian measures.** Let V be the Banach space of continuous on $[0, T]$ functions with $v(0) = 0$ and norm $\|v\|_V = \sup_{0<t<T} |v(t)|$. Then $\mu(A) = \mathbb{P}(w \in A)$ is a Gaussian measure on $(V, \mathcal{B}(V))$. Recall that μ is called the Wiener measure on V.

2. **Generalized fields.** Let $\{\mathfrak{B} = \mathfrak{B}(f),\ f \in L_2((0,T))\}$ be a collection of zero-mean Gaussian random variables such that

$$\mathbb{E}\big(\mathfrak{B}(f)\mathfrak{B}(g)\big) = \int_0^T f(s)g(s)ds. \tag{3.2.20}$$

For $0 < b \le T$, define the indicator function of the interval $[0,b]$:

$$\mathbf{1}_{[0,b]}(s) = \begin{cases} 1, & 0 \le s \le b, \\ 0, & \text{otherwise,} \end{cases} \tag{3.2.21}$$

Then

$$w(t) = \mathfrak{B}(\mathbf{1}_{[0,t]}). \tag{3.2.22}$$

3. **Orthogonal random measures.** Let $\boldsymbol{W}$ be a random set function on $\mathcal{B}([0,T])$ with the properties
   a) for every $A \in \mathcal{B}([0,T])$, $\boldsymbol{W}(A)$ is a Gaussian random variable with mean zero and the variance equal to the Lebesgue measure of A;
   b) for every $A, B \in \mathcal{B}([0,T])$ such that $A \cap B = \emptyset$, $\boldsymbol{W}(A)$ and $\boldsymbol{W}(B)$ are independent and $\boldsymbol{W}(A \cup B) = \boldsymbol{W}(A) + \boldsymbol{W}(B)$.

   Then (cf. (3.2.8) on page 103)

$$w(t) = \boldsymbol{W}([0,t]). \tag{3.2.23}$$

4. **Chaos expansion.** Let $\{\mathfrak{m}_k,\ k \ge 1\}$ be an orthonormal basis in $L_2((0,T))$ and let $\{\xi_k,\ k \ge 1\}$ be independent standard Gaussian random variables. Define $M_k(t) = \int_0^t \mathfrak{m}_k(s)ds$. Then

$$w(t) = \sum_{k\ge1} M_k(t)\xi_k. \tag{3.2.24}$$

5. **White noise.** Let $\{\mathfrak{m}_k,\ k \ge 1\}$ be an orthonormal basis in $L_2((0,T))$ and let $\{\xi_k,\ k \ge 1\}$ be independent standard Gaussian random variables. Then

$$\dot{w}(t) = \sum_{k\ge1} \mathfrak{m}_k(t)\xi_k \tag{3.2.25}$$

   is a Gaussian white noise process on $L_2((0,T))$ and

$$\mathbb{E}\dot{w}(t)\dot{w}(s) = \delta(t-s), \tag{3.2.26}$$

   where δ is the Dirac delta-function.

Exercise 3.2.34 (B)

(a) Let $\{\mathfrak{m}_k,\ k \geq 1\}$ be an orthonormal basis in $L_2((0,T))$ and let $\{\xi_k,\ k \geq 1\}$ be independent standard Gaussian random variables. Verify that

$$\mathfrak{B}(f) = \sum_{k\geq 1} \left(\int_0^T f(s)\mathfrak{m}_k(s)ds \right) \xi_k \tag{3.2.27}$$

is a generalized Gaussian field over $L_2((0,T))$ with the property (3.2.20).

(b) Verify (3.2.26). **Hint.** Fix t and interpret $\mathbb{E}\dot{w}(t)\dot{w}(s)$ as a distribution φ_t, acting on smooth functions according to the rule

$$\varphi_t(f) = \sum_{k\geq 1} \left(\int_0^T f(s)\mathfrak{m}_k(s)ds \right) \mathfrak{m}_k(t).$$

(c) Verify that $\dfrac{\partial^2 \min(t,s)}{\partial t \partial s} = \delta(t-s)$ in the sense of generalized functions.

The Wiener process is the usual driving force in stochastic ordinary differential equations with continuous solutions. While the same process can be a driving force in stochastic partial differential equations, there is also a possibility to have spacial dependence in the stochastic driving force. Spacial dependence in the Wiener process (or in any other process) can be introduced in at least two ways: (a) by considering a process with values in a linear topological space; (b) by considering a process indexed by both time and space variables. In the case of the Wiener process, these considerations lead, respectively, to the Q-cylindrical Brownian motion and to the Brownian sheet. In what follows, we investigate both objects and the connections between them.

To begin, we define a Q-cylindrical Gaussian process.

Definition 3.2.35 Let $x = x(t)$ be zero-mean Gaussian process with the covariance function $R(t,s) = \mathbb{E}\big(x(t)x(s)\big)$. A `Q-cylindrical x-process` on (or over) a Hilbert space H is a collection $\mathfrak{X}^{\mathrm{Q}} = \{X_f^{\mathrm{Q}}(t),\ f \in H,\ t \in [0,T]\}$ of zero-mean Gaussian random variables with the following property: there exists a bounded linear non-negative self-adjoint operator $\mathrm{Q} : H \to H$ such that

$$\mathbb{E}\Big(X_f^{\mathrm{Q}}(t)X_g^{\mathrm{Q}}(s)\Big) = (\mathrm{Q}f, g)_H\, R(t,s) \tag{3.2.28}$$

for all $f, g \in H$ and all $t, s \in [0,T]$. If Q is the identity operator, then X^{Q} is called a `cylindrical x-process` and denoted simply by X.

In particular, a `Q-cylindrical Brownian motion` $W^{\mathrm{Q}} = W^{\mathrm{Q}}(t),\ t \in [0,T]$, also known as the `Q-cylindrical Wiener process` on (or over) a Hilbert space H, is a collection $\{W_f^{\mathrm{Q}}(t),\ f \in H,\ t \in [0,T]\}$ of zero-mean Gaussian random variables such that

$$\mathbb{E}\Big(W_f^{\mathrm{Q}}(t)W_g^{\mathrm{Q}}(s)\Big) = (\mathrm{Q}f, g)_H \min(t,s). \tag{3.2.29}$$

A `cylindrical Brownian motion`, also known as the `cylindrical Wiener process`, W corresponds to Q equal to the identity operator:

$$\mathbb{E}\Big(W_f(t)W_g(s)\Big) = (f,g)_H \min(t,s). \tag{3.2.30}$$

Definition 3.2.36 A Q-cylindrical x-process $\mathfrak{X}^{\mathrm{Q}}$ over a Hilbert space H is called `regular` over H if there exists an H-valued process $X^{\mathrm{Q}} = X^{\mathrm{Q}}(t)$ such that $X_f^{\mathrm{Q}}(t) = (X^{\mathrm{Q}}(t), f)_H$, $t \geq 0, f \in H$. In that case, $X^{\mathrm{Q}} = X^{\mathrm{Q}}(t)$ is an alternative notation for $\mathfrak{X}^{\mathrm{Q}}$.

If W is a cylindrical Brownian motion over a separable Hilbert space H, then it is convenient to represent W as a formal series

$$W(t) = \sum_{k\geq 1} w_k(t)\mathfrak{m}_k, \tag{3.2.31}$$

where $\{\mathfrak{m}_k,\ k \geq 1\}$ is an orthonormal basis in H and $\{w_k,\ k \geq 1\}$ are independent standard Brownian motions. The following exercise outlines the rationale behind this representation.

Exercise 3.2.37 (C) Let H be a separable Hilbert space with an orthonormal basis $\{\mathfrak{m}_k,\ k \geq 1\}$.

(a) Show that if W is a cylindrical Brownian motion over H, then $\{W_{\mathfrak{m}_k}(t), k \geq 1\}$ are independent standard Brownian motions.
(b) Show that if $\{w_k,\ k \geq 1\}$ are independent standard Brownian motions, then

$$\{W_f(t) = \sum_{k\geq 1}(f,\mathfrak{m}_k)_H\, w_k(t),\ f \in H\}$$

is a cylindrical Brownian motion over H.

Similar to (3.2.31), every cylindrical x-process X over a separable Hilbert space H can be written as

$$X_f(t) = \sum_{k\geq 1}(\mathfrak{m}_k, f)_H\, x_k(t), \tag{3.2.32}$$

or

$$X(t) = \sum_{k\geq 1}\mathfrak{m}_k\, x_k(t), \tag{3.2.33}$$

where $\{\mathfrak{m}_k,\ k \geq 1\}$ is an orthonormal basis in H and $x_k,\ k \geq 1$, are independent copies of the process x. Since the series (3.2.32) does not converge in the space H, a cylindrical x-process is not an H-valued process, but, similar to Exercise 3.2.17 on page 108, we have the following result.

Exercise 3.2.38 (B)

(a) Let x be a Gaussian process and X, a cylindrical x-process over a separable Hilbert space H. Let $\tilde{H}$ be a Hilbert space such that $H \subset \tilde{H}$ and the inclusion operator $\mathrm{j} : H \to \tilde{H}$ is Hilbert-Schmidt. Show that (3.2.33) defines a Q-cylindrical x-process over $\tilde{H}$ with $\mathrm{Q} = \mathrm{jj}^*$, and X is an $\tilde{H}$-valued process. For a systematic way to construct the space $\tilde{H}$, see Problem 3.1.2 on page 96.
(b) Show that every Q-cylindrical process becomes regular when considered over a suitable extension of the original Hilbert space.

The following theorem shows that a Q-cylindrical Gaussian process over H is regular over H if and only if the operator Q is nuclear.

Theorem 3.2.39 *Let x be a Gaussian process and X^{Q}, a Q-cylindrical x-process over a separable Hilbert space H.*

(a) If the operator Q is nuclear, then there exists an H-valued Gaussian process $X^{\mathrm{Q}}(t)$ such that $X^{\mathrm{Q}}_f(t) = (X^{\mathrm{Q}}(t), f)_H$. This process is defined by

$$X^{\mathrm{Q}}(t) = \sum_{k \geq 1} \sqrt{q_k}\, x_k(t)\, \mathfrak{h}_k, \tag{3.2.34}$$

where x_k, $k \geq 1$, are independent copies of the process x, $\mathfrak{h}_k$, $k \geq 1$, are the orthonormal eigenfunctions of Q and q_k, $k \geq 1$, are the corresponding eigenvalues

(b) If there exists an H-valued Gaussian process $Y = Y(t)$ such that $X^{\mathrm{Q}}_f(t) = (Y(t), f)_H$, $f \in H$, then the operator Q is nuclear and $Y(t) = X^{\mathrm{Q}}(t)$, where $X^{\mathrm{Q}}(t)$ is defined by (3.2.34).

Proof

(a) Assume that Q is nuclear. Then Q is a compact operator (Exercise 3.1.20), and so H has an orthonormal basis consisting of the eigenfunctions of Q (Theorem 3.1.28(b)). Moreover, by Theorem 3.1.41(b), $\sum_{k\geq 1} q_k < \infty$ (note that if Q is non-negative and self-adjoint, then $\sqrt{\mathrm{Q}^*\mathrm{Q}} = \mathrm{Q}$). As a result, (3.2.34) defines $X^{\mathrm{Q}}(t)$ as an element of H for every t. Direct computations show that $\{(X^{\mathrm{Q}}(t), f)_H,\ f \in H,\ t \in [0, T]\}$, is a collection of zero-mean Gaussian random variables with the property (3.2.28).
(b) Assume that $X^{\mathrm{Q}}_f(t) = (Y(t), f)_H$ for an H-valued Gaussian process $Y = Y(t)$, and choose an orthonormal basis $\{\mathfrak{m}_k,\ k \geq 1\}$ in H. Then, by (3.2.28),

$$\sum_{k\geq 1} \mathbb{E}|X^{\mathrm{Q}}_{\mathfrak{m}_k}(t)|^2 = \Big(\mathbb{E}x^2(t)\Big) \sum_{k\geq 1} (\mathrm{Q}\mathfrak{m}_k, \mathfrak{m}_k)_H.$$

On the other hand,

$$\sum_{k\geq 1} \mathbb{E}|X^{\mathrm{Q}}_{\mathfrak{m}_k}(t)|^2 = \sum_{k\geq 1} \mathbb{E}(Y(t), \mathfrak{m}_k)_H^2 = \mathbb{E}\|Y(t)\|_H^2 < \infty.$$

By Theorem 3.1.41(b), the operator Q is nuclear, and therefore the process $X^{\mathrm{Q}}(t)$ can be defined by (3.2.34). Since

$$Y(t) = \sum_{k\geq 1}(Y(t), \mathfrak{m}_k)_H \mathfrak{m}_k = \sum_{k\geq 1} X^{\mathrm{Q}}_{\mathfrak{m}_k}(t)\mathfrak{m}_k = \sum_{k\geq 1}(X^{\mathrm{Q}}(t), \mathfrak{m}_k)_H \mathfrak{m}_k,$$

it follows the $Y(t) = X^{\mathrm{Q}}(t)$.

This completes the proof of Theorem 3.2.39.
Let us summarize the discussion as it applies to a Q-cylindrical Brownian motion over a separable Hilbert space H:

- If $\{q_k,\ k \geq 1\}$ is a bounded sequence of non-negative numbers and $\{\mathfrak{m}_k,\ k \geq 1\}$ is an orthonormal basis in H, then

$$W^{\mathrm{Q}}(t) = \sum_{k\geq 1} \sqrt{q_k}\,\mathfrak{m}_k\, w_k(t), \tag{3.2.35}$$

  is Q-cylindrical Brownian motion and the linear operator Q is defined by $\mathrm{Q}\mathfrak{m}_k = q_k\mathfrak{m}_k$. In particular, cylindrical Brownian motion has $q_k = 1$ for all k.
- If the operator Q is nuclear, then $W^{\mathrm{Q}}(t)$ has representation (3.2.35), where $\mathfrak{m}_k,\ k \geq 1$, are the eigenfunctions of Q, q_k are the corresponding eigenvalues, and $\sum_k q_k < \infty$.
- In general, not every Q-cylindrical Brownian motion can be written in the form (3.2.35). For example, (3.2.35) is impossible if the operator Q does not have a complete system of eigenfunctions.

3.2.3 Martingales

Let $\mathbb{F} = (\Omega, \mathcal{F}, \{\mathcal{F}_t\}_{t\geq 0}, \mathbb{P})$ be a stochastic basis with the usual assumptions (completeness of $\mathcal{F}_0$ and right-continuity of $\mathcal{F}_t$). Recall that a real-valued, $\mathcal{F}_t$-adapted process $M = M(t),\ t \geq 0$, with $\mathbb{E}|M(t)| < \infty$ for all $t \geq 0$, is called

- `submartingale`, if $\mathbb{E}\big(M(t)|\mathcal{F}_s\big) \geq M(s),\ t > s$;
- `martingale`, if $\mathbb{E}\big(M(t)|\mathcal{F}_s\big) = M(s),\ t > s$;
- `supermartingale`, $\mathbb{E}\big(M(t)|\mathcal{F}_s\big) \leq M(s),\ t > s$;

If M is a martingale and $\varphi = \varphi(x)$ is a convex function (for example $\varphi(x) = x^2$), such that $\mathbb{E}|\varphi(M(t))| < \infty,\ t \geq 0$, then, by Jensen's inequality, the process $X(t) = \varphi(M(t))$ is a sub-martingale.

A martingale M is called `square integrable` if $\mathbb{E}|M(t)|^2 < \infty$ for all $t \geq 0$. If M is a continuous square integrable martingale with $M(0) = 0$, then, by the `Doob-Meyer decomposition`, there exists a unique continuous non-decreasing process $\langle M\rangle$, called the `quadratic variation` of M, such that

$\langle M\rangle(0) = 0$ and the process $M^2 - \langle M\rangle$ is a martingale. For two continuous square integrable martingales M, N with $M(0) = N(0) = 0$, the `cross variation` process $\langle M, N\rangle$ is defined by

$$\langle M, N\rangle = \frac{1}{4}\left(\langle M + N\rangle - \langle M - N\rangle\right); \tag{3.2.36}$$

see, for example, Karatzas and Shreve [103, Sect. 1.5]. Note that

$$\langle M, M\rangle(t) = \langle M\rangle(t) \text{ and } \langle cM\rangle(t) = c^2\langle M\rangle(t),\ c \in \mathbb{R}.$$

If $M_1, \ldots, M_n$ are continuous square-integrable martingales, $M_i(0) = 0$, then

$$\left\langle \sum_{k=1}^{n} M_k \right\rangle(t) = \sum_{k,\ell=1}^{n} \langle M_k, M_\ell\rangle(t). \tag{3.2.37}$$

Exercise 3.2.40 (C)

(a) Verify that $\langle M, N\rangle$ is the unique process in the class of processes with bounded variation such that $MN - \langle M, N\rangle$ is a martingale. **Hint.** A square-integrable martingale with bounded variation is constant.
(b) Verify that if the sigma-algebras generated by the random variables $M(t)$, $t \geq 0$, and $N(t)$, $t \geq 0$, are independent, then $\langle M, N\rangle = 0$. **Hint.** Use the uniqueness statement in part (a).

Definition 3.2.41

(1) A cylindrical process X over a linear topological space V is called a `martingale (submartingale, supermartingale)` if the process $X_f(t)$, $t \geq 0$, is a martingale (submartingale, supermartingale) for every $f \in V$.
(2) A process $M = M(t)$ is called a continuous square-integrable martingale with values in a separable Hilbert space H (or an H-valued continuous, square-integrable martingale) if the M has the following properties:
   - $M(t) \in H$ and $\mathbb{E}\|M(t)\|_H^2 < \infty$ for every $t \geq 0$;
   - $\lim_{s\to 0} \|M(t+s) - M(t)\|_H = 0$ for all $t \geq 0$ and $\omega \in \Omega$, where, by convention, $M(t) = M(0)$ for $t < 0$;
   - for every $f \in H$, the process $M_f(t) = \big(f, M(t)\big)_H$ is a martingale.

Exercise 3.2.42 (C) Let H be a Hilbert space and let $f = f(t)$, $t \in \mathbb{R}$, be an H-valued function. Verify that the two conditions are equivalent:

1. $\lim_{s\to 0} \|f(t+s) - f(t)\|_H = 0$ for all $t \in \mathbb{R}$;
2. the norm $\|f(\cdot)\|$ is a continuous function and, for every $h \in H$, the function $f_h(t) = (f(t), h)_H$, $t \in \mathbb{R}$, is continuous.

Hint. See Problem 3.1.3(a), page 96.

Our next objective is to define and study the quadratic variation of a continuous H-valued martingale. As a motivation, consider an $\mathbb{R}^{\mathrm{d}}$-valued continuous, square-integrable martingale M. Fixing an orthonormal basis in $\mathbb{R}^{\mathrm{d}}$, we get a representation of M as a vector martingale $(M_1, \ldots, M_{\mathrm{d}})$. If $f = (f_1, \ldots, f_{\mathrm{d}})$ is a vector in $\mathbb{R}^{\mathrm{d}}$, then $M_f = \sum_{k=1} f_k M_k$ is a real-valued continuous, square-integrable martingale and, by (3.2.37),

$$\langle M_f \rangle(t) = \sum_{k,n=1}^{\mathrm{d}} f_k f_n \langle M_k, M_n \rangle(t). \tag{3.2.38}$$

We will now derive an alternative representation for $\langle M_f \rangle$.

The process $|M|^2 = \sum_{i=1}^{\mathrm{d}} M_i^2$ is a sub-martingale, and $|M|^2 - \sum_{i=1}^{\mathrm{d}} \langle M_i \rangle$ is a martingale. It is therefore natural to define

$$\langle M \rangle = \sum_{i=1}^{\mathrm{d}} \langle M_i \rangle \tag{3.2.39}$$

This definition seems to depend on the basis in $\mathbb{R}^{\mathrm{d}}$, but, on the other hand, $\langle M \rangle$ is such that $|M|^2 - \langle M \rangle$ is a martingale; therefore, by the uniqueness part of the Doob-Meyer decomposition, $\langle M \rangle$ must be the same in every basis.

Proposition 3.2.43 *If N_1 and N_2 are two real-valued continuous, square-integrable martingales, then the function $\langle N_1, N_2 \rangle$ is absolutely continuous with respect $\langle N_1 \rangle + \langle N_2 \rangle$:*

$$\langle N_1, N_2 \rangle(t) = \int_0^t \frac{d\langle N_1, N_2 \rangle(s)}{d\Big(\langle N_1 \rangle + \langle N_2 \rangle\Big)(s)}\, d\Big(\langle N_1 \rangle + \langle N_2 \rangle\Big)(s).$$

Proof It is known that the function $\langle N_1, N_2 \rangle$ is absolutely continuous with respect to both $\langle N_1 \rangle$ and $\langle N_2 \rangle$ (see, for example, Liptser and Shiryaev [138, Theorem 2.2.8]). Since both $\langle N_1 \rangle$ and $\langle N_2 \rangle$ are absolutely continuous with respect to $\langle N_1 \rangle + \langle N_2 \rangle$, the result follows,

For a continuous square-integrable martingale M with values in $\mathbb{R}^{\mathrm{d}}$, we use Proposition 3.2.43 to define the $\mathrm{d} \times \mathrm{d}$ symmetric matrix

$$Q_M(s) = \left(\frac{d\langle M_i, M_j \rangle(s)}{d\langle M \rangle(s)},\ i,j = 1, \ldots, \mathrm{d} \right).$$

Treating f as a column vector and writing $f^{\top}$ for the corresponding row vector, we can then re-write (3.2.38) as

$$\langle M_f \rangle(t) = \int_0^t \big(f^{\top} Q_M(s) f\big)\, d\langle M \rangle(s).$$

Once we think of $Q_M(s)$ as a matrix representation of some linear *operator* $\mathrm{Q}_M(s)$ on $\mathbb{R}^d$, and interpret $\langle M\rangle$ as the process that compensates $|M|^2$ to a martingale, we have all the ingredients for an *intrinsic* (coordinate-free) interpretation of (3.2.38). The following theorem carries out this plan, and also extends (3.2.38) to an infinite-dimensional setting.

Theorem 3.2.44 *Let M be a continuous square-integrable martingale with values in a separable Hilbert space H and $M(0) = 0$. Then*

(a) the process $\|M\|_H^2$ is a non-negative sub-martingale and there exists a unique continuous non-decreasing process $\langle M\rangle$ such that $\|M\|_H^2 - \langle M\rangle$ is a martingale.
(b) There exists a unique process $\mathrm{Q}_M = \mathrm{Q}_M(t)$, called the `correlation operator of the martingale` *$M$, with the following properties:*

- *for every t and ω, Q_M is a non-negative definite self-adjoint nuclear operator on H;*
- *for every $f, g \in H$*

$$\langle (M,f)_H, (M,g)_H\rangle(t) = \int_0^t \big(\mathrm{Q}_M(s)f, g\big)_H \, d\langle M\rangle(s). \tag{3.2.40}$$

Proof Let $\{\mathfrak{m}_k,\ k \geq 1\}$ be an orthonormal basis in H.

(a) We have $M(t) = \sum_{k\geq 1} M_k(t)\mathfrak{m}_k$ and $\|M(t)\|_H^2 = \sum_{k\geq 1} M_k^2(t)$, where $M_k = (M, \mathfrak{m}_k)_H$ is a continuous square-integrable martingale. Then we can take $\langle M\rangle(t) = \sum_{k\geq 1}\langle M_k\rangle(t)$. For a more detailed proof, see Rozovskii [199, Sect. 2.1.8].
(b) For fixed t,

$$\tilde{F}_{f,g}(t) = \langle (M,f)_H, (M,g)_H\rangle(t)$$

is a bi-linear form on H, and so $\tilde{F}_{f,g}(t) = (\tilde{\mathrm{Q}}(t)f, g)_H$. The operator $\tilde{\mathrm{Q}}(t)$ must be nuclear:

$$\mathrm{tr}(\tilde{\mathrm{Q}}(t)) = \sum_{k\geq 1}\langle (M,\mathfrak{m}_k)_H\rangle(t) = \sum_{k\geq 1}\langle M_k\rangle(t) = \langle M\rangle(t).$$

Then, for fixed f, g, look at $\tilde{F}_{f,g}(t)$ as a real-valued function of time and argue that this function is absolutely continuous with respect to $\langle M\rangle$. For fixed t, the corresponding Radon-Nykodim derivative defines a bi-linear form on H, and we get $\mathrm{Q}_M(t)$ from the relation

$$(\mathrm{Q}_M(t)f, g)_H = \frac{d\tilde{F}_{f,g}(t)}{d\langle M\rangle(t)}.$$

For details, see Métivier and Pellaumail [159, Theorem 14.3].

This completes the proof of Theorem 3.2.44.

Exercise 3.2.45 (B)

(a) Verify that (3.2.40) implies that, for every $t \geq 0$ and $\omega \in \Omega$, the operator $\mathrm{Q}_M(t)$ is self-adjoint, non-negative definite, and $\mathrm{tr}(\mathrm{Q}_M(t)) = 1$.
(b) Let M be a continuous square-integrable H-valued martingale, X, a separable Hilbert space, and $\mathrm{A} : H \to X$, a bounded linear operator. Show that the process $N(t) = \mathrm{A}M(t)$ is a continuous square-integrable X-valued martingale and

$$\mathrm{Q}_N(t) = \frac{\mathrm{A}\mathrm{Q}_M(t)\mathrm{A}^*}{\mathrm{tr}\big(\mathrm{A}\mathrm{Q}_M(t)\mathrm{A}^*\big)}.$$

Hint. See Exercise 3.2.20 on page 109.

To summarize, if H is a separable Hilbert space with an orthonormal basis $\{\mathfrak{m}_k,\ k \geq 1\}$ and M is an H-valued continuous square-integrable martingale, then, for each k, each process $M_k(t) = (M(t), \mathfrak{m}_k)_H$ is a real-valued continuous, square-integrable martingale and

$$M(t) = \sum_{k\geq 1} M_k(t)\,\mathfrak{m}_k, \quad \langle M\rangle(t) = \sum_{k\geq 1} \langle M_k\rangle(t),$$
$$\langle M_k, M_\ell\rangle(t) = \int_0^t \big(\mathrm{Q}_M(s)\mathfrak{m}_k, \mathfrak{m}_\ell\big)_H d\langle M\rangle(s). \tag{3.2.41}$$

Exercise 3.2.46 (C) Let W^{Q} be a Q-cylindrical Wiener process on a Hilbert space H, and assume that the operator Q is nuclear. Show that W^{Q} is an H-valued martingale with $\langle W^{\mathrm{Q}}\rangle(t) = t\,\mathrm{tr}(\mathrm{Q})$ and $\mathrm{Q}_{W^{\mathrm{Q}}}(t) = \mathrm{Q}/\mathrm{tr}(\mathrm{Q})$.

Theorem 3.2.47 (Burkholder-Davis-Gundy (BDG) Inequality) *If H is a separable Hilbert space and M is an H-valued continuous square-integrable martingale with $M(0) = 0$, then, for every $p > 0$, there exists a positive number C, depending only on p, such that*

$$\mathbb{E}\left(\sup_{t\leq T} \|M(t)\|_H\right)^p \leq C\mathbb{E}\big(\langle M\rangle(T)\big)^{p/2}. \tag{3.2.42}$$

Proof We derive the result from the finite-dimensional version. Fix an integer $N \geq 1$ and use the notations from (3.2.41) to define

$$\bar{M}^{(N)}(t) = (M_1(t), \ldots, M_N(t)).$$

Then $\bar{M}^{(N)}$ is a martingale with values in $\mathbb{R}^N$ and, as $N \nearrow \infty$,

$$|\bar{M}^{(N)}(t)| \nearrow \|M(t)\|_H, \qquad \langle \bar{M}^{(N)}\rangle(t) \nearrow \langle M\rangle(t). \tag{3.2.43}$$

By the finite-dimensional version of the BDG inequality applied to $\bar{M}^{(N)}$,

$$\mathbb{E}\left(\sup_{t\leq T}|\bar{M}^{(N)}(t)|\right)^{p} \leq C\mathbb{E}\left(\langle \bar{M}^{(N)}\rangle(T)\right)^{p/2}, \tag{3.2.44}$$

see, for example, Krylov [118, Theorem IV.4.1]. The key feature of the result is that number C in (3.2.44) depends only on p; in particular, C does not depend on N.

Then (3.2.42) follows after passing to the limit $N \to \infty$ in (3.2.44) and using the monotone convergence theorem together with (3.2.41) and (3.2.43). Note that $|\bar{M}^{(N)}(t)| \leq \sup_t |\bar{M}^{(N)}(t)|$, so that $\sup_t \|M(t)\|_H \leq \lim_N \sup_t |\bar{M}^{(N)}(t)|$. On the other hand, $\|M(t)\|_H \geq |\bar{M}^{(N)}(t)|$, and so $\sup_t \|M(t)\|_H \geq \lim_N \sup_t |\bar{M}^{(N)}(t)|$. That is,

$$\sup_t \|M(t)\|_H = \lim_N \sup_t |\bar{M}^{(N)}(t)|.$$

This completes the proof of Theorem 3.2.47.

Remark 3.2.48 It is not just a lucky coincidence that the constant in the BDG inequality does not depend on the dimension of the space. A general result of Kallenberg and Sztencel [99] shows that, for every continuous martingale M, there exists a continuous two-dimensional martingale M' such that $|M| = |M'|$ and $\langle M\rangle = \langle M'\rangle$. As a result, every inequality for a continuous two-dimensional martingale M involving only $|M|$ and $\langle M\rangle$ extends to any number of dimensions, including infinite, with the same constants.

3.2.4 Problems

Problem 3.2.1 outlines some computations related to Gaussian white noise on $L_2(\mathbb{R}^{\mathrm{d}})$ and introduces regularization of white noise. Problem 3.2.2 is yet another version of the Kolmogorov continuity criterion, this time in terms of Gaussian measures. Problem 3.2.3 is an exercise on finding the reproducing kernel Hilbert space of a Gaussian measure. Problem 3.2.4 is yet another look at the standard Brownian motion, via the Karhunen-Loève expansion and the limiting distribution of a certain SPDE. Problem 3.2.6 is an example of computing the covariance operator of a Gaussian measure. Problem 3.2.5 leads to the notion of the Cameron-Martin space. Problem 3.2.7 introduces and investigates homogeneous random fields. Problem 3.2.8 provides an example of a homogeneous field: a Euclidean free field. Problem 3.2.9 investigates some properties of the cross variation of two martingales. Problem 3.2.10 establishes a useful technical result: completeness of the spaces of continuous square-integrable martingales with values in a Hilbert space.

Problem 3.2.1 Let $\dot{W} = \dot{W}(x)$ be Gaussian white noise on $L_2(\mathbb{R}^d)$.

(a) For $f \in \mathcal{S}(\mathbb{R}^{\mathrm{d}})$, define the regular field $\dot{W}_{[f]}(x)$ (also known as a `regularization of white noise`) by

$$\dot{W}_{[f]}(x) = \dot{W}(f(x - \cdot)) = \int_{\mathbb{R}^{\mathrm{d}}} f(x-y)dW(y). \tag{3.2.45}$$

Show that, for every $g \in \mathcal{S}(\mathbb{R}^{\mathrm{d}})$,

$$\dot{W}_{[f]}(g) = \dot{W}(f \bullet g), \tag{3.2.46}$$

where $(f \bullet g)(y) = \int_{\mathbb{R}^{\mathrm{d}}} f(x-y)g(x)dx$.

(b) Let φ be an element of $\mathcal{S}(\mathbb{R}^{\mathrm{d}})$ such that $\varphi(x) = \varphi(-x)$, $\varphi(0) > 0$, and $\int_{\mathbb{R}^{\mathrm{d}}} \varphi(x)dx = 1$. For $\varepsilon > 0$, define $\varphi_\varepsilon(x) = \varepsilon^{-\mathrm{d}}\varphi(x/\varepsilon)$. Show that, for every $f \in \mathcal{S}(\mathbb{R}^{\mathrm{d}})$,

$$\lim_{\varepsilon \to 0} \mathbb{E}\Big(\dot{W}_{[\varphi_\varepsilon]}(f) - \dot{W}(f)\Big)^2 = 0. \tag{3.2.47}$$

(c) Let $W_r(x)$ be the Gaussian field defined by (3.2.6) on page 103. Show that, for every $f \in \mathcal{S}(\mathbb{R}^{\mathrm{d}})$,

$$\lim_{r \to 0} \mathbb{E}\Big(W_r(f) - \dot{W}(f)\Big)^2 = 0. \tag{3.2.48}$$

(d) How would you define a regularization of Gaussian white noise over $L_2(G)$ for $G \neq \mathbb{R}^{\mathrm{d}}$?

Problem 3.2.2 Using Theorem 2.1.7 on page 28, Prove the following version of Kolmogorov's continuity criterion: If $K = K(x, y)$ is a `positive-definite kernel` on $[0, 1]^{\mathrm{d}} \times [0, 1]^{\mathrm{d}}$ and there exist real numbers $\alpha < 1$ and $C > 0$ such that

$$K(x, x) + K(y, y) - 2K(x, y) \le C|x - y|^{2\alpha}$$

then, for every $\beta < \alpha$, there exits a centered Gaussian measure on the space $V = \mathcal{C}^\beta([0, 1])$ (functions on $[0, 1]^{\mathrm{d}}$ that are Hölder continuous of order β), such that

$$\int_V f(x)f(y)\,\mu(df) = K(x, y).$$

Problem 3.2.3 Describe the reproducing kernel Hilbert space of a finite-dimensional Gaussian vector.

Problem 3.2.4

(a) Show that the Karhunen-Loève expansion of the standard Brownian motion $w = w(t)$ on $[0, T]$ is

$$w(t) = \sqrt{2T} \sum_{k \geq 1} \xi_k \frac{\sin\left(\left(k - \frac{1}{2}\right) \pi t/T\right)}{\left(k - \frac{1}{2}\right) \pi}, \tag{3.2.49}$$

where ξ_k are independent standard Gaussian random variables. Reconcile this result with representation (3.2.24) on page 115.

(b) Show that the right-hand side of (3.2.49) appears as the limit in distribution, as $t \to \infty$, of the solution $u = u(t, x)$ of the stochastic heat equation

$$u_t(t, x) = u_{xx}(t, x) + \sqrt{2}\dot{W}(t, x),\ t > 0,\ x \in (0, 1),$$

with zero initial condition $u(0, x) = 0$ and boundary conditions $u(t, 0) = 0$, $u_x(t, 1) = 0$.

Problem 3.2.5 Let μ be the Wiener measure on the space V of continuous on $[0, 1]$ functions that are zero at $t = 0$. Find the corresponding covariance operator C_μ. Recall that the dual space of V is the collection of regular measures on $[0, 1]$; see Dunford and Schwartz [42, Theorem IV.6.3].

Problem 3.2.6 Let $\mathfrak{B}$ be a Gaussian white noise over a separable Hilbert space H, and let $\tilde{H}$ be a bigger Hilbert space such that the embedding operator $\mathrm{j} : H \to \tilde{H}$ is Hilbert-Schmidt.

(a) Verify that $\mathfrak{B}$ extends to a regular Gaussian field over $\tilde{H}$, with the nuclear covariance operator $\mathrm{K} = \mathrm{j}\mathrm{j}^*$.
(b) Verify that the corresponding Gaussian measure μ on $\tilde{H}$ has the covariance operator C_μ such that $C_\mu(f, g) = (\mathrm{K}f, g)_{\tilde{H}} = (\mathrm{j}^*f, \mathrm{j}^*g)_H$.
(c) Verify that H is a reproducing kernel Hilbert space with the reproducing kernel $K_H = C_\mu$.

Problem 3.2.7 Let $\mathfrak{B}$ be a zero-mean generalized Gaussian field over the space $\mathcal{S}(\mathbb{R}^{\mathrm{d}})$.

For a function $f \in \mathcal{S}(\mathbb{R}^{\mathrm{d}})$ and $h \in \mathbb{R}^{\mathrm{d}}$, define f^h by $f^h(x) = f(x + h)$. The field $\mathfrak{B}$ is called homogenous if, for every $N \geq 1$, $(f_1, \ldots, f_N) \subset \mathcal{S}(\mathbb{R}^{\mathrm{d}})$ and $h \in \mathbb{R}^{\mathrm{d}}$, the vectors $(\mathfrak{B}(f_1), \ldots, \mathfrak{B}(f_N))$ and $(\mathfrak{B}(f_1^h), \ldots, \mathfrak{B}(f_N^h))$ have the same distribution.

(a) Show that $\mathfrak{B}$ is homogeneous if and only if there exists a positive measure on μ on $(\mathbb{R}^{\mathrm{d}}, \mathcal{B}(\mathbb{R}^{\mathrm{d}}))$ with the following properties:

a. There exists a non-negative number p such that

$$\int_{\mathbb{R}^{\mathrm{d}}} \frac{\mu(dx)}{(1 + |x|^2)^p} < \infty; \tag{3.2.50}$$

b. For every $f, g \in \mathcal{S}(\mathbb{R}^{\mathrm{d}})$,

$$\mathbb{E}\big(\mathfrak{B}(f)\mathfrak{B}(g)\big) = \int_{\mathbb{R}^{\mathrm{d}}} \widehat{f}(y)\overline{\widehat{g}(y)}\mu(dy), \tag{3.2.51}$$

where $\widehat{f}, \widehat{g}$ are the Fourier transforms of f, g, and $\overline{\widehat{g}(y)}$ is the complex conjugate of $\widehat{g}(y)$.

The measure μ is called the `spectral measure` of $\mathfrak{B}$.

(b) Show that $\dot{W}$, the Gaussian white noise on $L_2(\mathbb{R}^{\mathrm{d}})$ (see (3.2.10) on page 104), is homogeneous and, up to a constant, its spectral measure is the Lebesque measure.

(c) We say that $\mathfrak{B}$ is `regular` if there exists a zero-mean Gaussian field $B = B(x)$ such that $\int_G \mathbb{E}B^2(x)dx < \infty$ for every compact set $G \subset \mathbb{R}^{\mathrm{d}}$ and $\mathfrak{B}(f) = \int_{\mathbb{R}^{\mathrm{d}}} f(x)B(x)dx$ for every $f \in \mathcal{S}(\mathbb{R}^{\mathrm{d}})$.

a. Show that a homogenous field over $\mathcal{S}(\mathbb{R}^{\mathrm{d}})$ is regular if and only if its spectral measure μ is finite: $\mu(\mathbb{R}^{\mathrm{d}}) < \infty$; equivalently, a finite μ means we can put $p = 0$ in (3.2.50).

b. Show that a non-trivial regular homogenous field over $\mathcal{S}(\mathbb{R}^{\mathrm{d}})$ CANNOT be extended to a regular field over $L_2(\mathbb{R}^{\mathrm{d}})$.

A Comment Strictly speaking, the random fields discussed in this problem are *translation invariant*; truly homogenous field would require a modification of the definition to include an arbitrary isometry of $\mathbb{R}^{\mathrm{d}}$ instead of only translations $f^h(x) = f(x+h)$ (see Wong [227, Sect. 4]).

Problem 3.2.8 Let $\mathfrak{X}$ be a homogeneous field over $\mathcal{S}(\mathbb{R}^{\mathrm{d}})$ with the spectral measure $\mu(dy) = dy/(1+|y|^2)$ (see Problem 3.2.7). Such a field is called the `Euclidean free field`

(a) For what d is $\mathfrak{X}$ regular? Find the corresponding representation of $\mathfrak{X}$.

(b) Verify that $\mathfrak{X}$ can be considered a Gaussian white noise over both $H^1(\mathbb{R}^{\mathrm{d}})$ and $H^{-1}(\mathbb{R}^{\mathrm{d}})$.

Problem 3.2.9 Let M, N be continuous, square-integrable martingales with values in a separable Hilbert space H, and, with $\{\mathfrak{m}_k,\ k \geq 1\}$ denoting an orthonormal basis in H,

$$M(t) = \sum_{k\geq 1} M_k(t)\mathfrak{m}_k,\ N(t) = \sum_{k\geq 1} N_k(t)\mathfrak{m}_k.$$

Show that

$$\langle M, N\rangle = \frac{1}{4}\Big(\langle M+N\rangle - \langle M+N\rangle\Big),\ \langle M, N\rangle = \sum_{k\geq 1}\langle M_k, N_k\rangle,$$

$$(M, N)_H - \langle M, N\rangle \text{ is a martingale,}$$

$$\langle M, N\rangle = 0 \ \text{ if } M, N \text{ are independent.}$$

Problem 3.2.10 Show that the space of continuous, square-integrable martingales with values in a separable Hilbert space and starting at zero is complete. In other words, let $M_n = M_n(t),\ n \geq 1,\ t \in [0,T]$ be real-valued continuous, square integrable martingales with values in a separable Hilbert space H, $M_n(0) = 0$, and such that

$$\lim_{m,n\to\infty} \mathbb{E}\langle M_m - M_n\rangle(T) = 0.$$

Show that there exists an H-valued continuous, square-integrable martingale M such that $M(0) = 0$ and

$$\lim_{n\to\infty} \mathbb{E} \sup_{0<t<T} \|M_n(t) - M(t)\|_H^2 = 0.$$

3.3 Stochastic Integration

3.3.1 Construction of the Integral

To begin, we review the construction of the stochastic Itô integral with respect to a real-valued continuous, square-integrable martingale. Many books, such as Karatzas and Shreve [103, Sect. 3.2], Krylov [118, Sect. III.10], Kunita [125, Sect. 3.2], Liptser and Shiryaev [138, Sect. 2.2], and Protter [192, Chap. II], provide the details of the construction and the proofs of the properties of the integral at different levels of generality. Below is the summary of the main results.

Let $M = M(t),\ t \geq 0$, be a real-valued continuous, square-integrable martingale with $M(0) = 0$, and let $f = f(t),\ t \geq 0$, be a real-valued continuous adapted process such that

$$\mathbb{E}\int_0^T f^2(t)\, d\langle M\rangle(t) < \infty. \tag{3.3.1}$$

The construction of the stochastic integral $M_f = \int_0^T f(t)dM(t)$ proceeds in the following three steps: (1) Integration of simple (can also be called elementary) functions; (2) An approximation result; (3) Integration of more general functions. We say that f a `simple predictable function` if there exist non-random times $0 = t_0 < t_1 < \ldots < t_N = T$ and *bounded*, $\mathcal{F}_{t_{n-1}}$-measurable random variables $f_n,\ n = 1, \ldots, N$ such that

$$f(t) = \sum_{\ell=1}^{N} f_\ell \mathbf{1}_{(t_{\ell-1},t_\ell]}(t),$$

where $\mathbf{1}_A(s)$ is the `indicator function` of the set A: $\mathbf{1}_A(s) = 1$ if $s \in A$ and $\mathbf{1}_A(s) = 0$ if $s \notin A$.

Here is a more detailed description of the three steps:

Step 1. If f is a `simple predictable function`, then *define*

$$M_f = \sum_{\ell=1}^{N} f_\ell \big(M(t_\ell) - M(t_{\ell-1})\big)$$

and *show* that

$$\mathbb{E}M_f^2 = \mathbb{E}\int_0^T f^2(t)\, d\langle M\rangle(t).$$

Step 2. If f is a continuous adapted process satisfying (3.3.1), *show* that there exists a sequence $\{f^{(n)},\ n \geq 1\}$ of simple predictable functions such that

$$\lim_{n\to\infty} \mathbb{E}\left(\int_0^T \big(f^{(n)}(t) - f(t)\big)^2 d\langle M\rangle(t)\right) = 0. \qquad (3.3.2)$$

Also show that, *for every* sequence $\{f^{(n)},\ n \geq 1\}$ of simple predictable functions satisfying (3.3.2), there is a unique square-integrable random variable M_f such that

$$\lim_{n\to\infty} \mathbb{E}\Big(M_f - M_{f^{(n)}}\Big)^2 = 0 \qquad (3.3.3)$$

Step 3. If f is a continuous adapted process satisfying (3.3.1), then *define*

$$\int_0^T f(t) dM(t) = M_f,$$

where M_f is the random variable from (3.3.3), and *show* that

$$\mathbb{E}\left(\int_0^T f(t) dM(t)\right)^2 = \mathbb{E}\int_0^T f^2(t)\, d\langle M\rangle(t).$$

It can be shown [192, Theorem II.21] that the resulting integral $\int_0^T f(s)dM(s)$ is the limit in probability of the integral sums

$$\sum_k f(t_k)(M(t_{k+1}) - M(t_k)),\ 0 = t_0 < t_1 < \ldots < t_{N-1} = T,$$

as $\Delta_N = \max_n(t_{n+1} - t_n)$ decreases to zero

Once we have $\int_0^T f(s)dM(s)$ (a random variable), we define $\int_0^t f(s)dM(s)$ (a random process), $0 < t < T$, by

$$\int_0^t f(s)dM(s) = \int_0^T f(s)\mathbf{1}_{(0,t]}(s)dM(s),$$

and prove that the process $M_f = M_f(t)$, $t \in [0,T]$, is a continuous square-integrable martingale with quadratic variation

$$\langle M_f\rangle(t) = \int_0^t f^2(s)\, d\langle M\rangle(s). \tag{3.3.4}$$

More generally, if $N = N(t)$ is another continuous square-integrable martingale and $g = g(t)$ is a continuous adapted process such that $\int_0^T g^2(t)d\langle N\rangle(t) < \infty$, then

$$\langle M_f, N_g\rangle(t) = \int_0^t f(s)g(s)\, d\langle M, N\rangle(s); \tag{3.3.5}$$

see Kunita [125, Theorem 2.3.2].

Further analysis shows that the space of admissible integrands (functions f) extends to the *predictable* processes (the closure of the set of simple predictable processes under the convergence (3.3.2)). After that, localization arguments (using suitable stopping times) show that M_f can still be defined when condition (3.3.1) is relaxed to $\mathbb{P}\left(\int_0^T f^2(t)\, d\langle M\rangle(t) < \infty\right) = 1$, and M is a continuous local martingale.

Our next step is to define the stochastic Itô integral with respect to Hilbert space-valued martingales. A popular approach is to *repeat* the above three-step construction when f is an *operator*; the resulting integral is then an element of a suitable Hilbert space. This approach is used, in particular, by Chow [24, Sect. 6.3], Da Prato and Zabczyk [31, Sect. 4.2], and Prévôt and Röckner [191, Sect. 2.3].

An alternative approach, used by Rozovskii [199, Sect. 2.2], is to *reduce* the construction to the one-dimensional case via expansions in suitable orthonormal bases. We will use this approach (in our opinion, it is less technical and more explicit), and start with a very special class of integrands so that the resulting integral is real-valued.

Let $M = M(t)$, $t \geq 0$, be a continuous square-integrable martingale with values in a separable Hilbert space H, $M(0) = 0$. Denote by Q_M the correlation operator of M (see Theorem 3.2.44 on page 122 for the definition of Q_M).

Definition 3.3.1 An H-valued process $f = f(t)$, $t \in [0,T]$, is called an `M-admissible integrand` if, for every $h \in H$, the process $x(t) = (f(t), h)_H$ is a continuous adapted process and

$$\mathbb{E}\int_0^T \big(\mathrm{Q}_M(t)f(t), f(t)\big)_H dt < \infty, \tag{3.3.6}$$

where Q_M is the correlation operator of M (see Theorem 3.2.44 on page 122).

Definition 3.3.2 Given an orthonormal basis $\mathfrak{M} = \{\mathfrak{m}_k,\ k \geq 1\}$ in H, and an M-admissible integrand f, we *define* the stochastic integral $M_f(t) = \int_0^t (f(s), dM(s))_H$ by

$$M_f(t) = \sum_{k\geq 1} \int_0^t f_k(s) dM_k(s), \tag{3.3.7}$$

where $f_k(t) = \big(f(t), \mathfrak{m}_k\big)_H,\ M_k(t) = \big(M(t), \mathfrak{m}_k\big)_H$.

It follows from Definition 3.3.2 that the resulting operation of stochastic integration is linear in both M and f: if a, b are real numbers, then

$$(aM + bN)_f = aM_f + bN_f,\ \ M_{af+bg} = aM_f + bM_g.$$

Other properties of the stochastic integral are established below.

Theorem 3.3.3 *The process M_f has the following properties:*

1. *M_f is a real-valued continuous square-integrable martingale and*

$$\langle M_f\rangle(t) = \int_0^t \big(\mathrm{Q}_M(s)f(s), f(s)\big)_H d\langle M\rangle(s). \tag{3.3.8}$$

In particular,

$$\mathbb{E}\left(\int_0^t \big(f(s), dM(s)\big)_H\right)^2 = \mathbb{E}\int_0^t \big(\mathrm{Q}_M(s)f(s), f(s)\big)_H d\langle M\rangle(s). \tag{3.3.9}$$

2. *The definition of M_f does not depend on the choice of the basis in H: if $\bar{\mathfrak{M}} = \{\bar{\mathfrak{m}}_k,\ k \geq 1\}$ is another orthonormal basis in H, $\bar{f}_k = (f, \bar{\mathfrak{m}}_k)_H$, $\bar{M}_k = \big(M, \bar{\mathfrak{m}}_k\big)_H$, then, for all $t \in [0, T]$,*

$$\mathbb{E}\sup_{0<t<T}\left(\sum_{k\geq 1}\int_0^t f_k(s)dM_k(s) - \sum_{k\geq 1}\int_0^t \bar{f}_k(s)d\bar{M}_k(s)\right)^2 = 0. \tag{3.3.10}$$

Proof To prove that M_f is a real-valued continuous, square-integrable martingale with quadratic variation (3.3.8), define

$$M_{f,K}(t) = \sum_{k=1}^{K}\int_0^t f_k(s)dM_k(s),\ K \geq 1.$$

It is a real-valued, continuous, square-integrable martingale. Then, keeping in mind that $\langle M_k, M_\ell\rangle(t) = \int_0^t \big(\mathrm{Q}_M(s)\mathfrak{m}_k, \mathfrak{m}_\ell\big)_H d\langle M\rangle(s)$ (see Theorem 3.2.44 on page 122), we use (3.2.37) on page 120 to find, for $K < N$,

$$\begin{aligned}\langle M_{f,N} - M_{f,K}\rangle(T) &= \sum_{k,\ell=K+1}^{N} \int_0^T f_k(s) f_\ell(s)\, d\langle M_k, M_\ell\rangle(s)\\ &= \sum_{k,\ell=K+1}^{N} \int_0^T f_k(s) f_\ell(s) \big(\mathrm{Q}_M(s)\mathfrak{m}_k, \mathfrak{m}_\ell\big)_H\, d\langle M\rangle(s).\end{aligned}$$

On the other hand,

$$\begin{aligned}\mathbb{E} \sum_{k,\ell=1}^{\infty} \int_0^T & f_k(s) f_\ell(s) \big(\mathrm{Q}_M(s)\mathfrak{m}_k, \mathfrak{m}_\ell\big)_H\, d\langle M\rangle(s)\\ & \mathbb{E} \int_0^T \big(\mathrm{Q}_M(t) f(t), f(t)\big)_H dt < \infty,\end{aligned}$$

which means

$$\lim_{K,N\to\infty} \mathbb{E}\langle M_{f,N} - M_{f,K}\rangle(T) = 0.$$

By the Burkholder-Davis-Gundy inequality with $p = 2$ (Theorem 3.2.47 on page 123),

$$\lim_{K,N\to\infty} \mathbb{E} \sup_{0<t<T} |M_{f,N}(t) - M_{f,K}(t)|^2 = 0.$$

After these computations, to conclude that M_f is a continuous square-integrable martingale we use completeness of the space of real-valued continuous square-integrable martingales (see, for example, Karatzas and Shreve [103, Proposition 1.5.23] or Kunita [125, Theorem 2.1.2]).

To prove (3.3.10), define

$$M_f^k(t) = \int_0^t f_k(s) dM_k(s), \quad \bar{M}_f^k(t) = \int_0^t \bar{f}_k(s) d\bar{M}_k(s). \tag{3.3.11}$$

Then, by equality (3.2.37) on page 120 and by the Burkholder-Davis-Gundy inequality with $p = 2$,

$$\begin{aligned}\mathbb{E} \sup_{0<t<T} & \left(\sum_{k\ge 1} \int_0^t f_k(s) dM_k(s) - \sum_{k\ge 1} \int_0^t \bar{f}_k(s) d\bar{M}_k(s) \right)^2\\ & \le C \sum_{k,\ell\ge 1} \mathbb{E}\langle M_f^k - \bar{M}_f^k, M_f^\ell - \bar{M}_f^\ell\rangle(T).\end{aligned}$$

By Theorem 3.2.44 on page 122,

$$
\begin{aligned}
&\langle M_f^k - \bar{M}_f^k, M_f^\ell - \bar{M}_f^\ell\rangle(t) \\
&= \int_0^t \Big(f_k(s)f_\ell(s)\big(\mathrm{Q}_M(s)\mathfrak{m}_k, \mathfrak{m}_\ell\big)_H + \bar{f}_k(s)\bar{f}_\ell(s)\big(\mathrm{Q}_M(s)\bar{\mathfrak{m}}_k, \bar{\mathfrak{m}}_\ell\big)_H\Big)\, d\langle M\rangle(s) \\
&- \int_0^t \Big(f_k(s)\bar{f}_\ell(s)\big(\mathrm{Q}_M(s)\mathfrak{m}_k, \bar{\mathfrak{m}}_\ell\big)_H + f_\ell(s)\bar{f}_k(s)\big(\mathrm{Q}_M(s)\mathfrak{m}_\ell, \bar{\mathfrak{m}}_k\big)_H\Big)\, d\langle M\rangle(s)
\end{aligned}
$$

Note that

$$
\sum_{\ell\geq 1} f_\ell(s)\big(\mathrm{Q}_M(s)\mathfrak{m}_k, \mathfrak{m}_\ell\big)_H = \big(\mathrm{Q}_M(s)\mathfrak{m}_k, f(s)\big)_H \text{ and}
$$

$$
\sum_{k\geq 1} f_k(s)\big(\mathrm{Q}_M(s)\mathfrak{m}_k, f(s)\big)_H = \big(\mathrm{Q}_M(s)f(s), f(s)\big)_H.
$$

Similarly,

$$
\sum_{\ell\geq 1} \bar{f}_\ell(s)\big(\mathrm{Q}_M(s)\mathfrak{m}_k, \bar{\mathfrak{m}}_\ell\big)_H = \big(\mathrm{Q}_M(s)\mathfrak{m}_k, f(s)\big)_H \text{ and}
$$

$$
\sum_{k\geq 1} \bar{f}_k(s)\big(\mathrm{Q}_M(s)\bar{\mathfrak{m}}_k, f(s)\big)_H = \big(\mathrm{Q}_M(s)f(s), f(s)\big)_H.
$$

As a result,

$$
\begin{aligned}
&\sum_{k,\ell\geq 1} \langle M_f^k - \bar{M}_f^k, M_f^\ell - \bar{M}_f^\ell\rangle(T) \\
&= \int_0^T \Big(\big(\mathrm{Q}_M(s)f(s), f(s)\big)_H + \big(\mathrm{Q}_M(s)f(s), f(s)\big)_H\Big)\, d\langle M\rangle(s) \\
&\quad - \int_0^T \Big(\big(\mathrm{Q}_M(s)f(s), f(s)\big)_H + \big(\mathrm{Q}_M(s)f(s), f(s)\big)_H\Big)\, d\langle M\rangle(s) = 0.
\end{aligned}
$$

This completes the proof of Theorem 3.3.3.

Exercise 3.3.4 (C)

(a) Show that

$$
\langle M_f, M_g\rangle(t) = \int_0^t \big(\mathrm{Q}_M(s)f(s), g(s)\big)_H\, d\langle M\rangle(s).
$$

An expression for $\langle M_f, N_g\rangle$ is impossible without introducing additional operators.

(b) Show that

$$\begin{aligned} &\mathbb{E} \sup_{0<t<T} \left| \int_0^t \big(f(s), dM(s)\big)_H \right| \\ &\quad \le C\mathbb{E} \left(\int_0^T \big(\mathrm{Q}_M(s)f(s), f(s)\big)_H \, d\langle M\rangle(s) \right)^{1/2}, \end{aligned} \tag{3.3.12}$$

and, if $\mathbb{E} \sup_{0<t<T} \|f(s)\|_H^2 < \infty$, then

$$\mathbb{E} \sup_{0<t<T} \left| \int_0^t \big(f(s), dM(s)\big)_H \right| \le \frac{1}{2} \mathbb{E} \sup_{0<t<T} \|f(s)\|_H^2 + C_1 \mathbb{E}\langle M\rangle(T). \tag{3.3.13}$$

Hint. Use the BDG inequality with $p = 1$; for (3.3.13), remember that the operator norm of Q_M is less than or equal to one and, for $a, b, \varepsilon > 0$, $ab \le \varepsilon a^2 + b^2/\varepsilon$.

Now, we consider a more general class of integrands. Let H, X be separable Hilbert spaces. As before, let $M = M(t)$, $t \ge 0$, be a continuous square-integrable martingale with values in H and with $M(0) = 0$. Denote by Q_M the correlation operator of M (see Theorem 3.2.44 on page 122 for the definition of Q_M).

Definition 3.3.5 A collection $\mathrm{B} = \mathrm{B}(t, \omega)$ of (random) operators from H to X is called an M-`admissible integrand` if

1. For every $h \in H$ and $x \in X$, the process $g(t) = \big(\mathrm{B}(t)h, x\big)_X$ is continuous and adapted.
2. For every $t \in [0, T]$ and $\omega \in \Omega$, the operator $\mathrm{B}(t, \omega)\mathrm{Q}_M(t, \omega)\mathrm{B}^*(t, \omega)$ (acting from X to X) is trace-class and

$$\mathbb{E} \int_0^T \mathrm{tr}\Big(\mathrm{B}(t)\mathrm{Q}_M(t)\mathrm{B}^*(t)\Big)\, d\langle M\rangle(t) < \infty. \tag{3.3.14}$$

Exercise 3.3.6 (C) Verify that if $X = \mathbb{R}$, then (3.3.14) is equivalent to (3.3.6).

Hint. The operator B in this case is given by $\mathrm{B}(t)h = (f(t), h)_H$, and $\mathrm{B}^*(t)r = rf(t)$, $r \in \mathbb{R}$.

Definition 3.3.7 Let H, X be separable Hilbert spaces, $M = M(t)$, a continuous, H-valued square-integrable martingale, $M(0) = 0$, and $\mathrm{B} : H \to X$, an M-admissible integrand. We *define* the stochastic integral $M_{\mathrm{B}}(t) = \int_0^t \mathrm{B}(s)dM(s)$ as the unique X-valued process such that, for every $x \in X$, $t \in [0, T]$,

$$(M_{\mathrm{B}}(t), x)_X = \int_0^t \big(\mathrm{B}^*(s)x, dM(s)\big)_H. \tag{3.3.15}$$

Exercise 3.3.8 (C) Verify that the process $f(t) = \mathrm{B}^*(t)x$ is an M-admissible integrand and thus we get the right-hand side of (3.3.15) from Definition 3.3.2 on page 131.

Theorem 3.3.9 *The process M_{B} defined by* (3.3.15) *is an X-valued continuous, square-integrable martingale such that, for all $x, y \in X$,*

$$\langle (M_{\mathrm{B}}, x)_X, (M_{\mathrm{B}}, y)_X \rangle(t) = \int_0^t \big(\mathrm{B}(s)\mathrm{Q}_M(s)\mathrm{B}^*(s)x, y\big)\, d\langle M\rangle(s).$$

In particular,

$$\mathbb{E}\left\| \int_0^t \mathrm{B}(s)dM(s) \right\|_X^2 = \mathbb{E}\int_0^t \mathrm{tr}\Big(\mathrm{B}(s)\mathrm{Q}_M(s)\mathrm{B}^*(s)\Big)\, d\langle M\rangle(s). \qquad (3.3.16)$$

Proof By Theorem 3.3.3 on page 131, the process $F_x(t) = (M_{\mathrm{B}}(t), x)_X$ is a real-valued, continuous, square-integrable martingale for every $x \in X$ and, by Exercise 3.3.4,

$$\langle F_x, F_y\rangle(t) = \int_0^t \big(\mathrm{Q}_M(s)\mathrm{B}^*(s)x, \mathrm{B}^*(s)y\big)_H\, d\langle M\rangle(s).$$

Also, if $\{\mathfrak{m}_k,\ k \geq 1\}$ is an orthonormal basis in X, then

$$\begin{aligned}\mathbb{E}\|M_{\mathrm{B}}(t)\|_X^2 &= \mathbb{E}\sum_{k\geq 1}\langle F_{\mathfrak{m}_k}\rangle(t) = \mathbb{E}\sum_{k\geq 1}\int_0^t \big(\mathrm{B}(s)\mathrm{Q}_M(s)\mathrm{B}^*(s)\mathfrak{m}_k, \mathfrak{m}_k\big)_X\, d\langle M\rangle(s)\\ &= \mathbb{E}\int_0^t \mathrm{tr}\big(\mathrm{B}(s)\mathrm{Q}_M(s)\mathrm{B}^*(s)\big)\, d\langle M\rangle(s) < \infty.\end{aligned}$$

This completes the proof of Theorem 3.3.9.

Exercise 3.3.10 (A) Find the correlation operator of the martingale M_{B}.

Hint. Use Exercise 3.2.45 on page 123, but keep in mind that $\mathrm{tr}\big(\mathrm{B}^*(s)\mathrm{Q}_M(s)\mathrm{B}^*(s)\big)$ can be zero.

We conclude this section with a brief discussion of integrals with respect to a Q-cylindrical Brownian motions W^{Q}. The main difference from the general construction is that, similar to the standard Brownian motion, every adapted, and not just predictable, process can be admissible integrand for W^{Q}. Otherwise, the general construction applies because W^{Q} is a square-integrable martingale on a possibly bigger Hilbert space, and the resulting construction does not depend on the choice of the bigger space; see Prévôt and Röckner [191, Sect. 2.5]. Still, the special structure of W^{Q} makes it possible to construct the integral directly without enlarging the underlying Hilbert space even if the operator Q is not trace-class; the details are outlined in Problem 3.3.1.

3.3.2 Itô Formula

We start by reviewing the finite-dimensional result. Let $A = A(t)$ be an $\mathbb{R}^{\mathrm{d}}$-valued adapted process with bounded variation (each component A_k can be written as a difference of two non-decreasing processes), $A(0) = 0$, $M = M(t)$, an $\mathbb{R}^{\mathrm{d}}$-valued continuous, square-integrable martingale, $M(0) = 0$, and $f = f(t,x)$, $t \geq 0$, $x \in \mathbb{R}^{\mathrm{d}}$, a function that is continuously differentiable in t and twice continuously differentiable in x. Define $X(t) = X(0) + A(t) + M(t)$. Then, with $f_t = \partial f/\partial t$, $D_k f = \partial f/\partial x_k$,

$$f(t,X(t)) = f(0,X(0)) + \int_0^t f_t(s,X(s))ds + \sum_{k=1}^{\mathrm{d}} \int_0^t D_k f(s,X(s))dX_k(s)$$
$$+\frac{1}{2}\sum_{k,\ell=1}^{\mathrm{d}} \int_0^t D_k D_\ell f(s,X(s))\, d\langle M_k, M_\ell\rangle(s). \tag{3.3.17}$$

This result is known as Itô formula; see, for example, Karatzas and Shreve [103, Theorem 3.6]. Of course, $dX_k = dA_k + dM_k$; $\int \cdots dA_k$ is the usual Lebesque-Stiltjes integral, and $\int \cdots dM_k$ is the stochastic Itô integral.

In particular, taking $f(t,x) = |x|^2$, we find

$$|X(t)|^2 = |X(0)|^2 + 2\sum_{k=1}^{\mathrm{d}} \int_0^t X_k(s)dX_k(s) + \sum_{k=1}^{\mathrm{d}} \langle M_k\rangle(t). \tag{3.3.18}$$

Equivalently, writing $(\cdot,\cdot)$ for the inner product in $\mathbb{R}^{\mathrm{d}}$ and using (3.2.39) on page 121 to define $\langle M\rangle$,

$$|X(t)|^2 = |X(0)|^2 + 2\int_0^t \big(X(s), dX(s)\big) + \langle M\rangle(t). \tag{3.3.19}$$

Exercise 3.3.11 (C) Verify that (3.3.18) follows from (3.3.17).

The main advantage of (3.3.19) over (3.3.18), beside a more compact form, is that (3.3.19) suggests that the result does not depend on the choice of the basis in $\mathbb{R}^{\mathrm{d}}$. Indeed, the inner product can be defined in an intrinsic way, and so can $\langle M\rangle$, as the process compensating $|M|^2$ to a martingale. Having coordinate-independent formulas is useful for possible extensions to an infinite-dimensional setting, so let us write (3.3.17) in a coordinate-free way. For that, we need to recall the intrinsic definitions of the first two derivatives of a function on $\mathbb{R}^{\mathrm{d}}$. We continue to use the notation $(\cdot,\cdot)$ for the inner product in $\mathbb{R}^{\mathrm{d}}$.

Given a smooth function g on $\mathbb{R}^{\mathrm{d}}$, the first derivative (the gradient) of g at a point $P \in \mathbb{R}^{\mathrm{d}}$, $Dg(P)$, is a linear mapping from $\mathbb{R}^{\mathrm{d}}$ to $\mathbb{R}$, and thus, by the Riesz representation theorem, can be identified with a vector $Dg(P)$ in $\mathbb{R}^{\mathrm{d}}$ so that

$$Dg(P)(x) = \big(Dg(P), x\big).$$

Similarly, the second derivative $D^2g(P)$ is a symmetric bi-linear mapping from $\mathbb{R}^{\mathrm{d}} \times \mathbb{R}^{\mathrm{d}}$ to $\mathbb{R}$, and thus can be identified with a symmetric linear operator on $\mathbb{R}^{\mathrm{d}}$ so that

$$D^2g(P)(x, y) = \big(D^2g(P)x, y\big).$$

In a given orthonormal basis $\{\mathfrak{e}_k, k = 1, \ldots d\}$, $Dg(P)$ is represented by the vector

$$\nabla g(P) = \big(Dg(P)(\mathfrak{e}_1), \ldots, Dg(P)(\mathfrak{e}_{\mathrm{d}})\big),$$

and $Dg(P)(\mathfrak{e}_k) = D_k g(P)$ is the *partial derivative* of g or the *derivative of g at P in the direction of $\mathfrak{e}_k$*. Similarly, $D^2g(P)$ is represented by the matrix of the second partial derivatives of g at P. For details, see Zeidler [232, Sect. 4.1].

With these notations, (3.3.17) becomes

$$\begin{aligned} f(t, X(t)) = f(0, X(0)) &+ \int_0^t f_t(s, X(s))ds + \int_0^t \big(Df(s, X(s)), dX(s)\big) \\ &+ \frac{1}{2}\int_0^t \mathrm{tr}\Big(D^2 f(s, X(s))\mathrm{Q}_M(s)\Big)\, d\langle M\rangle(s). \end{aligned} \tag{3.3.20}$$

Exercise 3.3.12 (C) Verify that (3.3.17) can indeed be written as (3.3.20).

Hint. See the discussion prior to Theorem 3.2.44 on page 122 about the operator Q_M.

Given a function g on a Hilbert space, we consider the `Fréchet derivatives` Dg, D^2g. Recall that Dg can be interpreted as an element of H and D^2g, as a bounded self-adjoint operator on H. For details about the Fréchet derivatives see Zeidler [232, Sects. 4.4 and 4.5]. Then (3.3.20) extends to processes with values in a separable Hilbert space.

Theorem 3.3.13 (Infinite-Dimensional Itô Formula) *Let H be a separable Hilbert space. Consider the real-valued function $F = F(t, h)$, $t \in [0, T]$, $h \in H$, and an H-valued process $X = X(t)$, and assume that*

1. *F, F_t, DF, D^2F exist and are continuous in $[0, T] \times H$, and are bounded and uniformly continuous on bounded sub-sets of $[0, T] \times H$.*
2. *There exists an H-valued adapted process B with $\mathbb{E}\int_0^T \|B(t)\|_H^2 dt < \infty$ and an H-valued continuous, square-integrable martingale M with correlation operator Q_M and $M(0) = 0$, such that*

$$X(t) = X(0) + \int_0^t B(s)ds + M(t). \tag{3.3.21}$$

Then

$$F(t,X(t)) = F(0,X(0)) + \int_0^t F_s(s,X(s))ds + \int_0^t \big(DF(s,X(s)), dX(s)\big)_H$$
$$+ \frac{1}{2}\int_0^t \operatorname{tr}\Big(\big(D^2F(s,X(s))\big)Q_M(s)\Big)\, d\langle M\rangle(s) \tag{3.3.22}$$

Proof Equality (3.3.20) suggests that (3.3.22) must be true as well holds; a complete proof is in Da Prato and Zabczyk [31, Theorem 4.1].

Exercise 3.3.14 (C) Verify that

$$\|X(t)\|_H^2 = \|X(0)\|_H^2 + 2\int_0^t \big(X(s), B(s)\big)_H ds$$
$$+2\int_0^t \big(X(s), dM(s)\big)_H + \langle M\rangle(t). \tag{3.3.23}$$

Hint. $F(t,h) = \|h\|_H^2$, $DF = 2h$, $(1/2)D^2F$ is the identity operator.

Let us take a closer look at (3.3.23), which looks identical to the final-dimension formula (3.3.19). The main reason for this similarity is that X, B, and M all take values in the same space H. On the other hand, most stochastic partial differential equations cannot be studied using a single space. For example, consider

$$u(t,x) = \int_0^t u_{xx}(s,x)ds + h(x)w(t),\ x \in \mathbb{R},$$

where w is a standard Brownian motion and h is a smooth function with compact support. This equation looks like (3.3.21) if we set $X(t) = u(t,\cdot)$, $B(t) = u_{xx}(t,\cdot)$, and $M(t) = hw(t)$, but it is clear that X and B cannot belong to the same space because the operator $u \to u_{xx}$ is not bounded. As a result, formula (3.3.22) does not look especially helpful.

On the other hand, if we assume that u is smooth as a function of x, then, for every fixed x, we can apply the one-dimensional Itô formula to $u^2(t,x)$ and get

$$u^2(t,x) = 2\int_0^t u(s,x)u_{xx}(s,x)ds + 2\int_0^t u(s,x)h(x)dW(s) + t. \tag{3.3.24}$$

We can then integrate with respect to x, but the term $\int_{\mathbb{R}} u(s,x)u_{xx}(s,x)dx$ should be interpreted not as the inner product in $L_2(\mathbb{R})$ but as the duality $[\cdot,\cdot]$ between the Sobolev spaces $H^1(\mathbb{R})$ and $H^{-1}(\mathbb{R})$ relative to the inner product in $L_2(\mathbb{R})$:

$$\|u(t,\cdot)\|_{L_2(\mathbb{R})}^2 = 2\int_0^t [u_{xx}(s,\cdot), u(s,\cdot)]ds + 2\int_0^t \big(u(s,\cdot), h\big)_{L_2(\mathbb{R})} + t. \tag{3.3.25}$$

Using the notion of the normal triple of Hilbert spaces, we will now implement this construction in a more general setting. The result, often known as the energy equality in Hilbert spaces, is an extension of the Itô formula (3.3.23) for the square of the norm to a normal triple of Hilbert space.

Theorem 3.3.15 (Energy Equality in Hilbert Spaces) *Let (V, H, V') be a normal triple of Hilbert spaces and $[\cdot, \cdot]$, the duality between V and V' relative to the inner product in H.*

Let x be an $\mathcal{F}_0$-measurable random element, and let $X = X(t)$, $B = B(t)$, and $M = M(t)$ be adapted processes with the following properties:

1. *x is H-valued and $\mathbb{E}\|x\|_H^2 < \infty$;*
2. *X is V-valued and $\mathbb{E}\int_0^T \|X(t)\|_V^2 dt < \infty$;*
3. *B is V'-valued and $\mathbb{E}\int_0^T \|B(t)\|_{V'}^2 dt < \infty$;*
4. *M is an H-valued continuous, square-integrable martingale;*
5. *For every $v \in V$ and every $t \in [0, T]$,*

$$\big(X(t), v\big)_H = (x, v)_H + \int_0^t [B(s), v]ds + \big(M(t), v\big)_H \tag{3.3.26}$$

on the same set of probability one. In other words, the equality $X(t) = x + \int_0^t B(s)ds + M(t)$ holds in $L_2\big(\Omega \times (0, T); V'\big)$.

Then

1. *$X \in L_2(\Omega; \mathcal{C}((0, T); H))$, that is, X has continuous trajectories as an H-valued process, and there is a number C, independent of T, such that*

$$\begin{aligned}\mathbb{E}\sup_{0<t<T} \|X(t)\|_H^2 \le C\Big(&\mathbb{E}\|x\|_H^2 + \mathbb{E}\int_0^T \|X(t)\|_V^2 dt \\ &+ \mathbb{E}\int_0^T \|B(t)\|_{V'}^2\, dt + \mathbb{E}\langle M\rangle(T)\Big);\end{aligned} \tag{3.3.27}$$

2. *The following equality holds for all $t \in [0, T]$ on the same set of probability one:*

$$\begin{aligned}\|X(t)\|_H^2 = \|x\|_H^2 &+ 2\int_0^t [B(s), X(s)]ds \\ &+ 2\int_0^t \big(X(s), dM(s)\big)_H + \langle M(t)\rangle.\end{aligned} \tag{3.3.28}$$

Proof The complete proof is long and difficult, and we will not present it here. An interested reader can find all the details in Rozovskii [199, Sect. 2.4] or Krylov and Rozovskii [122, Sect. 2.3].

In what follows, we explain why one should expect a result like this. First of all, recall that we are not distinguishing different modifications of the process, but in this case the distinction is crucial: it takes a lot of effort to show that X indeed has a modification that is continuous in H. On the other hand, under the assumptions of the theorem, continuity of X as an H-valued process is not at all surprising and is well known in the deterministic setting ($M = 0$), when the proof is simple (see Problem 3.1.3 on page 96.)

It is also easy to convince oneself that (3.3.28) should hold. Indeed, take an orthonormal basis $\{\mathfrak{m}_k,\ k \geq 1\}$ in H and assume that $\mathfrak{m}_k \in V$ (since V is dense in H, it is a reasonable assumption). Then write (3.3.26) with $v = \mathfrak{m}_k$ to get an expression for the Fourier coefficients of X:

$$X_k(t) = x_k + \int_0^t [B(s), \mathfrak{m}_k]ds + M_k(t).$$

After that apply the one-dimensional Itô formula to X_k^2:

$$X_k^2(t) = x_k^2 + 2\int_0^t [B(s), X_k(s)\mathfrak{m}_k]ds + 2\int_0^t X_k(s)dM_k(s) + \langle M_k\rangle(t).$$

To get (3.3.28), it now remain to sum over $k \geq 1$, keeping in mind the Parseval identity and the equality $X(t) = \sum_k X_k(t)\mathfrak{m}_k$.

Finally, (3.3.27) follows from (3.3.28) after taking the $\mathbb{E}\sup_{0<t<T}$ on both sides. Note that $|[B(s), X(s)]| \leq \|B(s)\|_{V'}^2 + \|X(s)\|_V^2$ and then all terms on the right-hand side, except the stochastic integral $\int_0^t (X(s), dM(s))_H$, become non-decreasing in t. For the stochastic integral, we use the Burkholder-Davis-Gundy inequality (see (3.3.13) on page 134) with $p = 1$, followed by the epsilon inequality:

$$\mathbb{E}\sup_{0<t<T}\left|\int_0^t (X(s), dM(s))_H\right| \leq \frac{1}{2}\mathbb{E}\sup_{0<t<T}\|X(t)\|_H^2 + C_1\mathbb{E}\langle M\rangle(T).$$

With this we end the proof of Theorem 3.3.15.

Exercise 3.3.16 (A) Find an expression for C in (3.3.27) in terms of the corresponding constant in the BDG inequality.

Let us summarize the main points about the Itô formula:

- There is a relatively straightforward extension of a finite-dimensional Itô formula to a separable Hilbert space.
- On the other hand, in a typical SPDE setting, there are several Hilbert spaces, and then the only available extension of the Itô formula is the energy equality (3.3.28), which is for the square of the norm.
- Both extensions are useful, and one does not immediately follow from the other.

3.3.3 Problems

Problems 3.3.1 and 3.3.2 discuss stochastic integration with respect to a Q-cylindrical Brownian motion. Problem 3.3.3 is an exercise on the infinite-dimensional Itô formula. Problem 3.3.4 suggests possible connections between the Itô formulas in one space and in the normal triple (energy equality).

Problem 3.3.1 Consider the Q-cylindrical Brownian motion

$$W^{\mathrm{Q}}(t) = \sum_{k\geq 1} \sqrt{q_k}\mathfrak{m}_k w_k(t),$$

where $\{\mathfrak{m}_k, k \geq 1\}$ is an orthonormal basis in a separable Hilbert space H and $\mathrm{Q}\mathfrak{m}_k = q_k\mathfrak{m}_k$ (see the discussion on page 119).

(a) Let $f = f(t)$ be an H-valued process such that, for every $h \in H$, the process $z(t) = (f(t), h)_H$ is adapted and $\mathbb{E}\int_0^T \big(\mathrm{Q}f(t), f(t)\big)_H dt < \infty$; call this process f `admissible`. Show that

$$W_f^{\mathrm{Q}}(t) = \sum_{k\geq 1} \left(\int_0^t f_k(s)dw_k(s)\right) \mathfrak{m}_k,$$

is a real-valued continuous, square-integrable martingale and, for all admissible f, g,

$$\langle W_f^{\mathrm{Q}}, W_g^{\mathrm{Q}}\rangle(t) = \int_0^t \big(\mathrm{Q}f(s), g(s)\big)_H ds. \tag{3.3.29}$$

(b) Let X be a separable Hilbert space and $\mathrm{B} = \mathrm{B}(t,\omega)$, a family of (random) linear operators such that, for every $h \in H$ and $x \in X$, the process $z(t) = \big(\mathrm{B}(t)h, x\big)_X$ is adapted and

$$\mathbb{E}\int_0^T \mathrm{tr}\big(\mathrm{B}(s)\mathrm{Q}\mathrm{B}^*(s)\big)ds < \infty.$$

Show that the X-valued process $W_{\mathrm{B}}^{\mathrm{Q}}(t)$, defined by

$$\big(W_{\mathrm{B}}^{\mathrm{Q}}(t), x\big)_X = \sum_{k\geq 1} \sqrt{q_k}\int_0^t (\mathrm{B}(s)\mathfrak{m}_k, x)_X dw_k(s)$$

is an X-valued continuous, square-integrable martingale, and

$$\Big\langle \big(W_{\mathrm{B}}^{\mathrm{Q}}(t), x\big)_X, \big(W_{\mathrm{B}}^{\mathrm{Q}}(t), y\big)_X \Big\rangle(t) = \int_0^t \big(\mathrm{B}(s)\mathrm{Q}\mathrm{B}^*(s)x, y\big)_X ds. \tag{3.3.30}$$

Problem 3.3.2 Recall that not every Q-cylindrical Brownian motion can be written as $W^{\mathrm{Q}} = \sum_{k\geq 1} \sqrt{q_k}\mathfrak{m}_k w_k(t)$ (for example, if the operator Q does not have any point spectrum). Nonetheless, show that stochastic integrals W_f^{Q} and $W_{\mathrm{B}}^{\mathrm{Q}}$ from Problem 3.3.1 can be defined for every Q-cylindrical Brownian motion over a separable Hilbert space H without assuming that $W^{\mathrm{Q}} = \sum_{k\geq 1} \sqrt{q_k}\mathfrak{m}_k w_k(t)$. Show that equalities (3.3.29) and (3.3.30) continue to hold.

Problem 3.3.3 Use Itô formula to establish a particular case of the BDG inequality: for $p \geq 2$,

$$\mathbb{E} \sup_{0<t<T} \left\| \int_0^t B(s)dM(s) \right\|^p$$
$$\leq C_p \mathbb{E} \left(\int_0^T \left(\mathrm{tr}\big(\mathrm{B}(s)\mathrm{Q}_M(s)\mathrm{B}^*(s)\big) \right) d\langle M\rangle(s) \right)^{p/2}.$$

Problem 3.3.4 Compare and contrast the infinite-dimensional Itô formula in a single Hilbert space (Theorem 3.3.13) and the corresponding energy equality (3.3.23) with the energy equality in a normal triple of Hillbert spaces (Theorem 3.3.15). Can one of the equalities (3.3.23), (3.3.28), be derived from the other?

Chapter 4
Linear Equations: Square-Integrable Solutions

4.1 A Summary of SODEs and Deterministic PDEs

There are many standard references on SODEs and even more standard references on deterministic PDEs. Here are a few of each, listed in a non-decreasing order of difficulty:

SODEs: Øksendal [181], Friedman [57], Krylov [118], Karatzas and Shreve [103], Stroock and Varadhan [214], Kunita [125].
Deterministic PDEs: Strauss [212], Evans [50], Rauch [193], John [97], Krylov [119, 121], Friedman [58].

No prior familiarity with any of these books is necessary to proceed.

4.1.1 Why Square-Integrable Solutions?

One of the main differences between ordinary differential equations (ODEs) and equations with partial derivatives (PDEs) is availability of a general existence/uniqueness theorem: there one for ODEs, but there is none for PDEs.

Consider a deterministic ordinary differential equation

$$y'(t) = f(t, y(t)),\ t > t_0,\ y(t_0) = y_0$$

If f is a continuous function of two variables, then a classical solution exists in some neighborhood of the point $(t_0, y(t_0))$. The solution is unique if f is Lipschitz continuous in the second argument. The solution exists globally in time if f is at most of linear growth in the second variable or otherwise does not allow a blow-up to happen. These existence and uniqueness results carry over to stochastic equations essentially without any changes, see Problem 4.1.1 on page 163 below.

S.V. Lototsky, B.L. Rozovsky, *Stochastic Partial Differential Equations*, Universitext, DOI 10.1007/978-3-319-58647-2_4

Existence of a continuous solution of an SODE implies that no a priori integrability conditions are necessary to define the solution. For example, given two continuous real-valued functions a and b and a standard Brownian motion w, consider the one-dimensional equation

$$dX(t) = a\big(X(t)\big)dt + b\big(X(t)\big)dw(t),\ 0 < t \le T,\ X(0) = X_0. \tag{4.1.1}$$

It is natural to define a solution of (4.1.1) as a random process $X = X(t)$ with the following properties:

1. The trajectories of X are continuous on $[0, T]$ with probability one;
2. The equality

$$X(t) = X_0 + \int_0^t a\big(X(s)\big)ds + \int_0^t b\big(X(s)\big)dw(s) \tag{4.1.2}$$

   holds with probability one for all $t \in [0, T]$ at once.

Note that if X is indeed continuous, then having equality (4.1.2) with probability one for each t separately is as good as having the equality for all t at once: just take the intersection of all the probability-one sets corresponding to the rational points in $[0, T]$. Note also that by assumption $a\big(X(t)\big)$ is a continuous function of t, and therefore the first integral on the right-hand side of (4.1.2) is well defined. As far as the second (stochastic) integral, the basic construction requires the integrand to be square-integrable in both time and probability space, which would impose an additional condition on the process X, namely,

$$\mathbb{E}\int_0^T b^2\big(X(t)\big)dt < \infty. \tag{4.1.3}$$

There are at least two ways to avoid condition (4.1.3):

1. An easy way: assume that b is bounded;
2. A (slightly) harder way: extend the construction of the stochastic integral to integrands that are square-integrable in time with probability one; see Krylov [118, Sect. III.4] or Liptser and Shiryaev [139, Sect. 4.2.6]. In this `path-wise construction` condition (4.1.3) is replaced with

$$\int_0^T b^2\big(X(t)\big)dt < \infty \text{ with probability one,} \tag{4.1.4}$$

   which holds automatically because of continuity of b and X.

As a result, to define a solution of equation (4.1.1), there is no need to impose any integrability restrictions on the process X.

The situation is completely different for partial differential equations, where there are very few universal existence and uniqueness results. One example is the Cauchy-Kovalevskaya theorem about local existence and uniqueness for real-analytic evolution equations. The theorem has many versions: see, for example, Evans [50, Theorem 4.6.2] or Rauch [193, Theorem 1.3.3]). Unfortunately, the theorem is not especially helpful because (a) the result is local, and (b) the result does not cover many specific examples, such as the initial value problem for heat equation. Accordingly, every partial differential equation usually requires individual analysis. In particular, one has to make some a priori assumptions about the solution before starting to look for one.

Let us consider the initial value problem for the heat equation:

$$u_t = u_{xx},\ t > 0,\ x \in \mathbb{R};\ \ u(0,x) = \varphi(x).$$

Taking the Fourier transform, we find $\widehat{u}_t = -y^2\widehat{u}$ or $\widehat{u}(t,y) = \widehat{\varphi}(y)e^{-ty^2}$. Assuming that we can indeed invert the Fourier transform, we get the familiar formula for the solution:

$$u(t,x) = \frac{1}{\sqrt{4\pi t}} \int_{-\infty}^{+\infty} e^{-(x-y)^2/(4t)}\varphi(y)dy. \tag{4.1.5}$$

This formula suggests that the functions of the type $f_c(x) = e^{cx^2}$, $c > 0$, have special significance for the heat equation. For example, if the initial condition grows at infinity slower than any function f_c, then the integral in (4.1.5) converges. Equality (4.1.5) also suggests that, for all practical purposes, requiring $|\varphi|$ not to grow too fast is the same as requiring $\int_{\mathbb{R}} |u(t,x)|^2 dx < \infty$ for all $t > 0$. More detailed analysis shows that the growth restriction on the solution of the heat equation is essential to have uniqueness; see Körner [112, Chap. 67].

Next, let us consider the stochastic equation

$$du = u_{xx}dt + \sigma u_x dw(t),\ t > 0,\ x \in \mathbb{R};\ \ u(0,x) = \varphi(x), \tag{4.1.6}$$

with deterministic initial condition $\varphi = \varphi(x)$. Then $d\widehat{u} = -y^2\widehat{u}\,dt + \sigma\,\mathrm{i}\,y\widehat{u}\,dw(t)$ or

$$\widehat{u}(t,y) = \widehat{\varphi}(y)e^{-ty^2(1-(\sigma^2/2))+\mathrm{i}\sigma y w(t)}.$$

In particular, $|\widehat{u}(t,y)| = |\widehat{\varphi}(y)|e^{-ty^2(1-(\sigma^2/2))}$. If we are to use the results about the deterministic heat equation, then we have to assume $\sigma^2 < 2$. Under the assumption that φ is non-random, condition

$$\int_{\mathbb{R}} |u(t,x)|^2 dx < \infty$$

with probability one is the same as

$$\int_{\mathbb{R}} \mathbb{E}|u(t,x)|^2 dx < \infty.$$

That is, solutions of (4.1.6) that are path-wise square-integrable in space are also square-integrable as random variables. To put it differently, if we do not require solutions of (4.1.6) to be square-integrable as random variables, the equation will not be well-posed path-wise.

If the initial condition in (4.1.6) is random, then it is not absolutely necessary to consider square-integrable solutions. For example, one can consider $\varphi(x,\omega) = \xi\, e^{-x^2}$, where ξ is independent of w and has Cauchy distribution with probability density function $f_\xi(x) = 1/\big(\pi(1+x^2)\big)$, although the solution will still be square-integrable when conditioned on φ.

Throughout this chapter, we will consider only square-integrable solutions.

4.1.2 *Classification of PDEs*

This section consists of two parts. In the first part, we describe a connection between a concrete elliptic (hyperbolic, parabolic) PDE and an ellipse (hyperbola, parabola). In the second part, we explain how classification into elliptic/hyperbolic/parabolic types works for abstract equations.

4.1.2.1 Second-Order PDEs in Two Variables and Conic Sections

Recall that the regular circular cone is the surface defined in the three-dimensional Cartesian coordinate system by the equation $z^2 = x^2 + y^2$. It is known that ellipse, hyperbola, and parabola can all be obtained by cutting this surface with a suitable plane, hence the name `conic section`. For example, the plane $2z = x + 2$ produces an ellipse, $y = 1$ produces a hyperbola, and $z = x + 1$, a parabola.

The general equation of a conic section in two variables (x, y) is

$$Ax^2 + 2Bxy + Cy^2 + ax + by = K, \tag{4.1.7}$$

where $A, B, C, a, b, K \in \mathbb{R}$ and $A^2 + B^2 + C^2 > 0$. Then a "typical" equation (4.1.7) defines

- an ellipse, if $AC - B^2 > 0$;
- a hyperbola, if $AC - B^2 < 0$;
- a parabola, if $AC - B^2 = 0$.

We have to say "typical" because there are degenerate cases, such as $x^2 + y^2 = -1$ or $x^2 - y^2 = 0$.

Next, consider the second-order partial differential equation with constant coefficients in two independent variables (x, y) for the unknown function $u = u(x, y)$:

$$Au_{xx} + 2Bu_{xy} + Cu_{yy} + au_x + bu_y = 0, \tag{4.1.8}$$

where $A, B, C, a, b, K \in \mathbb{R}$ and $A^2+B^2+C^2 > 0$. The condition $A^2 + B^2 + C^2 > 0$ means that at least one of the numbers A, B, C is not zero; for both (4.1.7) and (4.1.8), this condition ensures that the corresponding equation is indeed of second order. In Eq. (4.1.8), we make a linear change of variables

$$s = c_1 x + c_2 y, \ t = c_3 x + c_4 y,$$

with $c_1 c_4 \neq c_2 c_3$, so that we can solve for (x, y) in terms of (s, t):

$$x = \tilde{c}_1 s + \tilde{c}_2 t, \ y = \tilde{c}_3 s + \tilde{c}_4 t.$$

Then we define a new function $v = v(s, t)$ by

$$v(s, t) = u(\tilde{c}_1 s + \tilde{c}_2 t, \tilde{c}_3 s + \tilde{c}_4 t).$$

It turns out (and the reader is welcome to verify it by executing the linear change of variables in (4.1.8) and then carefully considering the particular cases) that one can choose the numbers c_1, c_2, c_3, c_4 in such a way that Eq. (4.1.8) becomes

$$\begin{array}{ll} \bullet\ v_{ss} + v_{tt} = a_1 v_s + b_1 v_t & \text{if } AC - B^2 > 0; \\ \bullet\ v_{ss} - v_{tt} = \bar{a}_1 v_s + \bar{b}_1 v_t & \text{if } AC - B^2 < 0; \\ \bullet\ v_t = v_{ss} + \tilde{a}_1 v_s & \text{if } AC - B^2 = 0. \end{array} \tag{4.1.9}$$

Now the connections between the ellipse and the Laplace (elliptic) equation, a hyperbola and the wave (hyperbolic) equation, and a parabola and the heat (parabolic) equation become clear: the conditions for Eq. (4.1.7) to define an ellipse (hyperbola, parabola) are the same as the conditions for Eq. (4.1.8) to be elliptic (hyperbolic, parabolic). Note also that

1. We omitted the long but elementary computations required to reduce both (4.1.7) and (4.1.8) to the corresponding canonical form;
2. The connection between (4.1.7) and (4.1.8) essentially stops with the sign condition on $AC - B^2$: even the change of variables required for the reduction to the canonical form is different for the two equations;
3. When the coefficients A, B, C in Eq. (4.1.8) are functions of x and y, a possibly non-linear change of variables can still reduce the equation to one of the three canonical forms (4.1.9). For details, see John [97, Sect. 2.3].

Next, we discuss classification of equations in an abstract setting, when the starting point is a normal triple of Hilbert spaces (V, H, V') and a linear operator A. Recall that $[v', v]$, $v' \in V'$, $v \in V$, denotes the duality between V and V' relative to the inner product in H. For a function $u = u(t)$ with values either in $\mathbb{R}^{\mathrm{d}}$ or in an infinite-dimensional space, we write $\dot{u}$ to denote the time derivative: $\dot{u} = du/dt$.

Definition 4.1.1 We say that an operator A is `acting in a normal triple` (V, H, V') if A is a bounded linear operator from V to V': there exists a number $c_0 > 0$ such that, for all $v \in V$,

$$\|\mathrm{A}v\|_{V'} \leq c_0 \|v\|_V.$$

Definition 4.1.2 An operator A acting in the normal triple (V, H, V') is called `elliptic` if there exists a positive number c_A such that

$$[\mathrm{A}v, v] \geq c_A \|v\|_V^2 \tag{4.1.10}$$

for all $v \in V$.

Problem 4.1.2 on page 165 below illustrates why the operator A satisfying (4.1.10) is natural to call elliptic.

Definition 4.1.3 Let A be an elliptic operator acting in the normal triple (V, H, V'). Then we have the following classification of equations:

- equation $\mathrm{A}u = f$, where $f \in V'$, is called `elliptic`;
- equation $\ddot{u} + \mathrm{A}u = \mathrm{A}_1 u + \mathrm{B}\dot{u} + f$, where $f \in L_2((0, T); V')$ and $\mathrm{A}_1 : V \to H$, $\mathrm{B} : H \to H$ are bounded linear operators, is called `hyperbolic`;
- equation $\dot{u} + \mathrm{A}u = \mathrm{A}_1 u + f$, where $f \in L_2((0, T); V')$, and $\mathrm{A}_1 : V \to H$ is a bounded linear operator, is called `parabolic`.

Let us now illustrate how condition (4.1.10) provides a link between concrete and abstract equations.

Example 4.1.4 Consider the following setting: $V = H_0^1((0, 1))$, $H = L_2((0, 1))$, $V' = H^{-1}((0, 1))$, $\mathrm{A} = -d^2/dx^2$, with zero boundary conditions. We think of V as the closure of the set $\mathcal{C}_0^\infty((0, 1))$ of smooth functions with compact support in $(0, 1)$ with respect to the norm

$$\|v\|_V = \left(\|v\|_H^2 + \|v'\|_H^2\right)^{1/2}.$$

Taking $v \in \mathcal{C}_0^\infty((0, 1))$, we see that

$$[\mathrm{A}v, v] = -\int_0^1 v''(x)v(x)dx = \int_0^1 |v'(x)|^2 dx = \|v'\|_H^2,$$

where the second equality follows after integration by parts. Next, we make the following observations: if $v \in \mathcal{C}_0^\infty((0,1))$, then $v(0) = 0$ and

$$v(x) = \int_0^x v'(y)dy;$$

$$|v(x)|^2 = \left(\int_0^x v'(y)dy\right)^2 \le \left(\int_0^x dy\right)\left(\int_0^x |v'(y)|^2 dy\right) \le x\,\|v'\|_H^2; \tag{4.1.11}$$

$$\|v\|_H^2 = \int_0^1 |v(x)|^2 dx \le \left(\int_0^1 x dx\right)\|v'\|_H^2 \le \frac{1}{2}\|v'\|_H^2.$$

Since $\|v\|_V^2 = \|v\|_H^2 + \|v'\|_H^2$, we conclude that

$$\|v\|_V^2 \le \frac{1}{2}\|v'\|_H^2 + \|v'\|_H^2 = \frac{3}{2}[\mathrm{A}v, v],$$

that is, the operator A is elliptic and (4.1.10) holds with $c_A = 2/3$.

An example of the corresponding hyperbolic equation is $\ddot{u} + \mathrm{A}u = 0$ or $u_{tt} = u_{xx}$, which is the wave equation. An example of the corresponding parabolic equation is $\dot{u} + \mathrm{A}u = 0$ or $u_t = u_{xx}$, which is the heat equation. In other words, we see how a concrete setting under condition (4.1.10) reduces abstract equations to familiar equations of the corresponding type.
This concludes Example 4.1.4.

Exercise 4.1.5 (C) Verify that the operator A in Example 4.1.4 is a bounded operator from V to V'.

Exercise 4.1.6 (A) Repeat Example 4.1.4 in the normal triple (H^1, H^0, H^{-1}) with the spaces $H^r = H^r((0,1))$, $r \in \mathbb{R}$; see Example 3.1.29 on page 87. Show that in this setting $[\mathrm{A}v, v] \ge \pi^2\|v\|_V^2$.

4.1.3 Proving Well-Posedness of Linear PDEs

Our main objective in dealing with an equation is to show that the equation is well posed. Well-posedness of an equation means that, given the input (initial conditions, boundary conditions, and driving force) the solution exists, is unique, and depends continuously on the input. In other words, we have to find Banach spaces for the input and the solution, and to construct a bounded linear operator (the `solution operator`) acting from the space of input to the solution space.

To construct a variational solution, and the corresponding solution operator, one can follow these steps:

1. Construct a sequence of approximate solution, for example, using Galerkin approximation.

2. Derive a uniform estimate on the approximate solutions.
3. Use a compactness argument to extract a converging subsequence and identifying the limit as the solution.
4. Prove uniqueness of the solution.

For examples of this approach, see Evans [50, Sects. 7.1,7.2].

A lot of information about the equation can be obtained from a priori bounds, or estimates, on the solution. To derive an `a priori bound (estimate)`, we assume that the solution exists and is as regular as we want, and then derive an upper bound on some norm of the solution in terms of some norm of the input data. Along the way, we ignore the rule of formal logic that a false assumption implies anything you want. The choice of the norms depends on the structure of the equation, our ultimate goals, and the amount of work we are willing to carry out.

Below, we use the heat equation on the line to demonstrate the derivation of an a priori estimate. Then we discuss four different methods to construct the solution operator for the equation: (a) an explicit formula for the solution, (b) the method of continuity, (c) the Galerkin approximation, or (d) the semigroup method.

Theorem 4.1.7 below provides an example of an a priori estimate for the heat equation on the line. To follow the argument, recall the Sobolev spaces $H^r(\mathbb{R})$ from page 79, together with the corresponding notations for the norms $\|\cdot\|_r$ and the duality $[\cdot,\cdot]_0$, in this case between $H^1(\mathbb{R})$ and $H^{-1}(\mathbb{R})$. We also use that, because of the L_2 isometry of the Fourier transform, $\|h\|_1^2 = \|h\|_0^2 + \|h'\|_0^2$; see Exercise 3.1.10 on page 80.

Theorem 4.1.7 *Let $u = u(t,x)$, $t > 0$, $x \in \mathbb{R}^d$, be a classical solution of*

$$u_t = u_{xx} + f,\ 0 < t \le T,\ x \in \mathbb{R}, \tag{4.1.12}$$

with initial condition $u(0,x) = \varphi(x)$. Assume that

- $\varphi \in C_0^\infty(\mathbb{R})$,
- $f \in C_0^\infty((0,+\infty)\times\mathbb{R})$,
- *$u(t,x)$ is continuously differentiable in x for all $t \ge 0$,*
- $\lim_{|x|\to\infty} |u(t,x)| = \lim_{|x|\to\infty} |u_x(t,x)| = 0$ *for all $t \ge 0$.*

Then

$$\begin{aligned}\sup_{0<t<T} \|u(t,\cdot)\|_0^2 &+ \int_0^T \|u(s,\cdot)\|_1^2 ds \\ &\le C(T)\left(\|\varphi\|_0^2 + \int_0^T \|f(t,\cdot)\|_{-1}^2 dt\right),\end{aligned} \tag{4.1.13}$$

where $C(T) = 4 + 8T(1 + 2Te^{2T})$.

Proof After multiplying both sides of the equation by $2u$, integrating, first with respect to x, then with respect to t, and noticing that, after integration by parts,

$$\int_{\mathbb{R}} u(s,x)u_{xx}(s,x)dx = -\int_{\mathbb{R}} u_x^2(s,x)dx = -\|u_x(s,\cdot)\|_0^2,$$

we find

$$\begin{aligned}\|u(t,\cdot)\|_0^2 &= \|\varphi\|_0^2 - 2\int_0^t \|u_x(s,\cdot)\|_0^2 ds + 2\int_0^t [f(s,\cdot),u(s,\cdot)]_0 ds\\ &\le \|\varphi\|_0^2 - 2\int_0^t (\|u(s,\cdot)\|_0^2 + \|u_x(s,\cdot)\|_0^2)ds + 2\int_0^t \|u(s,\cdot)\|_0^2 ds\\ &+ 2\int_0^t \|u(s,\cdot)\|_1 \, \|f(s,\cdot)\|_{-1} ds \le \|\varphi\|_0^2 - 2\int_0^t \|u(s,\cdot)\|_1^2 ds\\ &+ 2\int_0^t \|u(s,\cdot)\|_0^2 ds + \int_0^t \|u(s,\cdot)\|_1^2 ds + \int_0^t \|f(s,\cdot)\|_{-1}^2 ds,\end{aligned} \tag{4.1.14}$$

or

$$\|u(t,\cdot)\|_0^2 + \int_0^t \|u(s,\cdot)\|_1^2 ds \le \|\varphi\|_0^2 + 2\int_0^t \|u(s,\cdot)\|_0^2 + \int_0^T \|f(s,\cdot)\|_{-1}^2 ds. \tag{4.1.15}$$

To get (4.1.13) from (4.1.15), we first drop the second term on the left-hand side of (4.1.15) and use Gronwall's inequality (1.1.22) on page 9 to estimate $\sup_{0<t<T} U(t)$. Then we put the result into the right hand-side of (4.1.15) to get the estimate for $\int_0^T \|u(s,\cdot)\|_1^2 ds$.

This completes the proof of Theorem 4.1.7.

The proof of Theorem 4.1.7 illustrates some general features of deriving a priori estimates:

1. Integration by parts;
2. Use of the epsilon inequality: it happened in (4.1.14) with $\epsilon = 1$ during the transition from $2\|u\|_1\|f\|_{-1}$ to $\|u\|_1 + \|f\|_{-1}$;
3. Importance of Gronwall's inequality.

There are also some features that the proof of Theorem 4.1.7 DOES NOT show, for example:

1. We were somewhat lucky that $\|h\|_1^2 = \|h\|_0^2 + \|h'\|_0^2$; usually, we do not get exactly the norm we want, but only an equivalent norm. This happens, for example, if we want to estimate the $H^2(\mathbb{R})$ norm of u.
2. Usually, one never bothers with the precise value of the ε when applying the epsilon inequality, nor does one keep track of all the time dependence in Gronwall's inequality (so that $C(T)$ stays $C(T)$).

As a result, one might have to deal with multiple lines of computations where there are letters $C_1, C_2, \ldots$ (or N, or something else) denoting numbers whose value can change from line to line. These numbers contain all the constants that have been introduced in the calculations.

Finally, the reader is encouraged to think about the following:

1. What are a priori reasons to multiply both sides of the heat equation by u, and not, say u^2?
2. How do we know to interpret $\int fu\,dx$ as $[f,u]_0$ and not as $(f,u)_0$?

The following exercise is a generalization of Theorem 4.1.7. The result will be used later in this section.

Exercise 4.1.8 (C) Let $a = a(t,x)$, $b = b(t,x)$, and $c = c(t,x)$ be smooth bounded functions, with all the derivatives bounded, and $1 \le a(t,x) \le 2$ for all t, x. Let u, f, φ be the same as in Theorem 4.1.7, except that now u is a classical solution of the equation

$$u_t = au_{xx} + bu_x + cu + f,\ u|_{t=0} = \varphi. \tag{4.1.16}$$

Show that (4.1.13) still holds with a suitable $C(T)$.

Inequality (4.1.13) implies both uniqueness and continuous dependence on the input data for the heat equation (4.1.12) in the following Hilbert spaces: $u \in L_2\big((0,T); H^1(\mathbb{R})\big)$, $\varphi \in L_2(\mathbb{R})$, and $f \in L_2\big((0,T); H^{-1}(\mathbb{R})\big)$. Indeed, by linearity, if

$$u_t = u_{xx} + f,\ u|_{t=0} = \varphi,$$

$$v_t = v_{xx} + g,\ v|_{t=0} = \psi,$$

then $h = u - v$ satisfies $h_t = h_{xx} + (f - g)$, $h|_{t=0} = \varphi - \psi$, and therefore, by (4.1.13),

$$\int_0^T \|u - v\|_1^2 ds \le C(T)\left(\|\varphi - \psi\|_0^2 + \int_0^T \|f - g\|_{-1}^2 ds\right).$$

In particular, if $\varphi = \psi$ as elements of $L_2(\mathbb{R})$ and $f = g$ as elements of $L_2\big((0,T); H^{-1}(\mathbb{R})\big)$, then $u = v$ as elements of $L_2\big((0,T); H^1(\mathbb{R})\big)$.

The best way to construct the solution operator for Eq. (4.1.12) is using the closed-form solution (2.2.7) on page 43. If φ, f are as in Theorem 4.1.7, then taking the Fourier transform shows that $u(t,x)$ from (2.2.7), as a function of x, is an element of the Schwartz space $\mathcal{S}(\mathbb{R})$ and therefore satisfies assumptions of Theorem 4.1.7. Then (4.1.13) immediately extends the solution operator (2.2.7) from smooth input φ, f to $\varphi \in L_2(\mathbb{R}), f \in L_2\big((0,T); H^{-1}(\mathbb{R})\big)$. The solution constructed in this way is the weak variational solution [W2], defined on page 46.

In the end, we conclude that, for every $\varphi \in L_2(\mathbb{R}), f \in L_2\big((0,T); H^{-1}(\mathbb{R})\big)$, equation $u_t = u_{xx} + f$ has a unique solution $u \in L_2\big((0,T); H^1(\mathbb{R})\big)$ and (4.1.13) holds.

The next question is how to construct the solution operator if a closed-form solution is not available, for example, for Eq. (4.1.16). Below, we describe several methods. The first is the `method of continuity`.

Theorem 4.1.9 (Method of Continuity) *Let $\mathbb{H}$ and X be two Banach spaces. Suppose that we have a family of linear equations $\mathrm{A}_\lambda u = f$, $\lambda \in [0,1]$, such that (a) a priori bound $\|u\|_{\mathbb{H}} \leq C\|f\|_X$ holds for every solution u of every equation, with the number C and the spaces $\mathbb{H}, X$ independent of λ; (b) there exists a $\mu \in [0,1]$ such that equation $\mathrm{A}_\mu u = f$ has a solution, and (c) there exists a positive continuous at zero function $\varpi = \varpi(s)$ such that $\varpi(0) = 0$ and, for all $h \in \mathbb{H}$ and all $\lambda, \nu \in [0,1]$, $\|(\mathrm{A}_\lambda - \mathrm{A}_\nu)h\|_X \leq \varpi(|\lambda - \nu|)\, \|h\|_{\mathbb{H}}$.*

Then, for every $\lambda \in [0,1]$ and every $f \in X$, the equation $\mathrm{A}_\lambda u = f$ has a unique solution $u \in \mathbb{H}$ and $\|u\|_{\mathbb{H}} \leq C\|f\|_X$.

Proof Denote by $\Psi_\mu : f \mapsto u$ the solution operator for the equation $\mathrm{A}_\mu u = f$, that is, for every $f \in X$, $u = \Psi_\mu f$ is the solution of the equation. For a fixed $f \in X$, define the operator $\Phi : \mathbb{H} \to \mathbb{H}$ by $\Phi u = \Psi_\mu\Big(f + (\mathrm{A}_\mu - \mathrm{A}_\lambda)u\Big)$.

The next step is an exercise.

Exercise 4.1.10 (C)

(a) Verify that $\Phi u = u$ if and only if $\mathrm{A}_\lambda u = f$.
(b) Verify that $\|\Phi(u - v)\|_{\mathbb{H}} \leq C\,\varpi(|\mu - \lambda|)\, \|u - v\|_{\mathbb{H}}$.

As a result, the operator Φ is a contraction if $C\,\varpi(|\mu - \lambda|) < 1/2$, that is, if $|\mu - \lambda| \leq s_0$, where s_0 is such that $\varpi(s) < 1/(2C)$ for all $|s| \leq s_0$ (s_0 exists because by assumption ϖ is continuous at 0 and $\varpi(0) = 0$). Since the value of s_0 does not depend on μ or λ, we can get from μ to every $\lambda \in [0,1]$ in finitely many steps, thus solving $\mathrm{A}_\lambda u = f$ for all λ.
This completes the proof of Theorem 4.1.9
For more discussions of the method of continuity and related topics, see Krylov [119, Theorem 4.3.2] and [120, Sect. 2].

Example 4.1.11 Let us illustrate Theorem 4.1.9 for Eq. (4.1.16) when $\varphi = 0$. Define

$$\mathbb{H} = \Big\{v \in L_2\big((0,T); H^1(\mathbb{R})\big) : \; v(t) = \int_0^t g(s)ds, \; g \in L_2\big((0,T); H^{-1}(\mathbb{R})\big)\Big\},$$
$$X = L_2\big((0,T); H^{-1}(\mathbb{R})\big). \tag{4.1.17}$$

Problem 3.1.4 on page 97 shows that $\mathbb{H}$ is a Hilbert space with norm

$$\|v\|_{\mathbb{H}}^2 = \int_0^T \|g(s)\|_{H^{-1}(\mathbb{R})}^2 ds + \int_0^T \|v(s)\|_{H^1(\mathbb{R})}^2 ds.$$

Define A_λ by $\mathrm{A}_\lambda v = v_t - (1-\lambda)v_{xx} - \lambda(av_{xx} + bv_x + cv)$ and consider the family of equations $\mathrm{A}_\lambda u = f$, $u(0) = 0$, so that the usual heat equation $u_t = u_{xx} + f$ and Eq. (4.1.16) correspond to $\lambda = 0$ and $\lambda = 1$ respectively. After Theorem 4.1.7 and Exercise 4.1.8, it is not hard to verify conditions of Theorem 4.1.9 (with a linear function ϖ) and to deduce solvability of (4.1.16) with zero initial conditions. Note that $\|u_{xx}\|_{H^{-1}(\mathbb{R})} \leq C\|u\|_{H^1(\mathbb{R})}$. The solution constructed in this way is the weak variational solution [W2], defined on page 46). By the fundamental theorem of calculus in the normal triple $(H^1(\mathbb{R}), L_2(\mathbb{R}), H^{-1}(\mathbb{R}))$ (see Problem 3.1.3(b) on page 96), we also get $u \in \mathcal{C}((0,T); L_2(\mathbb{R}))$.
This concludes Example 4.1.11.

Exercise 4.1.12 (C)

(a) Construct the solution operator for (4.1.16) when $\varphi \neq 0$. **Hint.** Define $V = u - \bar{u}$, where u solves (4.1.16) and $\bar{u}$ solves the usual heat equation (4.1.12), both with the same initial condition φ. Then V satisfies (4.1.16) with zero initial condition and slightly different f.
(b) Verify that the operator $\mathbf{A} : f \mapsto f_t - (af_{xx} + bf_x + cf)$ defines a homeomorphism between the spaces $\mathbb{H}$ and X, and explain why the space $L_2\big((0,T); H^1(\mathbb{R})\big)$ will not work instead of $\mathbb{H}$.

Hint. The solution operator provides the inverse of $\mathcal{A}$. The time derivative f_t is not defined for a generic function from $L_2\big((0,T); H^1(\mathbb{R})\big)$. Remember that we assume a, b, c to be smooth bounded functions and $1 \leq a \leq 2$.

Let us now discuss another method of constructing the solution operator, the `Galerkin approximation`. The method can be traced to a 1915 paper by the Russian civil and mechanical engineer BORIS GRIGORIEVICH GALERKIN (1871–1945). The idea of the method is to approximate the original infinite-dimensional equation by a sequence of finite-dimensional equations, extract a limit using a compactness argument, and then identify the limit as the solution of the original equation. The method usually produces a variational solution, works for a large class of equations (linear or nonlinear, evolution or stationary), and can be used to compute the solution numerically. The approximation is often carried out by projecting the equation on an orthonormal basis.

Let us illustrate Galerkin approximation for Eq. (4.1.16). Take an orthonormal basis $\{\mathfrak{h}_k,\ k \geq 1\}$ in $L_2(\mathbb{R})$ such that the set $\{\mathfrak{h}_k,\ k \geq 1\}$ is dense in $H^1(\mathbb{R})$ and in $H^{-1}(\mathbb{R})$. A non-constructive argument for existence of such a basis is that $L_2(\mathbb{R})$ is separable, contains $H^1(\mathbb{R})$ as a dense subset, and is dense in $H^{-1}(\mathbb{R})$. For an explicit construction of such a basis, see Problem 4.1.3 on page 166.

Denote by Π^N the orthogonal projection in $L_2(\mathbb{R})$ on the linear subspace generated by $\{\mathfrak{h}_1, \ldots, \mathfrak{h}_N\}$, and denote by A the operator $a\partial^2/\partial x^2 + b\partial/\partial x + c$. Consider the family of equations

$$du^N/dt = \Pi^N \mathrm{A} u^N,\ t > 0,\ u^N(0) = \Pi^N \varphi. \tag{4.1.18}$$

With the notations $u^N(t) = \sum_{k=1}^N u_k^N(t)\mathfrak{h}_k$, $A^N = \Big([\mathrm{A}\mathfrak{h}_j, \mathfrak{h}_i],\ i,j = 1, \ldots, N\Big)$, (4.1.18) becomes a linear system of ordinary differential equations for the vector

$\bar{u}^N(t) = (u_1^N(t), \ldots, u_N^N(t))$:

$$d\bar{u}^N(t)/dt = A^N \bar{u}^N(t).$$

Next, we look at the sequence $\{u^N,\ N \geq 1\}$. Let $\mathbb{H}$ be the Hilbert space defined in (4.1.17). The same arguments as in the proof of a priori bound (4.1.13) show that $\sup_N \|u^N\|^2_{\mathbb{H}}$ is bounded by the right-hand side of (4.1.13). The weak sequential compactness of $\mathbb{H}$ (see page 86) implies that there is a subsequence $\{u^{N'},\ N' \geq 1\}$ converging weakly to some $u \in \mathbb{H}$. This weak convergence also implies that u is a weak variational solution of (4.1.16). Note also that, by the fundamental theorem of calculus in the normal triple $(H^1(\mathbb{R}), L_2(\mathbb{R}), H^{-1}(\mathbb{R}))$, we also get $u \in \mathcal{C}((0,T); L_2(\mathbb{R}))$.

For a more detailed example of the Galerkin method, see Evans [50, Theorem 7.1.3].

Here are several comments about the Galerkin approximation in a normal triple (V, H, V') (above, we saw a particular case with $V = H^1(\mathbb{R})$, $H = L_2(\mathbb{R})$, $V' = H^{-1}(\mathbb{R})$):

1. In general, there is no need to have the functions $\mathfrak{h}_k$ orthonormal in H, but they do have to form a dense set in V, H, and V'. [Later on, we will see that, in our analysis of the stochastic equations, it is necessary to have $\mathfrak{h}_k$ orthonormal in H.]
2. Sometimes, but not always, it is possible to choose the functions $\mathfrak{h}_k$ to be orthogonal not only in H but also in V and V'. While the construction of the solution does not rely on the property of the orthogonal projection $\|\Pi^N \cdot\| \leq \|\cdot\|$, this property can simplify some of the arguments.
3. As a finite-dimension object, the Galerkin approximation satisfies all the additional regularity assumptions necessary to derive the a priori estimate, but strictly speaking some work is required to show that the projections of the operators (e.g. $\Pi^N \mathrm{A}$) have the required properties uniformly in N.
4. For linear equations, the weak convergence of the approximation in the suitably constructed solution space $\mathbb{H}$ automatically implies that the limit is a variational solution of the equation.

Yet another method of constructing the solution operator is the `semigroup method`. For simplicity, we will only discuss one-parameter semigroups, corresponding to time-homogeneous equations, where operators/coefficients do not depend on time.

Recall that a `strongly continuous semigroup`, also known as a C_0-semigroup, is a family $\{\Phi(t),\ t \geq 0\}$ of bounded linear operators on a Banach space X such that $\Phi(0)$ is the identity operator I ($\Phi(0)x = x$), $\Phi(t)\Phi(s) = \Phi(t+s)$, $t, s \geq 0$ (the semigroup property), and $\lim_{t\to s} \|\Phi(t)x - \Phi(s)x\|_X = 0$ for all $s \geq 0$ and $x \in X$ (continuity). The `generator of the semigroup` Φ is the operator A defined by

$$\mathrm{A}x = \lim_{t \to 0+} \frac{\Phi(t)x - x}{t}$$

for those x for which the limit exists in the norm of X. It can be shown that the set of all such x is dense in X and that A is a closed operator, as an operator from X to X [50, Theorem 7.4.2]. Intuitively, $\Phi(t) = e^{t\mathrm{A}}$; in many cases this equality can be made rigorous, and, in fact, $e^{t\mathrm{A}}$ is sometimes used to denote the semigroup generated by A.

The idea of the semigroup method is that, given an evolution equation $\dot{u} = \mathrm{A}u + f$, if the operator A happens to be a generator of a strongly continuous semigroup $\Phi(t)$ on a Banach space X, then, for $u_0 \in X$ and $f \in L_1\big((0,T);X\big)$, we get the mild solution of the equation

$$u(t) = \Phi(t)u_0 + \int_0^t \Phi(t-s)f(s)ds \tag{4.1.19}$$

simply by definition of the mild solution. The mild solution is unique because the semigroup is unique. The following exercise presents some other properties of the mild solution.

Exercise 4.1.13 (B) Verify that, if $u_0 \in X$ and $f \in L_1((0,T);X)$, then $u \in \mathcal{C}((0,T);X)$ and $\sup_{0<t<T} \|u(t)\|_X \leq C(T)(\|u_0\|_X + \int_0^T \|f(t)\|_X dt)$.

Hint. The definition of Φ implies that there exists a bounded on $[0,T]$ function $C = C(t)$ such that $\|\Phi(t)x\|_X \leq C(t)\|x\|_X$ for $0 \leq t \leq T$.

The next results show that, the mild solution (4.1.19) is also a variational solution.

Proposition 4.1.14 *Let (V, H, V') be a normal triple of Hilbert spaces and let* A : $V \to V'$ *be a bounded linear operator. Assume that* A *is a generator of a strongly continuous semigroup Φ on H.*

If u is a mild solution of equation $\dot{u} = \mathrm{A}u + f$ then u is also a $w(V, V')$ variational solution as defined on page 48.

Proof This is left as an exercise for the reader.

The main tool in the semigroup method is the theorem providing necessary and sufficient conditions for an operator to generate a strongly continuous semigroup. The result is known as the `Hille-Yosida theorem`, after the American mathematician EINAR HILLE (1894–1980) and the Japanese mathematician KÔSAKU YOSIDA (1909–1990).

Theorem 4.1.15 (Hille-Yosida Theorem) *Let* A *be a linear operator defined on a dense subset of a Banach space X and such that* A *is closed as an operator from X to X. Then the operator* A *is a generator of a strongly continuous semigroup on X if and only if there exist a real number λ_0 and a positive number M such that, for every $\lambda > \lambda_0$, the inverse $(\lambda \mathrm{I} - \mathrm{A})^{-1}$ of $\lambda\mathrm{I} - \mathrm{A}$ is defined and is a bounded linear* *operator on X, and, for every positive integer n,*

$$\sup_{x\in X:\|x\|_X\leq 1} \|(\lambda \mathrm{I} - \mathrm{A})^{-n}x\|_X \leq \frac{M}{(\lambda - \lambda_0)^n}. \tag{4.1.20}$$

Moreover, condition (4.1.20) *for the generator is equivalent to the condition*

$$\sup_{x\in X:\|x\|_X\leq 1} \|\Phi(t)x\|_X \leq Me^{\lambda_0 t} \tag{4.1.21}$$

for the corresponding semigroup.

Proof See Dunford and Schwartz [42, Theorem VIII.1.13]. A particular case $\lambda_0 = 0$ is in Evans [50, Theorem 7.4.4]. Inequality (4.1.21) suggests that $\lambda_0 < 0$ is of special significance: if $\lambda_0 < 0$, then the semigroup and all the solutions of the corresponding evolution equation decay exponentially in time.
This concludes the proof of Theorem 4.1.15.
To construct a mild solution (4.1.19) of the evolution equation $\dot{u} = \mathrm{A}u + f$, $u(0) = 0$, we need to verify that the operator A satisfies conditions of Theorem 4.1.15. The technical part is proving that the operator is closed as an operator from X to X (here, Theorem 3.1.18 on page 83 can be useful), and verifying (4.1.20). To verify (4.1.20), note that if $M = 1$, then the simple inequality for the operator norms,

$$\|(\lambda \mathrm{I} - \mathrm{A})^{-n}\| \leq \|(\lambda \mathrm{I} - \mathrm{A})^{-1}\|^n$$

will work if we can show that

$$\|(\lambda \mathrm{I} - \mathrm{A})^{-1}\| \leq 1/(\lambda - \lambda_0). \tag{4.1.22}$$

In turn, both the existence of the inverse operator and the bound (4.1.22) will follow from

$$\|(\lambda \mathrm{I} - \mathrm{A})x\|_X \geq (\lambda - \lambda_0)\|x\|_X,\ \lambda > \lambda_0. \tag{4.1.23}$$

Indeed, if a linear operator B on X satisfies $\|\mathrm{B}x\|_X \geq b\|x\|_X$, then B is one-to-one and, replacing x with $\mathrm{B}^{-1}x$, we find $\|\mathrm{B}^{-1}x\|_X \leq (1/b)\|x\|_X$. In other words, to establish (4.1.20) it is enough to verify (4.1.23) on a dense subset of X.

Below, we verify conditions of Theorem 4.1.15 for the time-homogeneous version of (4.1.16).

Proposition 4.1.16 *Let* $a = a(x)$, $b = b(x)$, $c = c(x)$ *be smooth bounded functions on* $\mathbb{R}$, *with all derivatives also bounded, and* $1 \leq a(x) \leq 2$ *for all* $x \in \mathbb{R}$. *Define the operator* $\mathrm{A} : H^2(\mathbb{R}) \to L_2(\mathbb{R})$ *by* $\mathrm{A}f = af'' + bf' + cf$; *as always,* f' *means the (generalized) derivative of* f. *Then* A *generates a strongly continuous semigroup on* $X = L_2(\mathbb{R})$.

Proof The domain of A is the space $H^2(\mathbb{R})$ and is a dense subset of $L_2(\mathbb{R})$, for example, because the set of smooth functions with compact support is dense in both $H^2(\mathbb{R})$ and $L_2(\mathbb{R})$.

Next, we will verify that the operator A is closed as an unbounded operator from $L_2(\mathbb{R})$ to $L_2(\mathbb{R})$. [Note that $\mathrm{A} : H^2(\mathbb{R}) \to L_2(\mathbb{R})$ is obviously closed, but this is not what we need.] We will use Theorem 3.1.18 on page 83, with $X = L_2(\mathbb{R})$ and $H = H^2(\mathbb{R})$. Thus, we need to show that

$$\|f\|_2 \le C\big(\|\mathrm{A}f\|_0 + \|f\|_0\big),\ C > 0. \tag{4.1.24}$$

For brevity, we will use the notations $\|\cdot\|$ and $(\cdot,\cdot)$ for the norm and inner product in $L_2(\mathbb{R})$. First, we recall that an equivalent norm in $H^2(\mathbb{R})$ is $\|f\| + \|f''\|$, see Exercise 3.1.10 on page 80. Moreover, from (3.1.7) with $\alpha = 0$, $\beta = 1$, $\gamma = 2$, followed by epsilon inequality (1.1.19), we conclude that $\|f\|_1 \le \varepsilon\|f''\| + C(\varepsilon)\|f\|$ and therefore (4.1.24) will follow if we can show that

$$\|\mathrm{A}f\|^2 \ge C_1\|f''\|^2 + C_2\|f'\|^2 + C_3\|f\|^2 \tag{4.1.25}$$

with $C_1 > 0$. Indeed, in (4.1.25) we only want the coefficient of $\|f''\|$ to be positive: a negative coefficient of $\|f\|$ in (4.1.25) can be compensated by a sufficiently large positive C in (4.1.24); the terms $\|f'\|$ and $\|f\|$ make $\|f\|_1$ and therefore can be compensated by $\|f''\|$ with arbitrarily small coefficient.

To establish (4.1.25), we write

$$\|\mathrm{A}f\|^2 = \|af''\|^2 + 2(af'', bf') + \|bf'\|^2 + 2(cf, af'' + bf') + \|cf\|^2. \tag{4.1.26}$$

Since $a \ge 1$, we get $\|af''\|^2 \ge \|f''\|^2$. Next, we integrate by parts to conclude that

$$2(af'', bf') = -\int_{\mathbb{R}} \big(a(x)b(x)\big)' \big(f'(x)\big)^2 dx \ge C_3\|f'\|^2$$

with $C_3 = -\sup_x \big|\big(a(x)b(x)\big)'\big|$. Similarly,

$$\begin{aligned}2(cf, af'' + bf') &= -2\int_{\mathbb{R}} \big(a(x)c(x)\big)' \big(f'(x)\big)^2 dx - 2\int_{\mathbb{R}} \big(b(x)c(x)\big)' f^2(x)dx\\ &\ge C_4\|f'\|^2 + C_5\|f\|^2\end{aligned}$$

with $C_4 = -2\sup_x \big|\big(a(x)c(x)\big)'\big|$, $C_5 = -2\sup_x |(b(x)c(x))'|$, and (4.1.25) follows.

To verify (4.1.23), we need to show that there exists a $\lambda_0 \in \mathbb{R}$ such that, for all $\lambda > \lambda_0$ and all smooth compactly supported f,

$$\|\lambda f - (af'' + bf' + cf)\|^2 \ge (\lambda - \lambda_0)^2\|f\|^2. \tag{4.1.27}$$

To simplify notation, we continue to write $\|\cdot\|$ for the norm in $L_2(\mathbb{R})$ and $(\cdot,\cdot)$ for the inner product in $L_2(\mathbb{R})$. The left-hand side of (4.1.27) becomes

$$\|af''+bf'\|^2+\|(\lambda-c)f\|^2-2\big((\lambda-c)f,af''\big)-2\big((\lambda-c)f,bf'\big). \tag{4.1.28}$$

We will look at each term in (4.1.28) individually, assuming that $\lambda>0$ is sufficiently large.

The first term in (4.1.28) is non-negative and can be ignored: we are looking for a bound from below. For the second term, define

$$c_0=\sup_x|c(x)|$$

to get

$$\|(\lambda-c)f\|^2=\int_{\mathbb{R}}\big(\lambda-c(x)\big)^2f^2(x)dx\geq(\lambda-c_0)^2\|f\|^2. \tag{4.1.29}$$

After integration by parts, the third term in (4.1.28) becomes

$$2\int_{\mathbb{R}}\big(\lambda-c(x)\big)a(x)(f'(x))^2dx+2\int_{\mathbb{R}}\big(a(x)(\lambda-c(x))\big)'f(x)f'(x)dx \tag{4.1.30}$$

The first term in (4.1.30) is non-negative if $\lambda>c_0$ because of the assumption $a(x)\geq 1$. For the second term (4.1.30), integrate by parts:

$$2\int_{\mathbb{R}}\big(a(x)(\lambda-c(x))\big)'f(x)f'(x)dx=-\int_{\mathbb{R}}\big(a(x)(\lambda-c(x))\big)''f^2(x)dx. \tag{4.1.31}$$

If

$$a_1=\sup_x|a''(x)|,\quad c_1=\sup_x\big|\big(a(x)c(x)\big)''\big|,$$

then

$$-2\big((\lambda-c)f,af''\big)\geq-(\lambda a_1+c_2)\|f\|^2. \tag{4.1.32}$$

Similarly, for the last term in (4.1.28),

$$-2\big((\lambda-c)f,bf'\big)=\int_{\mathbb{R}}(b(x)(\lambda-c(x))'f^2(x)dx. \tag{4.1.33}$$

If

$$b_1=\sup_x|b'(x)|,\quad c_2=\sup_x\big|\big(b(x)c(x)\big)''\big|,$$

then

$$-2\big((\lambda - c)f, bf'\big) \geq -(\lambda b_1 + c_3)\|f\|^2. \tag{4.1.34}$$

We now combine (4.1.29), (4.1.32), and (4.1.34), and introduce new constants

$$A = c_0 + (a_1 + b_1)/2,\ B^2 = \max(A^2 + c_1 + c_2 - c_0^2, 0).$$

The result is

$$\|\lambda f - (af'' + bf' + cf)\|^2 \geq \big(\lambda^2 - (2c_0 + a_1 + b_1)\lambda + c_0^2 - c_1 - c_1\big)\|f\|^2$$
$$\geq \big((\lambda - A)^2 - B^2\big)\|f\|^2 \geq \big(\lambda - (A + B)\big)^2\|f\|^2,$$

if $\lambda > A + B$. Then (4.1.27) follows with $\lambda_0 = A + B$.
This concludes the proof of Proposition 4.1.16.

Exercise 4.1.17 (A) In the setting of Proposition 4.1.16 find

(a) the minimal smoothness conditions on a, b, c for the operator to satisfy conditions of Theorem 4.1.15;
(b) a condition on a, b, c that would ensure $\lambda_0 < 0$ in (4.1.20).

Hint. (a) Two continuous bounded derivatives for a and c and one continuous bounded derivative on b will work. (b) For example, $a = 1, b = 0, c = -1$ will give $\lambda_0 = -1$.

For two more examples of verifying (4.1.23) for certain classes of partial differential operators see Evans [50, Sect. 7.4.3]. Keep in mind that sometimes construction of the semigroup follows the construction and analysis of a variational solution, and the properties of the solution are used to verify conditions of Theorem 4.1.15.

To finish the section, let us summarize the four methods to construct the solution operator for equation $u_t = au_{xx} + bu_x + c$:

The explicit formula: available when a, b, c do not depend on x.

The method of continuity: requires a priori bound (4.1.13) and a separate analysis of the case $a = 1, b = c = 0$, gives variational solution in the Hilbert space $\mathbb{H}$ from (4.1.17).

The Galerkin approximation: requires a priori bound (4.1.13), gives variational solution in the Hilbert space $\mathbb{H}$ from (4.1.17), suggests a way to compute the solution numerically. There is no need to consider special cases, but there is a fair amount of work while passing to the limit.

The semigroup method: works best for time-independent coefficients, requires (4.1.27) instead of (4.1.13), gives a mild solution in $\mathcal{C}\big((0, T); L_2(\mathbb{R})\big)$ provided $f \in L_1\big((0, T); L_2(\mathbb{R})\big)$.

4.1.4 Well-Posedness of Abstract Equations

The objective of this section is to present the theorems about existence, uniqueness, and continuous dependence on the input for variational solutions of linear elliptic, hyperbolic, and parabolic equations in Hilbert spaces. We will work in the normal triple (V, H, V') of Hilbert space and denote by $[\cdot, \cdot]$ the duality between V and V' relative to the inner product in H.

Theorem 4.1.18 (Solvability of Elliptic Equations) *Let* (V, H, V') *be a normal triple of Hilbert spaces and let* $\mathrm{A} : V \to V'$ *be a bounded linear operator with the following property: there exists a positive number* c_A *such that, for all* $v \in V$, $[\mathrm{A}v, v] \geq c_A \|v\|_V^2$. *Then, for every* $f \in V'$, *there exists a unique* $u \in V$ *such that the equality* $\mathrm{A}u = f$ *holds in* V'. *In addition,*

$$\|u\|_V \leq \frac{1}{c_A} \|f\|_{V'}. \tag{4.1.35}$$

Proof Recall that $[\cdot, \cdot]$ is the duality between V' and V relative to the inner product in H. By assumption, the mapping $\mathrm{B} : V \times V \to \mathbb{R}$ defined by $\mathrm{B}(u, v) = [\mathrm{A}u, v]$ satisfies the conditions of the Lax-Milgram theorem (Theorem 3.1.23 on page 85). Since $\mathrm{A}u = f$ is equivalent to $[\mathrm{A}u, v] = [f, v] = \mathrm{B}(u, v)$ for all $v \in V$, and, in the setting of the normal triple, the spaces V and V' are duals of each other, existence and uniqueness of u follow. Next, if $\mathrm{A}u = f$, then $\|f\|_{V'} \|u\|_V \geq [f, u] = [\mathrm{A}u, u] \geq c_A \|u\|_V^2$, and (4.1.35) follows.
This concludes the proof of Theorem 4.1.18.
To discuss hyperbolic and parabolic equations, we start with a technical definition related to time-dependent operators.

Definition 4.1.19 Let X, Y be separable Banach spaces and let

$$\mathrm{A} = \{\mathrm{A}(t), 0 \leq t \leq T\}$$

be a family of mappings from X to Y. The family is called

- (X, Y)-`measurable` if, for every $x \in X$ and $\ell \in Y'$, the real-valued function $t \mapsto \ell(\mathrm{A}(t)x)$ is measurable
- (X, Y)-`uniformly bounded` if there exists a positive number C such that, for all $x \in X$ and all $t \in [0, T]$, $\|\mathrm{A}(t)x\|_Y \leq C\|x\|_X$.
- `differentiable` with derivative $\dot{\mathrm{A}}$ if there exists a family of operators $\dot{\mathrm{A}}$ such that, for every $x \in X$,

$$\lim_{r \to 0} \frac{\|\big(\mathrm{A}(t + r) - \mathrm{A}(t) - r\dot{\mathrm{A}}(t)\big)x\|_Y}{r} = 0.$$

For more information about measurability of Banach space-valued mappings, see Yosida [231, Sect. V.4].

Let (V, H, V') be a normal triple of Hilbert spaces, and let $\mathrm{A} = \{\mathrm{A}(t),\ 0 \le t \le T\}$, $\mathrm{A}_1 = \{\mathrm{A}_1(t),\ 0 \le t \le T\}$, $\mathrm{B} = \{\mathrm{B}(t),\ 0 \le t \le T\}$ be families of linear operators such that A is (V, V')-measurable and (V, V')-uniformly bounded, A_1 is (V, H)-measurable and (V, H)-uniformly bounded, and B is (H, H)-measurable and (H, H)-uniformly bounded.

Consider the equation

$$\ddot{u} = \mathrm{A}u + \mathrm{A}_1 u + \mathrm{B}\dot{u} + f,\ 0 < t \le T;\ \ u(0) = u_0,\ \dot{u}(0) = v_0. \tag{4.1.36}$$

Definition 4.1.20 (Solution of Hyperbolic Equations) The variational solution of (4.1.36) is two functions $u = u(t)$, $v = v(t)$ such that $u \in L_1((0,T);V)$, $v \in L_1((0,T);H)$,

$$u(t) = u_0 + \int_0^t v(s)ds \ \text{ in } L_1((0,T);H),$$

and

$$v(t) = v_0 + \int_0^t \mathrm{A}u(s)ds + \int_0^t \mathrm{A}_1 u(s)ds + \int_0^t \mathrm{B}v(s)ds + \int_0^t f(s)ds$$

in $L_1((0,T);V')$.

Given a Banach space X, denote by $H^1\big((0,T);X\big)$ the collection of functions $f \in L_2\big((0,T);X\big)$ such that $f(t) = f_0 + \int_0^t g(s)ds$ for some $f_0 \in X$, $g \in L_2\big((0,T);X\big)$. It is a Hilbert space with norm

$$\|f\|^2_{H^1((0,T);X)} = \|f_0\|^2_X + \|g\|^2_{L_2((0,T);X)}. \tag{4.1.37}$$

Theorem 4.1.21 (Solvability of Hyperbolic Equations) *Assume that (i)* $[\mathrm{A}u, v] = [u, \mathrm{A}v]$ *for all* $u, v \in V$, *(ii) there exist a positive number* c_A *and a real number* M *such that, for all* $u \in V$ *and all* $t \in [0,T]$,

$$[\mathrm{A}u, u] + c_A\|u\|^2_V \le M\|u\|^2_H,$$

and (iii) the family of operators A *is differentiable and the derivative* $\dot{\mathrm{A}}$ *is* (V, V')*-measurable and* (V, V')*-uniformly bounded.*

Then, for every $u(0) = u_0 \in V$, $\dot{u}(0) = v_0 \in H$, *and* $f = f_1 + f_2$ *with* $f_1 \in L_1\big((0,T);H\big)$, $f_2 \in H^1\big((0,T);V'\big)$, *Eq.* (4.1.36) *has a unique variational solution. Moreover,* $u \in \mathcal{C}\big((0,T);V\big)$, $v \in \mathcal{C}\big((0,T);H\big)$, *and*

$$\begin{aligned}\sup_{0<t<T}\|u(t)\|^2_V + \sup_{0<t<T}\|v(t)\|^2_H \le C(T)\Big(&\|u_0\|^2_V + \|v_0\|^2_H \\ &+ \|f_1\|^2_{L_1((0,T);H)} + \|f_2\|^2_{H^1((0,T);V')}\Big).\end{aligned} \tag{4.1.38}$$

Proof This is left to the reader as Problems 4.1.4 and 4.1.5.
Finally, we present the result for parabolic equations. Let (V, H, V') be a normal triple of Hilbert spaces and let $\mathrm{A} = \{\mathrm{A}(t), 0 \le t \le T\}$ be a family of (V, V')-measurable and (V, V')-uniformly bounded linear operators (see Definition 4.1.19). Consider the equation

$$\dot{u} = \mathrm{A}u + f,\ 0 < t \le T,\ \ u(0) = u_0. \tag{4.1.39}$$

Definition 4.1.22 (Solution of Parabolic Equations) The variational solution of (4.1.39) is the function $u = u(t)$ such that $u \in L_1((0,T); V)$ and

$$u(t) = u_0 + \int_0^t \mathrm{A}u(s)ds + \int_0^t f(s)ds \ \text{ in } L_1((0,T); V').$$

Theorem 4.1.23 (Solvability of Parabolic Equations) *Assume that there exist a positive number c_A and a real number M such that, for all $u \in V$ and all $t \in [0,T]$,*

$$[\mathrm{A}u, u] + c_A \|u\|_V^2 \le M\|u\|_H^2.$$

Then, for every $u(0) = u_0 \in H$ and $f \in L_2\big((0,T); V'\big)$ Eq. (4.1.39) has a unique variational solution. Moreover, $u \in L_2\big((0,T); V\big) \bigcap \mathcal{C}\big((0,T); H\big)$ and

$$\int_0^T \|u(t)\|_V^2 dt + \sup_{0<t<T} \|u(t)\|_H^2 \le C(T)\Big(\|u_0\|_H^2 + \|f\|_{L_2((0,T);V')}^2\Big). \tag{4.1.40}$$

Proof This is left to the reader as Problem 4.1.7.

4.1.5 Problems

Problem 4.1.1 provides some general conditions for global existence and uniqueness of strong solutions for SODEs. Problem 4.1.2 builds on the ideas from Example 4.1.4 on page 148. In particular, the problem introduces the Poincaré inequality and explains why the operator $-\boldsymbol{\Delta}$ in a smooth bounded domain with zero boundary conditions is an elliptic operator in the sense of Definition 4.1.2 on page 148. Problems 4.1.4–4.1.6 outline the proof of Theorem 4.1.21. Problem 4.1.7 outlines the proof of Theorem 4.1.23. Problems 4.1.8 and 4.1.9 invite the reader to take a critical look at Theorems 4.1.21 and 4.1.23. Problem 4.1.10 presents three concrete partial differential operators that often appear in elliptic, hyperbolic, and parabolic equations.

Problem 4.1.1 Consider stochastic ordinary differential equation

$$dX(t) = b(t, X(t))dt + \sigma(t, X(t))dw(t),\ 0 < t \le T,\ \ X(0) = x_0 \in \mathbb{R}^{\mathrm{d}}, \tag{4.1.41}$$

where $b : [0, +\infty) \times \mathbb{R}^{\mathrm{d}} \to \mathbb{R}^{\mathrm{d}}$, $\sigma : [0, +\infty) \times \mathbb{R}^{\mathrm{d}} \to \mathbb{R}^{\mathrm{d}\times m}$ are non-random measurable functions and w is a standard m-dimensional Brownian motion. Define the $\mathrm{d} \times \mathrm{d}$ matrix $a = a(t,x)$ by $a_{ij}(t,x) = \sum_{k=1}^{m} \sigma_{ik}(t,x)\sigma_{jk}(t,x)$. Recall that the generator associated with Eq. (4.1.41) is the partial differential operator

$$\mathrm{L}_X = \frac{1}{2}\sum_{i,j=1}^{\mathrm{d}} a_{ij}(t,x)\frac{\partial^2}{\partial x_i \partial x_j} + \sum_{i=1}^{\mathrm{d}} b_i(t,x)\frac{\partial}{\partial x_i}. \tag{4.1.42}$$

We say that Eq. (4.1.41) is globally well posed if, for every non-random initial condition $x_0 \in \mathbb{R}^{\mathrm{d}}$ and every $T > 0$, Eq. (4.1.41) has a unique probabilistically strong solution.

Below are two sufficient conditions for the equation to be globally well posed.
(a) Assume that the functions b and σ do not depend on t and are locally Lipschitz continuous in x, that is, for every compact subset K of $\mathbb{R}^{\mathrm{d}}$, there exists a number $C > 0$ such that, for all $x, y \in K$, $i, j = 1, \ldots, \mathrm{d}$,

$$|b_i(x) - b_i(y)| + |\sigma_{ij}(x) - \sigma_{ij}(y)| \le C|x - y|.$$

Show that if there exist a twice continuously differentiable real-valued function $F = F(x)$ and a real number c such that $\lim_{|x|\to\infty} F(x) = +\infty$ and, for all $x \in \mathbb{R}^{\mathrm{d}}$, $\mathrm{L}_X F(x) \le cF(x)$, then Eq. (4.1.41) is globally well posed. Moreover, for every real-valued function $f = f(x)$ satisfying $|f(x)| \le F(x) + A$ for some $A \in \mathbb{R}$, we have $\mathbb{E}f(X(t)) < \infty$ for all $t \ge 0$ and all non-random initial conditions x_0. In addition, if f is continuous and if

$$\sup_{x\in\mathbb{R}^{\mathrm{d}}} \frac{|f(x)|^p}{F(x) + A} < \infty$$

for some $p > 1$, then the function $v(t,x) = \mathbb{E}\big(f(X(t))|X(0) = x\big)$ is continuous in x for every $t \ge 0$.
(b) Assume that

1. For each $t \in [0, T]$ and all i, j, the functions $b_i(t,x)$ and $\sigma_{ij}(t,x)$ are continuous in x;
2. $$\int_0^T |b(t,x)|dt < \infty \quad \text{for} \quad \text{all} \quad x \in \mathbb{R}^{\mathrm{d}}; \tag{4.1.43}$$
3. For every $R > 0$ there exists a measurable in t function $K = K(t,R)$ such that

$$\int_0^T K(t,R)dt < \infty$$

and, for all $|x| < R,\ |y| < R$, we have

$$2\sum_{i=1}^{\mathrm{d}}(x_i - y_i)(b_i(t,x) - b_i(t,y)) + \sum_{i=1}^{\mathrm{d}}\sum_{j=1}^{m}|\sigma_{ij}(t,x) - \sigma_{ij}(t,y)|^2 \\ \le K(t,R)|x-y|^2; \tag{4.1.44}$$

4. For all $x \in \mathbb{R}^{\mathrm{d}}$ and almost all $t \in [0,T]$ we have

$$2\sum_{i=1}^{\mathrm{d}} x_i b_i(t,x) + \sum_{i=1}^{\mathrm{d}}\sum_{j=1}^{m}|\sigma_{ij}(t,x)|^2 \le K(t,1)(1+|x|^2). \tag{4.1.45}$$

Show that Eq. (4.1.41) is globally well posed.

Problem 4.1.2 Let G be a bounded domain in $\mathbb{R}^{\mathrm{d}}$ with a smooth boundary ∂G. Denote by $\mathcal{C}_0^\infty(G)$ the collection of smooth functions with compact support in G.

(a) Show that there exists a positive number C_π depending only on the domain G, such that, for all $u \in \mathcal{C}_0^\infty(G)$,

$$\|u\|^2_{L_2(G)} \le C_\pi \|\nabla u\|^2_{L_2(G)}, \tag{4.1.46}$$

where

$$\|\nabla u\|^2_{L_2(G)} = \sum_{i=1}^{\mathrm{d}} \int_G \left(\frac{\partial u}{\partial x_i}\right)^2 dx.$$

Show that one can take $C_\pi = 1/\lambda_1$, where $\lambda_1 > 0$ is the first eigenvalue of the operator $-\boldsymbol{\Delta}$ in G with zero boundary conditions.

A Comment The French mathematician JULES HENRI POINCARÉ (1854–1912) first studied inequalities of the type (4.1.46) in his 1890 paper on the eigenvalues of the Laplace operator. Later, in the 1930's, the German (later, American) mathematician KURT OTTO FRIEDRICHS (1901–1982) fully incorporated such inequalities in the theory of partial differential equations. Accordingly, inequality (4.1.46) is known as the `Poincaré inequality` (or sometimes, as the Poicaré-Friedrichs inequality—see Zeidler [233, Sect. 18.9]). We call the constant C_π in (4.1.46) the `Poincaré constant` for the domain G (and, similar to the notation of the fundamental group, use the subscript π in honor of Poincaré).

(b) [Problem 4.1.2 continues.] Let $H = L_2(G)$ and let $V = H_0^1(G)$ be the closure of $\mathcal{C}_0^\infty(G)$ with respect to the norm

$$\|u\|_V = \left(\|u\|_H^2 + \|\nabla u\|_H^2\right)^{1/2}$$

Define $V' = H^{-1}(G)$ as the dual of V relative to the inner product in H. Verify that $\mathrm{A} = -\boldsymbol{\Delta}$ with zero boundary conditions is an elliptic operator acting in the normal triple (V, H, V') and find the corresponding constant c_A in (4.1.10).

(c) Repeat part (b) in the normal triple (H^1, H^0, H^{-1}), where $H^r = H^r(G)$, $r \in \mathbb{R}$, are the Sobolev spaces on G from Example 3.1.29 on page 87. What differences do you notice?

(d) For each of the following three operators, find sufficient conditions on the functions a_{ij}, b_i, $\tilde{b}_i$, c to guarantee that the corresponding operator is elliptic in the normal triple $(H_0^1(G), L_2(G), H^{-1}(G))$ from part (a):

$$\mathbf{A}_1 u = -\sum_{i,j=1}^{\mathrm{d}} a_{ij}(x)\frac{\partial^2 u}{\partial x_i \partial x_j} + \sum_{i=1}^{\mathrm{d}} b_i(x)\frac{\partial u}{\partial x_i} + c(x)u;$$

$$\mathbf{A}_2 u = -\sum_{i=1}^{\mathrm{d}} \frac{\partial}{\partial x_i}\left(\sum_{j=1}^{\mathrm{d}} a_{ij}(x)\frac{\partial u}{\partial x_j} + \tilde{b}_i(x)u\right) + \sum_{i=1}^{\mathrm{d}} b_i(x)\frac{\partial u}{\partial x_i} + c(x)u;$$

$$\mathbf{A}_3 u = -\sum_{i,j=1}^{\mathrm{d}} \frac{\partial^2}{\partial x_i \partial x_j}\Big(a_{ij}(x)u\Big) - \sum_{i=1}^{\mathrm{d}} \frac{\partial}{\partial x_i}\Big(b_i(x)u\Big) + c(x)u. \tag{4.1.47}$$

Problem 4.1.3 Hermite polynomials (2.3.37) on page 61 are constructed using the standard normal distribution. A popular alternative construction is as follows:

$$e^{2xt-t^2} = \sum_{k=1}^{\infty} \frac{\bar{\mathrm{H}}_n(x)}{n!}\, t^n, \tag{4.1.48}$$

and $\bar{\mathrm{H}}_n$, $n \geq 0$, are also called `Hermite polynomials`.

(a) Comparing (4.1.48) and (2.3.37) on page 61, verify that

$$\mathrm{H}_n(x) = 2^{-n/2}\,\bar{\mathrm{H}}_n\big(x/\sqrt{2}\big).$$

(b) Similar to Problem 2.3.10 on page 71, verify that

$$\bar{\mathrm{H}}_n(x) = (-1)^n\, e^{x^2}\frac{d^n}{dx^n}e^{-x^2}$$

and

$$\int_{-\infty}^{+\infty} \bar{\mathrm{H}}_n(x)\,\bar{\mathrm{H}}_m(x)e^{-x^2}\,dx = \begin{cases} 0, & \text{if } m \neq n, \\ 2^n n!\sqrt{\pi}, & \text{if } m = n. \end{cases} \tag{4.1.49}$$

(c) Verify that the collection of `Hermite functions` $\{\mathfrak{w}_n,\ n \geq 1\}$ defined by

$$\mathfrak{w}_{k+1}(x) = \frac{1}{\sqrt{2^k k! \sqrt{\pi}}} \bar{\mathrm{H}}_k(x) e^{-x^2/2}, \quad k \geq 0, \tag{4.1.50}$$

is an orthonormal basis in $L_2(\mathbb{R})$ and is dense in every Sobolev space $H^r(\mathbb{R}),\ r \in \mathbb{R}$.

(d) Verify that $\mathfrak{w}_n$ is an eigenfunction of the operator $\Lambda = -d^2/dx^2 + x^2 + 1$:

$$-\mathfrak{w}_n''(x) + (x^2 + 1)\mathfrak{w}_n(x) = 2n\, \mathfrak{w}_n(x),\ n \geq 1. \tag{4.1.51}$$

(e) Verify that $\mathfrak{w}_n$ is an eigenfunction of the Fourier transform:

$$\frac{1}{\sqrt{2\pi}} \int_{\mathbb{R}} \mathfrak{w}_n(x) e^{-\mathrm{i}xy} dx = (-\mathrm{i})^{n-1} \mathfrak{w}_n(y). \tag{4.1.52}$$

For $n = 1$, this is the same as (1.1.12) on page 6.

Two Comments

(a) The corresponding orthonormal basis in $L_2(\mathbb{R}^{\mathrm{d}})$ is $\mathfrak{w}_{\boldsymbol{n}}, \boldsymbol{n} = (n_1, \ldots, n_{\mathrm{d}}), n_k \geq 1$:

$$\mathfrak{w}_{\boldsymbol{n}}(x) = \prod_{k=1}^{\mathrm{d}} \mathfrak{w}_{n_k}(x_k).$$

In place of (4.1.51) and (4.1.52) we get

$$-\boldsymbol{\Delta} \mathfrak{w}_{\boldsymbol{n}} + (|x|^2 + \mathrm{d})\mathfrak{w}_{\boldsymbol{n}} = 2|\boldsymbol{n}|\, \mathfrak{w}_{\boldsymbol{n}},\ |\boldsymbol{n}| = \sum_{k=1}^{\mathrm{d}} n_k,$$

and

$$\frac{1}{(2\pi)^{\mathrm{d}/2}} \int_{\mathbb{R}^{\mathrm{d}}} \mathfrak{w}_{\boldsymbol{n}}(x) e^{-\mathrm{i}xy} dx = (-\mathrm{i})^{|\boldsymbol{n}|-\mathrm{d}}\, \mathfrak{w}_{\boldsymbol{n}}(y).$$

(b) Given $\sigma > 0$, one can define

$$\mathrm{H}_n(x;\sigma) = (-1)^n\, e^{x^2/(2\sigma^2)} \frac{d^n}{dx^n} e^{-x^2/(2\sigma^2)},\ n \geq 0,$$

and the corresponding orthonormal basis in $L_2(\mathbb{R})$

$$w_{k+1}(x;\sigma) = C(k,\sigma)\mathrm{H}_k(x;\sigma)\, e^{-x^2/(4\sigma^2)}, k \geq 0.$$

Two particular choices of σ are especially popular: (a) $\sigma^2 = 1/2$, because of the connection with the standard Gaussian random variables, and (b) $\sigma^2 = 1$, because of the elegant equalities (4.1.51) and (4.1.52) satisfied by the Hermite functions.

Problem 4.1.4 Let (V, H, V') be a normal triple of Hilbert spaces. Consider a bounded linear operator $\mathrm{A} : V \to V'$ and assume that (i) $[\mathrm{A}u, v] = [u, \mathrm{A}v]$ for all $u, v \in V$, (ii) there exist a positive number c_A and a real number M such that, for all $u \in V$, $[\mathrm{A}u, u] + c_A\|u\|_V^2 \leq M\|u\|_H^2$.

Consider the equation $\ddot{u}(t) = \mathrm{A}u(t) + f(t),\ 0 < t \leq T,\ u(0) = u_0,\ \dot{u}(0) = v_0$.

(a) Show that the equation is hyperbolic in the sense of Definition 4.1.3 on page 148.
(b) Derive the following a priori estimate for the solution of the equation:

$$\sup_{0<t<T} \|u(t)\|_V^2 + \sup_{0<t<T} \|v(t)\|_H^2 \leq C(T)\left(\|u_0\|_V^2 + \|v_0\|_H^2 + \left(\int_0^T \|f(t)\|_H dt \right)^2 \right). \tag{4.1.53}$$

(c) Use (4.1.53) and Galerkin approximation to construct a solution operator for the equation.
(d) Under an additional assumption that $M = 0$ in (ii) (that is, $[\mathrm{A}u, u] + c_A\|u\|_V^2 \leq 0$), construct a mild solution for the equation using the semi-group method by showing that the operator $\bar{\mathrm{A}} : (u, v) \mapsto (v, \mathrm{A}u)$ generates a strongly continuous semigroup on $V \times H$.
(e) Compare the properties of the solutions constructed in parts (c) and (d).

Problem 4.1.5 Let (V, H, V') be a normal triple of Hilbert space and let $\mathrm{A} = \{\mathrm{A}(t),\ 0 \leq t \leq T\}$, $\mathrm{A}_1 = \{\mathrm{A}_1(t),\ 0 \leq t \leq T\}$, $\mathrm{B} = \{\mathrm{B}(t),\ 0 \leq t \leq T\}$ be families of linear operators such that A is (V, V')-measurable and (V, V')-uniformly bounded, A_1 is (V, H)-measurable and (V, H)-uniformly bounded, and B is (H, H)-measurable and (H, H)-uniformly bounded (see Definition 4.1.19).

Assume that

1. $[\mathrm{A}u, v] = [u, \mathrm{A}v]$ for all $u, v \in V$ and all $t \in [0, T]$;
2. There exist a positive number c_A and a real number M such that, for all $u \in V$ and all $t \in [0, T]$,

$$[\mathrm{A}u, u] + c_A\|u\|_V^2 \leq M\|u\|_H^2;$$

3. As a function of t, A is differentiable and the derivative is (V, V')-measurable and (V, V')-uniformly bounded.

Consider the hyperbolic equation

$$\ddot{u} = \mathrm{A}u + \mathrm{A}_1 u + \mathrm{B}\dot{u} + f. \tag{4.1.54}$$

(a) Show that the a priori estimate (4.1.53) holds.
(b) Verify that $\left(\int_0^T \|f(t)\|_H dt\right)^2$ on the right-hand-side of (4.1.53) can be replaced with $\|f\|^2_{H^1((0,T);V')}$ (see (4.1.37)).
(c) Complete the proof of Theorem 4.1.21, at least the existence part of it.

Problem 4.1.6

(a) Explain why a priori estimate (4.1.53) is not enough to claim uniqueness of the solution of the hyperbolic equation.
(b) Prove uniqueness of solution for the equation in Problem 4.1.5.

Problem 4.1.7 Let (V, H, V') be a normal triple of Hilbert spaces and let $\mathrm{A} = \{\mathrm{A}(t), 0 \le t \le T\}$ be a family of (V, V')-measurable and (V, V')-uniformly bounded linear operators (see Definition 4.1.19). Assume that there exist a positive number c_A and a real number M such that, for all $u \in V$ and all $t \in [0, T]$,

$$[\mathrm{A}u, u] + c_A\|u\|_V^2 \le M\|u\|_H^2.$$

Consider the parabolic equation

$$\dot{u} = \mathrm{A}u + f,\ 0 < t \le T,\ \ u(0) = u_0.$$

(a) Prove Theorem 4.1.23 by deriving (4.1.40) as an a priori bound and then constructing the solution using the Galerkin approximation.
(b) Assume that A does not depend on time. Show that A is a generator of a strongly continuous semigroup in H.

Problem 4.1.8 Consider the heat equation $u_t = u_{xx},\ 0 < t \le 1,\ x \in \mathbb{R}$, with initial condition

$$u_0(x) = \begin{cases} 1, & \text{if } |x| \le 1, \\ 0, & \text{if } |x| > 1, \end{cases}$$

in the normal triple $(H^1(\mathbb{R}), L_2(\mathbb{R}), H^{-1}(\mathbb{R}))$. By Theorem 4.1.23, the solution satisfies

$$u \in L_2\big((0,1); H^1(\mathbb{R})\big) \bigcap \mathcal{C}\big((0,1); L_2(\mathbb{R})\big).$$

In particular, for almost all t, $u(t,\cdot) \in H^1(\mathbb{R})$. On the other hand, we know from the explicit formula for the solution that, for all $t > 0$, $u(t,\cdot) \in \mathcal{S}(\mathbb{R})$. Does it mean that, in exchange for generality in Theorem 4.1.23, we are getting a very weak result?

Problem 4.1.9 Compare and contrast Theorems 4.1.21 and 4.1.23 in terms of assumptions and conclusions. For example, apply the suitable theorem to equations $u_{tt} = a(t,x)u_{xx}$ and $u_t = a(t,x)u_{xx}$.

Problem 4.1.10 Consider the operators (4.1.47) either in $\mathbb{R}^d$ or in a smooth bounded domain with zero boundary conditions. In each of the six cases (three operators, each in two different regions), find conditions on the coefficients so that

(a) The corresponding equation $u_t + \mathrm{A}u = 0$ is parabolic.
(b) The the corresponding operator $-\mathrm{A}$ generates a strongly continuous semigroup on $L_2(\mathbb{R}^d)$ or $L_2(G)$.
(c) The solution of the equation $u_t + \mathrm{A}u = 0$ can be constructed using each of the following three methods: continuity, Galerkin, semigroup.
(d) We get $\lambda_0 < 0$ in (4.1.20).

4.2 Stochastic Elliptic Equations

4.2.1 Existence and Uniqueness of Solution

Of the three main types of equations (elliptic, hyperbolic, parabolic), the stochastic elliptic equations seem to be the least studied. In fact, to the best of our knowledge, the book by Rozanov [198] provides the only systematic treatment of stochastic elliptic equations.

In the abstract Hilbert space setting, existence of the solution of a stochastic elliptic equation with additive noise is provided by Definition 3.2.19 on page 109. Let us recall the setting: if $\mathfrak{B}$ is a zero-mean generalized Gaussian field over a Hilbert space H and $\mathrm{A} : X \to Y$ is a bounded linear operator, then $\mathfrak{X} = \mathrm{A}\mathfrak{B}$ is a zero-mean generalized Gaussian field over Y such that $\mathfrak{X}(y) = \mathfrak{B}(\mathrm{A}^*y)$; recall that A^* is the adjoint operator acting from Y to H in such a way that $(\mathrm{A}h, y)_Y = (h, \mathrm{A}^*y)_H$.

Definition 4.2.1 Let $\mathfrak{B}$ is a zero-mean generalized Gaussian field over a Hilbert space H and let $\mathrm{B} : Y \to H$ be a bounded linear operator. Then a zero-mean generalized Gaussian random field $\mathfrak{X}$ over Y is a solution of the equation

$$\mathrm{B}\mathfrak{X} = \mathfrak{B}. \tag{4.2.1}$$

if, for every $h \in H$, defined by

$$\mathfrak{X}(\mathrm{B}^*h) = \mathfrak{B}(h).$$

Then we have the following result.

Theorem 4.2.2 *Let $\mathfrak{B}$ is a zero-mean generalized Gaussian field over a Hilbert space H and let* $\mathrm{B} : Y \to H$ *be a bounded linear bijection. Then a zero-mean generalized Gaussian random field $\mathfrak{X}$ over Y defined by*

$$\mathfrak{X}(y) = \mathfrak{B}\big((\mathrm{B}^{-1})^* y\big)$$

is the unique solution of the equation (4.2.1).

Proof This follows from Definition 4.2.1 and the equality $(\mathrm{B}^{-1})^* = (\mathrm{B}^*)^{-1}$.

Example 4.2.3 (Euclidean Free Field) The solution of the equation

$$\sqrt{1-\boldsymbol{\Delta}}\, u = \dot{W}$$

in $\mathbb{R}^{\mathrm{d}}$ is known as the `Euclidean free field`, or sometimes also as the `Markov free field`. By Theorem 4.2.2, with $\mathrm{B} = \sqrt{1-\boldsymbol{\Delta}} := \Lambda$, $H = H^0(\mathbb{R}^{\mathrm{d}}) = L_2(\mathbb{R}^{\mathrm{d}})$, $Y = H^1(\mathbb{R}^{\mathrm{d}})$ (see Example 3.1.9, page 79), we conclude that

$$u(f) = \dot{W}((\Lambda^{-1})^* f); \ \ \mathbb{E}\big(u(f)u(g)\big) = \big((\Lambda^{-1})^* f, (\Lambda^{-1})^* g\big)_0.$$

Note that $(\Lambda^{-1})^*$ is acting from $H^1(\mathbb{R}^{\mathrm{d}})$ to $H^0(\mathbb{R}^{\mathrm{d}})$ and, for $f \in H^1(\mathbb{R}^{\mathrm{d}})$, $g \in H^0(\mathbb{R}^{\mathrm{d}})$,

$$((\Lambda^{-1})^* f, g)_0 = (f, \Lambda^{-1} g)_1 = (\Lambda f, \Lambda\Lambda^{-1} g)_0 = (\Lambda f, g)_0,$$

that is, $(\Lambda^{-1})^* = \Lambda$ (for more explanations see (3.1.12) on page 80, as well as Exercise 3.1.35 on page 90). Therefore, for $f, g \in H^1(\mathbb{R}^{\mathrm{d}})$,

$$\mathbb{E}\big(u(f)u(g)\big) = (\Lambda f, \Lambda g)_0 = (f, g)_1,$$

that is, the Euclidean free field u can be interpreted as a Gaussian white noise on $H^1(\mathbb{R}^{\mathrm{d}})$.

This concludes Example 4.2.3.

A complete theory of stochastic elliptic regularity, similar to the deterministic equations, has not been developed even for Gaussian noise. The main difficulties lie in

1. Description of the noise;
2. Description of the boundary values of the solution.

There are several ways to describe a Gaussian field $\mathfrak{X}$ in $G \subseteq \mathbb{R}^{\mathrm{d}}$, for example:

1. Using the homogenous field construction (see Problem 3.2.7, page 126). This description works only in the whole space $\mathbb{R}^{\mathrm{d}}$.

2. Using a Fourier series

$$\mathfrak{X}(f) = \sum_{k\geq 1} q_k f_k \xi_k, \tag{4.2.2}$$

where f_k are the Fourier coefficients of f in some orthonormal basis in $L_2(G)$, $G \subseteq \mathbb{R}^{\mathrm{d}}$.

3. Using a correlation kernel

$$\mathbb{E}\Big(\mathfrak{X}(f)\mathfrak{X}(g)\Big) = \iint\limits_{G\times G} f(x)g(y)q(x,y)dxdy, \tag{4.2.3}$$

or even more generally, a correlation measure:

$$\mathbb{E}\Big(\mathfrak{X}(f)\mathfrak{X}(g)\Big) = \iint\limits_{G\times G} f(x)g(y)\mu(dx,dy). \tag{4.2.4}$$

For a regular field $\mathfrak{X}(f) = \int_G X(x)f(x)dx$, one can also use the correlation function $\mathbb{E}\big(X(x)X(y)\big)$.

While we discuss some of these representations and connections between them in Sect. 3.2.1 (see also Blömker [11]), many of the questions remain open. For example, there is in general no known way to connect the numbers q_k in (4.2.2) and the function q in (4.2.3).

As far as boundary conditions, in this book we mostly restrict ourselves to the zero Dirichlet conditions; the study of more general boundary value problems is one of the main objectives of [198].

Accordingly, most of the information about square-integrable solutions of stochastic elliptic equations is contained in specific examples. We will discuss one example in the following section; more examples are among the problems.

4.2.2 An Example and Further Directions

In this section we look at the elliptic equation with Gaussian white noise as the free term:

$$\mathbf{A}u(x) = \dot{W}(x),\ x \in G, \tag{4.2.5}$$

where $G \subset \mathbb{R}^{\mathrm{d}}$ is a bounded domain with sufficiently regular boundary or a smooth closed manifold (without boundary), and $\mathbf{A}$ is a second-order partial differential operator on G. To emphasize ideas over technicalities, we are not making precise assumptions about the operator $\mathbf{A}$ and the set G beyond the following:

1. The Sobolev spaces $H^r(G)$ and the Hölder spaces $\mathcal{C}^{n,\alpha}(G)$ are defined on G;
2. The deterministic equation $\mathbf{A}v = f$ is well-posed in these spaces.

The norm $\|\cdot\|_r$ in $H^r(G)$ is defined using the Fourier series in the eigenfunction of the Laplace operator on G; see Example 3.1.29 on page 87.

We show that, for the solution u of (4.2.5), the following holds:

1. $\mathbb{E}\|u\|_\gamma^2 < \infty$ if and only if $\gamma < 2 - d/2$.
2. If $\mathrm{d} = 1, 2$ or 3, then u is a regular field on $L_2(G)$, that is, $\mathbb{E}\|u\|^2_{L_2(G)} < \infty$ and

$$u(f) = \int_G u(x)f(x)dx, \; f \in L_2(G). \tag{4.2.6}$$

3. If u is a regular field, then the function $u = u(x)$ from (4.2.6) has, with probability one, the following regularity:
   - u is almost $\mathcal{C}^{3/2}(G)$ if $\mathrm{d} = 1$ (the derivative of u is almost $\mathcal{C}^{1/2}(G)$);
   - u is almost $\mathcal{C}^{1}(G)$ if $\mathrm{d} = 2$;
   - u is almost $\mathcal{C}^{3/8}(G)$ if $\mathrm{d} = 3$.

The first step in the study of Eq. (4.2.5) is to establish the Sobolev space regularity of the Gaussian white noise on $L_2(G)$.

Proposition 4.2.4 *Let $\dot{W}$ be a Gaussian white noise on $L_2(G)$ and $G \subset \mathbb{R}^{\mathrm{d}}$ is a bounded domain with sufficiently regular boundary or a smooth closed manifold. Then*

$$\dot{W} \in L_2\big(\Omega; H^{-\gamma}(G)\big) \text{ if and only if } \gamma > \frac{d}{2}. \tag{4.2.7}$$

Proof If G is a bounded domain with sufficiently regular boundary or a smooth compact manifold, then there is an orthonormal basis $\mathfrak{h}_k$, $k \geq 1$, in $L_2(G)$ consisting of the eigenfunctions of the Laplace operator on G (with zero boundary conditions if there is a boundary):

$$\boldsymbol{\Delta}\mathfrak{h}_k = -\lambda_k^2\mathfrak{h}_k, \; k \geq 1,$$

and $\lambda_k^2 > 0$. Then

$$\dot{W} = \sum_{k\geq 1} \mathfrak{h}_k \, \xi_k,$$

ξ_k are iid standard Gaussian random variables, and

$$\mathbb{E}\|\dot{W}\|_{-\gamma}^2 = \sum_{k\geq 1} \lambda_k^{-2\gamma}; \tag{4.2.8}$$

see Example 3.1.29 on page 87. It is known that

$$0 < \lim_{k\to\infty} \frac{\lambda_k}{k^{1/\mathrm{d}}} < \infty, \tag{4.2.9}$$

see, for example, Safarov and Vassiliev [201, Sect. 1.2]. Then, with (4.2.9) and (4.2.9) in mind, the series in (4.2.8) converges if and only if $2\gamma/\mathrm{d} > 1$, and (4.2.7) follows.

This completes the proof of Proposition 4.2.4.

Now we are ready to establish the Sobolev space regularity of the solution of (4.2.5).

Theorem 4.2.5 *Let* $\mathbf{A}$ *be a second-order partial differential operator on* $G \subset \mathbb{R}^{\mathrm{d}}$. *Suppose that, for some* $\gamma > \mathrm{d}/2$ *and every* $f \in H^{\gamma}(G)$, *the equation* $\mathbf{A}v = f$ *has a unique solution* $v \in H^{\gamma+2}(G)$. *Then equation* $\mathbf{A}u = \dot{W}$ *has a unique solution*

$$u \in L_2\big(\Omega; H^{-\gamma+2}(G)\big).$$

In particular, $\mathbb{E}\|u\|^2_{L_2(G)} < \infty$ *if and only if* $\mathrm{d} = 1, 2,$ *or* 3.

Proof The result follows directly from the assumptions and the Sobolev space regularity of $\dot{W}$, established in Proposition 4.2.4. Note that $\mathbb{E}\|u\|^2_{L_2(G)} < \infty$ means $-\gamma + 2 \geq 0$ or $\gamma \leq 2$, which, given $\gamma > \mathrm{d}/2$, is possible if and only if $\mathrm{d} < 4$.

This completes the proof of Theorem 4.2.5.

Finally, we establish the Hölder continuity of the solution in dimensions $\mathrm{d} = 1, 2$ and 3. Recall that both $\dot{W}(f)$ and $\int_G f(x)dW(x)$ can denote the action of $\dot{W}$ on $f \in L_2(G)$, and we have

$$\mathbb{E}|\dot{W}(f)|^2 = \int_G f^2(x)dx. \tag{4.2.10}$$

Theorem 4.2.6 *In addition to conditions of Theorem 4.2.5 assume that the solution of the equation* $\mathbf{A}u = \dot{W}$ *is a regular Gaussian field, that is,* (4.2.6) *holds. Then, possibly with additional regularity assumptions of the coefficients of* $\mathbf{A}$, *the function* $u = u(x)$ *in* (4.2.6) *has the following regularity with probability one:*

d = 1 *u is continuously differentiable and* u' *is Hölder continuous of every order less than* $1/2$;

d = 2 *u is Hölder continuous of every order less than* 1;

d = 3 *u is Hölder continuous of every order less than* $3/8$;

Proof Unlike the result on the Sobolev space regularity, this theorem cannot be proved by citing a corresponding deterministic result because $\dot{W}$ has no Hölder regularity. The Sobolev embedding theorems could have helped if we had the results of Theorem 4.2.5 in $H^r_p(G)$ for all $p \geq 2$ and not just for $p = 2$.

We are not aware of a complete proof of the result as stated in the theorem. Below, we follow Buckdahn and Pardoux [16, Lemma 2.1] and give the proof when

$\mathbf{A} = \boldsymbol{\Delta}$ Problem 4.2.8 on page 182 provides the main ingredient for the general proof.

Consider the equation

$$-\boldsymbol{\Delta}u(x) = \dot{W}(x)$$

in a bounded domain G with zero boundary conditions.

If $\mathbf{d} = \mathbf{1}$, then, with $G = (0, 1)$, the solution is

$$u(x) = x\int_0^1 w(t)dt - \int_0^x w(t)dt,$$

where w is a standard Brownian motion. This can be verified by direct computation; see also (2.3.34) on page 60. Then the first derivative of the solution $u'(x) = \int_0^1 w(t)dt - w(x)$ has the same regularity as the standard Brownian motion, that is, Hölder continuous of every order less than $1/2$.

If $\mathbf{d} = \mathbf{2}, \mathbf{3}$, then the solution is

$$u(x) = \dot{W}\Big(K(x, \cdot)\Big) = \int_G K(x, z)dW(z), \tag{4.2.11}$$

where K is Green's function of the (negative) Laplacian $-\boldsymbol{\Delta}$ in G with zero boundary conditions.

For $r > 0$, define

$$K_{\mathrm{d}}(r) = \begin{cases} \dfrac{1}{2\pi}\ln r, & \text{if d} = 2, \\[2ex] \dfrac{1}{4\pi\, r}, & \text{if d} = 3, \end{cases} \tag{4.2.12}$$

and

$$U_{\mathrm{d}}(x) = \dot{W}\Big(K_{\mathrm{d}}(|x - \cdot|)\Big) = \int_G K_{\mathrm{d}}(|x - z|)dW(z). \tag{4.2.13}$$

It is known (see, for example John [97, Sect. 4.3]) that the function K from (4.2.11) has a representation

$$K(x, y) = K_{\mathrm{d}}(|x - y|) - \bar{K}(x, y), \tag{4.2.14}$$

where the function $\bar{K}(x, y)$ is smooth in $G \times G$ and, for each fixed $y \in G$, satisfies

$$\boldsymbol{\Delta}\bar{K}(x, y) = 0,\ x \in G;\ \bar{K}(x, y)|_{x\in\partial G} = K_{\mathrm{d}}(|x - y|). \tag{4.2.15}$$

Then, with (4.2.13) in mind, (4.2.11) becomes

$$u(x) = U_{\mathrm{d}}(x) - \int_G \bar{K}(x,z)dW(z). \tag{4.2.16}$$

Since the function $\bar{K}$ has no singularities in G, the regularity of u is determined by the regularity of U_{d}.

Note that U_{d} is a Gaussian field. By Corollary 2.1.8 on page 28, the proof of the theorem will be complete once we verify that, for every sufficiently small positive δ, for example, $\delta \in (0, 1/8)$, there exists a number C_{d}, depending only on δ and the domain G, such that, for all $x, y \in G$,

$$\mathbb{E}|U_{\mathrm{d}}(x) - U_{\mathrm{d}}(y)|^2 \le \begin{cases} C_2|x-y|^{2-\delta}, & \text{if d} = 2, \\ C_3|x-y|^{3/4-\delta}, & \text{if d} = 3. \end{cases} \tag{4.2.17}$$

The proof of (4.2.17) is by direct computation. First, we note that, by (4.2.10) and (4.2.13),

$$\begin{aligned} \mathbb{E}|U_{\mathrm{d}}(x) - U_{\mathrm{d}}(y)|^2 &= \mathbb{E}\Big|\dot{W}\Big(K_{\mathrm{d}}(|x-\cdot|) - K_{\mathrm{d}}(|y-\cdot|)\Big)\Big|^2 \\ &= \int_G \Big|K_{\mathrm{d}}(|x-z|) - K_{\mathrm{d}}(|y-z|)\Big|^2 dz; \end{aligned} \tag{4.2.18}$$

then we use the explicit expression for K_{d} to estimate the right-hand side of (4.2.18).

Here is the general approach to estimating expressions of this type. We start with a little calculus exercise. Let $f = f(t)$, $t > 0$, be a continuously differentiable function. Then

$$\frac{d}{d\theta} f\big(t\theta + s(1-\theta)\big) = (t-s)f'\big(t\theta + s(1-\theta)\big),$$

or, after integrating both sides with respect to θ from 0 to 1,

$$f(t) - f(s) = (t-s)\int_0^t f'\big(t\theta + s(1-\theta)\big)d\theta.$$

Next, assume that the function $|f'(t)|$ is convex:

$$|f'\big(t\theta + s(1-\theta)\big)| \le \theta|f'(t)| + (1-\theta)|f'(s)|;$$

a sufficient condition for this convexity is $d^2|f'(t)|/dt^2 > 0$. Then, since $\int_0^1 \theta d\theta = \int_0^1 (1-\theta)d\theta = 1/2$,

$$|f(t) - f(s)| \le |t-s|\big(|f'(t)| + |f'(s)|\big).$$

Therefore, for every $\gamma \in (0,1)$,

$$\begin{aligned} |f(t)-f(s)| &= |f(t)-f(s)|^{\gamma}|f(t)-f(s)|^{1-\gamma} \\ &\le |t-s|^{\gamma}|f(t)-f(s)|^{1-\gamma}\big(|f'(t)|+|f'(s)|\big)^{\gamma}. \end{aligned} \tag{4.2.19}$$

We now use (4.2.19) with

- $f(t)=K_{\mathrm{d}}(t)$; note that $|K'_{\mathrm{d}}(t)|$ is convex for both d = 2 and d = 3;
- $t=|x-z|$, $s=|y-z|$; note that $|t-s| = \big||x-z|-|y-z|\big| \le |x-y|$.

If d = 2, then $|K'_{\mathrm{d}}(r)| = 1/(2\pi r)$. We take

$$\gamma = 1-\frac{\delta}{2}$$

to find from (4.2.18) and (4.2.19) that (4.2.17) holds:

$$\begin{aligned} &(2\pi)^2\,\mathbb{E}|U_{\mathrm{d}}(x)-U_{\mathrm{d}}(y)|^2 \\ &\le |x-y|^{2-\delta}\int_G \big|\ln|x-z|-\ln|y-z|\big|^{\delta}\left(\frac{1}{|x-z|}+\frac{1}{|y-z|}\right)^{2-\delta} dz \\ &\le \bar{C}_2\left(\int_G \big|\ln|x-z|-\ln|y-z|\big|^{q\delta}\,dz\right)^{1/q} \qquad (4.2.20) \\ &\times\left(\int_G \frac{dz}{|x-z|^{(2-\delta)p}}+\int_G\frac{dz}{|y-z|^{(2-\delta)p}}\right)^{1/p} \\ &\le C_2|x-y|^{2-\delta}, \qquad (4.2.21) \end{aligned}$$

where we get (4.2.20) by applying the Hölder inequality with $1<p<2/(2-\delta)$ and $q=(p-1)/p$, followed by the inequality $(a+b)^{(2-\delta)p} \le \bar{C}_2(a^{(2-\delta)p}+b^{(2-\delta)p})$; then (4.2.21) is a consequence of integrability at zero of the functions $\ln|x|$ and $|x|^{-\alpha}$ for $\alpha<2$ and $x\in\mathbb{R}^2$.

If d = 3, then $|K'_{\mathrm{d}}(r)| = 1/(4\pi r^2)$. We take

$$\gamma = \frac{3}{8}-\frac{\delta}{2}$$

to find from (4.2.18) and (4.2.19) that (4.2.17) holds:

$$\begin{aligned} &(4\pi)^2\,\mathbb{E}|U_{\mathrm{d}}(x)-U_{\mathrm{d}}(y)|^2 \\ &\le |x-y|^{(3/4)-\delta}\int_G\left|\frac{1}{|x-z|}-\frac{1}{|y-z|}\right|^{(5/4)+\delta}\left(\frac{1}{|x-z|^2}+\frac{1}{|y-z|^2}\right)^{(3/4)-\delta} dz \end{aligned}$$

$$\leq \bar{C}_3 \left(\int_G \frac{dz}{|x-z|^{(5/2)+2\delta}} + \int_G \frac{dz}{|y-z|^{(5/2)+2\delta}} \right)^{1/2} \tag{4.2.22}$$

$$\times \left(\int_G \frac{dz}{|x-z|^{3-4\delta}} + \int_G \frac{dz}{|y-z|^{3-4\delta}} \right)^{1/2}$$

$$\leq C_3 |x-y|^{(3/4)-\delta}, \tag{4.2.23}$$

where we get (4.2.22) by applying the Cauchy-Schwartz inequality to the integral, followed by two instances of $(a+b)^n \leq c(n)(a^n + b^n)$, $n > 1$; then (4.2.23) is a consequence of integrability at zero of the function $|x|^{-\alpha}$ for $\alpha < 3$ and $x \in \mathbb{R}^3$. *This concludes the proof of Theorem 4.2.6.*

Corollary 4.2.7 *The solution of the equation*

$$-\boldsymbol{\Delta}\, u = \dot{W} \text{ in } G \subset \mathbb{R}^{\mathrm{d}},\ \mathrm{d} = 2, 3,$$

with zero boundary conditions, has the following `probabilistic representation:`

$$u(x) = U_{\mathrm{d}}(x) - \mathbb{E}\Big(U_{\mathrm{d}}\big(w^x(\tau_x)\big) \big| \mathcal{F}^W \Big), \tag{4.2.24}$$

where

- U_{d} *is the function defined by* (4.2.13)*;*
- $w^x = w^x(t)$, $t \geq 0$, $x \in G$, *is a family of standard* d*-dimensional Brownian motions independent of* $\dot{W}$ *and satisfying* $w^x(0) = x$*;*
- $\tau_x = \inf\{t > 0 : w^x(t) \notin G\}$ *is the first exit time of* w^x *from* G*;*
- $\mathcal{F}^W$ *is the sigma-algebra generated by* $\dot{W}$.

Proof This follows from (4.2.16) and the probabilistic representation of the solution of (4.2.15)

$$\bar{K}(x,y) = \mathbb{E}K_{\mathrm{d}}\big(w^x(\tau_x), y\big). \tag{4.2.25}$$

Exercise 4.2.8 (B) Formulas (4.2.16) and (4.2.25) hold for all $\mathrm{d} \geq 2$, with $K_{\mathrm{d}}(t) = c_{\mathrm{d}} t^{2-\mathrm{d}}$, where c_{d} is a number depending only on d. What goes wrong if one writes (4.2.24) for $\mathrm{d} > 3$? **Hint.** The function K_{d} is square-integrable at 0 if and only if $(2\mathrm{d}-4) - (\mathrm{d}-1) < 1$ or $\mathrm{d} < 4$.

Markov Property of the Solution Recall that there are three types of the Markov property for random fields: strong, germ, and global (see pages 111– 113). All these properties are a version of conditional independence of the values of the field inside and outside of a set given the values of the field on the boundary of the set. Here are a few results, without proofs, about various Markov properties for elliptic equations.

- Walsh [223, Proposition 9.6] shows that the solution of $\boldsymbol{\Delta} u = \nabla \cdot \dot{W}$ in $\mathbb{R}^{\mathrm{d}}$, $\mathrm{d} \geq 3$, satisfies the sharp Markov property relative to bounded open sets.
- Rozanov [198, Sect. III.2.2] shows that the solution of $\mathbf{A}u = \dot{W}$ with suitable boundary conditions has the global Markov property.
- Donati-Martin and Nualart [40] (see also [39]) consider the equation

$$\boldsymbol{\Delta} u + f(u) = \dot{W} \tag{4.2.26}$$

with zero boundary conditions in a bounded domain with sufficiently smooth boundary in $\mathbb{R}^{\mathrm{d}}$, $\mathrm{d} = 1, 2, 3$. They show that the solution has the germ Markov property relative to bounded open sets if and only if the function f has the form $f(u) = au+b$, $a, b \in \mathbb{R}$; further restrictions on a are necessary to ensure existence and uniqueness of the solution. For another proof of this result see Nualart [175, Theorem 4.2.4]. Note that if $\mathrm{d} = 1, 2, 3$, then the solution of (4.2.26) is a regular field, making the germ Markov property relative to bounded open sets equivalent to the global Markov property [92, Remark 1.25].

To conclude this section, let us mention some other questions that can be studied in connection with stochastic elliptic equations.

Variational Formulation of the Solution If the solution is a regular field, then a more traditional definition of the solution is possible and is the subject of Problem 4.2.1 below.

Non-Dirichlet Deterministic Boundary Conditions One example is the work by Nualart and Tindel [178] on elliptic equations with reflection.

Random Boundary Conditions This is the subject of the book [198].

Regular Solutions in Dimensions d ≥ 4 A typical linear equation considered in the literature (for example, Martínez and Sanz-Solé [155]), is

$$\boldsymbol{\Delta} u = \dot{W}^{\mathrm{Q}},$$

in a bounded domain with zero boundary conditions, where $\dot{W}^{Q}$ is a colored noise on $L_2(G)$, usually described using the correlation kernel (4.2.3). Under suitable conditions on the kernel, the solution is a regular, Hölder-continuous random field. Analysis is similar to the proof of Theorem 4.2.6.

Nonlinear Equations If the coefficients are deterministic and noise is additive, then many results extend from linear equations to nonlinear. In fact, most of the papers on the subject of stochastic elliptic equations consider equations of the type

$$\boldsymbol{\Delta} u + f(x, u) = \dot{W}^{\mathrm{Q}} \tag{4.2.27}$$

with a suitable (non-random) function f. Of course, the solution has to be a regular field for the equation to make sense.

Numerical Methods Finite-difference schemes for equations of the type (4.2.27) were studied by Gyöngy and Martínez [69] and Martínez and Sanz-Solé [155].

Equations Driven by Fractional Noise Instead of the Gaussian white noise $\dot{W}$, corresponding to the zero mean Gaussian random field $W = W(x)$ with covariance $\mathbb{E}\big(W(x)W(y)\big) = \prod_{i=1}^{\mathrm{d}} \min(x_i, y_i)$, $x_i, y_i \geq 0$, one can consider fractional Gaussian noise $\dot{W}^H$, corresponding to the zero mean Gaussian field $W^H = W^H(x)$ with covariance

$$\mathbb{E}\big(W^H(x)W^H(y)\big) = \prod_{i=1}^{\mathrm{d}} \big(|x_i|^{2H_i} + |y_i|^{2H_i} - |x_i - y_i|^{2H_i}, H_i \in (0,1),$$

$x_i, y_i \geq 0$. Sanz-Solé and Torrecilla [205] study existence, uniqueness, regularity, and numerical approximation of the solution for the equation $\boldsymbol{\Delta} u = f\big(u(x)\big) + \dot{W}^H(x)$ in a smooth bounded domain in $\mathbb{R}^{\mathrm{d}}$ with zero boundary conditions, when $H_i \geq 1/2$ and $\sum_{i=1}^{\mathrm{d}} H_i > \mathrm{d} - 2$.

4.2.3 Problems

Problem 4.2.1 is about various definitions of the variational solution for a stochastic elliptic equation. Problems 4.2.2 outlines an alternative approach to the study of regularity of white noise. Problem 4.2.3 shows that the Gaussian white noise on $L_2(\mathbb{R}^{\mathrm{d}})$ is invariant under the Fourier transform and thus has no global Sobolev space regularity. Problem 4.2.4 provides another example of an equation with a closed-form solution and another representation of the Lévy Brownian motion. Problems 4.2.5–4.2.7 are about the Euclidean free field. Problem 4.2.8 extends the probabilistic representation of the solution in Corollary 4.2.7 to more general equations. Problem 4.2.9 invites the reader to investigate another class (in fact, two classes) of stochastic elliptic equations in $\mathbb{R}^{\mathrm{d}}$.

Problem 4.2.1 Consider the equation $\mathbf{A}u = \dot{W}$ with zero boundary conditions in $G \subseteq \mathbb{R}^{\mathrm{d}}$, where A is one of the operators $\mathbf{A}_i$, $i = 1, 2, 3$, from (4.1.47) on page 166. Assuming that the solution u is a regular field, define a variational solution in each case.

Problem 4.2.2 For $f \in \mathcal{C}_0^{\infty}(\mathbb{R})$ define

$$\Lambda f(x) = -f''(x) + (1 + x^2)f(x)$$

and let $\widetilde{H}^r$ be the closure of $\mathcal{C}_0^{\infty}(\mathbb{R})$ with respect to the norm

$$\|f\|_r^2 = \|\Lambda^{r/2} f\|_{L_2(\mathbb{R})}^2.$$

(a) Show that

$$\|f\|_r^2 = \sum_{k\geq 1}(2k)^r f_k^2$$

where $f_k = \int_{\mathbb{R}} f_k(x)\mathfrak{w}_k(x)dx$ and $\mathfrak{w}_k$ is the Hermite function (4.1.50) on page 167, and that $\mathcal{S}(\mathbb{R}) = \bigcap_{r>0}\widetilde{H}^r$ (at the very least establish equality of sets; as an extra challenge, you can compare the topologies as well).

(b) Show that $\dot{W}$, the Gaussian white noise on $L_2(\mathbb{R})$, is an element of $L_2(\Omega;\widetilde{H}^{-r})$ if and only if $r > 1$.

(c) Extend the results to $\mathbb{R}^{\mathrm{d}}$, $\mathrm{d} \geq 2$.

(d) Compare and contrast the spaces $\widetilde{H}^r$ with the usual Sobolev spaces $H^r(\mathbb{R}^{\mathrm{d}})$.

Problem 4.2.3 Let $\dot{W}$ be a Gaussian white noise on $L_2(\mathbb{R}^{\mathrm{d}})$.

(a) Show that the Fourier transform of $\dot{W}$ is defined as a generalized function and is also a Gaussian white noise on $L_2(\mathbb{R}^{\mathrm{d}})$. Conclude that $\dot{W}$ does not belong to any Sobolev space $H^\gamma(\mathbb{R}^{\mathrm{d}})$.

(b) Show that $\dot{W}$ is locally in $H^\gamma(\mathbb{R}^{\mathrm{d}})$ for every $\gamma < -\mathrm{d}/2$. More precisely, let φ be a smooth function with compact support in $\mathbb{R}^{\mathrm{d}}$, and define $\varphi\dot{W}$ as a generalized random field over $L_2(\mathbb{R}^{\mathrm{d}})$ such that

$$(\varphi\dot{W})(f) = \dot{W}(\varphi f).$$

The objective is to show that $\varphi\dot{W} \in H^\gamma(\mathbb{R}^{\mathrm{d}})$ for every $\gamma < -\mathrm{d}/2$.

Problem 4.2.4 Investigate the equation

$$\boldsymbol{\Delta} u = \dot{W} \tag{4.2.28}$$

in $G = \{x \in \mathbb{R}^{\mathrm{d}} : x \neq 0\}$, with the boundary condition $u(0) = 0$. In particular,

(a) Show that the solution is a regular field if $\mathrm{d} = 1, 2, 3$.

(b) Show that if $\mathrm{d} = 3$, the solution can be written as $u(x) = \sigma L(x)$, where $\sigma > 0$ and $L = L(x)$ is the `Lévy Brownian motion` on $\mathbb{R}^3$ (see Exercise 2.1.10 on page 29).

(c) Investigate the properties of the solution when $\mathrm{d} > 3$.

Problem 4.2.5 For each of the following equations on $\mathbb{R}^{\mathrm{d}}$, $\mathrm{d} \geq 1$,

(a) Interpret the solution u as a generalized Gaussian field on a suitable Hilbert space and compute $\mathbb{E}\big(u(f)u(g)\big)$;

(b) Determine whether the solution is a regular field.

$$\sqrt{m^2 - \boldsymbol{\Delta}}\, u = \dot{W},\ m \geq 0; \tag{4.2.29}$$

$$\boldsymbol{\Delta} u = \nabla \cdot \dot{\boldsymbol{W}}, \tag{4.2.30}$$

$$(1 - \boldsymbol{\Delta})u = \mathfrak{B}, \tag{4.2.31}$$

where $\dot{W}$ is a Gaussian white noise on $L_2(\mathbb{R}^d)$, $\dot{\boldsymbol{W}} = (\dot{W}_1, \ldots, \dot{W}_d)$, with $\dot{W}_i$ independent Gaussian white noises on $L_2(\mathbb{R}^d)$, and $\mathfrak{B}$ is a Gaussian white noise on $H^1(\mathbb{R}^d)$.

A Comment The `Euclidean free field` can be defined as the solution of (4.2.29) with $m > 0$ and $d \geq 1$ [172], or the solution of (4.2.30) for $d \geq 3$ [223, Proposition 9.4]. It also turns out that (4.2.30) is essentially the same as (4.2.29) with $m = 0$.

Problem 4.2.6 Show that the solution u of the equation $\sqrt{1 - \boldsymbol{\Delta}}\, u = \dot{W}$ in $\mathbb{R}^d$ is a Gaussian white noise on $H^{-1}(\mathbb{R}^d)$.

A Comment In Example 4.2.3 on page 171 we interpreted u as a Gaussian white noise on $H^1(\mathbb{R}^d)$. The reader is encouraged to think about these two different interpretations of the same object. To some extend, something similar happens in a normal triple (V, H, V') of Hilbert spaces: an element of the dual space of V can be an element of V (when the duality is relative to the inner product in V), or an element of V' (when the duality is relative to the inner product in H).

Problem 4.2.7 Find as many different constructions of the Euclidean free field as possible (in this text and other references), and investigate applications of this object in physics.

Problem 4.2.8 For $d = 2, 3$, let $\mathbf{A}$ be a partial differential operator that is the generator of the family of diffusion processes $X^y = X^y(t)$, $y \in \mathbb{R}^d$, $X^y(0) = y$, and let $c = c(x)$ be a continuous non-negative function. Denote by $K_A = K_A(x, y)$ the fundamental solution of the operator $-\mathbf{A} + c$ in $\mathbb{R}^d$, that is, for every smooth compactly supported f, the function

$$U(x) = -\int_{\mathbb{R}^d} K(x, y) f(y) dy$$

is a classical solution of

$$\mathbf{A}U(x) - c(x)U(x) = f(x), \; x \in \mathbb{R}^d.$$

Let G be a smooth bounded domain in $\mathbb{R}^d$, and, for $x \in G$,

$$\tau_x = \inf\{t > 0 : X^x(t) \notin G\}.$$

Let $\dot{W}$ be a Gaussian white noise on $L_2(G)$, and assume that $\dot{W}$ is independent of all the processes X^y. Define

$$U_A(x) = \int_G K_A(x, y) dW(y)$$

and let u be the solution of

$$\mathbf{A}u - cu = \dot{W} \text{ in } G$$

with zero boundary conditions. Verify the following `probabilistic representation` of u:

$$u(x) = U_A(x) - \mathbb{E}\left(U_A\big(X^x(\tau_x)\big)e^{-\int_0^{\tau_x} c(X^x(t))dt} \,\middle|\, \mathcal{F}^W \right),$$

where $\mathcal{F}^W$ is the sigma-algebra generated by $\dot{W}$.

Problem 4.2.9 Let $\dot{W}$ be a Gaussian white noise on $L_2(\mathbb{R}^d)$ and $\boldsymbol{\Delta}$, the Laplace operator. Consider the equation

$$(m^2 - \boldsymbol{\Delta})^{\gamma/2}u = \dot{W}$$

for $m \geq 0$ and $\gamma \in \mathbb{R}$. Is the solution u ever a regular field? Can it have any Hölder space regularity? Then answer the same questions for the solution of the equation

$$(m^2 + c^2|x|^2 - \boldsymbol{\Delta})^{\gamma/2}u = \dot{W},\ c \geq 0.$$

4.3 Stochastic Hyperbolic Equations

4.3.1 Existence and Uniqueness of Solution

Given two Banach spaces X, Y, recall the following notions for the time-dependent operators and functions:

1. (X, Y)-measurable, (X, Y)-uniformly bounded, and differentiable families of operators: Definition 4.1.19, page 161.
2. The space $H^1\big((0, T); X\big)$: page 162.

Now that we consider stochastic equations, we need to modify the corresponding definitions to include possible dependence on the elementary outcome. We fix the stochastic basis $(\Omega, \mathcal{F}, \{\mathcal{F}_t\}_{t\geq 0}, \mathbb{P})$ with the usual assumptions.

Definition 4.3.1 Let X, Y be separable Banach spaces and let

$$\mathrm{A} = \{\mathrm{A}(\omega, t), \omega \in \Omega,\ 0 \leq t \leq T\}$$

be a family of mappings from X to Y. The family is called

- (X, Y)-`adapted` if, for every $x \in X$ and $\ell \in Y'$, the real-valued process $t \mapsto \ell(\mathrm{A}(t)x)$ is $\mathcal{F}_t$-adapted;
- $L_\infty(X, Y)$-`uniformly bounded` if there exists a positive (non-random) number C such that, for all $x \in X$, all $t \in [0, T]$, and all $\omega \in \Omega$, $\|\mathrm{A}(t)x\|_Y \le C\|x\|_X$.

Take a collection $\{w_k,\ k \ge 1\}$ of independent standard Brownian motions and consider the following stochastic equation:

$$\begin{aligned}
\ddot{u} &= \mathrm{A}(t)u + \mathrm{A}_1(t)u + \mathrm{B}(t)\dot{u} + f(t)\\
&\quad + \sum_{k\ge 1}\big(\mathrm{M}_k(t)u + \mathrm{N}_k(t)\dot{u} + g_k(t)\big)\dot{w}_k(t),\ 0 < t \le T;\\
u(0) &= u_0,\ \dot{u}(0) = v_0,
\end{aligned} \tag{4.3.1}$$

in the normal triple (V, H, V') of Hilbert spaces. This equation is constructed by taking the deterministic version (4.1.36) on page 162 and adding the corresponding stochastic term. Similar to the deterministic case, the reason for considering both A and A_1 becomes clear during the derivation of the a priori bound. To keep formulas from getting too congested, we will often omit the time dependence in all operators. Recall that $[v, u]$, $v \in V'$, $u \in V$, denotes the duality between V and V' relative to the inner product in H.

To define a square-integrable variational solution of equation (4.3.1), we make the following assumptions:

[HP1] The families of operators $\mathrm{A} = \{\mathrm{A}(t),\ 0 \le t \le T\}$, $\mathrm{A}_1 = \{\mathrm{A}_1(t),\ 0 \le t \le T\}$, and $\mathrm{M}_k = \{\mathrm{M}_k(t),\ 0 \le t \le T\}$ are (V, V')-adapted and $L_\infty(V, V')$-uniformly bounded.

[HP2] The families of operators $\mathrm{B} = \{\mathrm{B}(t),\ 0 \le t \le T\}$ and $\mathrm{N}_k = \{\mathrm{N}_k(t),\ 0 \le t \le T\}$ are (H, V')-adapted and $L_\infty(H, V')$-uniformly bounded.

[HP3] The initial conditions are $\mathcal{F}_0$-measurable (equivalently, independent of all w_k), $u_0 \in L_2(\Omega, V)$, $v_0 \in L_2(\Omega; H)$; the process f and each of the processes g_k are $\mathcal{F}_t$-adapted and are elements of $L_2\big(\Omega \times (0, T); V'\big)$.

[HP4] $\sum_{k\ge 1}\mathbb{E}\int_0^T \|g_k(t)\|_{V'}^2 dt < \infty$.

[HP5] There exists a positive number C_o such that, for every $v \in L_2\big(\Omega \times (0, T); V\big)$ and $h \in L_2\big(\Omega \times (0, T); H\big)$,

$$\begin{aligned}
&\sum_{k\ge 1}\mathbb{E}\int_0^T \|\mathrm{M}_k(t)v\|_{V'}^2 dt < \infty,\\
&\sum_{k\ge 1}\mathbb{E}\int_0^T \|\mathrm{N}_k(t)h\|_{V'}^2 dt < \infty.
\end{aligned} \tag{4.3.2}$$

Definition 4.3.2 (Solution of Stochastic Hyperbolic Equations) The variational solution of (4.3.1) is two $\mathcal{F}_t$-adapted processes $u = u(t)$, $v = v(t)$ such that $u \in$

$L_2(\Omega \times (0,T); V)$, $v \in L_2(\Omega \times (0,T); H)$,

$$u(t) = u_0 + \int_0^t v(s)ds \quad \text{in } L_2(\Omega \times (0,T); H),$$

and

$$\begin{aligned} v(t) =& v_0 + \int_0^t \mathrm{A}u(s)ds + \int_0^t \mathrm{A}_1 u(s)ds + \int_0^t \mathrm{B}v(s)ds + \int_0^t f(s)ds \\ &+ \sum_{k\geq 1} \int_0^t \big(\mathrm{M}_k u(s) + \mathrm{N}_k v(s) + g_k(s)\big)dw_k(s) \end{aligned} \tag{4.3.3}$$

in $L_2(\Omega \times (0,T); V')$.

To construct a solution of (4.3.1), we have to modify the assumptions as follows (the precise conditions on the input are in the statement of the theorem below):

[HP1'] The family of operators $\mathrm{A} = \{\mathrm{A}(t),\ 0 \leq t \leq T\}$ is (V, V')-adapted and $L_\infty(V, V')$-uniformly bounded.

[HP2'] The families of operators $\mathrm{A}_1 = \{\mathrm{A}_1(t),\ 0 \leq t \leq T\}$ and $\mathrm{M}_k = \{\mathrm{M}_k(t),\ 0 \leq t \leq T\}$ are (V, H)-adapted and $L_\infty(V, H)$-uniformly bounded.

[HP3'] The families of operators $\mathrm{B} = \{\mathrm{B}(t),\ 0 \leq t \leq T\}$ and $\mathrm{N}_k = \{\mathrm{N}_k(t),\ 0 \leq t \leq T\}$ are (H, H)-adapted and $L_\infty(H, H)$-uniformly bounded.

[HP4'] There exists a positive number C_o such that, for every $v \in L_2\big(\Omega \times (0,T); V\big)$ and $h \in L_2\big(\Omega \times (0,T); H\big)$,

$$\sum_{k\geq 1} \mathbb{E} \int_0^T \|\mathrm{M}_k(t)v\|_H^2 dt \leq C_o \mathbb{E} \int_0^T \|v(s)\|_V^2 ds,$$

$$\sum_{k\geq 1} \mathbb{E} \int_0^T \|\mathrm{N}_k(t)h\|_H^2 dt \leq C_o \mathbb{E} \int_0^T \|h(s)\|_H^2 ds.$$

Before proceeding, the reader is encouraged to think about the differences between the two sets of conditions. After that, the reader can try to state and even prove the following theorem about existence and uniqueness of solution, because all the necessary technical tools have been already presented. On the other hand, the reader who ignored Problems 4.1.4 and 4.1.5 might find the following a bit overwhelming.

Theorem 4.3.3 (Solvability of Hyperbolic Equations) *In addition to* [HP1']–[HP5'] *assume that*

(i) $[\mathrm{A}u, v] = [u, \mathrm{A}v]$ *for all* $u, v \in V$, *all* $t \in [0,T]$, *and all* $\omega \in \Omega$,

(ii) there exist a positive number c_A *and a real number* M *such that, for all* $u \in V$, *all* $t \in [0,T]$, *and all* $\omega \in \Omega$,

$$[\mathrm{A}u, u] + c_A\|u\|_V^2 \leq M\|u\|_H^2, \tag{4.3.4}$$

and

(iii) the family of operators A *is differentiable for all* $\omega \in \Omega$ *and the derivative of* A *is* (V, V')*-adapted and* $L_\infty(V, V')$*-uniformly bounded.*

Then, for every $\mathcal{F}_0$*-measurable initial conditions* $u(0) = u_0 \in L_2(\Omega; V)$, $\dot{u}(0) = v_0 \in L_2(\Omega; H)$, *and* $\mathcal{F}_t$*-adapted free terms* $f \in L_2\big(\Omega \times (0,T); H\big)$ *and* g_k *satisfying* $\sum_k \|g_k\|^2_{L_2(\Omega\times(0,T);H)} < \infty$, *Eq. (4.3.1) has a solution. The solution is unique in* $L_2\big(\Omega; \mathcal{C}((0,T); V)\big) \times L_2\big(\Omega; \mathcal{C}((0,T); H)\big)$ *and*

$$\|u(t)\|^2_{L_2(\Omega;\mathcal{C}((0,T);V))} + \|v(t)\|^2_{L_2(\Omega;\mathcal{C}((0,T);H))} \le C(T)\Big(\|u_0\|^2_{L_2(\Omega;V)} + \|v_0\|^2_{L_2(\Omega;H)} + \|f\|^2_{L_2(\Omega\times(0,T);H)} + \sum_{k\ge1}\|g_k\|^2_{L_2(\Omega\times(0,T);H)}\Big). \tag{4.3.5}$$

Proof Below, we present three steps: (1) derivation of the a priori bound (4.3.5); (2) construction of the solution using the Galerkin approximation; (3) proof of continuity of the solution. The proof of uniqueness is similar to the deterministic case (see Problem 4.1.6 on page 169) and is left to the reader.

Step 1: The a priori Estimate (pages 186–188) Inequality (4.3.5) is equivalent to

$$\mathbb{E}\sup_{0<t<T}\|u(t)\|^2_V + \mathbb{E}\sup_{0<t<T}\|v(t)\|^2_H \le C(T)\Big(\mathbb{E}\|u_0\|^2_V + \mathbb{E}\|v_0\|^2_H + \mathbb{E}\int_0^T \|f(t)\|^2_H dt + \sum_{k\ge1}\mathbb{E}\int_0^T \|g_k(t)\|^2_H dt\Big). \tag{4.3.6}$$

To derive (4.3.6), we pretend that $v \in V$ and apply the energy equality (Theorem 3.3.15, page 139) to the process v. The result is

$$\begin{aligned}\|v(t)\|^2_H &= \|v_0\|^2_H + 2\int_0^t \big(\mathrm{A}u(s), v(s)\big)_H ds + 2\int_0^t \big(\mathrm{A}_1 u(s), v(s)\big)_H ds\\ &\quad + 2\int_0^t \big(\mathrm{B}v(s), v(s)\big)_H ds + \int_0^t \big(f(s), v(s)\big)_H ds\\ &\quad +2\sum_k \int_0^t \Big(\big(\mathrm{M}_k u(s), v(s)\big)_H + \big(\mathrm{N}_k v(s), v(s)\big)_H + \big(g_k(s), v(s)\big)_H\Big) dw_k(s)\\ &\quad + \sum_k \int_0^t \|\mathrm{N}_k u(s) + \mathrm{M}_k v(s) + g_k(s)\|^2_H ds.\end{aligned}$$

Next, we need to take $\mathbb{E}\sup_{0<t<T}$ on both sides, term by term and analyze the individual terms on the right-hand side. Because of the supremum preceding the expectation, we cannot disregard the stochastic integral.

Below is an outline of the main calculations; the value of the number C is changing from line to line.

By the product rule and the symmetry of the operator A,

$$\frac{d}{dt}[\mathrm{A}u, u] = 2[\mathrm{A}u, v] + [\dot{\mathrm{A}}u, u]. \tag{4.3.7}$$

Therefore,

$$2\int_0^t \big(\mathrm{A}u(s), v(s)\big)_H ds = [\mathrm{A}u(t), u(t)] - [\mathrm{A}u(0), u(0)] - \int_0^t [\dot{\mathrm{A}}u(s), u(s)]ds.$$

By assumption (4.3.4),

$$[\mathrm{A}u(t), u(t)] \leq -c_A\|u(t)\|_V^2 + M\|u(t)\|_H^2.$$

The term $-c_A\|u(t)\|_V^2$ goes to the left, while $u(t) = u_0 + \int_0^t v(s)ds$ implies

$$\|u(t)\|_H^2 \leq \|u_0\|_H^2 + \int_0^t \|v(s)\|_H^2 ds$$

and the last integral is removed by the Gronwall inequality.

Similarly,

$$\int_0^t |[\dot{\mathrm{A}}u(s), u(s)]|ds \leq C\int_0^t \|u(s)\|_V^2 ds,$$

$$\int_0^t |(\mathrm{A}_1u(s), v(s))|ds \leq C\int_0^t \|u(s)\|_V^2 ds + C\int_0^t \|v(s)\|_H^2 ds,$$

$$\int_0^t |\big(\mathrm{B}v(s), v(s)\big)_H|ds \leq C\int_0^t \|v(s)\|_H^2 ds,$$

and again the Gronwall inequality removes the integrals.

If $f \in L_2\big(\Omega \times (0,T); H\big)$, then

$$\int_0^t |\big(f(s), v(s)\big)_H|ds \leq \int_0^t \|f(s)\|_H^2 ds + \int_0^t \|v(s)\|_H^2 ds.$$

For each of the three stochastic integrals, we first apply the BDG inequality with $p = 1$, then we take $\sup_t \|v(t)\|_H$ out of the integral and finally apply the epsilon inequality. Condition [HP4′] ensures convergence of the sums, and the Gronwall inequality removes the integrals of u and v. For example,

$$\begin{aligned}
\mathbb{E}\sup_t\left|\int_0^t\sum_k\big(\mathrm{M}_k u(s),v(s)\big)_H dw_k(s)\right| &\leq C\mathbb{E}\left(\int_0^T\sum_k\big(\mathrm{M}_k u(s),v(s)\big)_H^2 ds\right)^{1/2}\\
&\leq C\mathbb{E}\left(\int_0^T\sum_k\|\mathrm{M}_k u(s)\|_H^2\,\|v(s)\|_H^2 ds\right)^{1/2}\\
&\leq C\mathbb{E}\left(\sup_t\|v(t)\|_H^2\sum_k\int_0^T\|\mathrm{M}_k u(s)\|_H^2 ds\right)^{1/2}\\
&\leq \varepsilon\mathbb{E}\sup_t\|v(t)\|_H^2 + C\mathbb{E}\sum_k\int_0^T\|\mathrm{M}_k u(s)\|_H^2 ds\\
&\leq \varepsilon\mathbb{E}\sup_t\|v(t)\|_H^2 + C\mathbb{E}\int_0^T\|u(s)\|_V^2 ds.
\end{aligned}$$

For the Itô correction term,

$$\begin{aligned}
&\sum_k\int_0^t\|\mathrm{N}_k u(s)+\mathrm{M}_k v(s)+g_k(s)\|_H^2 ds\\
&\qquad\leq 3\sum_k\int_0^t\Big(\|\mathrm{N}_k u(s)\|_H^2+\|\mathrm{M}_k v(s)\|_H^2+\|g_k(s)\|_H^2\Big)ds,
\end{aligned}$$

and condition [HP5′], together with the Gronwall inequality, provides the necessary bounds. This completes the proof of the a priori bound.

Step 2: The Galerkin Approximation (pages 188–189) Let $\{\mathfrak{h}_k,\ k\geq 1\}$ be an orthonormal basis in H that is also a dense set in V and V'; Π^N is the orthogonal in H projection onto the the linear span of $\{\mathfrak{h}_1,\dots,\mathfrak{h}_N\}$. For $y\in V'$,

$$\Pi^N y=\sum_{k=1}^N[y,\mathfrak{h}_k]\mathfrak{h}_k.$$

The sequence $(u^N,v^N),\ N\geq 1$, approximating the solution satisfies

$$\begin{aligned}
du^N &= v^N dt,\\
dv^N &= \big(\Pi^N \mathrm{A}u^N+\Pi^N\mathrm{A}_1 u^N+\Pi^N\mathrm{B}v^N+\Pi^N f\big)dt\\
&\quad+\sum_{k=1}^N\big(\Pi^N\mathrm{M}_k u^N+\Pi^N\mathrm{N}_k v^N+\Pi^N g_k\big)dw_k,\ 0<t\leq T,
\end{aligned}\tag{4.3.8}$$

$u^N(0) = \Pi^N u_0$, $v^N(0) = \Pi^N v_0$. Writing $u^N(t) = \sum_{k=1}^N \bar{u}^N(t)\mathfrak{h}_k$, $v^N(t) = \sum_{k=1}^N \bar{v}^N(t)\mathfrak{h}_k$ we interpret (4.3.8) as a $2N$-dimensional linear SODE for the functions $\bar{u}_k^N, \bar{v}_k^N$ and conclude existence and uniqueness of (u^N, v^N) for every N. Also, every $h \in \Pi^N H$ satisfies $h = \Pi^N h$ and therefore

$$[\Pi^N \mathrm{A}h, h_1)] = (\mathrm{A}h, h_1)_H = [\Pi^N \mathrm{A}h_1, h], \quad \|\Pi^N \mathrm{A}_1 h\|_H \leq \|\mathrm{A}_1 h\|_H,$$
$$\|\Pi^N \mathrm{B}h\|_H \leq \|\mathrm{B}h\|_H, \quad \|\Pi^N \mathrm{M}_k h\|_H \leq \|\mathrm{M}_k h\|_H, \quad \|\Pi^N \mathrm{N}_k h\|_H \leq \|\mathrm{N}_k h\|_H.$$

Consequently, the pair (u^N, v^N) satisfies (4.3.6) with $C(T)$ independent of N (to be absolutely sure, the reader should take a closer look at $\|\Pi^N u_0\|_V$: it is not necessarily true that $\|\Pi^N u_0\|_V \leq \|u_0\|_V$, but the convergence $\lim_{N\to\infty} \|u_0 - \Pi^N u_0\|_V = 0$ is good enough).

Define the Hilbert space

$$\mathbb{H} = L_2\big(\Omega \times (0,T); V\big) \times L_2\big(\Omega \times (0,T); H\big).$$

The a priori estimate means that the sequence $\{(u^N, v^N,\ N \geq 1)\}$ is weakly compact in $\mathbb{H}$ and therefore has a weakly converging sub-sequence. The limit (u, v) of this subsequence is, by definition, a solution we want. (Again, to be completely sure, the reader should write out explicitly what weak convergence in $\mathbb{H}$ means and why the limit is indeed a solution of (4.3.1) in the sense of Definition 4.3.2.) Note also that

$$u \in L_2\big(\Omega; L_\infty((0,T); V)\big) \bigcap L_2\big(\Omega; \mathcal{C}((0,T); H)\big),$$
$$v \in L_2\big(\Omega; L_\infty((0,T); H)\big) \bigcap L_2\big(\Omega; \mathcal{C}((0,T); V')\big).$$

Indeed, the a priori estimate ensures boundedness of u in V, while the equality $u(t) = u_0 + \int_0^t v(s)ds$ ensures continuity in H. Similarly, the a priori estimate ensures boundedness of v in H and (4.3.3) implies continuity of v in V'.

Step 3: Continuity of the Solution (pages 189–191) To prove continuity of u in V and of v in H, we follow the argument of Lions and Magenes [133, End of Section 3.8.4]. We need to show that, for each t, the sample trajectories of u and v satisfy

$$\lim_{r\to 0} \big(\|u(t+h) - u(t)\|_V + \|v(t+h) - v(t)\|_H\big) = 0. \tag{4.3.9}$$

To begin, we apply the energy equality (Theorem 3.3.15, page 139) to v on the interval $(t, t+r)$ for fixed $t > 0$ and r in a small neighborhood of zero. The result is

$$-\big([\mathrm{A}(t+r)u(t+r), u(t+r)] - [\mathrm{A}(t)u(t), u(t)]\big) + \|v(t+r)\|_H^2 - \|v(t)\|_H^2$$
$$= \int_t^{t+r} (\ldots)ds + \sum_k \int_t^{t+r} (\ldots)_k\, dw_k(s). \tag{4.3.10}$$

Now we have to keep track of time dependence in the operator A to consider expressions of the type $[\mathrm{A}(t_1)u(t_2), v(t_3)]$. Accordingly, for $x, y \in V$, we take the number M from (4.3.4) and define

$$a(t, x, y) = -[\mathrm{A}(t)x, y] + M(x, y)_H.$$

Then (4.3.10) and continuity of u in H imply that the (real-valued) function $F = F(t)$ defined by

$$F(t) = a\big(t, u(t), u(t)\big) + \|v(t)\|_H^2$$

is continuous in t.

Next, for a fixed $t > 0$ and for r in a small neighborhood of zero, define the function

$$\Phi(r) = a\big(t + r, u(t + r) - u(t), u(t + r) - u(t)\big) + \|v(t + r) - v(t)\|_H^2.$$

Note that (4.3.4) implies

$$0 \le c_A \|u(t + r) - u(t)\|_V^2 + \|v(t + r) - v(t)\|_H^2 \le \Phi(r).$$

Therefore, we will have (4.3.9) if we show that

$$\lim_{r \to 0} \Phi(r) = 0. \tag{4.3.11}$$

To prove (4.3.11), we expand all the inner products, rearrange some terms, and get

$$\begin{aligned}
\Phi(r) &= F(t + r) + F(r) - 2\big(v(t_r), v(t)\big)_H \\
&\quad - 2\Big(a\big(t + r, u(t + r), u(t)\big) - a\big(t, u(t + r), u(t)\big)\Big) \\
&\quad + \Big(a\big(t + r, u(t), u(t)\big) - a\big(t, u(t), u(t)\big)\Big) \\
&\quad - 2a\big(t, u(t + r), u(t)\big).
\end{aligned} \tag{4.3.12}$$

By continuity of F, $\lim_{r \to 0} F(t + r) = F(t)$. By the definition of a,

$$\begin{aligned}
&|a\big(t + r, u(t + r), u(t)\big) - a\big(t, u(t + r), u(t)\big)| \\
&\qquad\qquad \le \|\big(\mathrm{A}(t + r) - \mathrm{A}(t)\big)u(t + r)\|_{V'} \|u(t)\|_V, \\
&|a\big(t + r, u(t), u(t)\big) - a\big(t, u(t), u(t)\big)| \le \|\big(\mathrm{A}(t + r) - \mathrm{A}(t)\big)u(t)\|_{V'} \|u(t)\|_V
\end{aligned}$$

By the mean-value theorem,

$$\|\big(\mathrm{A}(t+r)-\mathrm{A}(t)\big)u(t+r)\|_{V'} = r\|\dot{\mathrm{A}}(t+\tilde{r})u(t)\|_{V'} \le Cr\|u(t+r)\|_{V}.$$

As a result,

$$\begin{aligned}&|a\big(t+r,u(t+r),u(t)\big)-a\big(t,u(t+r),u(t)\big)|\\&+|a\big(t+r,u(t),u(t)\big)-a\big(t,u(t),u(t)\big)| \le Cr\sup_t \|u(t)\|_V^2 \to 0,\ r\to 0.\end{aligned}$$

It remains to pass to the limit $r \to 0$ in the expressions $2\big(v(t+r),v(t)\big)_H$ and $2a\big(t,u(t+r),u(t)\big) = 2[\mathrm{A}(t)u(t),u(t+r)]$. If indeed

$$\begin{aligned}&\lim_{r\to 0}\big(v(t_r),v(t)\big)_H = \|v(t)\|_H^2,\\&\lim_{r\to 0}[\mathrm{A}(t)u(t),u(t+r)] = [\mathrm{A}(t)u(t),u(t)] = a\big(t,u(t),u(t)\big),\end{aligned}\tag{4.3.13}$$

then, passing to the limit $r \to 0$ in (4.3.12) yields

$$\lim_{r\to 0}\Psi(r) = 2F(t) - 2\Big(\|v(t)\|_H^2 + a\big(t,u(t),u(t)\big)\Big) = 0,$$

completing the proof.

With t fixed, the convergence in (4.3.13) would follow from the weak continuity of v in H (when the dual of H is H) and weak continuity of u in V (when the dual of V is V'). It turns out that both the weak continuity of u in V and of v in H follow from a general result somewhat similar to Problem 3.1.3(b): if a Hilbert space X is continuously embedded into a Hilbert space Y, then weak continuity in Y and boundedness in X imply weak continuity in X; see Lions and Magenes [133, Lemma 3.8.1]. Retracing our steps, from (4.3.13) we get (4.3.11), and from (4.3.11), (4.3.9). Note that our disregard for different modifications of the same process was a major simplification; a completely rigorous proof of continuity should follow the same lines as in Krylov and Rozovskii [122, Sect. 2.3]. We leave it to the reader as a challenge: to the best of our knowledge, nobody has ever done it.

This concludes the proof of Theorem 4.3.3.

Exercise 4.3.4 (B) Verify that one can always take $M = 0$ in (4.3.4).

Hint. Replace A with $\mathrm{A}+M\mathrm{I}$ and A_1, with $\mathrm{A}_1 - M\mathrm{I}$.

There are several advantages of considering the abstract evolution equation (4.3.1) rather than a concrete equation such as $u_{tt} = (a(t,x)u_x)_x + u\dot{W}(t,x)$:

1. The ability to write a very general equation in a relatively compact form;
2. The possibility to apply the result to the same equation in different normal triples (V,H,V');

3. The ability to circumvent various issues related to hard analysis, such as verification of ellipticity for specific partial differential operators, and to concentrate on the general features of the equation.

There are also some obvious disadvantages:

1. The conclusions of the theorem are usually not sharp for concrete equations;
2. Verification of the conditions can be rather difficult;
3. Essentially only one type of the representation of the noise is allowed by the theorem. For example, it is not at all clear how to apply the theorem to the equation $u_{tt} = u_{xx} + \dot{W}^{Q}(t,x)$ if all we know about $\dot{W}^{Q}(t,x)$ is that

$$\mathbb{E}\dot{W}^{Q}(t,x)\dot{W}^{Q}(s,y) = \delta(t-s)\varphi(|x-y|),$$

where the function φ has compact support and $\varphi(r) \sim r^{-1/2}$ as $r \to 0$.

We illustrate some of these points below. The main concrete example covered by Theorem 4.3.3 is equation

$$\begin{aligned} u_{tt}(t,x) &= \sum_{i,j=1}^{\mathrm{d}} \big(a_{ij}(t,x)u_{x_i}(t,x)\big)_{x_j} + \sum_{i=1}^{\mathrm{d}} \big(b_i(t,x)u_{x_i}(t,x) + \tilde{b}_i(t,x)u_{tx_i}(t,x)\big) \\ &\quad + c(t,x)u(t,x) + \tilde{c}(t,x)u_t(t,x) + f(t,x) \\ &\quad + \sum_{k\geq 1} \Big(\sigma_{ik}(t,x)u_{x_i}(t,x) \\ &\quad + \nu_k(t,x)u(t,x) + \tilde{\nu}(t,x)u_t(t,x) + g_k(t,x)\Big)\dot{w}_k(t) \end{aligned} \tag{4.3.14}$$

when $x \in G$ and $G = \mathbb{R}^{\mathrm{d}}$ or G is a smooth bounded domain with zero boundary conditions. Other concrete examples can be more exotic equations such as $u_{tt} = \sqrt{\boldsymbol{\Delta}^2 - \boldsymbol{\Delta}}\, u + \sqrt{1-\boldsymbol{\Delta}}u_t\dot{W}$.

To apply the theorem to Eq. (4.3.14) in the normal triple of Sobolev spaces $(H^{r+1}(G), H^{r}(G), H^{r-1}(G))$, $r \in \mathbb{R}$, it is enough to assume that all the coefficients are uniformly bounded, measurable, and adapted, and

- $a_{ij}(t,\cdot)$, $\partial a_{ij}(t,\cdot)/\partial t$, $b_i(t\cdot)$, $\tilde{b}_i(t,\cdot)$, $c(t,\cdot)$, $\tilde{c}(t,\cdot)$, $\nu_k(t,\cdot)$, $\tilde{\nu}_k(t,\cdot)$, $\sigma_{ik}(t,\cdot) \in \mathcal{C}^{|r|+1}(G)$,
- $\sum_{i,j=1}^{\mathrm{d}} a_{ij}(t,x)y_iy_j \geq c_A|y|^2$.
- $\sum_k \big(\sup_{\omega,t,x} |D_x^m\sigma_{ik}(t,x)| + \sup_{\omega,t,x} |D_x^m\nu_k(t,x)| + \sup_{\omega,t,x} |D_x^m\tilde{\nu}_k(t,x)|\big) < \infty$, $m = 0,\ldots,|r|+1$

We also assume that initial conditions u_0 and v_0 are $\mathcal{F}_0$-measurable and satisfy $u_0 \in L_2(\Omega; H^{r+1}(G))$, $v_0 \in L_2(\Omega; H^{r}(G))$. In particular, if the coefficients and the input are as regular as we want, then we can take r as large as we want, and, together with the Sobolev embedding theorem, Theorem 4.3.3 almost immediately implies that the function $u = u(t)$ is continuously differentiable in time and infinitely differentiable in space. Proving existence, uniqueness and regularity of the solution of (4.3.14) from scratch would require at least as much work as the proof of Theorem 4.3.3.

To illustrate some difficulties related to applying Theorem 4.3.3 to a concrete equation, let us consider the wave equation driven by a Q-cylindrical Wiener process:

$$u_{tt} = \boldsymbol{\Delta} u + u_t \dot{W}^{\mathrm{Q}}. \tag{4.3.15}$$

One can think of (4.3.15) as the wave equation with random damping.

Exercise 4.3.5 (C) Integrating by parts, verify that if $u_{tt} = u_{xx} - au_t$, $x \in \mathbb{R}$, then

$$\frac{d}{dt}\int_{\mathbb{R}} (u_t^2 + u_x^2)dx = -a\int_{\mathbb{R}} u_t^2 dx.$$

Let us further assume that

$$\dot{W}^{\mathrm{Q}}(t,x) = \sum_{k\geq 1} \sqrt{q_k}\mathfrak{m}_k(x)\dot{w}_k(t),$$

where $\{\mathfrak{m}_k,\ k \geq 1\}$ is an orthonormal basis in $L_2(G)$. Without this assumption we will be unable to apply Theorem 4.3.3. On the other hand, as long as the operator Q is nuclear, the assumption leads to no loss of generality, although it might be hard to find the numbers q_k explicitly.

In other words, we make Eq. (4.3.15) a particular case of (4.3.1) with $f = g_k = 0$, $\mathrm{A} = \boldsymbol{\Delta}$, $\mathrm{A}_1 = \mathrm{B} = \mathrm{M}_k = 0$, and $\mathrm{N}_k u = \sqrt{q_k}\,\mathfrak{m}_k$. To satisfy condition [HP4′], that is

$$\sum_k \|\mathrm{N}_k h\|_r^2 \leq C_o\|h\|_r^2, \tag{4.3.16}$$

we need $\mathfrak{m}_k$ to be a point-wise multiplier in H^r. With the notation

$$\|\mathfrak{m}_k\|_{r\to r} = \sup_{h\in H^r(G):\|h\|_r\leq 1} \|\mathfrak{m}_k h\|_r,$$

inequality (4.3.16) becomes

$$\sum_k q_k\|\mathfrak{m}_k\|_{r\to r}^2 < \infty.$$

While this convergence can certainly be achieved by making q_k tend to zero fast enough, more precise bounds on q_k are connected with the theory of point-wise multipliers in Sobolev spaces and with the asymptotic behavior of the sup-norms of the functions $\mathfrak{m}_k$.

Let us consider the easiest case $r = 0$. Then $\|\mathfrak{m}_k\|_{0\to 0} = \sup_{x\in G}|\mathfrak{m}_k(x)|$. First we consider (4.3.15) in a smooth bounded domain $G \subset \mathbb{R}^{\mathrm{d}}$ with zero boundary conditions. As a further simplification, we assume that $\mathfrak{m}_k$, $k \geq 1$ are normalized ($\|\mathfrak{m}_k\|_{L_2(G)} = 1$) eigenfunctions of the Laplacian $\boldsymbol{\Delta}$: $\boldsymbol{\Delta}\mathfrak{m}_k = -\lambda_k\mathfrak{m}_k$. Then it is known that

- $\sup_{x\in G}|\mathfrak{m}_k(x)| \leq C\lambda_k^{(\mathrm{d}-1)/4}$: Grieser [66];
- $\lambda_k \sim k^{2/\mathrm{d}}$: Safarov and Vassiliev [201, Sect. 1.2].

Therefore, $\|\mathfrak{m}_k\|_{0\to 0} \leq Ck^{(\mathrm{d}-1)/(2\mathrm{d})}$, and Eq. (4.3.15) in a smooth bounded domain with zero boundary conditions and initial conditions $u_0 \in H^1(G)$, $v_0 \in L_2(G)$ will have a solution $u \in L_2\big(\Omega; \mathcal{C}((0,T); H^1(G))\big)$ as long as $\sum_{k\geq 1} q_k k^{(\mathrm{d}-1)/\mathrm{d}} < \infty$, for example, if $q_k = 1/k^2$. Note that the rate of decay of q_k is essentially independent of d.

Now let us consider Eq. (4.3.15) in $\mathbb{R}$. This time, we take $\mathfrak{m}_k(x) = \mathfrak{w}_k(x)$, the Hermite functions (4.1.50) on page 167. It is known that $\sup_x|\mathfrak{w}_n| \leq Cn^{-1/12}$: the standard reference is Hille and Phillips [77, Sect. 21.3]. As a result, Eq. (4.3.15) in $\mathbb{R}$ with initial conditions $u_0 \in H^1(\mathbb{R})$, $v_0 \in L_2(\mathbb{R})$ will have a solution $u \in L_2\big(\Omega; \mathcal{C}((0,T); H^1(\mathbb{R}))\big)$ as long as $\sum_{k\geq 1} q_k k^{-1/6} < \infty$, for example, if $q_k = 1/k$. By the Sobolev embedding theorem, the function u will have sample trajectories that are continuous in time and $1/2$-Hölder continuous in space.

Note that Theorem 4.3.3 does not satisfactorily cover equations driven by space-time white noise: even for the equation

$$u_{tt} = u_{xx} + \dot{W}(t,x),\ x \in [0,\pi],$$

with zero boundary conditions, we find $g_k = \sqrt{2/\pi}\sin(kx)$, which means that condition

$$\sum_k \|g\|^2_{L_2((\Omega\times[0,T];H))} < \infty$$

holds for $H = H^r((0,\pi))$ and $r < -1/2$. Therefore, Theorem 4.3.3 ensures only $u(t,\cdot) \in H^{r+1}((0,\pi))$, which, with $r+1 < 1/2$, does not even imply continuity of u, much less Hölder continuity as derived from the closed-form solution in Sect. 2.3.2.

4.3.2 Further Directions

Each of the following equations is hyperbolic:

$$u_t = u_x,\ t > 0,\ x \in G \subseteq \mathbb{R}; \tag{4.3.17}$$

$$u_{tx} = 0,\ t > 0,\ x \in G \subseteq \mathbb{R}; \tag{4.3.18}$$

$$u_{tt} = u_{xx},\ t > 0,\ x \in G \subseteq \mathbb{R}. \tag{4.3.19}$$

Accordingly, each of the equations has a stochastic counterpart and a generalization to more than two independent variables.

Recall that (4.3.18) and (4.3.19) are essentially the same, as one turns into another after a rotation of the (t, x) plane by $\pi/4$: see our study of the stochastic wave equation on page 57. The transport equation (4.3.17) can be thought of as "one-half" of the wave equation (4.3.19) by factoring the operator

$$\frac{\partial^2}{\partial t^2} - \frac{\partial^2}{\partial x^2} = \left(\frac{\partial}{\partial t} - \frac{\partial}{\partial x}\right)\left(\frac{\partial}{\partial t} + \frac{\partial}{\partial x}\right).$$

In other words, if we agree that the wave equation (4.3.19) is hyperbolic, then we should agree that Eqs. (4.3.17) and (4.3.18) are natural to call hyperbolic as well. The interesting developments start once we try to go beyond two independent variables, as the corresponding extensions in the three cases are very different.

Our discussion has been focused on the (abstract version of) wave equation (4.3.19): if $x \in \mathbb{R}^{\mathrm{d}}$, then, instead of u_{xx} we consider $\boldsymbol{\Delta} u$ or more generally, $\mathrm{A}u$, where $-\mathrm{A}$ is an elliptic operator. In other words, we still have an equation that is second-order in time.

In contrast, the natural extension of (4.3.18) to $\mathrm{d} \geq 3$ independent variables is

$$\frac{\partial^{\mathrm{d}} u}{\partial t_1 \partial t_2 \ldots \partial t_{\mathrm{d}}} = 0, \tag{4.3.20}$$

and this equation has order d. The reason for denoting the independent variables by t_k will become clear in a moment. This extension is especially convenient in connection with equations driven by multi-parameter white noise. As we saw in (3.2.19), page 108, given a Gaussian field $W = W(t_1, \ldots, t_{\mathrm{d}})$,

$$\dot{W} = \frac{\partial^{\mathrm{d}} W}{\partial t_1 \partial t_2 \ldots \partial t_{\mathrm{d}}}$$

in the sense of distributions, and therefore the solution of

$$\frac{\partial^{\mathrm{d}} u}{\partial t_1 \partial t_2 \ldots \partial t_{\mathrm{d}}} = \dot{W} \tag{4.3.21}$$

is $u(t_1, \ldots, t_{\mathrm{d}}) = W(t_1, \ldots, t_{\mathrm{d}})$. We can now take the idea further. To simplify the notations, write

$$\boldsymbol{t} = (t_1, \ldots, t_{\mathrm{d}}),\ \int_{\boldsymbol{0}}^{\boldsymbol{t}} f(\boldsymbol{s}) d\boldsymbol{s} = \int_0^{t_{\mathrm{d}}} \ldots \int_0^{t_1} f(s_1, \ldots, s_{\mathrm{d}}) ds_1, \ldots ds_{\mathrm{d}}.$$

In other words, we think of $\mathbf{t}$ as a d-dimensional time parameter, with all the components t_k having the same significance (unlike the d-dimensional wave equation, where, out of d + 1 variables, scalar time t plays a special role). Then, we generalize (4.3.21) to

$$u(\mathbf{t}) = u(\mathbf{0}) + \int_{\mathbf{0}}^{\mathbf{t}} b\big(\mathbf{s}, u(\mathbf{s})\big)d\mathbf{s} + \int_{\mathbf{0}}^{\mathbf{t}} \sigma\big(\mathbf{s}, u(\mathbf{s})\big)dW(\mathbf{s}). \tag{4.3.22}$$

Equation (4.3.22) describes a **multi-parameter** process, but, except for the bold symbols, looks exactly like a stochastic ordinary differential equation and can be studied in exactly the same way, once the questions of measurability and adaptedness are worked out. For details see Khosnevisan [108, Sect. 7.4]. In particular, global Lipschitz continuity and boundedness of $b(\mathbf{s}, \cdot)$ and $\sigma(\mathbf{s}, \cdot)$ imply existence and uniqueness of solution of (4.3.22) [108, Theorem 7.4.3.1].

There is no need to stop with existence and uniqueness of solution, and indeed many other problems that have been studied for stochastic ordinary differential equations have also been studied for Eq. (4.3.22):

- **Large deviations** (in the spirit of Freidlin and Wentzell [56]): Boufoussi et al. [15], Eddahbi [45];
- **Martingale problems**: Ondreját [183];
- **Potential theory** (study of probability that the solution hits a certain set): Dalang and Nualart [33];
- **Regularity of the probabilistic law of the solution** (Malliavin calculus): Millet Sanz-Solé [168], Rovira and Sanz-Solé [197], Sanz-Solé and Torrecilla-Tarantino [206];
- **Support theorem** (in the spirit of Stroock and Varadhan [213]): Boufoussi [14], Millet and Sanz-Solé [167], Rovira and Sanz-Solé [196].

Now let us discuss the transport equation (4.3.17). The stochastic version of this equation,

$$u_t = u_x + a(t,x) + \sigma(t,x)\dot{W}(t,x),$$

where $\dot{W}$ is space-time white noise, is a popular model in the study of term structure of interest rates [22, Sect. 7.2.1], [216]. Making noise multiplicative, such as $u_t = u_x\dot{W}$ or even $u_t = u_x + u\dot{W}$ is impossible in the framework of square-integrable solutions: u is not regular enough to act as a point-wise multiplier for $\dot{W}$. As a result, further generalizations of the equation are carried out in the Stratonovich formulation. In our study of the stochastic heat equation $u_t = au_{xx} + u_x\dot{w}(t)$, we saw that, when $a = 1/2$, the equation is equivalent to $u_t = u_x \circ \dot{w}(t)$; see (2.3.12) on page 54. For more, see Chow [24, Chap. 2] and Kunita [125, Sect. 6.1].

There is another direction in which Eq. (4.3.17) can be generalized, and that is `hyperbolic systems` [50, Sect. 11.1]. An example is

$$u_t = u_x + v_x,\ v_t = u_x - v_x,\ t > 0,\ x \in G \subseteq \mathbb{R},$$

although most of the physically relevant examples are non-linear. Stochastic counterparts of these hyperbolic systems have been studied, for example, by Chow [25] and Marcus and Mizel [154].

As a final comment, we mention that the stochastic version of the wave equation (4.3.19) can be extended beyond the setting of our Theorem 4.3.3, for example by introducing dependence on the past [111], or by considering a different random perturbation, such as Lévy noise [110], [185, Chap. 13].

4.3.3 Problems

Problem 4.3.1 gives alternative conditions on the function f and the operator B in Theorem 4.3.3. Problem 4.3.2 connects stochastic parabolic equations with an infinite system of independent second-order SODEs. Problem 4.3.3 lists several specific hyperbolic equations in one space variable.

Problem 4.3.1 Prove Theorem 4.3.3 under the following modifications of the conditions:

(a) $f \in L_2\big(\Omega; H^1((0,T);V')\big)$ instead of $f \in L_2(\Omega \times (0,T);H)$. The space $L_2\big(\Omega; H^1((0,T);V')\big)$ is the collection of V'-valued processes f such that $f(t) = f(0) + \int_0^t g(s)ds$, $f_0 \in L_2(\Omega;V')$, $g \in L_2(\Omega \times (0,T);V')$; it is a Hilbert space with norm

$$\|f\|^2_{H^1(\Omega\times(0,T);V')} = \mathbb{E}\|f_0\|^2_{V'} + \mathbb{E}\|g\|^2_{L_2((0,T);V')}.$$

(b) $(\mathrm{B}v, v)_H \le C\|v\|^2_H$ instead of $|(\mathrm{B}v, v)_H| \le C\|v\|^2_H$. In particular, equation

$$u_{tt} = u_{xx} + u_{txx} + \dot{W}(t,x)$$

becomes admissible.

Problem 4.3.2 Let w_k, $k \ge 1$, be independent standard Brownian motions and let $a_k, b_k, c_k\sigma_k$, $k \ge 1$, be real numbers. Consider a collection of processes u_k, $k \ge 1$, defined by

$$a_k\ddot{u}_k(t) + b_k\dot{u}_k(t) + c_k u_k(t) = \sigma_k \dot{w}_k(t),\ 0 < t \le T. \tag{4.3.23}$$

Find conditions on the numbers a_k, b_k, σ_k and the initial data $u_k(0)$, $\dot{u}_k(0)$ to have

$$\text{(a)}\ \mathbb{E}\sup_{0<t<T}\sum_{k\ge1}|u_k(t)|^2 < \infty,\quad \text{(b)}\ \mathbb{E}\sup_{0<t<T}\sum_{k\ge1}|\dot{u}_k(t)|^2 < \infty. \tag{4.3.24}$$

Problem 4.3.3 Investigate the following equations, both in $\mathbb{R}$ and on $(0, \pi)$:

$$u_{tt} = u_{xx} + au_x + bu + (cu + \sigma u_x + \nu u_t + g)\dot{W}^Q(t, x), \tag{4.3.25}$$

$$u_{tt} = u_{xx} + u_{txx} + (au + bu_x + \nu u_t + g)\dot{W}^Q(t, x), \tag{4.3.26}$$

$$u_{tt} = -u_{xxxx} + au_{xx} + (bu + cu_x + \sigma u_{xx} + \nu u_t + g)\dot{W}^Q(t, x), \tag{4.3.27}$$

$$u_{tt} = u_{xx} + u_{txxx} + (au + bu_x + \nu u_t + g)\dot{W}^Q(t, x), \tag{4.3.28}$$

$$u_{tt} - au_{ttxx} = bu_{xx} + (cu + \sigma u_x + \nu u_t + g)\dot{W}^Q(t, x). \tag{4.3.29}$$

4.4 Stochastic Parabolic Equations

4.4.1 Existence and Uniqueness of Solution

We fix the stochastic basis $(\Omega, \mathcal{F}, \{\mathcal{F}_t\}_{t\geq 0}, \mathbb{P})$ with the usual assumptions, and a collection $\{w_k, k \geq 1\}$ of independent standard Brownian motions on this basis. Our objective in this section is to study stochastic evolution equation

$$\dot{u}(t) = \mathrm{A}(t)u(t) + f(t) + \sum_{k\geq 1} \big(\mathrm{M}_k(t)u(t) + g_k(t)\big)\dot{w}_k(t) \tag{4.4.1}$$

$0 < t \leq T$, $u(0) = u_0$, in the normal triple (V, H, V') of Hilbert spaces.

To begin, recall Definition 4.3.1:

Let X, Y be separable Banach spaces and let $\mathrm{A} = \{\mathrm{A}(\omega, t), \omega \in \Omega, 0 \leq t \leq T\}$ be a family of mappings from X to Y. The family is called (X, Y)-adapted if, for every $x \in X$ and $\ell \in Y'$, the real-valued process $t \mapsto \ell(\mathrm{A}(t)x)$ is $\mathcal{F}_t$-adapted. The family is called $\mathrm{L}_\infty(X, Y)$-uniformly bounded if there exists a positive non-random number C such that, for all $x \in X$, all $t \in [0, T]$, and all $\omega \in \Omega$, $\|\mathrm{A}(t, \omega)x\|_Y \leq C\|x\|_X$.

Also, recall that $[u, v]$, $u \in V$, $v \in V'$, denotes the duality between V and V' relative to the inner product in H.

To define a square-integrable variational solution of equation (4.4.1), we make the following assumptions:

[PR1] The families of operators $\mathrm{A} = \{\mathrm{A}(t),\ 0 \leq t \leq T\}$ and $\mathrm{M}_k = \{\mathrm{M}_k(t),\ 0 \leq t \leq T\}$ are (V, V')-adapted and $L_\infty(V, V')$-uniformly bounded.

[PR2] The initial condition u_0 is $\mathcal{F}_0$-measurable (equivalently, independent of all w_k), $u_0 \in L_2(\Omega, H)$, the process f and each of the processes g_k are $\mathcal{F}_t$-adapted and are elements of $L_2\big(\Omega \times (0, T); V'\big)$.

[PR3] $\sum_{k\geq 1} \mathbb{E} \int_0^T \|g_k(t)\|_{V'}^2 dt < \infty$.

[PR4] There exists a positive number C_o such that, for every $v \in L_2\big(\Omega \times (0, T); V\big)$

$$\sum_{k\geq 1} \mathbb{E} \int_0^T \|\mathrm{M}_k(t)v(t)\|_{V'}^2 dt < \infty. \tag{4.4.2}$$

Definition 4.4.1 (Solution of Stochastic Parabolic Equations) The variational solution of (4.4.1) is an $\mathcal{F}_t$-adapted process $u = u(t)$, such that $u \in L_2\big(\Omega\times(0,T);V\big)$ and

$$u(t) = u_0 + \int_0^t \mathrm{A}u(s)ds + \int_0^t f(s)ds + \sum_{k\geq 1}\int_0^t \big(\mathrm{M}_k u(s) + g_k(s)\big)dw_k(s) \tag{4.4.3}$$

in $L_2\big(\Omega \times (0,T); V'\big)$.

To construct a solution of (4.4.1), we have to modify the assumptions about the operators A and M_k as follows:

[PR1′] The family of operators $\mathrm{A} = \{\mathrm{A}(t),\ 0 \leq t \leq T\}$ is (V, V')-adapted and $L_\infty(V, V')$-uniformly bounded.
[PR2′] For each k, the family of operators $\mathrm{M}_k = \{\mathrm{M}_k(t),\ 0 \leq t \leq T\}$ is (V,H)-adapted and $L_\infty(V,H)$-uniformly bounded.
[PR3′] (`stochastic parabolicity condition`) There exist a positive number c_A and a real number M such that, for all $\omega \in \Omega$, $t \in [0,T]$, and $v \in V$,

$$2[\mathrm{A}(t)v, v] + \sum_{k\geq 1} \|\mathrm{M}_k(t)v\|_H^2 + c_A\|v\|_V^2 \leq M\|v\|_H^2. \tag{4.4.4}$$

The conditions on u_0, f, g_k are in the statement of Theorem 4.4.3 below.

Exercise 4.4.2 (C) Verify that inequalities $\|\mathrm{A}v\|_{V'} \leq C\|v\|_V$ and (4.4.4) together imply existence of $C_M > 0$ such that, for all ω, t, v,

$$\sum_{k\geq 1} \|\mathrm{M}_k v\|_H^2 \leq C_M\|v\|_V^2. \tag{4.4.5}$$

Hint. Remember that $|\,[\mathrm{A}(t)v, v]| \leq \|\mathrm{A}v\|_{V'}\|v\|_V$ and $\|v\|_H \leq C\|v\|_V$.

Theorem 4.4.3 (Solvability of Stochastic Parabolic Equations) *Assume that* [PR1′]–[PR3′] *hold.*

Then, for every $\mathcal{F}_0$-measurable initial condition $u(0) = u_0 \in L_2(\Omega; H)$, and $\mathcal{F}_t$-adapted free terms f, g_k with $f \in L_2\big(\Omega \times (0,T); V'\big)$ and g_k satisfying $\sum_k \|g_k\|^2_{L_2(\Omega\times(0,T);H)} < \infty$, Eq. (4.4.1) has a unique variational solution. The solution is an element of $L_2\big(\Omega; \mathcal{C}\big((0,T);H\big)\big)$ and

$$\begin{aligned}\|u\|^2_{L_2(\Omega;\mathcal{C}((0,T);H))} + \|u\|^2_{L_2(\Omega\times(0,T);V)} \leq C(T)\Big(&\|u_0\|^2_{L_2(\Omega;H)} \\ &+ \|f\|^2_{L_2(\Omega\times(0,T);V')} + \sum_{k\geq 1}\|g_k\|^2_{L_2(\Omega\times(0,T);H)}\Big).\end{aligned} \tag{4.4.6}$$

Proof There are two steps: (1) derivation of the a priori bound (4.4.6); (2) construction of the solution using the Galerkin approximation. Uniqueness of the solution and continuity of the solution in H follow from Theorem 3.3.15 on page 139.

Step 1: The A Priori Estimate (pages 200–203) An equivalent form of (4.4.6) is

$$\begin{aligned}\mathbb{E}\sup_{0<t<T}\|u(t)\|_H^2 + \mathbb{E}\int_0^T \|u(t)\|_V^2 dt \le C(T)\Big(\mathbb{E}\|u_0\|_H^2 \\ + \mathbb{E}\int_0^T \|f(t)\|_{V'}^2 dt + \sum_{k\ge1}\mathbb{E}\int_0^T \|g_k(t)\|_H^2 dt\Big).\end{aligned} \tag{4.4.7}$$

To derive (4.4.7), we apply the energy equality (Theorem 3.3.15) to the process u:

$$\begin{aligned}\|u(t)\|_H^2 = \|u_0\|_H^2 + 2\int_0^t [\mathrm{A}u(s), u(s)]ds + 2\int_0^t [f(s), u(s)]ds \\ + 2\sum_k \int_0^t \Big(\big(\mathrm{M}_k u(s), u(s)\big)_H + \big(g_k(s), u(s)\big)_H\Big)dw_k(s) \\ + \sum_k \int_0^t \|\mathrm{M}_k u(s) + g_k(s)\|_H^2 ds.\end{aligned} \tag{4.4.8}$$

Note that, under assumptions [PR1′]–[PR3′], the solution u, if exists, satisfies the conditions of Theorem 3.3.15.

Let us look at the individual terms on the right-hand side of (4.4.8), starting with the Itô correction term:

$$\begin{aligned}\sum_k \int_0^t \|\mathrm{M}_k u(s) + g_k(s)\|_H^2 ds = \sum_k \int_0^t \|\mathrm{M}_k u(s)\|_H^2 ds \\ + \sum_k \int_0^t \|g_k(s)\|_H^2 ds + 2\sum_k \int_0^t \big(\mathrm{M}_k u(s), g_k(s)\big)_H ds.\end{aligned} \tag{4.4.9}$$

The term $\sum_k \int_0^t \|\mathrm{M}_k u(s)\|_H^2 ds$ combines with $2\int_0^t [\mathrm{A}u(s), u(s)]ds$ in (4.4.8), and then condition (4.4.4) results in

$$\begin{aligned}2\int_0^t [\mathrm{A}u(s), u(s)]ds + \sum_k \int_0^t \|\mathrm{M}_k u(s)\|_H^2 ds \\ \le -c_A \int_0^t \|u(s)\|_V^2 ds + M\int_0^t \|u(s)\|_H^2 ds.\end{aligned}$$

Then the term $-c_A\int_0^t \|u(s)\|_V^2 ds$ goes to the left so that (4.4.8) becomes

$$\|u(t)\|_H^2 + c_A \int_0^t \|u(s)\|_V^2 ds \le \cdots .$$

Thus, (4.4.6) will follow from (4.4.8) once we can manipulate the remaining terms containing u on the right-hand side of (4.4.8) to one of the following:

- $C\int_0^t \|u(s)\|_H^2 ds$; this integral is then removed by the Gronwall inequality because of the term $\|u(t)\|_H^2$ on the left-hand side of (4.4.8);
- $\varepsilon \int_0^t \|u(t)\|_V^2 dt$ or $\varepsilon \sup_s \|u(s)\|_H^2$ with sufficiently small ε.

As always, we allow the constant C to change from line to lime. The last term on the right-hand side of (4.4.9) is OK:

$$\begin{aligned}
\sum_k \int_0^t |(\mathrm{M}_k u(s), g_k(s))_H| ds &\le \sum_k \int_0^t \|\mathrm{M}_k u(s)\|_H \, \|g_k(s)\|_H ds \\
&\le \sum_k \left(\int_0^t \|\mathrm{M}_k u(s)\|_H^2 ds\right)^{1/2} \left(\int_0^t \|g_k(s)\|_H^2 ds\right)^{1/2} \\
&\le \varepsilon \sum_k \int_0^t \|\mathrm{M}_k u(s)\|_H^2 ds + C \sum_k \int_0^t \|g_k(s)\|_H^2 ds \\
&\le C_M \varepsilon \int_0^t \|u(s)\|_V^2 ds + C \sum_k \int_0^t \|g_k(s)\|_H^2 ds;
\end{aligned} \tag{4.4.10}$$

the last inequality follows from (4.4.5). Similarly, the term $\int_0^t [f(s), u(s)] ds$ is OK:

$$\int_0^t [f, u] ds \le \int_0^t \|f\|_{V'} \, \|u\|_V \, ds \le \varepsilon \int_0^t \|u\|_V^2 ds + C \int_0^t \|f\|_{V'}^2 ds. \tag{4.4.11}$$

The remaining term is the stochastic integral. To estimate $\mathbb{E}\sup_t$, we use the BDG inequality with $p = 1$:

$$\begin{aligned}
&\mathbb{E}\sup_t \left| \sum_k \int_0^t \Big((\mathrm{M}_k u(s), u(s))_H + (g_k(s), u(s))_H \Big) dw_k(s) \right| \\
&\le C\mathbb{E} \left(\int_0^T \sum_k \Big((\mathrm{M}_k u(s), u(s))_H + (g_k(s), u(s))_H \Big)^2 ds \right)^{1/2} \\
&\le C\mathbb{E} \left(\int_0^T \sum_k \Big((\mathrm{M}_k u(s), u(s))_H^2 + (g_k(s), u(s))_H^2 \Big) ds \right)^{1/2} \\
&\le C\mathbb{E} \left(\int_0^T \sum_k \|\mathrm{M}_k u\|_H^2 \|u\|_H^2 ds \right)^{1/2} + C\mathbb{E} \left(\int_0^T \sum_k \|g_k\|_H^2 \|u\|_H^2 ds \right)^{1/2};
\end{aligned}$$

beside the BDG and the Cauchy-Schwarz, we use $(a+b)^2 \le 2a^2 + 2b^2$ and $\sqrt{a+b} \le \sqrt{a} + \sqrt{b}$, $a, b \ge 0$.

To process expressions of the type

$$\mathbb{E}\left(\int_0^T \sum_k B_k(t)\|u(t)\|_H^2 dt\right)^{1/2},$$

take $\sup_t \|u(t)\|_H$ outside the integral and use the epsilon inequality:

$$\mathbb{E}\left(\int_0^T \sum_k B_k(t)\|u(t)\|_H^2 dt\right)^{1/2} \le \varepsilon \mathbb{E} \sup_t \|u(t)\|_H^2 + C \sum_k \mathbb{E} \int_0^T B_k(t)dt.$$

This works well if $B_k = \|g_k\|_H^2$, but if $B_k = \|\mathrm{M}_k u\|_H^2$, then, after applying (4.4.5), we get $C\int_0^T \|u\|_V^2 dt$ with a constant C that we cannot make arbitrarily small. In other words, we get

$$\begin{aligned} &\mathbb{E} \sup_t \left| \sum_k \int_0^t \Big(\big(\mathrm{M}_k u(s), u(s)\big)_H + \big(g_k(s), u(s)\big)_H \Big) dw_k(s) \right| \\ &\le \varepsilon \mathbb{E} \sup_t \|u(t)\|_H^2 + C\mathbb{E} \int_0^T \|u(t)\|_V^2 dt + C \sum_k \mathbb{E} \int_0^T \|g_k(t)\|_H^2 dt, \end{aligned} \tag{4.4.12}$$

and there is no clear way to proceed.

Since the problem appears to be with the stochastic integral, let us go back to (4.4.8) and simply take $\mathbb{E}$ on both sides. This will eliminate the stochastic integral. Looking over (4.4.9)–(4.4.11) and choosing a suitable ε, we get the following inequality:

$$\begin{aligned} \mathbb{E}\|u(T)\|_H^2 + \frac{c_A}{2}\mathbb{E}\int_0^T \|u(t)\|_V^2 dt &\le \mathbb{E}\|u_0\|_H^2 + C\mathbb{E}\int_0^T \|u(t)\|_V^2 dt \\ &+ C\mathbb{E}\int_0^T \|f(t)\|_{V'}^2 dt + C\sum_{k\ge 1}\mathbb{E}\int_0^T \|g_k(t)\|_H^2 dt. \end{aligned} \tag{4.4.13}$$

If we ignore the second term on the left-hand side of (4.4.13), then, by the Gronwall inequality,

$$\mathbb{E}\|u(T)\|_H^2 \le C(T)\Big(\mathbb{E}\|u_0\|_H^2 + \mathbb{E}\int_0^T \|f(t)\|_{V'}^2 dt + \sum_{k\ge 1}\mathbb{E}\int_0^T \|g_k(t)\|_H^2 dt\Big).$$

Putting this back into (4.4.13), we get the key estimate for the second term on the left-hand side of (4.4.13):

$$\mathbb{E}\int_0^T \|u(t)\|_V^2 dt \le C(T)\Big(\mathbb{E}\|u_0\|_H^2 + \mathbb{E}\int_0^T \|f(t)\|_{V'}^2 dt + \sum_{k\ge 1}\mathbb{E}\int_0^T \|g_k(t)\|_H^2 dt\Big). \tag{4.4.14}$$

Now we go back to (4.4.8) and this time take $\mathbb{E}\sup_t$ on both sides. With (4.4.14) taking care of the term $C\mathbb{E}\int_0^T \|u(t)\|_V^2 dt$ in (4.4.12), we complete the proof of a priori estimate (4.4.7).

Step 2: The Galerkin Approximation (pages 203–204) We define the Hilbert space $\mathbb{H}$ by

$$\begin{aligned}\mathbb{H} = \Big\{ & u \in L_2\big(\Omega\times(0,T);V\big) : u(t) = u_0 + \int_0^t F(s)ds + \sum_{k\ge 1}\int_0^t G_k(s)dw_k(s),\\ & u_0 \in L_2(\Omega;H),\ F\in L_2\big(\Omega\times(0,T);V'\big),\ G_k \in L_2\big(\Omega\times(0,T);H\big),\\ & \sum_k \|G_k\|^2_{L_2(\Omega\times(0,T);H)} < \infty,\ u, F, G_k \text{ are } \mathcal{F}_t\text{- adapted}\Big\};\\ & \|u\|^2_{\mathbb{H}} = \|u_0\|^2_{L_2(\Omega;H)} + \|u\|^2_{L_2(\Omega\times(0,T);V)}\\ & + \|F\|^2_{L_2(\Omega\times(0,T);V')} + \sum_k \|G_k\|^2_{L_2(\Omega\times(0,T);H)}.\end{aligned} \tag{4.4.15}$$

According to a priori estimate (4.4.6), $\mathbb{H}$ is the natural solution space for the equation

$$du = (\mathrm{A}u + f)dt + \sum_k \big(\mathrm{M}_k u + g_k\big)dw_k.$$

Exercise 4.4.4 (C) Convince yourself that $\mathbb{H}$ is indeed a Hilbert space and that F and G_k are uniquely determined by u. Write an expression for the inner product in $\mathbb{H}$.

Let $\{\mathfrak{h}_k,\ k\ge 1\}$ be an orthonormal basis in H that is also a dense set in V and V'; Π^N is the orthogonal in H projection onto the the linear span of $\{\mathfrak{h}_1,\dots,\mathfrak{h}_N\}$. For $v\in V'$,

$$\Pi^N v = \sum_{k=1}^N [v,\mathfrak{h}_k]\mathfrak{h}_k.$$

The sequence u^N, $N \geq 1$, approximating the solution satisfies

$$\begin{aligned} du^N &= \big(\Pi^N \mathrm{A} u^N + \Pi^N f\big)dt \\ &+ \sum_{k=1}^{N} \big(\Pi^N \mathrm{M}_k u^N + \Pi^N g_k\big)dw_k,\ 0 < t \leq T, \end{aligned} \tag{4.4.16}$$

$u^N(0) = \Pi^N u_0$. Writing $u^N(t) = \sum_{k=1}^{N} \bar{u}^N(t)\mathfrak{h}_k$, we interpret (4.4.16) as an N-dimensional linear SODE for the functions $\bar{u}^N_k$ and conclude existence and uniqueness of u^N for every N. Also, if $v \in \Pi^N H$, the $v = \Pi^N v$ and therefore

$$[\Pi^N \mathrm{A} v, v] = (\mathrm{A} v, v)_H = [\mathrm{A} v, v],\ \|\Pi^N \mathrm{M}_k v\|_H \leq \|\mathrm{M}_k v\|_H.$$

As a result, the operators $\Pi^N \mathrm{A}$ and $\Pi^N \mathrm{M}_k$ satisfy (4.4.4) and (4.4.5) uniformly in N:

$$2[\Pi^N \mathrm{A} v, v] + \sum_{k=1}^{N} \|\Pi^N \mathrm{M}_k v\|_H^2 \leq 2[\mathrm{A} v, v] + \sum_{k} \|\mathrm{M}_k v\|_H^2 \leq -c_A \|v\|_V^2 + M\|v\|_H^2,$$

$$\sum_{k=1}^{N} \|\Pi^N \mathrm{M}_k v\|_H^2 \leq C_M \|v\|_V^2.$$

Consequently, u^N satisfies (4.4.7) with $C(T)$ independent of N [While it is not necessarily true that $\|\Pi^N f\|_{V'} \leq \|f\|_{V'}$, convergence $\lim_{N\to\infty} \|f - \Pi^N f\|_{V'} = 0$ is enough to have a uniform bound]. In other words, the sequence $\{u^N,\ N \geq 1\}$ is weakly compact in $\mathbb{H}$ and therefore has a weakly converging sub-sequence. The limit u of this subsequence is, by definition, the variational solution we want.

Theorem 3.3.15 on page 139 (the energy equality in a normal triple of Hilbert spaces) implies both uniqueness of the solution and existence of a continuous in H modification of u.

This completes the proof of Theorem 4.4.3.

The stochastic parabolicity condition (4.4.4) is the abstract version of the condition $2a > \sigma^2$ for the equation $du = au_{xx}dt + \sigma u_x dw(t)$ we derived earlier using the Fourier transform (page 53).

Exercise 4.4.5 (C) Consider equation

$$du = \sum_{i,j=1}^{\mathrm{d}} a_{ij}(\omega, t, x) u_{x_i} u_{x_j} dt + \sum_{k\geq 1} \sum_{i=1}^{\mathrm{d}} \sigma_{ik}(\omega, t, x) u_{x_i} dw_k,$$

$0 < t \le T$, $x \in G \subseteq \mathbb{R}^{\mathrm{d}}$, and assume that there exists a $c > 0$ such that, for all $t \in [0, T]$, $x \in G$, $\omega \in \Omega$, and $y \in \mathbb{R}^{\mathrm{d}}$,

$$\sum_{i,j=1}^{\mathrm{d}} \left(a_{ij} - \frac{1}{2} \sum_{k \ge 1} \sigma_{ik}\sigma_{jk} \right) y_i y_j \ge c \sum_{i=1}^{\mathrm{d}} y_i^2. \tag{4.4.17}$$

Show that condition (4.4.4) holds in each of the following normal triples of Hilbert space:

- $(H^{\gamma+1}(G), H^{\gamma}(G), H^{\gamma-1}(G))$,
- $(W^{n+2,r,\mathrm{d}}, W^{n+1,r,\mathrm{d}}, W^{n,r,\mathrm{d}})$ (see Exercise 3.1.6 on page 78),

Assume all the necessary smoothness and boundedness of the coefficients. **Hint.** First, consider the normal triple $(H^1(\mathbb{R}^{\mathrm{d}}), L_2(\mathbb{R}^{\mathrm{d}}), H^{-1}(\mathbb{R}^{\mathrm{d}}))$ and use that $\|v\|_1^2 = \|v\|_0^2 + \|v_{x_1}\|_0^2 + \ldots + \|v_{x_{\mathrm{d}}}\|_0^2$.

In our study of equations with a closed form solution we saw that the right-hand side of (4.4.17) can be zero: for example, equation $du = u_{xx}dt + \sqrt{2}u_x dw(t)$, $u(0,x) = e^{-x^2}$ is perfectly fine and has solution $u(t,x) = \exp(-(x+\sqrt{2}w(t))^2)$ (see (2.3.11) on page 54). In the language of abstract equations, this observation suggests that we should be able to allow (4.4.4) to hold with $c_A = 0$. We thus arrive at the notion of `degenerate` stochastic parabolic equations, which we will discuss next.

Fix the stochastic basis $(\Omega, \mathcal{F}, \{\mathcal{F}_t\}_{t \ge 0}, \mathbb{P})$ with the usual assumptions and a collection $\{w_k, k \ge 1\}$ of independent standard Brownian motions on this basis, and consider the following equation

$$\dot{u}(t) = \mathrm{A}(t)u(t) + f(t) + \sum_{k \ge 1} \big(\mathrm{M}_k(t)u(t) + g_k(t)\big)\dot{w}_k(t) \tag{4.4.18}$$

$0 < t \le T$, $u(0) = u_0$, in the normal triple (V, H, V') of Hilbert spaces.

Definition 4.4.6 Equation (4.4.18) is called `degenerate parabolic` in the normal triple (V, H, V') if

1. The family of operators $\mathrm{A} = \{\mathrm{A}(t),\ 0 \le t \le T\}$ is (V, V')-adapted and $L_\infty(V, V')$-uniformly bounded (see page 198 for the definition).
2. For each k, the family of operators $\mathrm{M}_k = \{\mathrm{M}_k(t),\ 0 \le t \le T\}$ is (V, H)-adapted and $L_\infty(V, H)$-uniformly bounded.
3. There exists a real number M such that, for all $\omega \in \Omega$, $t \in [0, T]$, and $v \in V$,

$$2[\mathrm{A}(t)v, v] + \sum_{k \ge 1} \|\mathrm{M}_k(t)v\|_H^2 dt \le M\|v\|_H^2. \tag{4.4.19}$$

Note that the definition covers deterministic equations such as $u_t = \big(1 + \sin(x)\big)u_{xx}$.

Remark 4.4.7 If the stochastic parabolicity condition (4.4.4) holds, then equation (4.4.18) is sometimes called non-degenerate parabolic or super-parabolic.

Exercise 4.4.8 (C) Let $a = a(t,x)$ be a bounded measurable non-negative function. Consider $u_t = (au_x)_x$ in the normal triple $(H^1(\mathbb{R}), L_2(\mathbb{R}), H^{-1}(\mathbb{R}))$. Show that the equation is degenerate parabolic in the sense of Definition 4.4.6.

While Definition 4.4.1 on page 199 defines a solution for a wide class of equations, including degenerate parabolic, constructing a solution is a different story. To get a better idea of what to expect, let us go back to equation

$$du = u_{xx}dt + \sqrt{2}u_x dw(t),\ 0 < t \le T,\ x \in \mathbb{R}, \tag{4.4.20}$$

with non-random initial condition $u(0,x) = u_0(x)$. Simple integration by parts shows that

$$\mathbb{E}\int_{\mathbb{R}} u^2(t,x)dx \le \int_{\mathbb{R}} u_0^2(x)dx. \tag{4.4.21}$$

This is not bad, but certainly not enough to get an $H^1(\mathbb{R})$-valued solution. On the other hand, assuming that u_0 and the solution u are as regular as we want, we can take $\partial/\partial x$ on both sides of the equation and conclude that $v(t,x) = u_x(t,x)$ also satisfies $dv = v_{xx}dt + \sqrt{2}v_x dw(t)$ and therefore

$$\mathbb{E}\int_{\mathbb{R}} u_x^2(t,x)dx \le \int_{\mathbb{R}} \big(u_0'(x)\big)^2 dx \tag{4.4.22}$$

(as usual, u_0' is the derivative of u_0). This is better, as now we have an a priori bound on the $H^1(\mathbb{R})$ norm of u:

$$\mathbb{E}\|u(t,\cdot)\|_1^2 \le \|u_0\|_1^2,\ 0 \le t \le T, \tag{4.4.23}$$

and can use it to construct a solution.

Exercise 4.4.9 (C)

(a) Verify (4.4.21) and (4.4.22).
(b) Derive an analogue of (4.4.23) for equation $u_t = (1 + \sin x)u_{xx}$.

Hint. In (b), for $v = u_x$, you will get $v_t = (1+\sin x)v_{xx} + (\cos x)v_x$, which will require additional integrations by parts and Gronwall's inequality and will produce a $C(T)$ on the right-hand side of (4.4.23).

We will construct a solution of (4.4.20) by **elliptic regularization**. Take $\varepsilon > 0$ and consider the equation

$$du^{\varepsilon} = (1+\varepsilon)u_{xx}^{\varepsilon}dt + \sqrt{2}u_x^{\varepsilon}dw(t),\ u^{\varepsilon}(0,x) = u_0(x), \tag{4.4.24}$$

in the normal triple $(H^2(\mathbb{R}), H^1(\mathbb{R}), L_2(\mathbb{R}))$. Assume that $u_0 \in H^1(\mathbb{R})$. For $\varepsilon > 0$ the equation is not degenerate and has a unique solution. A priori bound (4.4.23) allows us to extract a converging subsequence of u^{ε}, $\varepsilon \to 0$, and to identify the limit as a solution of (4.4.20). Uniqueness follow from (4.4.23). The details are left to the reader.

Exercise 4.4.10 Show that, for Eq. (4.4.24),

(a) condition (4.4.17) holds with $c_A = \varepsilon$, and therefore
(b) there is a unique solution

$$u^{\varepsilon} \in L_2\big(\Omega \times (0,T); H^2(\mathbb{R})\big) \bigcap L_2\big(\Omega; \mathcal{C}((0,T); H^1(\mathbb{R}))\big);$$

(c) If $\varepsilon > 0$, then

$$\mathbb{E}\|u^{\varepsilon}(t,\cdot)\|_1^2 \le \|u_0\|_1^2, \tag{4.4.25}$$

and, after integrating in time,

$$\|u^{\varepsilon}\|^2_{L_2(\Omega\times(0,T);H^1(\mathbb{R}))} \le T\|u_0\|_1^2. \tag{4.4.26}$$

(d) Using weak sequential compactness of $L_2(\Omega\times(0,T); H^1(\mathbb{R}))$ and passing to the limit $\varepsilon \to 0$ (if necessary, along a subsequence), argue existence of the solution of (4.4.20) for every $u_0 \in H^1(\mathbb{R})$.
(e) Use (4.4.23) to claim uniqueness.

Let us summarize what we learned about Eq. (4.4.20).

Proposition 4.4.11 *Consider the equation*

$$du = u_{xx}dt + \sigma u_x dw(t),\ 0 < t \le T,\ x \in \mathbb{R},\ u(0,x) = u_0(x). \tag{4.4.27}$$

(a) If $|\sigma| < \sqrt{2}$ (non-degenerate case), then, for every $u_0 \in L_2(\mathbb{R})$ the equation has a unique solution such that $u \in L_2\big(\Omega; \mathcal{C}((0,T); L_2(\mathbb{R}))\big)$ and

$$\mathbb{E} \sup_{0<t<T} \|u(t,\cdot)\|^2_{L_2(\mathbb{R})} + \mathbb{E}\int_0^T \|u(t,\cdot)\|^2_{H^1(\mathbb{R})} \le C(T)\|u_0\|^2_{L_2(\mathbb{R})}.$$

(b) If $|\sigma| = \sqrt{2}$ (degenerate case), then, for every $u_0 \in H^1(\mathbb{R})$ the equation has a unique solution such that, for all $0 \le t \le T$,

$$\mathbb{E}\|u(t,\cdot)\|^2_{H^1(\mathbb{R})} \le \|u_0\|^2_{H^1(\mathbb{R})}.$$

Proposition 4.4.11 shows that there are two main differences between the non-degenerate and degenerate equation (4.4.27):

- **At the level of the results,** the solution in the non-degenerate case is, in some sense better than the initial condition, whereas there is no improvement for degenerate equations: we start in $H^1(\mathbb{R})$, we stay in $H^1(\mathbb{R})$ for all $t > 0$;

- **At the level of the proof,** to establish existence of solution in the degenerate case, we need to go beyond the original normal triple (H^1, L_2, H^{-1}).

To prove an analogue of Proposition 4.4.11(b) for the abstract equation (4.4.18), we have to apply the elliptic regularization in abstract setting. For that, we need something to go in place of the space $H^2(\mathbb{R})$ and something to replace the operator $\varepsilon\partial^2/\partial x^2$. Turns out, a `Hilbert scale` provides all the necessary ingredients.

Given a normal triple (V, H, V') of Hilbert spaces, let $\{H^r,\ r \in \mathbb{R}\}$ be the Hilbert scale such that $H^1 = V$, $H^0 = H$, $H^{-1} = V'$ and let Λ be a positive-definite self-adjoint operator generating the scale (see Theorems 3.1.11 and 3.1.12 on page 80). We will use $\|\cdot\|_r$ to denote the norm in H^r and $[\cdot,\cdot]_r$ to denote the duality between H^{r+1} and H^{r-1}. Note that $\|\cdot\|_0 = \|\cdot\|_H$ and $\|\cdot\|_1 = \|\cdot\|_V$.

The following is the general result about existence and uniqueness of the solution for degenerate parabolic equation (4.4.18).

Theorem 4.4.12 *Assume that*

1. *For* $\gamma = 0$ *and for* $\gamma = 1$, *the operator* A *is* $(H^{\gamma+1}, H^{\gamma-1})$*-adapted and* $L_\infty((H^{\gamma+1}, H^{\gamma-1})$*-uniformly bounded, each operator* M_k *is* $(H^{\gamma+1}, H^\gamma)$*-adapted and* $L_\infty((H^{\gamma+1}, H^\gamma)$*-uniformly bounded, and there exists a positive number* C_M *such that, for all* $t \in [0,T]$, $\omega \in \Omega$, *and* $v \in H^\gamma$,

$$\sum_{k\geq 1} \|\mathrm{M}_k(t)v\|_\gamma^2 \leq C_M \|v\|_\gamma^2.$$

2. *Condition* (4.4.19) *is satisfied in the normal triple* (H^2, H^1, H^0)*: there exists a real number* M *such that, for all* $t \in [0,T]$, $\omega \in \Omega$ *and* $v \in H^2$,

$$2[\mathrm{A}(t)v, v]_1 + \sum_{k\geq 1} \|\mathrm{M}_k(t)v\|_1^2 dt \leq M\|v\|_1^2. \tag{4.4.28}$$

3. $u_0 \in L_2(\Omega; H^1)$, $f \in L_2\big(\Omega \times (0,T); H^1\big)$,

$$\sum_{k\geq 1} \|g_k\|^2_{L_2(\Omega\times(0,T);H^2)} < \infty,$$

u_0 *is* $\mathcal{F}_0$*-measurable,* f *and each* g_k *are* $\mathcal{F}_t$*-adapted.*

Then Eq. (4.4.18) *has a unique solution* $u = u(t)$ *and*

$$\sup_{0<t<T} \mathbb{E}\|u(t)\|_1^2 \leq C(T)\Big(\mathbb{E}\|u_0\|_1^2 + \mathbb{E}\int_0^T \|f(t)\|_1^2 dt + \sum_{k\geq 1} \mathbb{E}\int_0^T \|g_k(t)\|_2^2 dt\Big). \tag{4.4.29}$$

Proof For $\varepsilon > 0$, consider the elliptic regularization of (4.4.18),

$$du^\varepsilon = \big((\mathrm{A} - \varepsilon\Lambda^2)u + f\big)dt + \sum_k (\mathrm{M}_k u + g_k)dw_k, \tag{4.4.30}$$

$0 < t \le T$, $u^{\varepsilon}(0) = u_0$. We consider Eq. (4.4.30) in the normal triple (H^2, H^1, H^0). By Theorem 4.4.3, the equation has a unique solution for every $\varepsilon > 0$.

Exercise 4.4.13

(a) Verify conditions of Theorem 4.4.3 for Eq. (4.4.30).
(b) Verify that (4.4.28) implies

$$\mathbb{E}\|u^{\varepsilon}(t)\|_1^2 \le C(T)\Big(\mathbb{E}\|u_0\|_1^2 + \mathbb{E}\int_0^T \|f(t)\|_1^2 dt + \sum_{k\ge 1}\mathbb{E}\int_0^T \|g_k(t)\|_2^2 dt\Big). \tag{4.4.31}$$

The key is that $C(T)$ does not depend on ε.

Hint. For (b), repeat the derivation of the a priori bound in the proof of Theorem 4.4.3. Not all of derivations are necessary because you are not taking $\sup_t$ before the expectation and therefore there is no need to deal with the stochastic integral. Use $|(f, u)_1| \le \|f\|_1\|u\|_1$, $|(\mathrm{M}_k u, g_k)_1| = [g_k, \mathrm{M}_k u]_1$.

Then we take a subsequence u^{ε_n} converging weakly in $L_2\big(\Omega \times [0, T]; H^1\big)$ and identify the limit with the solution of (4.4.18). Inequality (4.4.31) is preserved in the limit, leading to (4.4.29) and uniqueness of the solution.
This concludes the proof of Theorem 4.4.12.

Exercise 4.4.14 (C) Consider the equation

$$du = \sum_{i,j=1}^{\mathrm{d}} a_{ij}(\omega, t, x) u_{x_i} u_{x_j} dt + \sum_{k\ge 1}\sum_{i=1}^{\mathrm{d}} \sigma_{ik}(\omega, t, x) u_{x_i} dw_k, \tag{4.4.32}$$

$0 < t \le T$, $x \in G \subseteq \mathbb{R}^{\mathrm{d}}$. Assume that, for all $t \in [0, T]$, $x \in G$, $\omega \in \Omega$, and $y \in \mathbb{R}^{\mathrm{d}}$,

$$\sum_{i,j=1}^{\mathrm{d}} \left(a_{ij} - \frac{1}{2}\sum_{k\ge 1}\sigma_{ik}\sigma_{jk}\right) y_i y_j \ge 0. \tag{4.4.33}$$

Show that condition (4.4.28) holds in the each of the normal triples of Hilbert space

- $(H^{\gamma+1}(G), H^{\gamma}(G), H^{\gamma-1}(G))$,
- $(W^{n+2,r,\mathrm{d}}, W^{n+1,r,\mathrm{d}}, W^{n,r,\mathrm{d}})$ (see Exercise 3.1.6 on page 78).

Assume all the necessary smoothness and boundedness of the coefficients.
This concludes Exercise 4.4.14.

The conclusion is that solvability of a degenerate parabolic equation requires much higher regularity of the input than comparable solvability of a non-degenerate equation. In fact, Theorem 4.4.12 effectively says that, to be solvable in the normal triple (H^1, H^0, H^{-1}), the operators in the equation must be acing in the triple

(H^2, H^1, H^0). For equations of the type (4.4.32), this requirement means higher regularity of the coefficients. We illustrate this point on the following simple example. Consider the equation

$$du = (u_{xx} + f(x))dt + (\sigma(x)u_x + g(x))dw(t),\ 0 < t \le T,\ x \in \mathbb{R},$$

in the normal triple $(H^1(\mathbb{R}), L_2(\mathbb{R}), H^{-1}(\mathbb{R}))$. For simplicity assume that $u(0,\cdot), f, g, \sigma$ are non-random. Below is a comparison of the conditions for the non-degenerate and degenerate situations.

Input	Non-degenerate: $\sigma^2(x) \le c < 2$	Degenerate: $\sigma^2(x) \le 2$
$u(0,\cdot)$	$L_2(\mathbb{R})$	$H^1(\mathbb{R})$
f	$H^{-1}(\mathbb{R})$	$H^1(\mathbb{R})$
σ	bounded and measurable	Multiplier in $H^1(\mathbb{R})$
g	$L_2(\mathbb{R})$	$H^2(\mathbb{R})$

We conclude the section by observing that Exercise 4.4.14 and the Sobolev embedding theorem imply the following result.

Theorem 4.4.15 *Assume that*

1. *the functions $a_{ij}, b_i, c, \sigma_{ij}, \nu_k$ are predictable, bounded, measurable, infinitely differentiable in x, with all the derivatives also bounded;*
2. *the functions f, g_k are predictable, grow polynomially in x, infinitely differentiable in x, with all the derivatives also of polynomial growth in x;*
3. $\sum_{k\ge1} \sup_{\omega,t,x} |D^n \nu_k(\omega,t,x)| < \infty$ *for all $n \ge 0$, where D^n is n-th partial derivative, $D^0\nu_k = \nu_k$;*
4. $\sum_{k\ge1} \sup_{\omega,t,x} |D^n \sigma_{ik}(\omega,t,x)| < \infty$ *for all $n \ge 0$ and $i = 1,\dots,\mathrm{d}$;*
5. *for every $n \ge 0$ there exists an $r \in \mathbb{R}$ such that*

$$\sum_{k\ge1} \sup_{\omega,t} \|g_k(\omega,t,\cdot)\|^2_{W^{n,r,\mathrm{d}}} < \infty;$$

6. *the initial condition u_0 is $\mathcal{F}_0$-measurable, grows polynomially in x, infinitely differentiable in x, with all the derivatives also of polynomial growth in x;*
7. *condition* (4.4.33) *is satisfied.*

Then Eq. (4.4.32) *has a unique classical solution that grows polynomially in x.*

Proof The idea is to apply Theorem 4.4.12 in the normal triple

$$(W^{n+2,r,\mathrm{d}}, W^{n+1,r,\mathrm{d}}, W^{n,r,\mathrm{d}})$$

with $n > (d+4)/2$ and suitable r, and then use the Sobolev embedding theorem to conclude that the solution is twice continuously differentiable in x. The details are left to the reader.
This concludes the proof of Theorem 4.4.15.

4.4.2 A Change of Variables Formula

In the course of our study of closed-form solutions, we discovered that

1. change of the unknown function

$$v(t,x) = u(t,x)\exp\left(-\int_0^t h(s)dw(s) + \frac{1}{2}\int_0^t h^2(s)ds\right) \tag{4.4.34}$$

transforms equation $du = u_{xx}dt + h(t)\,u\,dw(t)$ to $v_t = v_{xx}$;
2. change of the unknown function

$$v(t,x) = u\big(t, x - \sigma w(t)\big) \tag{4.4.35}$$

transforms equation $du = u_{xx}dt + \sigma u_x dw(t)$ to $v_t = (1-(\sigma^2/2))v_{xx}$.

The following exercise shows that something similar is possible for equations with variable coefficients.

Exercise 4.4.16 (C) Let u be a classical solution of the equation $du = u_{xx}dt + \nu(x)udw(t)$ and let

$$v(t,x) = u(t,x)H(t,x),\ H(t,x) = \exp\left(-\nu(x)w(t) + \frac{t}{2}\nu^2(x)\right) \tag{4.4.36}$$

Using Itô formula, show that v satisfies

$$v_t = v_{xx} - 2(H_x/H)v_x + H(1/H)_{xx}v. \tag{4.4.37}$$

Our objective in this section is to find a change of variables that transforms a stochastic bi-linear equation

$$\begin{aligned} u_t(t,x) &= \sum_{i,j=1}^{\mathrm{d}} a_{ij}(\omega,t,x)u_{x_ix_j} + \sum_{i=1}^{\mathrm{d}} b_i(\omega,t,x)u_{x_i} + c(\omega,t,x)u \\ &+ \sum_{k\geq 1}\left(\left(\sum_{i=1}^{\mathrm{d}}\sigma_{ik}(\omega,t,x)u_{x_i}\right) + \nu_k(\omega,t,x)u\right)\dot{w}_k(t),\ x\in\mathbb{R}^{\mathrm{d}}, \end{aligned} \tag{4.4.38}$$

with sufficiently regular in (t, x) coefficients, to a linear PDE with random coefficients

$$v_t = \sum_{i,j=1}^{\mathrm{d}} A_{ij}(\omega, t, x) v_{x_i x_j} + \sum_{i=1}^{\mathrm{d}} B_i(\omega, t, x) v_{x_i} + C(\omega, t, x) v. \tag{4.4.39}$$

Here is the result.

Theorem 4.4.17 *Assume that*

1. *the functions $a_{ij}, b_i, c, \sigma_{ik}, \nu_k$ are $\mathcal{F}_t$-adapted, bounded, continuous in t, and infinitely differentiable in x, with all the derivatives also bounded;*
2. *Eq.* (4.4.38) *is degenerate parabolic, that is, condition* (4.4.33) *on page 209 holds);*
3. *$\sum_{k\geq 1} \sup_{\omega,t,x} |D^n \nu_k(\omega, t, x)| < \infty$ for all $n \geq 0$, where D^n is n-th partial derivative, $D^0 \nu_k = \nu_k$.*
4. *$\sum_{k\geq 1} \sup_{\omega,t,x} |D^n \sigma_{ik}(\omega, t, x)| < \infty$ for all $n \geq 0$ and $i = 1, \ldots, \mathrm{d}$.*

Define the functions $\Psi = (\Psi_1(t, x), \ldots, \Psi_{\mathrm{d}}(t, x))$ and $H = H(t, x)$ by

$$\Psi_i(t, x) = x_i - \sum_{k\geq 1} \int_0^t \sigma_{ik}(s, \Psi(s, x)) dw_k(s), \ i = 1, \ldots, \mathrm{d};$$

$$H(t, x) = \exp\left(-\sum_{k\geq 1} \int_0^t \nu_k(s, \Psi(s, x)) dw_k(s) \right. \tag{4.4.40}$$

$$\left. + \frac{1}{2} \sum_{k\geq 1} \int_0^t |\nu_k(s, \Psi(s, x))|^2 ds \right),$$

and the functions $G = G(t, x)$ and $\Phi = (\Phi_1(t, x), \ldots, \Phi_{\mathrm{d}}(t, x))$ by

$$G(t, x) = \exp\left(\sum_{k\geq 1} \int_0^t \nu_k(s, x) dw_k(s) - \frac{1}{2} \sum_{k\geq 1} \int_0^t |\nu_k(s, x)|^2 ds \right),$$

$$\Phi_m(t, x) = x_m$$

$$+ \int_0^t \left(\sum_{i,j=1}^{\mathrm{d}} \sum_{k\geq 1} \sigma_{ik} \frac{\partial}{\partial x_i} \left(\sigma_{jk} \frac{\partial \Phi_m}{\partial x_j} \right) - \frac{1}{2} \sum_{i,j=1}^{\mathrm{d}} \sum_{k\geq 1} \sigma_{ik} \sigma_{jk} \frac{\partial^2 \Phi_m}{\partial x_i \partial x_j} \right) ds \tag{4.4.41}$$

$$+ \int_0^t \sum_{k\geq 1} \sum_{i=1}^{\mathrm{d}} \sigma_{ik} \frac{\partial \Phi_m}{\partial x_i} dw_k(s), \ m = 1, \ldots, \mathrm{d}.$$

Let $u = u(t, x)$ *be a classical solution of* (4.4.38) *and define*

$$v(t,x) = u(t, \Psi(t,x))H(t,x). \tag{4.4.42}$$

Then

(a) The functions Ψ *and* Φ *are infinitely differentiable in* x *and*

$$\Psi(t, \Phi(t,x)) = x, \;\; \Phi(t, \Psi(t,x)) = x \tag{4.4.43}$$

for all $t \geq 0$, $x \in \mathbb{R}^{\mathrm{d}}$, *and* $\omega \in \Omega$.

(b) The function v *is a classical solution of* (4.4.39) *with suitable functions* A_{ij}, B_i, C.

(c) $u(t,x) = v(t, \Phi(t,x))G(t,x)$.

Proof Direct computations using the Itô-Wentzell formula (Theorem 2.1.16 on page 34). The computations also lead to explicit expressions for the functions A_{ij}, B_i, C; see Rozovskii [199, Sect. 5.2.2].
This concludes the proof of Theorem 4.4.17.

A more detailed proof of Theorem 4.4.17 involves analysis of the functions $|\Psi$ and Φ. Both Ψ and Φ are solutions of stochastic differential equations. For Ψ, the equation is ordinary, and the existence and uniqueness of solution are well-known (Problem 4.1.1 on page 163). For Φ, the equation is with partial derivatives, and is a degenerate parabolic equation (in fact, it is **fully degenerate**: we have an equality in (4.4.33)). Moreover, the initial condition is not a square-integrable function. As a result, existence and uniqueness of Ψ are not at all obvious, and follow from Theorem 4.4.12 in the normal triple $(W^{n+2,r,\mathrm{d}}, W^{n+1,r,\mathrm{d}}, W^{n,r,\mathrm{d}})$ with $r < -(d+2)/2$ and $n \geq 2$ (see Exercise 3.1.6 on page 78 for the definition of the spaces $W^{n,r,\mathrm{d}}$). By the Sobolev embedding theorem, taking $n > N + (\mathrm{d}/2)$ ensures that Φ has N continuous partial derivatives in x. To show that Ψ is smooth actually requires some effort as well: differentiating (4.4.41) according to the chain rule, it is natural to expect that

$$\frac{\partial \Psi_i(t,x)}{\partial x_m} = \delta_{im} - \int_0^t \sum_{\ell=1}^{\mathrm{d}} \frac{\partial \sigma_{ik}}{\partial x_\ell}(s, \Psi(s,x)) \frac{\partial \Psi_\ell}{\partial x_m}(s,x)dw_k(s), \tag{4.4.44}$$

which is a linear equation (and similarly for higher-order derivatives), but justification is required.

The change of variables (4.4.42) is better understood as a composition of two transformations. First, we replace u with

$$\tilde{u}(t,x) = u(t,x)/G(t,x).$$

According to Exercise 4.4.16, the function $\tilde{u}$ satisfies an equation of the form:

$$d\tilde{u} = (\ldots)dt + \sum_k \sum_{i=1}^{\mathrm{d}} \tilde{\sigma}_{ik}\tilde{u}_{x_i}dw_k.$$

In other words, by going from u to $\tilde{u}$, we remove the $\nu_k\, u\, dw_k$ terms. The second change of variables,

$$v(t,x) = \tilde{u}(t, \Psi(t,x)),$$

finishes the job by removing the $\tilde{\sigma}_{ik}\, \tilde{u}_{x_i}\, dw_k$ terms.

Finally, note that the classical solution of (4.4.38) is a variational solution with two continuous derivative in x. Given the conditions of the theorem, existence of such a solution will follow from Theorem 4.4.12 if the initial condition $u(0,x)$ belongs to $H^n(\mathbb{R}^{\mathrm{d}})$ for sufficiently large n; $u(0,\cdot) \in C_0^\infty(\mathbb{R}^{\mathrm{d}})$ or $u(0,\cdot) \in \mathcal{S}(\mathbb{R}^{\mathrm{d}})$ will certainly work. The change of variables does not affect the initial condition.

To conclude our discussion of the change of variables, let us have a closer look at Part (a) of Theorem 4.4.17. The solution Ψ of the stochastic ordinary differential equation in (4.4.40), when considered as a function of both t and x, is called a `stochastic flow`. Its `inverse flow` Φ satisfies a fully degenerate stochastic parabolic equation. While the stochastic flow in (4.4.40) is somewhat special (it contains no drift), the connection between Ψ and Φ is remarkable enough to justify an extension to more general flows.

Theorem 4.4.18 (Direct and Inverse Stochastic Flows) *Assume that the functions $b = \big(b_i(\omega,t,x),\ i = 1\ldots,\mathrm{d}\big)$ and $\sigma = \big(\sigma_{ik}(\omega,t,x),\ i = 1,\ldots,\mathrm{d},\ k \geq 1\big)$ have the following properties:*

1. *The components of b and σ are $\mathcal{F}_t$-adapted, bounded, continuous in t, and infinitely differentiable in x, with all the derivatives also bounded;*
2. $\sum_{k\geq 1} \sup_{\omega,t,x} |D^n\sigma_{ik}(\omega,t,x)| < \infty$ *for all $n \geq 0$ and $i = 1,\ldots,\mathrm{d}$.*

Define the functions $\Psi = (\Psi_1(t,x),\ldots,\Psi_{\mathrm{d}}(t,x))$ and $\Phi = (\Phi_1(t,x),\ldots,\Phi_{\mathrm{d}}(t,x))$ by

$$\Psi_i(t,x) = x_i + \int_0^t b_i(s,\Psi(s,x))ds + \sum_{k\geq 1}\int_0^t \sigma_{ik}(s,\Psi(s,x))dw_k(s),\ i = 1,\ldots,\mathrm{d};$$

$$\begin{aligned}\Phi_m(t,x) = x_m + \int_0^t \Bigg(&\sum_{i,j=1}^{\mathrm{d}}\sum_{k\geq 1} \sigma_{ik}\frac{\partial}{\partial x_i}\left(\sigma_{jk}\frac{\partial \Phi_m}{\partial x_j}\right) \\ &- \frac{1}{2}\sum_{i,j=1}^{\mathrm{d}}\sum_{k\geq 1}\sigma_{ik}\sigma_{jk}\frac{\partial^2\Phi_m}{\partial x_i\partial x_j} - \sum_{i=1}^{\mathrm{d}} b_i\frac{\partial \Phi_m}{\partial x_i}\Bigg)ds \\ &- \int_0^t \sum_{k\geq 1}\sum_{i=1}^{\mathrm{d}} \sigma_{ik}\frac{\partial \Phi_m}{\partial x_i}dw_k(s),\ m = 1,\ldots,\mathrm{d}.\end{aligned} \tag{4.4.45}$$

Then the functions Ψ and Φ are infinitely differentiable in x and

$$\Psi(t, \Phi(t,x)) = x, \ \Phi(t, \Psi(t,x)) = x \tag{4.4.46}$$

for all $t \geq 0$, $x \in \mathbb{R}^{\mathrm{d}}$, and $\omega \in \Omega$.

Proof The mechanical part is to verify (4.4.46) using the Itô-Wentzell formula (Theorem 2.1.16 on page 34). The easy technical part is to realize that the fully degenerate parabolic equation for Φ_m must be solved in the normal triple $(W^{n+2,r,\mathrm{d}}, W^{n+2,r,\mathrm{d}}, W^{n+2,r,\mathrm{d}})$ of weighted Sobolev spaces with integer n arbitrarily large and with $r < -(\mathrm{d}+2)/2$; the Sobolev embedding theorem then implies that Φ is infinitely differentiable. The medium-hard technical part is justifying an analogue of (4.4.44) to ensure that Ψ is smooth. The hard technical part is making sure that every sample trajectory of Ψ and Φ is infinitely differentiable in x. We leave the details to the reader. The details are left to the reader.
This concludes the proof of Theorem 4.4.18.

4.4.3 Probabilistic Representation of the Solution, Part I: Method of Characteristics

In our study of closed-form solutions we derived the following representation of the solution of the equation

$$u_t = au_{xx} + \sigma u_x \dot{w}, \ x \in \mathbb{R}, \ t > 0, \ u(0,x) = u_0(x), \ 2a \geq \sigma^2 : \tag{4.4.47}$$

$$u(t,x) = \mathbb{E}\Big(u_0\big(x + \sqrt{2a - \sigma^2}\widetilde{w}(t) + \sigma w(t)\big)\big|\mathcal{F}_t^w\Big), \tag{4.4.48}$$

where $\widetilde{w}$ is a standard Brownian motion independent of w and $\mathcal{F}_t^w$ is the sigma-algebra generated by $w(s)$, $s \leq t$ (see (2.3.11) on page 54). One can interpret (4.4.48) as a `probabilistic representation` of the solution of (4.4.48) by averaging the initial condition along the `stochastic characteristic` $X(t,x) = x + \sqrt{2a-\sigma^2}\,\widetilde{w}(t) + \sigma w(t)$.

The objective of this section is to derive a similar representation for the solution of a more general equation

$$\begin{aligned} u_t(t,x) &= \sum_{i,j=1}^{\mathrm{d}} a_{ij}(t,x)u_{x_ix_j}(t,x) + \sum_{i=1}^{\mathrm{d}} b_i(t,x)u_{x_i}(t,x) + c(t,x)u(t,x) \\ &+ f(t,x) + \sum_{k\geq 1}\Big(\sum_{i=1}^{\mathrm{d}} \sigma_{ik}(t,x)u_{x_i}(t,x) \\ &+ \nu_k(t,x)u(t,x) + g_k(t,x)\Big)\dot{w}_k(t), \\ &0 < t \leq T, \ x \in \mathbb{R}^d; \ u(0,x) = u_0(x). \end{aligned} \tag{4.4.49}$$

It will be essential for our analysis to have randomness in the equation coming only from the Brownian motions w_k; all other objects (coefficients a, b, c, σ, ν, free terms f, g, and the initial condition u_0) are assumed to be non-random.

To get an idea of what to expect, we have to review the original method of characteristics for deterministic first-order PDEs: the special structure of Eq. (4.4.47) (more precisely, the fact that the coefficients a and σ do not depend on time) make (4.4.48) somewhat misleading.

Recall that, for the first-order partial differential equation $a(x,y)u_x + b(x,y)u_y = 0$, a `characteristic curve`, or simply a `characteristic`, is a solution of the system of two ordinary differential equations $\dot{x}(t) = a\big(x(t), y(t)\big)$, $\dot{y}(t) = b\big(x(t), y(t)\big)$. If $\big(x(t), y(t)\big)$ is a characteristic curve and u is a solution of the PDE, then, by the chain rule, $(d/dt)u\big(x(t), y(t)\big) = 0$. If the values of u are known on some curve $h(x,y) = 0$ that is not a characteristic curve, then the value of u can be determined at every point (x_0, y_0) by finding the intersection (x^*, y^*) of the characteristic curve through the point (x_0, y_0) with the curve $h(x,y) = 0$; then $u(x_0, y_0) = u(x^*, y^*)$.

As an illustration, consider the initial value problem for the transport equation $u_t + cu_x = 0$, $u(0,x) = \varphi(x)$. The characteristic equation is $dt/ds = 1$, $dx/ds = c$ (with t already in use, we parameterize the characteristics with s). To find $u(t_0, x_0)$ for $t_0 > 0$, we find the characteristic curve through the point (t_0, x_0): $t(s) = t_0 + s$, $x(s) = x_0 + cs$. The curve intersects the line $t = 0$ when $s = -t_0$, and then $x^* = x_0 - ct_0$. This means

$$u(t_0, x_0) = \varphi(x^*) = \varphi(x_0 - ct_0). \tag{4.4.50}$$

The result is the familiar formula for the solution: $u(t,x) = u(0, x - ct)$. The important feature of this derivation is that, to find the solution of the initial-value problem, we run the characteristic curve backward in time: from the current time $t = t_0$ to the initial time $t = 0$. This time-reversal, that is, running the characteristic curve backward in time to solve an initial-value problem, will be essential for our analysis of equation (4.4.49).

As a further illustration of time reversal in the method of characteristics, let us review the probabilistic representation of solution for deterministic parabolic equations. We start with the time homogeneous case:

$$v_t = \frac{1}{2}\sum_{i,j=1}^{\mathrm{d}} a_{ij}(x)v_{x_ix_j} + \sum_{i=1}^{\mathrm{d}} b_i(x)v_{x_i} + c(x)v + f(t,x), \tag{4.4.51}$$

$t > 0,\ x \in \mathbb{R}^{\mathrm{d}},\ v(0,x) = v_0(x).$

Theorem 4.4.19 *Assume that*

- *the functions a_{ij}, b_i, c are non-random, bounded, and uniformly Lipschitz continuous in x;*
- *the functions f, v_0 are non-random, continuous, and of polynomial growth in x;*
- *the function v is a classical solution of* (4.4.51) *and is of polynomial growth in x.*

Then

(a) there exists a square matrix $\sigma = (\sigma_{ik}(x),\ i,k = 1,\dots,\mathrm{d}$ with Lipschitz continuous entries such that $\sum_{k=1}^{\mathrm{d}} \sigma_{ik}\sigma_{jk} = a_{ij}$;
(b) For every $x \in \mathbb{R}^{\mathrm{d}}$, the system of stochastic ordinary differential equations

$$X_i(t,x) = x_i + \int_0^t b_i\big(X(s,x)\big)ds + \sum_{k=1}^{\mathrm{d}} \int_0^t \sigma_{ik}\big(X(s,x)\big)dw_k(s) \qquad (4.4.52)$$

has a unique strong solution [as usual, $(w_1,\dots,w_{\mathrm{d}})$ are independent standard Brownian motions], and can serve as a stochastic characteristic for equation (4.4.51)*:*

$$\begin{aligned} v(t,x) = \mathbb{E}\Big(& v_0\big(X(t,x)\big)e^{\int_0^t c(X(s,x))ds} \\ & + \int_0^t e^{\int_0^s c(X(r,x))dr} f\big(t-s, X(s,x)\big)ds\Big). \end{aligned} \qquad (4.4.53)$$

Proof Part (a) is a multi-dimensional analogue of Lipschitz continuity of $f(t) = \sqrt{t}$ on every interval away from zero; for details, see, for example, Stroock and Varadhan [214, Theorem 5.2.2].

In Part (b), solvability of (4.4.52) follows from global Lipschitz continuity of the coefficients. To establish (4.4.53), fix $T > 0$, use t to denote the time variable, apply Itô formula to the function

$$F(t,x) = v(T-t, X(t,x))e^{\int_0^t c(X(s,x))ds},\ 0 \le t \le T,$$

take the expectation on both sides, and notice that $F(0,x) = v(T,x)$, $F(T,x) = v_0(X(T,x))e^{\int_0^T c(X(s,x))ds}$. The result is equality (4.4.53). We need polynomial growth assumption to be sure that all expectations are finite, and we need continuity of f and v_0 to make sense out of expressions such as $f\big(t-s, X(s,x)\big)$.
This concludes the proof of Theorem 4.4.19.
Note that

(i) Representation (4.4.53) is not **canonical** in the sense that representation of the solution is in terms of $f(t-s,x)$ rather than $f(s,x)$;
(ii) The proof cannot be carried out if the functions a_{ij}, b_i depend on time;

(iii) The same characteristic equation represents the solution v at the point x for all $t \geq 0$: if we got $v(t_1, x)$ from (4.4.53) and now want $v(t_1, x)$ for some $t_2 > t_1$, we only need to solve (4.4.52) on $[t_1, t_2]$.

Items (i) and (ii) can be corrected by reversing the time and considering a **backward** equation for the characteristic:

$$Y_i(s, x; t) = x_i + \int_s^t b_i\big(Y(r, x; t)\big)dr + \sum_{k=1}^{\mathrm{d}} \int_s^t \sigma_{ik}\big(Y(r, x; t)\big) * dw_k(r), \tag{4.4.54}$$

$s \leq t$, where, for fixed $t > 0$ and $0 \leq s \leq t$,

$$\int_s^t F(r) * dw(r) = \int_0^{t-s} F(t-r)d\big(w(t) - w(t-r)\big). \tag{4.4.55}$$

Equation (4.4.54) is a backward-in-time stochastic equation, which is reduced to the usual forward-in-time equation via a time change. [This is very different from backward stochastic differential equations, which are forward in time and seek an adapted solution given a terminal condition.]

Note that, for fixed $t > 0$ and $r \in [0, t]$, the process $\bar{w}(r) = w(t) - w(t-r)$ is a standard Brownian motion. Then

$$\begin{aligned} v(t, x) = \mathbb{E}\Big(& v_0\big(Y(0, x; t)\big)e^{\int_0^t c\left(Y(s,x;t)\right)ds} \\ & + \int_0^t e^{\int_s^t c\left(Y(r,x;t)\right)dr} f\big(s, Y(s, x; t)\big)ds\Big). \end{aligned} \tag{4.4.56}$$

The resulting representation is canonical and easily extends to time-dependent coefficients a, b, c, but there is a price to pay: now, a new equation (4.4.54) must be solved for every new t at which the solution of (4.4.53) is represented. Indeed, changing t to t_1 changes the terminal condition in (4.4.54), meaning that the equation must be solved from scratch.

Exercise 4.4.20 (C)

(a) Verify (4.4.56) by reversing time in (4.4.53).
(b) Write an analogue of (4.4.56) when the coefficients a, b, c in (4.4.51) depend on t.

If $f = 0$ and $c = 0$, then there is a complete analogy between (4.4.56) and (4.4.50): to find the solution of the initial value problem at point (t_0, x_0), $t_0 > 0$, we run the corresponding characteristic backward in time from the point (t_0, x_0) and evaluate the initial condition at the point where the characteristic hits the hyperplane $t = 0$ in the (t, x) space.

If we insist on keeping (4.4.52) as a characteristic, then define

$$U(t,x) = \mathbb{E}\left(v_0(X(T-t))e^{\int_t^T c(X(s))ds} + \int_t^T e^{\int_t^s c(X(r))dr} f(s, X(s))ds \right). \tag{4.4.57}$$

The function U satisfies

$$-U_t = \frac{1}{2}\sum_{i,j=1}^{d} a_{ij}(x)U_{x_ix_j} + \sum_{i=1}^{d} b_i(x)U_{x_i} + c(x)v + f(t,x), \tag{4.4.58}$$

$0 \le t < T$, $x \in \mathbb{R}^d$, $U(T,x) = v_0(x)$ (a backward parabolic equation): see Karatzas and Shreve [103, Theorem 5.7.6].

Exercise 4.4.21 (B) Find a backward in time deterministic PDE whose solution can be represented using (4.4.54), and write the corresponding representation.

Some of the problems that can be addressed using probabilistic representations are

1. Maximum principle for the PDE. For example, if the initial condition v_0 and the free term f are non-negative, then (4.4.53) implies that the solution of (4.4.51) remains non-negative for all $t > 0$;
2. Relaxing conditions on the coefficients in the equations. For example, the right-hand side of (4.4.53) makes sense even if the stochastic equation (4.4.52) has a unique weak, rather than strong, solution.
3. Numerical solution of the PDE by the Monte-Carlo simulations.

Let us summarize what we discussed so far:

(a) The canonical representation by the method of characteristics is of the forward-backward or backward-forward type: it requires one of the equations (either the PDE or the ODE) to be backward in time;
(b) In this canonical representation, a different characteristic is necessary for every point at which the solution is represented.
(c) Forward-forward and backward-backward representations are not canonical, but can have a computational advantage because one characteristic equation works for all t.

Now we are ready to represent the solution of (4.4.49) using the method of characteristics.

Theorem 4.4.22 *Assume that*

- *the functions $a_{ij}, b_i, c, \sigma_{ik}, \nu_k$ are non-random, bounded, continuous, and uniformly Lipschitz continuous in x;*
- *the functions f, v_0, g_k are non-random, continuous, and of polynomial growth in x;*
- *the function u is a classical solution of* (4.4.49) *and is of polynomial growth in x;*

- *the following technical conditions hold:*

$$\sum_{k\geq 1} \sup_{t,x} |\nu_k(t,x)| < \infty, \tag{4.4.59}$$

$$\sum_{k\geq 1} \sup_{t,x} |g_k(t,x)| < \infty, \tag{4.4.60}$$

$$2a_{ij}(t,x) = \sum_{k\geq 1} \sigma_{ik}(t,x)\sigma_{jk}(t,x) + \sum_{\ell=1}^{\mathrm{d}} \widetilde{\sigma}_{i\ell}(t,x)\widetilde{\sigma}_{j\ell}(t,x), \tag{4.4.61}$$

and the functions $\widetilde{\sigma}_{i\ell}(t,x)$ are non-random, bounded, continuous, and uniformly Lischitz continuous in x.

For fixed $t > 0$ and $x \in \mathbb{R}^{\mathrm{d}}$, consider the following stochastic ordinary differential equation in $\mathbb{R}^{\mathrm{d}}$:

$$\begin{aligned} X_i(s,x;t) = x_i &+ \int_s^t B_i\big(r, X(r,x;t)\big)dr + \int_s^t \sum_{k\geq 1} \sigma_{ik}\big(r, X(r,x;t)\big) * dw_k(r) \\ &+ \sum_{\ell=1}^{\mathrm{d}} \int_s^t \widetilde{\sigma}_{i\ell}\big(r, X(r,x;t)\big) * d\widetilde{w}_\ell(r);\ 0 < s \leq t,\ i = 1,\dots,\mathrm{d} \end{aligned} \tag{4.4.62}$$

where $B_i = b_i - \sum_{k\geq 1}\sigma_{ik}\nu_k$, $\widetilde{w}_k$, $k \geq 1$, are independent standard Wiener processes independent of w_k, $k \geq 1$, and $*dw$ is defined in (4.4.55).

Then (4.4.62) *has a unique strong solution and*

$$\begin{aligned} u(t,x) = \mathbb{E}\Bigg(&\int_0^t f\big(s, X(s,x;t)\big)\gamma(s,x;t)ds \\ &+ \sum_{k\geq 1} \int_0^t g_k\big(s, X(s,x;t)\big)\gamma(s,x;t) * dw_k(s) \\ &+ u_0\big(X(0,x;t)\big)\gamma(0,x;t) \,\Big|\mathcal{F}_t^w \Bigg), \end{aligned} \tag{4.4.63}$$

where $\mathcal{F}_t^w$ is the σ-algebra generated by $w_k(s)$, $k \geq 1$, $0 < s < t$, and

$$\begin{aligned} \gamma(s,x;t) = \exp\Bigg(&\int_s^t c\big(r, X(r,x;t)\big)dr \\ &+ \sum_{k\geq 1} \int_s^t \nu_k\big(r, X(r,x;t)\big) * dw_k(r) - \frac{1}{2}\int_s^t \sum_{k\geq 1} \nu_k^2\big(r, X(r,x;t)\big)dr \Bigg). \end{aligned}$$

Proof We begin with some general comments. The result is of the forward-backward type: we have a representation of the solution of a forward in time SPDE using a backward in time SODE. Representation (4.4.63) holds for all $t \geq 0$ and $x \in \mathbb{R}^{\mathrm{d}}$, but to get it, we have to fix t and x and then solve (4.4.62) from $s = t$ to $s = 0$. Having the matrix $2a_{ij} - \sum_k \sigma_{ik}\sigma_{jk}$ uniformly positive definite is sufficient for (4.4.61) to hold: see Stroock and Varadhan [214, Theorem 5.2.2]. Conditions (4.4.59) and (4.4.60) ensure convergence of all infinite sums. Existence and uniqueness of the strong solution of (4.4.62) follows from boundedness and uniform Lipschitz continuity of the coefficients. Since a weak solution will also work, conditions on the coefficients can be relaxed. Polynomial growth of u_0, u, f, g_k is necessary to take expectations after applying the Itô formula.

The proof of (4.4.63) is carried out by reducing the SPDE to a deterministic equation and applying (4.4.56). Here is an outline:

- Since t is fixed in (4.4.63), it will be convenient to set $t = T$ in (4.4.63), and use t as a the time variable.
- Define

$$\begin{aligned} Y(T,x) = & \int_0^T f\big(s, X(s,x;T)\big)\gamma(s,x;T)ds \\ & + \sum_{k\geq 1}\int_0^T g_k\big(s, X(s,x;T)\big)\gamma(s,x;T) * dw_k(s) \\ & + u_0\big(X(0,x;T)\big)\gamma(0,x;T). \end{aligned} \tag{4.4.64}$$

- Let $h = (h_1(s), \ldots, h_N(s)), 0 < s < T$, be a collection of smooth functions; N is arbitrary but finite if there are infinitely many Wiener processes in the SPDE (4.4.49); otherwise, N is the number of those Wiener processes.
- Define

$$\mathcal{E}_h(t) = \exp\left(\sum_{k=1}^{N}\left(\int_0^t h_k(s)dw_k(s) - \frac{1}{2}\int_0^t h_k^2(s)ds\right)\right)$$

  and write $\mathcal{E}_h(T) = \mathcal{E}_h$. Note that $\mathcal{E}_h(t) = \mathbb{E}(\mathcal{E}_h|\mathcal{F}_t^w)$ and $d\mathcal{E}_h(t) = \mathcal{E}_h(t)h_k(t)dw_k(t)$.
- The following result will be necessary: If $\xi \in L_2(\Omega)$, ξ is $\mathcal{F}_T^w$-measurable and $\mathbb{E}(\xi\mathcal{E}_h) = 0$ for every h as above, then $\mathbb{E}\xi^2 = 0$ [181, Lemma 4.3.2]. In other words, any $\mathcal{F}_T^w$-measurable square-integrable random variable ξ is completely characterized by the collection of numbers $\xi_h = \mathbb{E}\big(\xi\mathcal{E}_h\big)$:

$$\mathbb{E}\big(\xi\mathcal{E}_h\big) = \mathbb{E}\big(\xi\mathcal{E}_h\big) \text{ for all } h \iff \xi = \eta \text{ with probability one.} \tag{4.4.65}$$

We refer to this result as `completeness of the system` $\mathcal{E}_h$. To some extend, this is an analogue of saying that equality of the Fourier transforms implies equality of functions.

- By (4.4.65), equality (4.4.63) is equivalent to

$$\mathbb{E}\big(u(T,x)\mathcal{E}_h\big) = \mathbb{E}\big(\mathcal{E}_h \mathbb{E}(Y(T,x)|\mathcal{F}_T^w)\big). \tag{4.4.66}$$

So, all we need is to establish (4.4.66).

- To prove (4.4.66), define $U_h(s,x) = \mathbb{E}\big(u(s,x)\mathcal{E}_h\big)$. Since u is $\mathcal{F}_t^w$-adapted, we have

$$U_h(t,x) = \mathbb{E}\big(u(t,x)\mathcal{E}_h\big) = \mathbb{E}\big(u(t,x)\mathbb{E}(\mathcal{E}_h|\mathcal{F}_t^w)\big) = \mathbb{E}\big(u(t,x)\mathcal{E}_h(t)\big).$$

By the Itô formula

$$\frac{\partial U_h}{\partial t} = \sum_{i,j=1}^{\mathrm{d}} a_{ij}\frac{\partial^2 U_h}{\partial x_i x_j} + \sum_{i=1}^{\mathrm{d}} b_i \frac{\partial U_h}{\partial x_i} + c\, U_h + f$$
$$+ \sum_{k=1}^{N}\left(\sum_{i=1}^{\mathrm{d}} \sigma_{ik}\frac{\partial U_h}{\partial x_i} + \nu_k U_h + g_k h_k\right)$$

with initial condition $U_h|_{t=0} = u_0$.

- Introduce a new probability measure $d\mathbb{P}'_T = \mathcal{E}_h d\mathbb{P}_T$, where $\mathbb{P}_T$ is the restriction of $\mathbb{P}$ to $\mathcal{F}_T^w$. Then, with $\mathbb{E}'$ denoting the corresponding expectation,

$$U_h(T,x) = \mathbb{E}' Y(T,x), \tag{4.4.67}$$

where $Y(t,x)$ is defined in (4.4.64) and where we use the Girsanov theorem [103, Theorem 3.5.1] and the probabilistic representation (4.4.56) for deterministic PDEs. In particular, it is the Girsanov theorem that produces the modified drift in (4.4.49).

- To complete the proof, we note that

$$\mathbb{E}' Y(T,x) = \mathbb{E}\big(\mathcal{E}_h Y(T,x)\big) = \mathbb{E}\Big(\mathbb{E}\big(\mathcal{E}_h Y(T,x)\big|\mathcal{F}_T^w\big)\Big),$$

recall that $U_h(T,x) = \mathbb{E}\big(u(T,x)\mathcal{E}_h\big)$, and get (4.4.66) from (4.4.67).

The concludes the proof of Theorem 4.4.22.

Exercise 4.4.23 (C)

(a) In (4.4.49), assume that $f \geq 0$, $u_0 \geq 0$, and $g_k = 0$. Show that $u \geq 0$ (this is a version of the `maximum principle` for (4.4.49)).
(b) Verify that setting σ, ν, and g to zero results in (4.4.56).

4.4.4 Probabilistic Representation of the Solution, Part II: Measure-Valued Solutions and the Filtering Problem

Measure-valued solutions were first introduced on page 47. These solutions are considered for a special type of linear parabolic equations involving formal adjoint operators and lead to a new type of the existence/uniqueness result (because the collection of measures is not a Hilbert space). In this section, we construct measure-valued solutions for stochastic parabolic equations and apply the result to the problem of optimal nonlinear filtering of diffusion processes.

If A is a partial differential operator in $\mathbb{R}^d$, then its formal adjoint $\mathrm{A}^\top$ is defined by the equality

$$(\mathrm{A}f, g)_{L_2(\mathbb{R}^d)} = (f, \mathrm{A}^\top g)_{L_2(\mathbb{R}^d)}$$

for all smooth compactly supported functions f, g. For example, if, in $\mathbb{R}$, $\mathrm{A}f(x) = a(x)f''(x)$, then $\mathrm{A}^\top g(x) = (a(x)g(x))''$.

Exercise 4.4.24 (C) Verify that if

$$\mathrm{A}f = \frac{1}{2}\sum_{i,j=1}^{\mathrm{d}} a_{ij}(\omega,t,x)\frac{\partial^2 f}{\partial x_i \partial x_j} + \sum_{i=1}^{\mathrm{d}} b_i(\omega,t,x)\frac{\partial f}{\partial x_i} + c(\omega,t,x)f, \tag{4.4.68}$$

then

$$\mathrm{A}^\top g = \frac{1}{2}\sum_{i,j=1}^{\mathrm{d}} \frac{\partial^2\big(a_{ij}(\omega,t,x)f\big)}{\partial x_i \partial x_j} - \sum_{i=1}^{\mathrm{d}} \frac{\partial\big(b_i(\omega,t,x)f\big)}{\partial x_i} + c(\omega,t,x)f. \tag{4.4.69}$$

Our objective is to study measure-valued solutions for stochastic partial differential equations of the form

$$u_t = \mathrm{A}^\top u + \sum_{k\geq 1} \mathrm{M}_k^\top u\, \dot{w}_k(t),\ 0 < t \leq T, \tag{4.4.70}$$

where A is from (4.4.68) and

$$\begin{aligned} \mathrm{M}_k u &= \sum_{i=1}^{\mathrm{d}} \sigma_{ik}(\omega,t,x)\frac{\partial u}{\partial x_i} + \nu_k(\omega,t,x)u, \\ \mathrm{M}_k^\top &= -\sum_{i=1}^{\mathrm{d}} \frac{\partial\big(\sigma_{ik}(\omega,t,x)u\big)}{\partial x_i} + \nu_k(\omega,t,x)u. \end{aligned} \tag{4.4.71}$$

To define the measure-valued solution of (4.4.70) we assume that

- all coefficients are measurable and $\mathcal{F}_t$-adapted;
- the initial condition $u(0,\cdot) = \mu_0$ is a finite positive measure on $\mathbb{R}^{\mathrm{d}}$;
- $\sup\limits_{\omega,t,x} \sum_k \sigma_{ik}^2(\omega,t,x) < \infty$ for all $i = 1,\ldots,\mathrm{d}$, and $\sup\limits_{\omega,t,x} \sum_k \nu_k^2(\omega,t,x) < \infty$

Definition 4.4.25 A collection of random measures $\mu_t = \mu_t(\omega, dx)$ on $\mathbb{R}^{\mathrm{d}}$ with the Borel sigma-algebra is called a `measure-valued solution` of (4.4.70) if

1. μ_t is a positive measure for every $(\omega,t) \in \Omega \times [0,T]$: $\mu_t[A] \geq 0$ for all Borel subsets A of $\mathbb{R}$;
2. μ_t is a finite measure on $\mathbb{R}^d$ for every $(\omega,t) \in \Omega \times [0,T]$:

$$\mu_t[\mathbb{R}^{\mathrm{d}}] := \int_{\mathbb{R}^{\mathrm{d}}} \mu_t(dx) < \infty;$$

3. For every $T > 0$, $\int_0^T \mathbb{E}|\mu_t[\mathbb{R}^d]|^2 dt < \infty$;
4. For every bounded measurable function f, $\mu_t[f] = \int_{\mathbb{R}^{\mathrm{d}}} f(x)\mu_t(dx)$ is an $\mathcal{F}_t$-adapted continuous process;
5. For every smooth compactly supported function $f = f(x)$,

$$\mu_t[f] = \mu_0[f] + \int_0^t \mu_s[(\mathrm{A}f)(s,\cdot)]ds + \sum_{k\geq 1} \int_0^t \mu_s[(\mathrm{M}_k f)(s,\cdot)]dw_k(s). \tag{4.4.72}$$

The following theorem is the main result about existence, uniqueness, and representation of the measure-valued solution for equation (4.4.70).

Theorem 4.4.26 *Assume that*

- *the functions* $a_{ij}, b_i, \sigma_{ik}, \nu_k$ *are adapted, bounded, continuous in* (t,x), *and uniformly Lipschitz continuous in x;*
- *the initial condition* u_0 *is a probability measure* μ_0 *on* $\mathbb{R}^{\mathrm{d}}$;
- *the following technical conditions hold:*

$$\sum_{k\geq 1} \sup_{\omega,t,x} |\nu_k(\omega,t,x)| < \infty, \tag{4.4.73}$$

$$a_{ij}(\omega,t,x) = \sum_{k\geq 1} \sigma_{ik}(\omega,t,x)\sigma_{jk}(\omega,t,x) + \sum_{\ell=1}^{\mathrm{d}} \widetilde{\sigma}_{i\ell}(\omega,t,x)\widetilde{\sigma}_{j\ell}(\omega,t,x) \tag{4.4.74}$$

and the functions $\widetilde{\sigma}_{i\ell}(\omega,t,x)$ *are adapted, bounded, continuous, and uniformly Lischitz continuous in x.*

Define the process $X = X(t) \in \mathbb{R}^{\mathrm{d}}$ *by*

$$X_i(t) = X_i(0) + \int_0^t B_i\,(s, X(s))\,ds + \int_0^t \sigma_{ik}\,(s, X\,(s))\,dw_k\,(s) \\ + \int_0^t \widetilde{\sigma}_{ik}\,(s, X\,(s))\,d\widetilde{w}_k\,(s) \tag{4.4.75}$$

where $B_i = b_i - \sum_{k\geq 1}\sigma_{ik}\nu_k$ *and* $\widetilde{w}_k$, $k \geq 1$, *are independent standard Wiener processes, independent of* $\{\mathcal{F}_t\}_{0<t\leq T}$, *the initial condition* $X(0)$ *is independent of* $w, \widetilde{w}$ *and has distribution* μ_0. *Also define*

$$\phi(t) = \exp\Bigg(\int_0^t c(s, X(s))ds + \int_0^t \nu_k(s, X(s))dW_k(s) \\ - \frac{1}{2}\int_0^t \sum_{k\geq 1}\nu_k^2(s, X(s))ds\Bigg).$$

Then there exists a unique measure-valued solution of (4.4.70) and, for every bounded measurable function $f = f(x)$,

$$\mu_t[f] = \mathbb{E}\Big(f\big(X(t)\big)\phi(t)|\mathcal{F}_t\Big). \tag{4.4.76}$$

Proof We begin with some general comments. Having the matrix $a_{ij} - \sum_k \sigma_{ik}\sigma_{jk}$ uniformly positive definite is sufficient for (4.4.74) to hold: see Stroock and Varadhan [214, Theorem 5.2.2]. Condition (4.4.73) ensures convergence of the corresponding infinite sums. Boundedness and uniform Lipschitz continuity of the coefficients ensure existence and uniqueness of the strong solution of (4.4.75). Since a weak solution will also work, conditions on the coefficients can be relaxed.

The proof is based on the following computation. Using Itô formula,

$$d\phi(t) = \phi(t)\Big(c dt + \sum_k \nu_k dw_k\Big),$$

$$df(X(t)) = \sum_i \Big(\big(b_i - \sum_k \sigma_{ik}\nu_k\big)f_{x_i}\,dt + \sum_k f_{x_i}\sigma_{ik}dw_k \\ + \sum_\ell f_{x_i}\widetilde{\sigma}_{i\ell}\,d\widetilde{w}_\ell + \frac{1}{2}\sum_j a_{ij}f_{x_ix_j}\Big),$$

$$d\Big(\phi(t)f(X(t))\Big) = f d\phi + \phi\,df + \sum_{i,k}\phi f_{x_i}\sigma_{ik}\nu_k \\ = \phi\,(\mathrm{A}f)\,dt + \phi\sum_k(\mathrm{M}_k f)\,dw_k + \phi\sum_{i,\ell} f_{x_i}\widetilde{\sigma}_{i\ell}d\widetilde{w}_\ell, \tag{4.4.77}$$

or

$$\phi(t)f(X(t)) = f(X(0)) + \int_0^t \phi(s)\mathrm{A}f(X(s))ds$$
$$+\sum_k \int_0^t \phi(s)\,\mathrm{M}_k f(X(s))\,dw_k(s) + \sum_{i,\ell} \int_0^t \phi(s)f_{x_i}(X(s))\widetilde{\sigma}_{i\ell}d\widetilde{w}_\ell(s) \tag{4.4.78}$$

Next, we take conditional expectation $\mathbb{E}(\cdot|\mathcal{F}_t)$ on both sides of (4.4.78). The result is

$$\begin{aligned}\mathbb{E}\Big(\phi(t)f(X(t))\big|\mathcal{F}_t\Big) = \mu_0[f] + \int_0^t \mathbb{E}\Big(\phi(s)\,\mathrm{A}f(X(s))\big|\mathcal{F}_s\Big)ds \\ +\sum_k \int_0^t \mathbb{E}\Big(\phi(s)\,\mathrm{M}_k f(X(s))\big|\mathcal{F}_s\Big)dw_k(s).\end{aligned} \tag{4.4.79}$$

To get from (4.4.78) to (4.4.79), we use the following results:

1. because $\tilde{w}_\ell$ is independent of $\mathcal{F}_t$,

$$\mathbb{E}\Big(\int_0^t \phi(s)f_{x_i}(X(s))\widetilde{\sigma}_{i\ell}d\widetilde{w}_\ell(s)\Big|\mathcal{F}_t\Big) = 0$$

   (see Liptser and Shiryaev [139, Corollary 2 to Theorem 5.13]);
2. by a version of the Fubini theorem [139, Lemma 8.3],

$$\mathbb{E}\Big(\int_0^t \phi(s)\,\mathrm{A}f(X(s))ds\big|\mathcal{F}_t\Big) = \int_0^t \mathbb{E}\Big(\phi(s)\,\mathrm{A}f(X(s))\big|\mathcal{F}_s\Big)ds$$

3. by a different version of the Fubini theorem [139, Theorem 5.14],

$$\mathbb{E}\Big(\int_0^t \phi(s)\,\mathrm{M}_k f(X(s))dw_k(s)\big|\mathcal{F}_t\Big) = \int_0^t \mathbb{E}\Big(\phi(s)\,\mathrm{M}_k f(X(s))\big|\mathcal{F}_s\Big)dw_k(s).$$

Now let us compare (4.4.79) and (4.4.72). The left-hand sides are the same. If μ_t is a measure-valued solution of (4.4.70), then (4.4.72) holds, and, by matching the terms on the right-hand side in (4.4.79) and (4.4.72), we conclude that μ_t satisfies (4.4.76). In other words, every measure-valued solution of (4.4.70) must satisfy (4.4.76), and, since the right-hand side of (4.4.76) does not depend on μ, the uniqueness of the measure-valued solution follows.

To prove existence, there are two options:

(i) Assume μ_0 has a smooth density and that all the coefficients are smooth, and reduce the problem to the Hilbert space setting. Then argue that the Hilbert space solution is the same as the density of the measure-valued solution. In general, use a limiting procedure, approximating the coefficients and initial condition with smooth objects.

(ii) Argue that there exists a measure μ_t such that, for every smooth function f, relation (4.4.76) holds. The construction of μ_t is similar to the construction of the regular conditional distribution.

For details and further discussions, see Rozovskii [199, Sect. 5.3].

This concludes the proof of Theorem 4.4.26.

Let us now discuss a connection between formula (4.4.76) and nonlinear filtering of diffusion processes. Consider two diffusion processes (X, Y) defined by

$$
\begin{aligned}
dX_i(t) &= b_i(t, X(t))dt + \sum_{\ell=1}^{m} \sigma_{i\ell}(t, X(t))d\widetilde{w}_\ell(t) + \sum_{k=1}^{n} \rho_{ik}(t, X(t))dw_k(t),\\
dY_k(t) &= h(t, X(t))dt + dw_k(t),
\end{aligned}
\tag{4.4.80}
$$

where $0 < t \leq T$, $i = 1, \ldots, \mathrm{d}$, $k = 1, \ldots, n$, $w_k, \widetilde{w}_\ell$ are independent standard Brownian motions, $X(0) \in \mathbb{R}^{\mathrm{d}}$ is a random variable independent of $w_k, \widetilde{w}_\ell$, $Y(0) = 0$. Thinking of X as an unobservable process and Y as observations, the `filtering problem` is to find the conditional distribution of $X(t)$ given the observations $Y(s)$, $s \leq t$. The problem is called nonlinear because Eq. (4.4.80) is nonlinear.

We assume that the coefficients b, σ, ρ, h are bounded, continuous in (t, x), and uniformly Lipschitz continuous in x. The coefficients can also be random, but the randomness at time t can only come through the dependence on $Y(s)$, $s \leq t$; for more details, see Rozovskii [199, Sect. 6.1]. We define the following operators:

$$
\begin{aligned}
\mathrm{A} &= \frac{1}{2} \sum_{i,j=1}^{\mathrm{d}} \left(\sum_{\ell=1}^{n} \sigma_{i\ell}\sigma_{j\ell} + \sum_{k=1}^{m} \rho_{ik}\rho_{jk} \right) \frac{\partial^2}{\partial x_i \partial x_j} + \sum_{i=1}^{\mathrm{d}} b_i \frac{\partial}{\partial x_i},\\
\mathrm{M}_k &= h_k + \sum_{i=1}^{\mathrm{d}} \rho_{ik} \frac{\partial}{\partial x_i}.
\end{aligned}
\tag{4.4.81}
$$

Here is the main result about the filtering problem, describing the conditional distribution of X as a measure-valued solution of a linear stochastic parabolic equations.

Theorem 4.4.27 *If the processes X and Y are defined by* (4.4.80), *then*

(a) for every $0 < t \leq T$, the conditional distribution of $X(t)$ given $Y(s)$, $0 \leq s \leq t$ has the following representation:

$$
\mathbb{P}\big(X(t) \in G | Y(s), 0 \leq s \leq t\big) = \frac{\mu_t[G]}{\mu_t[\mathbb{R}^{\mathrm{d}}]}, \tag{4.4.82}
$$

where G is a Borel subset of $\mathbb{R}^{\mathrm{d}}$;

(b) *the family of measures* μ_t, $0 \le t \le T$, *is the measure-valued solution of the equation*

$$du = \mathrm{A}^{\top} u\, dt + \sum_{k=1}^{m} \mathrm{M}_k^{\top} u\, dY_k, \tag{4.4.83}$$

with initial condition μ_0 *equal to the probability distribution of* $X(0)$.

Proof Define the function

$$Z(t) = \exp\left(\sum_{k=1}^{m} \int_0^t h_k(s, X(s))dY(s) - \frac{1}{2}\sum_{k=1}^{m} \int_0^t h_k^2(s, X(s))ds \right) \tag{4.4.84}$$

and a new probability measure $\bar{\mathbb{P}}$ by

$$Z(T)\, d\bar{\mathbb{P}} = d\mathbb{P}. \tag{4.4.85}$$

Under the new measure $\bar{\mathbb{P}}$,

- The process Y is an m-dimensional standard Brownian motion, independent of $\tilde{w}_\ell$ (this is Girsanov's theorem);
- The process X satisfies

$$dX_i = \Big(b_i - \sum_{k=1}^{n} \rho_{ik} h_k\Big)dt + \sum_{k=1}^{n} \rho_{ik} dY_k + \sum_{\ell=1}^{m} \sigma_{i\ell}\, d\widetilde{w}_\ell \tag{4.4.86}$$

  (in the original equation for X, replace dw_k with $dY_k - h_k dt$);
- The distribution of $X(0)$ does not change.

If $\bar{\mathbb{E}}$ denotes the expectation with respect to $\bar{\mathbb{P}}$, then direct computations show that, for every bounded measurable function $f = f(x)$, $x \in \mathbb{R}^{\mathrm{d}}$,

$$\mathbb{E}\big(f(X(t))|Y(s),\ 0 \le s \le t\big) = \frac{\bar{\mathbb{E}}\big(Z(t)f(X(t))|Y(s),\ 0 \le s \le t\big)}{\bar{\mathbb{E}}\big(Z(t)|Y(s),\ 0 \le s \le t\big)}. \tag{4.4.87}$$

Indeed, the martingale property of Z under $\bar{\mathbb{P}}$ allows us to replace $Z(t)$ with $Z(T)$ on the right-hand side of (4.4.87) [details are in Exercise 4.4.29 below]. Writing $Z = Z(T), f = f\big(X(t)\big)$, and $\mathcal{Y}$ for the sigma-algebra generated by $Y(s),\ 0 \le s \le t$, we now need to show that

$$\bar{\mathbb{E}}(Z|\mathcal{Y})\, \mathbb{E}(f|\mathcal{Y}) = \bar{\mathbb{E}}(f\, Z|\mathcal{Y}). \tag{4.4.88}$$

Note that

$$\bar{\mathbb{E}}(Z|\mathcal{Y})\,\mathbb{E}(f|\mathcal{Y}) = \bar{\mathbb{E}}\big(Z\mathbb{E}(f|\mathcal{Y})\big|\mathcal{Y}\big).$$

Then, for every $\mathcal{Y}$-measurable random variable ξ,

$$\bar{\mathbb{E}}\big(\xi Z\mathbb{E}(f|\mathcal{Y})\big) = \mathbb{E}(\xi\mathbb{E}(f|\mathcal{Y})) = \mathbb{E}(\xi f)$$

and

$$\bar{\mathbb{E}}\big(\xi\bar{\mathbb{E}}(f\,Z|\mathcal{Y})\big) = \bar{\mathbb{E}}(\xi Zf|\mathcal{Y}) = \mathbb{E}(\xi f),$$

which implies (4.4.88).

Applying (4.4.76) with $\bar{\mathbb{E}}$ instead of $\mathbb{E}$, with the sigma-algebra generated by $Y(s)$, $0 \le s \le t$ in place of $\mathcal{F}_t$, and with $w = Y$, $\sigma = \rho$, $\nu = h$, and $\phi = Z$, we conclude that

$$\bar{\mathbb{E}}\big(Z(t)f(X(t))|Y(s),\ 0 \le s \le t\big) = \mu_t[f],$$

where μ_t is the measure-valued solution of (4.4.83).
This concludes the proof of Theorem 4.4.27.

Exercise 4.4.28 (B)

(a) Verify that, in general, $\mu_t[\mathbb{R}^d] \neq 1$ for $t > 0$.
(b) Verify that $\mu_t[\mathbb{R}^d] = 1$ for all $t \ge 0$ if $h_k = 0$.

Hint. In both cases, integrate (4.4.83) over $\mathbb{R}^d$ and note that $\mathrm{A}1 = 0$ [$\mathrm{A}1$ is the operator A applied to the constant function $f(x) = 1$], and $\mathrm{M}_k 1 = 0$ if $h_k = 0$.

Exercise 4.4.29 (B) Verify that

$$\bar{\mathbb{E}}\big(f(X(t))Z(t)|Y(s), 0 \le s \le t\big) = \bar{\mathbb{E}}\big(f(X(t))Z(T)|Y(s), 0 \le s \le t\big).$$

Hint. Condition on the sigma-algebra $\mathcal{F}_t^{X,Y}$ generated by both $Y(s)$ and $X(s)$, $0 \le s \le t$ and argue that $Z(t)$ is a martingale with respect to $\bar{\mathbb{P}}$ and $\mathcal{F}_t^{X,Y}$.

4.4.5 Further Directions

Of the three types (elliptic, hyperbolic, parabolic), parabolic equations have been studied the most. This could be because of the applications of such equations to nonlinear filtering of diffusion processes and to the study particle systems. This could also be because the heat semigroup is analytically much better than the wave semigroup, and, unlike their elliptic counterparts, parabolic equations can use the full power of the Itô calculus. Given the vast literature on the subject of stochastic parabolic equations, we will only mention a few selected topics.

Our Theorems 4.4.3 and 4.4.12 give more-or-less complete description of the Hilbert space theory for linear stochastic parabolic equations. The main drawback of this theory is inability to provide optimal regularity of the solution. For example, consider the equation driven by space-time white noise

$$u_t = u_{xx}dt + \dot{W}(t,x),\ x \in (0,\pi)$$

with zero initial and boundary conditions. Our study of the closed-form solutions showed that u is almost $1/2$ Hölder continuous in space and almost $1/4$ Hölder continuous in time; see (2.3.19), page 56. On the other hand, applying Theorem 4.4.3 in the normal triple $\big(H^{\gamma+1}((0,\pi)), H^{\gamma}((0,\pi)), H^{\gamma-1}((0,\pi))\big)$ with $\gamma < -1/2$ does not even yield continuity of u; with $g_k(x) = \sqrt{2/\pi}\sin(kx)$, we cannot take $\gamma \geq 1/2$. It would be nice if we could extend Theorem 4.4.3 to Sobolev spaces $H^{\gamma}_p((0,\pi))$ for $p > 2$: then the Sobolev embedding theorem will give the expected continuity in x. This extension is possible, but requires a lot of additional analytical developments: see Krylov [120]. It is also possible to study stochastic parabolic equations directly in the Hölder spaces [161]. For the semigroup approach see Da Prato and Zabzcyk [31].

Beside the usual Gaussian white noise, popular random perturbations for parabolic equations are fractional Brownian motion and Lévy-type noise. In this connection, we mention several papers dealing with fractional Brownian motion [41, 156, 179, 217] and the book by Peszat and Zabczyk [185] dealing with the Lévy-type noise.

There are several books dedicated exclusively to nonlinear filtering of diffusion processes: Bain and Crisan [5], Kallianpur [100], and Xiong [230]. The basic SPDE aspects of the problem are also discussed in Krylov [120, Sect. 8.1] and Rozovskii [199, Chap. 6]. With all these references in mind, our discussion of the filtering problem does not go beyond the very basic result of Theorem 4.4.27.

4.4.6 Problems

Problem 4.4.1 investigates one particular aspect of Theorem 4.4.3: a possibility to relax the regularity condition for the stochastic part of the equation. Problem 4.4.2 suggests an alternative conclusion for Theorem 4.4.3. Problem 4.4.3 investigates possible modifications of the stochastic parabolicity condition. Problem 4.4.4 is about a specific degenerate parabolic equation. Problem 4.4.5 connects stochastic parabolic equations with an infinite system of independent Ornstein-Uhlenbeck processes. Problem 4.4.5 lists several specific parabolic equations in one space variable.

Problem 4.4.1 Consider the equation

$$du = u_{xx}dt + g(x)dw(t),\ 0 < t \le T,\ x \in \mathbb{R}.$$

Show that if $g \in H^{-1}(\mathbb{R})$, then we cannot guarantee that the solution u will be an element of $L_2\big(\Omega \times (0,T); H^1(\mathbb{R})\big)$.

Problem 4.4.2 Recall the space $\mathbb{H}$ ((4.4.15), page 203). Show that, for Eq. (4.4.1), the solution operator $(u_0, f, \{g_k\}) \mapsto u$ is a homeomorphism from $L_2(\Omega; H) \times L_2\big(\Omega \times (0,T); V'\big) \times L_2\big(\Omega \times (0,T); \ell_2(H)\big)$ to $\mathbb{H}$. With ℓ_2 denoting the space of square-integrable sequences,

$$\|\{g_k\}\|^2_{L_2(\Omega\times(0,T);\ell_2(H))} = \sum_{k\ge1} \mathbb{E} \int_0^T \|g_k(t)\|_H^2 dt.$$

Problem 4.4.3 Investigate the possibility of replacing condition (4.4.4) with one of the following:

$$\begin{aligned} 2\int_0^t [\mathrm{A}(s)v(s), v(s)]ds &+ \sum_{k\ge1}\int_0^t \|\mathrm{M}_k v(s)\|_H^2 ds \\ &\le -c_A \int_0^t \|v(s)\|_V^2 ds + M\int_0^t \|v(s)\|_H^2 ds; \end{aligned}$$

$$\begin{aligned} 2\int_0^t \mathbb{E}[\mathrm{A}(s)v(s), v(s)]ds &+ \sum_{k\ge1}\int_0^t \mathbb{E}\|\mathrm{M}_k(s)v(s)\|_H^2 ds \\ &\le -c_A \int_0^t \mathbb{E}\|v(s)\|_V^2 ds + M\int_0^t \mathbb{E}\|v(s)\|_H^2 ds. \end{aligned}$$

Of course, the continuity conditions on A and M_k should also be modified accordingly.

Problem 4.4.4 Create an analogue of the table on page 210 for the equation

$$\begin{aligned} u_t = &\big(a(t,x)u_x\big)_x + b(t,x)u_x + c(t,x)u + f(t,x) \\ &+ \big(\sigma(t,x)u_x + \nu(t,x)u + g(t,x)\big)\dot W^{\mathrm{Q}}(t,x),\ 0 < t \le T,\ x \in \mathbb{R}, \end{aligned}$$

in the normal triple $(H^{r+1}(\mathbb{R}), H^r(\mathbb{R}), H^{r-1}(\mathbb{R}))$, $r \in \mathbb{R}$, where W^{Q} is the Q-cylindrical Brownian motion, written in the basis of Hermite functions $\mathfrak{w}_k$ (see (4.1.50) on page 167) as follows:

$$W^{\mathrm{Q}}(t,x) = \sum_{k\ge1} \sqrt{q_k}\, w_k(t)\, \mathfrak{w}_k(x).$$

Problem 4.4.5 Let $w_k,\ k \geq 1$, be independent standard Brownian motions and let $a_k, b_k, \sigma_k,\ k \geq 1$, be real numbers. Consider a collection of processes $u_k,\ k \geq 1$, defined by

$$a_k \dot{u}_k(t) = b_k u_k(t) + \sigma_k \dot{w}_k(t),\ 0 < t \leq T. \tag{4.4.89}$$

Find conditions on the numbers a_k, b_k, σ_k and the initial data $u_k(0)$ to have

$$\mathbb{E} \sup_{0<t<T} \sum_{k \geq 1} |u_k(t)|^2 < \infty. \tag{4.4.90}$$

Problem 4.4.6 Investigate the following equations, both in $\mathbb{R}$ and on $(0, \pi)$:

$$u_t = u_{xx} + au_x + bu + (cu + \sigma u_x + g)\dot{W}^{\mathrm{Q}}(t,x), \tag{4.4.91}$$

$$u_t = -u_{xxxx} + au_{xx} + (bu + cu_x + \sigma u_{xx} + g)\dot{W}^{\mathrm{Q}}(t,x), \tag{4.4.92}$$

$$u_t = au_{xxx} + bu_{xx} + cu_x + (\sigma u_x + \nu u + g)\dot{W}^{\mathrm{Q}}(t,x), \tag{4.4.93}$$

$$u_t - au_{txx} = bu_{xx} + (cu + \sigma u_x + g)\dot{W}^{\mathrm{Q}}(t,x), \tag{4.4.94}$$

$$u_t - au_{txx} = u_{xxx} + (bu + \sigma u_x + g)\dot{W}^{\mathrm{Q}}(t,x). \tag{4.4.95}$$

Chapter 5
The Polynomial Chaos Method

Separation of variables is a powerful idea in the study of partial differential equations, and the polynomial chaos method is a particular implementation of this idea for stochastic equations. While the elementary outcome ω is typically never mentioned explicitly in the notation of random objects, it is a variable that can potentially be separated from other variables, and the objective of this chapter is to outline a systematic approach to doing just that. Along the way, it quickly becomes clear that many ideas are closely connected to another modern branch of stochastic analysis, namely, Malliavin Calculus, and we explore these connections throughout.

Section 5.1 introduces the orthonormal basis in the space of square-integrable functionals of a Gaussian process, and Sect. 5.2 applies the results to stochastic elliptic equations. Section 5.3 repeats many of the constructions of Sect. 5.1 in the time-dependent setting, and then Sects. 5.4 and 5.5 apply the results to stochastic parabolic equations. Specific examples are discussed in Sect. 5.6. Since may of the constructions do not specifically rely on Gaussian nature of the randomness, Sect. 5.7 presents the corresponding results in the distribution-free setting.

5.1 Stationary Wiener Chaos

5.1.1 Cameron-Martin Basis

Let $\mathbb{F} = (\Omega, \mathcal{F}, \mathbb{P})$ be a probability space. The only assumption on this probability space needed and assumed in this chapter is that $\mathbb{F}$ is "reach enough" to support an infinite sequence of independent standard Gaussian random variables

$$\{\xi_k, k \geq 1\}. \tag{5.1.1}$$

S.V. Lototsky, B.L. Rozovsky, *Stochastic Partial Differential Equations*, Universitext, DOI 10.1007/978-3-319-58647-2_5

Definition 5.1.1 Let Ξ be the σ-algebra generated by $\{\xi_k, k \geq 1\}$ and completed by sets of probability zero

Without loss of generality, we assume that $\Xi \subset \mathcal{F}$.

Exercise 5.1.2 (B) Let X be a Hilbert space and let $\{x_k, k \geq 1\}$ be a sequence with $x_k \in X$. Let $w = \sum_{k\geq 1} x_k \xi_k$, where ξ_k are from (5.1.1) and $\sum_{k\geq 1} \|x_k\|_X^2 < \infty$. Prove that w is an X-valued Gaussian random element.

Next, let $\mathfrak{B}$ be a Gaussian white noise over a real separable Hilbert space $\mathcal{U}$ and $\{\mathfrak{u}_k, k \geq 1\}$ be an orthonormal basis in $\mathcal{U}$. Then (cf. [175, Chap. 1] or Definition 3.2.10 on page 105)

$$\mathfrak{B}(f) = \sum_{k\geq 1} f_k \xi_k,\ f \in \mathcal{U}, \tag{5.1.2}$$

where $\xi_k = \mathfrak{B}(\mathfrak{u}_k)$ is a collection of i.i.d. standard Gaussian random variables, $f_k = (f, \mathfrak{u}_k)_{\mathcal{U}}$ and, for $f, g \in \mathcal{U}$

$$\mathbb{E}\left(\mathfrak{B}(f)\,\mathfrak{B}(g)\right) = (f, g)_{\mathcal{U}}\,. \tag{5.1.3}$$

In the future, the space $L_2(\Xi) := L_2(\Omega, \Xi, \mathbb{P})$ will be referred to as the `Gaussian chaos space`. It is a Hilbert space with scalar product $(\xi, \eta) = \mathbb{E}(\xi\eta)$.

Exercise 5.1.3 (A) Definition 5.1.1 and Exercise 5.1.2 provide two views on the constructions of Gaussian white noise. Are these two constructions equivalent?

Exercise 5.1.4 (C) Confirm that the variance of $\mathfrak{B}(f)$ is $\|f\|_{\mathcal{U}}^2$.

The following construction provides an important and popular example of the Gaussian field $\mathfrak{B}$.

Example 5.1.5 Let G be a smooth bounded domain in $\mathbb{R}^{\mathrm{d}}$, $\{\mathfrak{h}_k,\ k \geq 1\}$, an orthonormal basis in $L_2(G)$, and $\{\xi_k, k \geq 1\}$, a sequence of independent standard Gaussian random variables. Then a formal series

$$\dot{W}(x) = \sum_{k\geq 1} \mathfrak{h}_k(x)\xi_k, \tag{5.1.4}$$

is called `spatial Gaussian white noise` on $L_2(G)$. The variance of $\dot{W}(x)$ is

$$\mathbb{E}\left|\dot{W}(x)\right|^2 = \sum_{k\geq 1} \mathfrak{h}_k^2(x). \tag{5.1.5}$$

Therefore,

$$\mathbb{E}\left\|\dot{W}\right\|_{L_2(G)}^2 = \infty. \tag{5.1.6}$$

Property (5.1.6) indicates that the spatial Gaussian white noise is quite "rough". This is not unexpected. Indeed the derivative of standard Brownian motion $w(t)$ is also rough.

Let H be a real separable Hilbert space with the inner product $(\cdot,\cdot)_H$ and

$$\mathfrak{B} = \sum_{k\geq 1} \mathfrak{u}_k\,\xi_k, \tag{5.1.7}$$

where $\{\mathfrak{u}_k, k \geq 1\}$ is a complete orthonormal basis in H. Then $\mathfrak{B}$ is a (Gaussian) **white noise** on H and, for $f \in H$,

$$\mathfrak{B}(f) = \sum_{k\geq 1} (f, \mathfrak{u}_k)_H\,\xi_k \tag{5.1.8}$$

and

$$\mathbb{E}\,|\mathfrak{B}(f)|^2 = \sum_{k\geq 1} (f, \mathfrak{u}_k)_H^2 = \|f\|_H^2\,. \tag{5.1.9}$$

If X is another Hilbert space, then $L_2(\Xi, X)$ stands for the Hilbert space $L_2(\Omega, \Xi, \mathbb{P}; X)$ with the norm

$$\|x\|^2_{L_2(\Xi,X)} = \mathbb{E}\,\|x\|_X^2\,. \tag{5.1.10}$$

In the cxample below we use the notational agreement following Definition 3.2.10.

Example 5.1.6

(a) Let $H := \mathbb{L}_2[0,\pi]$, $\mathfrak{u}_k := \sqrt{\frac{2}{\pi}}\sin(kx)$. Then Gaussian white noise $\mathfrak{B} = \mathfrak{B}(x)$ on H is

$$\mathfrak{B}(x) = \sum_{k=1}^{\infty} \xi_k \sqrt{\frac{2}{\pi}}\sin(kx)\,,$$

where $\{\xi_k, k \geq 1\}$ is a sequence of independent standard Gaussian random variables, and, for $f \in \mathbb{L}_2[0,\pi]$

$$\mathfrak{B}(f) = \int_0^{\pi} f(x)\,\mathfrak{B}(x)\,dx = \sum_{k=1}^{\infty} \xi_k \sqrt{\frac{2}{\pi}} \int_0^{\pi} f(x)\sin(kx)\,dx.$$

Clearly, $\mathbb{E}\,\|\mathfrak{B}\|^2_{\mathbb{L}_2[0,\pi]} = \infty$, but $\mathbb{E}|\mathfrak{B}(f)|^2 = \|f\|^2_{\mathbb{L}_2[0,\pi]} < \infty$.

(b) More generally, assume that $H := \mathbb{L}_2(\mathcal{O})$, where $\mathcal{O}$ is an open set in $\mathbb{R}^{\mathrm{d}}$ and $\mathfrak{u}_k = \mathfrak{u}_k(x)$, $x \in \mathcal{O}$, is an orthonormal basis in $\mathbb{L}_2(\mathcal{O})$. Then, for $f \in \mathbb{L}_2(\mathcal{O})$,

$$\mathfrak{B}(f) = \sum_{k=1}^{\infty} \xi_k \int_{\mathcal{O}} f(x)\, \mathfrak{u}_k(x)\, dx.$$

(c) If $H = \mathbb{R}$, then $\dot{W}(f) = f\xi_1$, because $\mathbb{R} = L_2(S)$, where S is the space consisting of a single point.

Next we will introduce a complete orthonormal basis in the space $L_2(\Xi)$. This basis is usually referred to as `Cameron-Martin basis` ([20]).

The following exercise shall re-acquaint the reader with the notion of Hermite polynomials, which will play a crucial role in this section.

Exercise 5.1.7 (A) Prove that Hermite polynomials $\mathrm{H}_n(x)$, originally defined by (2.3.37) on page 61, have an alternative representation

$$\mathrm{H}_n(x) = (-1)^n e^{x^2/2} \frac{d^n}{dx^n} e^{-x^2/2}. \tag{5.1.11}$$

Denote $\mathrm{h}_n(x) = \mathrm{H}_n(x)/\sqrt{n!}$. Since

$$\frac{1}{\sqrt{2\pi}} \int_{-\infty n}^{\infty} \mathrm{h}_n(x) e^{-x^2/2} dx = 1,$$

we will refer to $\mathrm{h}_n(x)$ as `normalized Hermite polynomials`.

The following definition is technical but still very important for the construction of an orthonormal basis in $L_2(\Xi)$.

Definition 5.1.8 Let $\mathcal{J}^1$ be the collection of multi-indices $\boldsymbol{\alpha}$ with $\boldsymbol{\alpha} = (\alpha_1, \alpha_2, \ldots)$ so that each α_k is a non-negative integer and $|\boldsymbol{\alpha}| := \sum_{k\geq 1} \alpha_k < \infty$. For $\boldsymbol{\alpha}, \boldsymbol{\beta} \in \mathcal{J}^1$, define

$$\boldsymbol{\alpha} + \boldsymbol{\beta} = (\alpha_1 + \beta_1, \alpha_2 + \beta_2, \ldots) \text{ and } \quad \boldsymbol{\alpha}! = \prod_{k\geq 1} \alpha_k!.$$

The multi-index consisting of all zeroes will be denoted by $(\mathbf{0})$.

Let us write

$$\xi_{\boldsymbol{\alpha}} = \prod_{k\geq 1} \mathrm{h}_{\alpha_k}(\xi_k) \text{ and } \eta_{\boldsymbol{\alpha}} = \mathbb{E}(\eta \mathrm{h}_{\boldsymbol{\alpha}}), \tag{5.1.12}$$

where $\eta \in L_2(\Xi)$. For example, if $\boldsymbol{\alpha} = (0, 0, ...)$, then $\xi_{\boldsymbol{\alpha}} = 1$ and if $\boldsymbol{\alpha} = (0, 2, 0, 1, 3, 0, 0, \ldots)$, then

$$\xi_{\boldsymbol{\alpha}} = \frac{\mathrm{H}_2(\xi_2)}{\sqrt{2!}} \cdot \xi_4 \cdot \frac{\mathrm{H}_3(\xi_5)}{\sqrt{3!}}.$$

By (5.1.11) $H_0(x) = 1$ and, by (2.3.39) on page 61,

$$\mathbb{E}h_{\alpha}^2 = 1 \text{ for all } \boldsymbol{\alpha} \in \mathcal{J}^1 \tag{5.1.13}$$

and

$$\mathbb{E}h_{\alpha}h_{\beta} = 0 \text{ for all } \boldsymbol{\alpha} \neq \beta. \tag{5.1.14}$$

Before proceeding farther, we shall introduce a notation that will be used often in what follows.

Notations

(a) *Let* $\boldsymbol{\alpha} = (\alpha_1, \alpha_2, ...)$ *and* $\mathbf{z} = (z_1, z_2, ...)$ *. Then*

$$\frac{\partial^{|\alpha|}}{\partial \mathbf{z}^{\alpha}} = \frac{\partial^{|\alpha|}}{\partial z_1^{\alpha_1} \partial z_2^{\alpha_2} ...}. \tag{5.1.15}$$

(b) Let $\mathbf{z} = (z_1, z_2, ...)$ be a sequence of real numbers such that $\sum_{k=1}^{\infty} z_k^2 < \infty$. Define

$$\mathcal{E}(\mathbf{z}) = \exp\left\{ \sum_{k=1}^{\infty} z_k \xi_k - \frac{1}{2} \sum_{k=1}^{\infty} z_k^2 \right\}. \tag{5.1.16}$$

The random function $\mathcal{E}(\mathbf{z})$ is often referred to as `stochastic exponent.` It is useful in the analysis of non-linear functions of Gaussian random variables and, in particular, functionals of Brownian motion.

Lemma 5.1.9 *For* $\boldsymbol{\alpha} \in \mathcal{J}^1$,

$$\xi_{\alpha} = \frac{\partial^{|\alpha|}}{\partial \mathbf{z}^{\alpha}} \mathcal{E}(\mathbf{z})\,|_{\mathbf{z}=0}. \tag{5.1.17}$$

The proof of the Lemma follows from formula (2.3.37) on page 61, and we leave it to the interested reader.

Exercise 5.1.10 (A) Consider a polynomial in $(s, x) \in (0, \infty) \times \mathbb{R}$ given by

$$H_n(s, x) = \frac{(-s)^n}{n!} \exp\left\{ \frac{x^2}{2s} \right\} \frac{\partial^n}{\partial x^n} \exp\left\{ -\frac{x^2}{2s} \right\}, \; n = 0, 1, ...$$

Prove that

(a)

$$\sum_{n=0} \lambda^n H_n(s, x) = \exp\left\{ \lambda x - \frac{\lambda^2 s}{2} \right\}.$$

(b) For $n \geq 2$,

$$\mathrm{H}_{n+2}(s,x) = \frac{x}{n+2}\mathrm{H}_{n+1}(s,x) - \frac{s}{n+2}\mathrm{H}_n(s,x). \tag{5.1.18}$$

(c) Show that the system $\mathrm{H}_n(s,x)$, $n = 0, 1, ...$introduced in Example 5.1.10 is a complete orthonormal system in the space $L_2\left(\mathbb{R}^1, \mathcal{B}\left(\mathbb{R}^1\right), \mathcal{N}_s\right)$, where $\mathcal{N}_s(dx) = \left\{\exp\left\{-x^2/2s\right\}/\sqrt{2\pi s}\right\} dx$ is Gaussian measure on $\mathcal{B}\left(\mathbb{R}^1\right)$.

Exercise 5.1.11 (A) Prove that the collection of random variables $\{\exp\{\xi_i - 1/2\}, i \geq 1\}$ is a total subset in $L_2(\Xi)$.

The following fundamental theorem is due to American mathematicians Robert H. Cameron (1908–1989) and William T. Martin (1911–2004). It is routinely referred to as `Cameron-Martin Theorem.`

Theorem 5.1.12 *The collection* $\{\xi_\alpha, \ \boldsymbol{\alpha} \in \mathcal{J}^1\}$ *is a complete orthonormal system in* $L_2(\Xi)$.

Proof By formula (2.3.37) and Lemma 5.1.9,

$$\mathcal{E}(\mathbf{z}) = \sum_{n=0}^{\infty}\prod_{k\geq 1}\frac{\mathrm{H}_n(\xi_k)}{n!}z_k = \sum_{\alpha\in\mathcal{J}^1}\frac{\mathbf{z}^\alpha}{\sqrt{\alpha!}}\xi_\alpha. \tag{5.1.19}$$

It follows from (5.1.13) and (5.1.14) that $\{\xi_\alpha, \ \boldsymbol{\alpha} \in \mathcal{J}^1\}$ is an orthonormal system. Let us consider a random element $f \in L_2(\Xi)$. Suppose that

$$\mathbb{E}(f\,\xi_\alpha) = 0 \text{ for all } \boldsymbol{\alpha} \in \mathcal{J}^1.$$

For $\mathbf{z} = (z_1, ...z_n)$, $z_i \in \mathbb{R}^1$, we write

$$\mathcal{E}_n(\mathbf{z}) = \prod_{k=1}^{n}\sum_{m=0}^{\infty}\frac{\mathrm{H}_m(\xi_k)}{m!}z_k.$$

By our assumptions,

$$\mathbb{E}\left(f\prod_{k=1}^{n}\sum_{m=0}^{\infty}\frac{\mathrm{H}_m(\xi_k)}{m!}z_k\right) = 0.$$

Consider now $\mathbb{E}(f|\xi_1, ...\xi_n)$. By the definition of the conditional expectation and Exercise 5.1.11,

$$\mathbb{E}\left(\mathbb{E}(f|\xi_1, ...\xi_n)\mathcal{E}_n(\mathbf{z})\right) = \mathbb{E}\left(f\mathcal{E}_n(\mathbf{z})\right) = 0.$$

Therefore, $\mathbb{E}(f|\xi_1, ...\xi_n) = 0$. Now, letting n tend to infinity, we get that $f = 0$.

Example 5.1.13 Assume that $L_2(\mathbb{W}, \mathbb{R}) = L_2(\xi, \mathbb{R})$, where ξ is a standard Gaussian random variable $\boldsymbol{\alpha} = (n, 0, 0, \ldots) \in \mathcal{J}^1, n = 0, 1, 2, \ldots$, and $v = v(\xi) \in L_2(\xi, \mathbb{R})$. By the Cameron-Martin Theorem,

$$v(\xi) = \sum_{n=0}^{\infty} \mathbb{E}\left(v(\xi)\,\xi_{(n)}\right)\xi_{(n)} = \sum_{n \in \mathcal{J}^1} \mathbb{E}v(\xi)\,\xi_{(n)},$$

where $\xi_{(n)} = \mathrm{h}_n(\xi) = \mathrm{H}_n(\xi)/\sqrt{n!}$.

Theorem 5.1.12 implies that for any random function $f \in L_2(\Xi)$ the Cameron-Martin expansion

$$f = \sum_{\alpha \in \mathcal{J}^1} f_\alpha \xi_\alpha. \tag{5.1.20}$$

holds in $L_2(\Xi)$ and with probability 1. Therefore, analysis of functionals of Gaussian random variables in many settings could be efficiently reduced to the analysis of the deterministic coefficients f_α.

In the future we will often refer to expansions of the form (5.1.20) as `Cameron-Martin expansions` or `Wiener Chaos expansion`.

Definition 5.1.14 For $n \geq 0$, the n-th chaos space $\mathcal{H}_n$ is the closed subset of $L_2(\Xi)$ generated by finite linear combinations of the random variables $\{\xi_\alpha, |\boldsymbol{\alpha}| = n\}$.

We will also refer to $\mathcal{H}_n$ as the nth `Gaussian chaos`. Obviously, $\mathcal{H}_0$ consists of constants, $\mathcal{H}_1$ contains only Gaussian random variables, etc. This leads to the following orthogonal decomposition

$$L_2(\Xi) = \bigotimes_{n=0}^{\infty} \mathcal{H}_n.$$

5.1.2 Elementary Operators on Cameron-Martin Basis

In this section we will introduce `creation` and `annihilation` operators on the elements of the on the elements of Cameron-Martin basis. Let $\boldsymbol{\epsilon}(k)$ be a multi-index of length $|\boldsymbol{\epsilon}(k)| = 1$ such that the entry #k is its only non-zero entry.

Definition 5.1.15 `Creation operator` $\delta_{\xi_k}(\cdot)$ and `annihilation operator` $\mathbb{D}_{\xi_k}(\cdot)$ are defined as follows:

$$\delta_{\xi_k}(\xi_\alpha) = \sqrt{\alpha_k + 1}\,\xi_{\alpha+\epsilon(k)} \tag{5.1.21}$$

and

$$\mathbb{D}_{\xi_k}(\xi_\alpha) = \sqrt{\alpha_k}\,\xi_{\alpha-\epsilon(k)}. \tag{5.1.22}$$

Remark 5.1.16 Historically, the creation and annihilation operators were introduced in Quantum Physics (see, for example, [63] and [209]). Later on, these operators were greatly generalized and connected with the fundamental operators of Malliavin calculus. See, for example, [175, 204], and the developments below in this chapter.

Creation operator is a special case of another operator, called `Wick product`, cf. (2.3.40) on page 62.

It is readily checked that

$$\mathbb{E}\left|\delta_{\xi_k}(\xi_\alpha)\right|^2 = (\alpha_k + 1) \text{ and } \sum_k \mathbb{E}\left|\mathbb{D}_{\xi_k}(\xi_\alpha)\right|^2 = |\boldsymbol{\alpha}|. \tag{5.1.23}$$

Exercise 5.1.17 (C) Prove that

$$\mathbb{E}\left[\delta_{\xi_k}(\xi_\alpha)\,\delta_{\xi_n}(\xi_\beta)\right] = \sqrt{(\alpha_k+1)(\beta_n+1)}\,I_{\{\alpha+\epsilon(k)=\beta+\epsilon(n)\}}.$$

Exercise 5.1.18 (C) Assume that the entries $\alpha_{k_1}, \ldots, \alpha_{k_m}$ of the multiindex $\boldsymbol{\alpha}$, are positive. Compute $\mathbb{D}_{\xi_{k_m}}\left(\ldots\mathbb{D}_{\xi_{k_2}}\left(\mathbb{D}_{\xi_{k_1}}(\xi_\alpha)\right)\right)$.

Exercise 5.1.19 (C) Prove that for any k and $\boldsymbol{\alpha}$ the elements of the Cameron-Martin basis are eigenfunctions of the operator $\delta_{\xi_k}\mathbb{D}_{\xi_k}$ and

$$\delta_{\xi_k}\left(\mathbb{D}_{\xi_k}(\xi_\alpha)\right) = \alpha_k \xi_\alpha. \tag{5.1.24}$$

Let us denote $\mathbb{L}_{\xi_k}(\cdot) = \delta_{\xi_k}\left(\mathbb{D}_{\xi_k}(\cdot)\right)$. It follows from Exercise 5.1.19 that for every k and $\boldsymbol{\alpha}$,

$$\mathbb{L}_{\xi_k}(\xi_\alpha) = = |\alpha_k|\xi_\alpha. \tag{5.1.25}$$

Operator $\mathbb{L}_{\xi_k}$ is often called the `Ornstein-Uhlenbeck operator` associated with the random variable ξ_k. Similarly, $\mathbb{L}(\cdot) = \sum_{k\geq 1}\mathbb{L}_{\xi_k}(\cdot)$ is often called the `Ornstein-Uhlenbeck operator` (associated with the system of random variables $(\xi_k, k \geq 1)$).

Exercise 5.1.20 (C) Prove that the elements of the Cameron-Martin basis $\{\xi_\alpha, \boldsymbol{\alpha} \in \mathcal{J}^1\}$ are eigenfunctions of the operator $\mathbb{L}$:

$$\mathbb{L}(\xi_\alpha) = |\boldsymbol{\alpha}|\,\xi_\alpha. \tag{5.1.26}$$

Formula (5.1.26) provides an alternative description of the Cameron-Martin basis. Specifically, the Cameron-Martin basis consists of eigenfunctions of the Ornstein-Uhlenbeck operator $\mathbb{L}$.

Now we shall introduce an alternative way to describe a multi-index $\boldsymbol{\alpha} \in \mathcal{J}^1$.

Definition 5.1.21 Let $\boldsymbol{\alpha} \in \mathcal{J}^1$ and $|\boldsymbol{\alpha}| = n > 0$. An ordered n-tuple $K_{\boldsymbol{\alpha}} = \{k_1, \ldots, k_n\}$, where $k_1 \leq k_2 \leq \ldots \leq k_n$ characterize the locations and the values of the non-zero elements of $\boldsymbol{\alpha}$, will be called `characteristic set` of $\boldsymbol{\alpha}$.

In this definition, k_1 is the index of the first non-zero element of $\boldsymbol{\alpha}$. If $\alpha_{k_1} > 1$, then it is followed by $\max(0, \alpha_{k_1} - 1)$ of entries with the same value. The next entry after that is the index of the second non-zero element of α, followed by $\max(0, \alpha_{k_2} - 1)$ of entries with the same value, and so on.

For example, if $n = 7$ and $\boldsymbol{\alpha} = (1, 0, 2, 0, 0, 1, 0, 3, 0, \ldots)$, then the non-zero elements of a are $\alpha_1 = 1$, $\alpha_3 = 2$, $\alpha_6 = 1$, $\alpha_8 = 3$. As a result, $K_{\alpha} = \{1, 3, 3, 6, 8, 8, 8\}$, that is, $k_1 = 1$, $k_2 = k_3 = 3$, $k_4 = 6$, $k_5 = k_6 = k_7 = 8$.

Exercise 5.1.22 (A) Let $\boldsymbol{\alpha} \in \mathcal{J}^1$ and it's characteristic set $K_{\boldsymbol{\alpha}} = \{k_1, \ldots, k_n\}$. Prove that

$$\xi_{\alpha} = \delta_{\xi_{k_n}}\left(\ldots \delta_{\xi_{k_2}}\left(\delta_{\xi_{k_1}}\left(1/\sqrt{\alpha!},\right)\right)\ldots\right). \tag{5.1.27}$$

Formula (5.1.27) shows that the elements of Cameron-Martin basis ξ_{α} can be obtained by multiple sequential applications of creation operators to the constant $1/\sqrt{\alpha!}$.

In conclusion of this section, let us discuss actions of Wick product on the elements of Cameron-Martin basis. An extension of formula (2.3.40) on page 62 to the elements of Cameron-Martin basis yields the following relation:

$$\xi_{\alpha} \diamond \xi_{\beta} = \sqrt{\left(\frac{(\boldsymbol{\alpha} + \boldsymbol{\beta})!}{\boldsymbol{\alpha}!\boldsymbol{\beta}!}\right)}\xi_{\boldsymbol{\alpha}+\boldsymbol{\beta}}. \tag{5.1.28}$$

Let us denote

$$\mathrm{H}_{\alpha} = \sqrt{\alpha!}\xi_{\alpha}. \tag{5.1.29}$$

In the future we will refer to $\{\mathrm{H}_{\alpha}, \boldsymbol{\alpha} \in \mathcal{J}^1\}$ as the `unnormalised Cameron-Martin basis`

Exercise 5.1.23 (C) Prove that

$$\mathrm{H}_{\alpha} \diamond \mathrm{H}_{\beta} = \mathrm{H}_{\alpha+\beta}. \tag{5.1.30}$$

Clearly,

$$\delta_{\xi_k}(\xi_{\alpha}) = \sqrt{\alpha_k + 1}\,\xi_{\alpha+\epsilon(k)} = \xi_{\alpha} \diamond \xi_{\epsilon(k)}. \tag{5.1.31}$$

Wick product extends by linearity to the elements of $L_2(\Xi)$: for $f, g \in L_2(\Xi)$,

$$f \diamond g = \sum_{\alpha,\beta \in \mathcal{J}^1} f_\alpha g_\beta \sqrt{\left(\frac{(\alpha+\beta)!}{\alpha!\beta!}\right)} \xi_{\alpha+\beta}. \quad (5.1.32)$$

Exercise 5.1.24 (B) Rewrite expansion (5.1.32) in terms of unnormalised Cameron-Martin basis.

An important and now obvious property of the Wick product follows from (5.1.32):

$$\mathbb{E}(f \diamond g) = (\mathbb{E}f)(\mathbb{E}g) < \infty. \quad (5.1.33)$$

for $f, g \in L_2(\Xi)$.

Exercise 5.1.25 (A) Let $\alpha \in \mathcal{J}^1$ be a multi-index with $|\alpha| = n \geq 1$ with characteristic set $K_\alpha = \{k_1, \dots, k_n\}$. Verify that

$$\xi_\alpha = \frac{\xi_{k_1} \diamond \xi_{k_2} \diamond \cdots \diamond \xi_{k_n}}{\sqrt{\alpha!}}. \quad (5.1.34)$$

Formula (5.1.34) is a simple analog of the well-known result of Itô [90].

5.1.3 Elements of Stationary Malliavin Calculus

In this section we will discuss the basics of general stochastic calculus, which is often referred to as Malliavin calculus (see, for example, Nualart [175]). In particular, Malliavin calculus covers Itô's calculus and its extensions to non-adapted (anticipating) processes. The latter feature makes it possible to apply Malliavin calculus to spatial random fields, in particular to solutions of elliptic SPDEs.

As before, let $\mathfrak{B}$ be the Ξ-measurable Gaussian white noise on a separable Hilbert space H. Everywhere below, $\mathfrak{B}$ will be identified with its `chaos expansion`

$$\mathfrak{B} = \sum_{k \geq 1} \mathrm{u}_k \, \xi_k, \quad (5.1.35)$$

where $\{\xi_k, k \geq 1\}$ is a sequence of independent standard Gaussian random variables. $(\mathrm{u}_k, k \geq 1)$ is an orthonormal basis in Hilbert space H.

An important example of the white noise $\mathfrak{B}$ is spatial white noise $\dot{W}(x)$ in a Hilbert space $\mathbb{L}_2(G)$, where G is a domain in $\mathbb{R}^{\mathrm{d}}$. If $\mathfrak{u}_k = \mathfrak{u}_k(x)$ is an orthonormal basis in $\mathbb{L}_2(G)$, then the spatial white noise on G is given by

$$\dot{W}(x) = \sum_{k\geq 1} \mathfrak{u}_k(x)\,\xi_k. \tag{5.1.36}$$

Now, let us define the `Malliavin derivative` $\mathbb{D}_{\mathcal{B}}$ on the elements of the Cameron-Martin basis $\{\xi_\alpha,\ \boldsymbol{\alpha} \in \mathcal{J}^1\}$ as follows:

$$\mathbb{D}_{\mathfrak{B}}\xi_\alpha = \sum_{k\geq 1} \sqrt{\alpha_k}\xi_{\alpha-\epsilon(k)}\mathfrak{u}_k, \tag{5.1.37}$$

where $\boldsymbol{\epsilon}(k)$ is the multi-index of length 1 with all coordinates being zero except for the entry #k equal to 1.

Remark 5.1.26 Note that the Malliavin derivative $\mathbb{D}_{\mathfrak{B}}$ is a "linear combination" of annihilation operators $\mathbb{D}_{\xi_k}$ given by

$$\mathbb{D}_{\mathfrak{B}}\xi_\alpha = \sum_{k\geq 1} \mathbb{D}_{\xi_k}(\xi_\alpha)\,\mathfrak{u}_k \in H, \tag{5.1.38}$$

and, by (5.1.23),

$$\mathbb{E}\,\|\mathbb{D}_{\mathfrak{B}}\xi_\alpha\|_H^2 = |\alpha|\,. \tag{5.1.39}$$

In particular, if $\mathfrak{B} = \dot{W}(x)$, then

$$\mathbb{D}_{\dot{W}(x)}\xi_\alpha = \sum_{k\geq 1} \sqrt{\alpha_k}\xi_{\alpha-\epsilon(k)}\mathfrak{u}_k(x) \in L_2(G).$$

Exercise 5.1.27 (B) Consider the stochastic exponent

$$\mathcal{E}(\mathbf{z}) = e^{\sum_k (z_k\xi_k - (1/2)z_k^2)} = \prod_{k\geq 1}\sum_{n=0}^{\infty} \frac{\mathrm{H}_n(\xi_k)}{n!} z_k^n = \sum_{\alpha\in\mathcal{J}^1} \frac{\mathbf{z}^\alpha}{\sqrt{\alpha!}}\xi_\alpha.$$

and specify the requirements on $\mathbf{z}$ to ensure that $\mathbb{D}_{\mathfrak{B}}\mathcal{E}(\mathbf{z}) \in \mathbb{L}_2(\Xi; H)$.

Our next goal is to extend the Malliavin derivative $\mathbb{D}_{\mathfrak{B}}$ from the elements of Cameron-Martin basis to the space $L_2(\Xi)$. In view of (5.1.39), if $v \in \mathbb{L}_2(\Xi)$, then its Malliavin derivative $\mathbb{D}_{\mathfrak{B}}v$ does not necessarily belong to $\mathbb{L}_2(\Xi; H)$. However, due to (5.1.39), for $v \in \mathbb{L}_2(\Xi)$,

$$\mathbb{E}\,|\mathbb{D}_{\mathfrak{B}}v|^2 = \sum_{\alpha\in\mathcal{J}^1} |\alpha|\,v_\alpha^2. \tag{5.1.40}$$

With this in view, let us introduce the following subspace of $L_2(\varXi)$:

$$L_2^1(\varXi) = \left\{ v \in L_2(\varXi) : \sum_{\alpha \in \mathcal{J}^1} |\alpha| \, v_\alpha^2 < \infty \right\}. \tag{5.1.41}$$

Definition 5.1.28 For $v \in L_2^1(\varXi)$, the Malliavin derivative $\mathbb{D}_{\mathfrak{B}} v$ is

$$\mathbb{D}_{\mathfrak{B}} v = \sum_{\alpha \in \mathcal{J}^1} \sum_{k \geq 1} \sqrt{\alpha_k} \xi_{\alpha - \epsilon(k)} v_\alpha \mathfrak{u}_k. \tag{5.1.42}$$

Exercise 5.1.29 (C) Prove that if $v \in \mathbb{L}_2(\varXi)$, then

$$\mathbb{E}\left(\mathbb{D}_{\mathfrak{B}}(v)\, \xi_\alpha\right) = \sum_{k \geq 1} \sqrt{\alpha_{k+1}} v_{\alpha + \epsilon(k)} \mathfrak{u}_k. \tag{5.1.43}$$

and

$$\left\| \mathbb{E}\left(\mathbb{D}_{\mathfrak{B}}(v)\, \xi_\alpha\right) \right\|_H^2 < \infty. \tag{5.1.44}$$

Exercise 5.1.30 (C) Prove that Malliavin derivative $\mathbb{D}_{\mathfrak{B}}$ is a continuous linear operator from $L_2^1(\varXi)$ to $\mathbb{L}_2(\varXi)$.

The above construction can be easily extended to Gaussian random variables taking values in a Hilbert space X. Firstly, let us "upgrade" the domain of $\mathbb{D}_{\mathfrak{B}}$ from $L_2^1(\varXi)$ to

$$L_2^1(\varXi; X) = \left\{ v \in L_2(\varXi; X) : \sum_{\alpha \in \mathcal{J}^1} |\alpha| \, \|v_\alpha\|_X^2 < \infty \right\}, \tag{5.1.45}$$

where X is a Hilbert space. Secondly, for $v \in L_2^1(\varXi; X)$, let us define the Malliavin derivative as follows:

$$\mathbb{D}_{\mathfrak{B}}(v) = \sum_{\alpha \in \mathcal{J}^1} \sum_{k \geq 1} \sqrt{\alpha_k} \xi_{\alpha - \epsilon(k)} v_\alpha \otimes \mathfrak{u}_k. \tag{5.1.46}$$

Note that the "extended" version of the Malliavin derivative takes values in the tensor product Hilbert space $X \otimes H$.

Theorem 5.1.31 *Malliavin derivative $\mathbb{D}_{\mathfrak{B}}$ is a continuous linear operator from $L_2^1(\varXi; X)$ to $L_2(\varXi; X \otimes H)$. Moreover, if $v \in L_2^1(\varXi; X)$, then*

$$\left\| \mathbb{D}_{\mathfrak{B}}(v) \right\|_{L_2(\varXi; X \otimes H)}^2 = \|v\|_{L_2^1(\varXi; X)}^2 .$$

Proof Similar to (5.1.43),

$$\mathbb{E}\left(\mathbb{D}_{\mathfrak{B}}(v)\,\xi_\alpha\right) = \sum_{k\geq 1} \sqrt{\alpha_{k+1}} v_{\alpha+\epsilon(k)} \otimes \mathfrak{u}_k.$$

Therefore,

$$\|\mathbb{D}_{\mathfrak{B}}(v)\|^2_{L_2(\Xi;X\otimes H)} = \sum_{\alpha\in\mathcal{J}^1}\sum_{k\geq 1} |\alpha|\,\|v_\alpha\|^2_X = \|v\|^2_{L_2^1(\Xi;X)} < \infty.$$

The following exercise should persuade a sceptical reader that calling the operator $\mathbb{D}_{\mathfrak{B}}$ a *derivative* is not far-fetched or misleading.

Exercise 5.1.32 (B) If $F(x_1, ..., x_n)$ is a polynomial, $h_i \in X$, and $v = F(h_1\xi_1, ..., h_n\xi_n)$, then

$$\mathbb{D}_{\mathfrak{B}}(v) = \sum_{k=1}^{n} \frac{\partial F}{\partial x_k}(h_1\xi_1, ..., h_n\xi_n)\, h_k \otimes \mathfrak{u}_k. \tag{5.1.47}$$

The statement of the Exercise 5.1.32 is a rather important fact. It has been used in the literature as a starting point for the development of Malliavin calculus (see e.g. Nualart [175, 176]).

Next, we would like to define the "*anti-derivative*" associated with the Malliavin derivative $\mathbb{D}_{\mathfrak{B}}$. In other words, we want to find an operator that is dual to $\mathbb{D}_{\mathfrak{B}}$ in an appropriate sense. This operator is usually referred to as the Malliavin `divergence operator` is and denoted by $\delta_{\mathfrak{B}}$. Getting a little bit ahead of the story, we should say that, in the context of the Itô calculus, the divergence operator is simply the familiar Itô integral.

Firstly, let us try to build an anti-derivative (inverse) to $\mathbb{D}_{\mathfrak{B}}$ using creation operators δ_{ξ_k} (see Definition 5.1.15) as building blocks. With this in mind, let us introduce a the following simple but useful extension of the creation operator with values in the Hilbert space H

For $u \in H$, let us define

$$\delta_{u\xi_k}(\xi_\alpha) = u\delta_{\xi_k}(\xi_\alpha) = \sqrt{\alpha_k + 1}\,\xi_{\alpha+\epsilon(k)} \tag{5.1.48}$$

If $\{\mathfrak{u}_k, k \geq 1\}$ is a basis in H, then $\delta_{\mathfrak{u}_k\xi_k}(\xi_\alpha) = \mathfrak{u}_k\sqrt{\alpha_k + 1}\,\xi_{\alpha+\epsilon(k)}$ could be interpreted as is the creation operator "in the direction" of the vector $\mathfrak{u}_k$. It follows from (5.1.48) that

$$\delta_{\mathfrak{u}_k\xi_k}(\xi_\alpha) := \sqrt{\alpha_k + 1}\,\mathfrak{u}_k\xi_{\alpha+\epsilon(k)} = \mathfrak{u}_k\delta_{\xi_k}(\xi_\alpha). \tag{5.1.49}$$

Let us consider now the class of functions f such that

$$f = \sum_{k\geq 1} f_k \otimes \mathfrak{u}_k \in L_2(\Xi; X \otimes H). \tag{5.1.50}$$

Write $f_{k,\alpha} = \mathbb{E}\left(f_k \xi_\alpha\right)$. Then, by the Cameron-Martin theorem, f can be expanded as follows:

$$f = \sum_{\alpha \in \mathcal{J}^1} \sum_{k \geq 1} f_{k,\alpha} \otimes \mathfrak{u}_k \xi_\alpha. \tag{5.1.51}$$

Now, it is intuitively clear that the divergence operator $\delta_{\mathfrak{B}}$ (anti-derivative to $\mathbb{D}_{\mathfrak{B}}$) should be defined as follows:

$$\begin{aligned}\delta_{\mathfrak{B}}(f) &= \sum_{\alpha \in \mathcal{J}^1} \sum_{k \geq 1} f_{k,\alpha} \otimes \mathfrak{u}_k \delta_{\xi_k}(\xi_\alpha) = \\ &= \sum_{\alpha \in \mathcal{J}^1} \sum_{k \geq 1} f_{k,\alpha} \otimes \mathfrak{u}_k \sqrt{\alpha_k + 1}\, \xi_{\alpha + \epsilon(k)} \\ &= \sum_{\beta \in \mathcal{J}^1} \sum_{k \geq 1} \sqrt{\beta_k} f_{k,\beta - \epsilon(k)} \otimes \mathfrak{u}_k \xi_\beta.\end{aligned} \tag{5.1.52}$$

Of course, the manipulations we performed above were formal. An important question now is the assumptions on f to guarantee $\delta_{\mathfrak{B}}(f) \in L_2(\Xi; X \otimes H)$. An explicit answer to this question is provided next.

Theorem 5.1.33 *If $f \in L_2(\Xi; X \otimes H)$ and*

$$\sum_{\alpha \in \mathcal{J}^1} \sum_{k \geq 1} |\alpha| \, \|f_{k,\alpha}\|_X^2 < \infty, \tag{5.1.53}$$

then $\delta_{\mathfrak{B}}(f) \in L_2(\Xi; X)$ and

$$\left(\delta_{\mathfrak{B}}(f)\right)_\alpha = \sum_{k \geq 1} \sqrt{\alpha_k} f_{k,\alpha - \epsilon(k)}. \tag{5.1.54}$$

Proof Since

$$\sum_{\alpha \in \mathcal{J}^1} \sum_{k \geq 1} \sqrt{\alpha_k + 1} f_{k,\alpha} \xi_{\alpha + \epsilon(k)} = \sum_{\alpha} \sum_{k \geq 1} \sqrt{\alpha_k} f_{k,\alpha - \epsilon(k)}\, \xi_\alpha,$$

we have

$$\mathbb{E}\left(\delta_{\mathfrak{B}}(f)\, \xi_\alpha\right) = \mathbb{E}\left(\sum_{\beta} \sum_{k \geq 1} f_{k,\beta - \epsilon(k)} \sqrt{\beta_k}\, \xi_\beta \xi_\alpha\right) = \sum_{k \geq 1} \sqrt{\alpha_k} f_{k,\alpha - \epsilon(k)}. \tag{5.1.55}$$

Therefore,

$$\mathbb{E}\left\|\delta_{\mathfrak{B}}(f)\right\|^2 = \sum_{\alpha} \sum_{k \geq 1} (\alpha_k + 1) \left\|f_\alpha\right\|_X^2 < \infty. \tag{5.1.56}$$

Example 5.1.34 Suppose that $\mathfrak{u}_k = \mathfrak{u}_k(x)$ is an orthonormal basis in $\mathbb{L}_2(G)$ and $\mathfrak{B}$ is spatial white noise $\dot{W}(x)$ (see (5.1.36)). Consider $f \in \mathbb{L}_2(\Xi, G)$. Denote

$$f_k = \int_{\mathbb{L}_2(G)} f(x)\,\mathfrak{u}_k(x)\,dx,$$

$f_{k,\alpha} = \mathbb{E}(f_k \xi_\alpha)$, and assume that

$$\sum_{\alpha\in\mathcal{J}^1}\sum_{k\geq 1} |\alpha|\, \|f_{k,\alpha}\|_X^2 < \infty.$$

Then, one can expand $f(x)$ as follows:

$$f(x) = \sum_{k\geq 1}\sum_{\alpha\in\mathcal{J}^1} f_{k,\alpha}\mathfrak{u}_k(x)\,\xi_\alpha$$

and, by (5.1.52),

$$\delta_{\dot{W}(x)}(f) = \sum_{\alpha\in\mathcal{J}^1}\sum_{k\geq 1} \sqrt{\alpha_k+1} f_{k,\alpha}\xi_{\alpha+\epsilon(k)}. \tag{5.1.57}$$

Recall now that, by Exercise 5.1.19, for any ξ_k, the elements of the Cameron-Martin basis are eigenfunctions of the operator $\delta_{\xi_k}\mathbb{D}_{\xi_k}$ and

$$\delta_{\xi_k}\mathbb{D}_{\xi_k}(\xi_\alpha) = |\alpha|\xi_\alpha. \tag{5.1.58}$$

This important equality extends to Malliavin derivative $\mathbb{D}_{\mathfrak{B}}$ and divergence operator $\delta_{\mathfrak{B}}$. The operator

$$\mathbb{L}_{\mathfrak{B}} = \delta_{\mathfrak{B}}\mathbb{D}_{\mathfrak{B}} \tag{5.1.59}$$

is often called `Ornstein-Uhlenbeck operator.` The following simple result provides an explicit description of the action of Ornstein-Uhlenbeck operator on elements of the space

$$L_2^2(\Xi;X) = \{v : \sum_{\alpha\in\mathcal{J}^1} |\alpha|^2\|v_\alpha\|_X^2 < \infty.\}$$

Corollary 5.1.35 *If* $v \in L_2^2(\Xi;X)$, *then* $\mathbb{L}_{\mathfrak{B}}(v) \in L_2(\Xi;X)$ *and*

$$\mathbb{L}_{\mathfrak{B}}(v) = \sum_{\alpha\in\mathcal{J}^1} |\alpha|\, v_\alpha\xi_\alpha.$$

Proof For $v \in L_2^2(\Xi; X)$, the Malliavin derivative

$$\mathbb{D}_{\mathfrak{B}}(v) = \sum_{\alpha\in\mathcal{J}^1}\sum_{k\geq 1}\sqrt{\alpha_k}\xi_{\alpha-\epsilon(k)}v_\alpha \otimes \mathfrak{u}_k$$
$$= \sum_{\alpha\in\mathcal{J}^1}\sum_{k\geq 1}\sqrt{\alpha_k+1}v_{\alpha+\epsilon(k)} \otimes \mathfrak{u}_k\xi_\alpha.$$

Recall that for $f \in L_2(\Xi; X\otimes H)$, $f = \sum_{\alpha\in\mathcal{J}^1}\sum_{k\geq 1} f_{k,\alpha}\otimes \mathfrak{u}_k\xi_\alpha$ and

$$\delta_{\mathfrak{B}}(f) = \sum_{\alpha\in\mathcal{J}^1}\sum_{k\geq 1}\sqrt{\alpha_k+1}f_{k,\alpha}\xi_{\alpha+\epsilon(k)}.$$

Now, let us take $f_{k,\alpha} = \sqrt{\alpha_k+1}v_{\alpha+\epsilon(k)}$. Then

$$\mathbb{L}_{\mathfrak{B}}(v) = \delta_{\mathfrak{B}}\left(\sum_{\alpha\in\mathcal{J}^1}\sum_{k\geq 1}\sqrt{\alpha_k+1}v_{\alpha+\epsilon(k)} \otimes \mathfrak{u}_k\xi_\alpha\right)$$
$$= \sum_{\alpha\in\mathcal{J}^1}\sum_{k\geq 1}(\alpha_k+1)\, v_{\alpha+\epsilon(k)}\xi_{\alpha+\epsilon(k)} = \sum_{\alpha\in\mathcal{J}^1}|\alpha|\, v_\alpha\xi_\alpha$$

and we are done.

It is, of course, a standard fact that integrals and derivatives are dual operators. We have also noticed that creation and annihilation operators δ_{ξ_k} and $\mathbb{D}_{\xi_k}$ are dual. Next we will show that in some sense it holds also for Malliavin derivative $\mathbb{D}_{\mathfrak{B}}$ and divergence operator $\delta_{\mathfrak{B}}$.

Theorem 5.1.36 *Let* $f = \sum_{k\geq 1} f_k\mathfrak{u}_k$ *be an element of* $L_2(\Xi; X\otimes H)$ *and*

$$\sum_\alpha\sum_{k\geq 1}\alpha_k\left\|f_{k,\alpha}\right\|_X^2 < \infty.$$

Suppose also that $\varphi \in L_2^1(\Xi)$. *Then* $\mathbb{D}_{\mathfrak{B}}(f) \in L_2(\Xi; X\otimes H)$ *and* $\mathbb{D}_{\mathfrak{B}}$ *is dual to* $\delta_{\mathfrak{B}}$ *in that*

$$\mathbb{E}(\varphi\delta_{\mathfrak{B}}(f)) = \mathbb{E}(\mathbb{D}_{\mathfrak{B}}(\varphi), f)_H.$$

Proof To begin with, let us assume that $\varphi = \xi_\alpha$. Let $f \in L_2(\Xi; X\otimes H)$. It follows from the definition of the divergence operator that

$$\mathbb{E}(\xi_\alpha\delta_{\mathfrak{B}}(f)) = \sum_{k\geq 1}\sqrt{\alpha_k}f_{k,\alpha-\epsilon(k)}. \tag{5.1.60}$$

On the other hand,

$$\mathbb{D}_{\mathfrak{B}}\left(\xi_\alpha\right) = \sum_{k\geq 1} \sqrt{\alpha_k}\xi_{\alpha-\epsilon(k)}\mathfrak{u}_k.$$

Thus,

$$\mathbb{E}\left(\mathbb{D}_{\mathfrak{B}}\left(\xi_\alpha\right), f\right)_H = \sum_{k\geq 1} \sqrt{\alpha_k}\mathbb{E}\left(\xi_{\alpha-\epsilon(k)}f_k\right) = \sum_{k\geq 1} \sqrt{\alpha_k}f_{k,\alpha-\epsilon(k)}$$

Therefore we proved that

$$\mathbb{E}\left(\xi_\alpha\delta_{\mathfrak{B}}\left(f\right)\right) = \mathbb{E}\left(\mathbb{D}_{\mathfrak{B}}\left(\xi_\alpha\right), f\right)_H.$$

An immediate consequence of (5.1.28) and (5.1.52) is the following identity:

$$\delta_{\mathfrak{B}}(\xi_\alpha h\otimes \mathfrak{u}_k) = h\xi_\alpha \diamond \xi_k,\ h\in X. \tag{5.1.61}$$

5.1.4 Problems

Problem 5.1.1 is an invitation to investigate the Wick product as a bi-linear operator on $L_2(\mathfrak{B})$. Problem 5.1.2 address a basic question about completeness of polynomials. Problem 5.1.3 illustrates a challenge related to extension of chaos approach to non-Gaussian setting.

Problem 5.1.1 Starting with one-dimensional case $\mathfrak{B} = \xi \sim \mathcal{N}(0,1)$, and then proceeding to infinite dimensions,

(a) Give an example showing that the Wick product $\diamond$ is not a bounded operator from $L_2(\Xi)\times L_2(\Xi)$ to $L_2(\Xi)$
(b) Derive a sufficient condition on η and ζ so that $\eta\diamond\zeta\in L_2(\Xi)$.
(c) Is there a sufficient condition on η so that $\eta\diamond\zeta\in L_2(\Xi)$ for all $\zeta\in L_2(\Xi)$?

Problem 5.1.2 Consider the `Stieltjes-Wigert weight function`

$$\varphi(x) = \frac{1}{\sqrt{\pi}}e^{-(\ln x)^2},\ x>0.$$

(a) Confirm that

$$s_n = \int_0^{+\infty} x^n\varphi(x)dx = e^{(n+1)^2/4},\ n=0,1,2,\ldots.$$

(b) Confirm that

$$\int_0^{+\infty} x^n \sin(2\pi \ln x)\, \varphi(x)dx = 0$$

for every $n = 0, 1, 2, \ldots$ and so the polynomials are not complete in $L_2((0, +\infty), \varphi(x)dx)$.

(c) Confirm that the integral

$$\int_0^{\infty} e^{ax}\varphi(x)dx$$

diverges for every $a > 0$, but the integral

$$\int_0^{\infty} \frac{\ln \varphi(x)}{1+x^2}\, dx$$

and the series

$$\sum_{k\geq 1} \frac{1}{\sqrt[2k]{S_{2k}}}$$

both converge.

The book [1] can provide further information about this and similar problems.

Problem 5.1.3 Let $\xi_1, \xi_2, \ldots$ be iid random variables with zero mean and unit variance, and let $\{\mathfrak{m}_k(t),\ k \geq 1,\ t \in [0, T]$, be an orthonormal basis in $L_2((0, T))$. Define the process

$$W(t) = \sum_{k\geq 1} \xi_k M_k(t),$$

where $M_k(t) = \int_0^t \mathfrak{m}_k(s)ds$.

(a) Verify that $W = W(t)$ is a `wide-sense Wiener process`, that is,

$$W(0) = 0,\ \mathbb{E}W(t) = 0,\ \mathbb{E}\big(W(t)W(s)\big) = \min(t, s);$$

cf. [140, Definition 15.1.2].

(b) Give a sufficient condition on the distribution of ξ_k for the sample trajectories of W to be continuous with probability one.

[The Kolmogorov continuity criterion implies that the sample trajectories of W are continuous if the distribution of ξ_k has finite moments of sufficiently high order, and this includes all standard distributions (exponential, uniform, Poisson, Binomial). As a result, generalized polynomial chaos framework requires further modification to study evolution equations driven by processes with jumps; cf. [38, Chap. 13].]

(c) Find an example of a distribution of ξ_k resulting in discontinuous trajectories of W.

5.2 Stationary SPDEs

5.2.1 Definitions and Basic Examples

The objective of this section is to introduce *stationary stochastic PDEs.* Roughly speaking, by stationary stochastic PDEs we understand SPDEs that do not involve time variable. For example, elliptic SPDEs can be characterized as stationary SPDEs. In this section we will discuss a large class of stationary SPDEs of the form

$$\mathbf{A}u + \delta_{\mathfrak{B}}(\mathbf{M}u) = f, \tag{5.2.1}$$

which includes equations with additive and/or multiplicative noise. The important examples of equations from this class include:

1. Poisson equation with a simple random potential

$$\begin{aligned} \Delta V(x) + V(x) \diamond \xi &= f(x), \; x \in (0,1) \\ V(0) = V(1) &= 0, \end{aligned} \tag{5.2.2}$$

   where ξ is a standard Gaussian random variable and $\diamond$ is the Wick product.
2. Poisson equation with a more complicated random potential

$$\Delta v(x) + \delta_{\dot{W}(x)} v(x) = f(x), \tag{5.2.3}$$

   where $\delta_{\dot{W}(x)}$ denotes the divergence operator in the sense of Malliavin calculus.
3. Poisson equations in random medium:

$$\nabla (A_\epsilon(x) \diamond \nabla u(x)) = f(x), \tag{5.2.4}$$

   where $A_\epsilon(x) := \left(a(x) + \epsilon \dot{W}(x)\right)$, $a(x)$ is a deterministic positive-definite matrix, and ϵ is a real number. Of course, since $a(x)$ is deterministic, $a(x) \diamond \nabla u(x) = a(x) \nabla u(x)$).

Equations (5.2.2)–(5.2.4) are random perturbation of the corresponding deterministic equation. An important feature of these types of perturbations, which is a consequence of the property (5.1.32) of the Wick product, is that the stochastic equations are *unbiased* in that they preserve the mean dynamics.

Exercise 5.2.1 (C) For Eq. (5.2.3), confirm that the function $u_0(x) := \mathbb{E}u(x)$ solves the deterministic Poisson equation

$$\nabla\left(a(x)\nabla u_0(t,x)\right) = \mathbb{E}f(x).$$

The main approach to solving the equation discussed above and further generalizations of these equations is based on utilization of the Cameron-Martin expansion of the solutions. For example, let us construct a solution of equation (5.2.2) by using the Cameron-Martin expansion

$$u(x) = \sum_{n=0}^{\infty} u_n(x)\,\mathrm{h}_n(\xi),$$

where $u_n(x) = \mathbb{E}\left(u(x)\,\mathrm{h}_n(\xi)\right)$ (see (5.1.7) and related notations). Then, proceeding as if all the series converged, (5.1.28) implies

$$u(x) \diamond \xi = \sum_{n=0}^{\infty} u_n(x)\,\mathrm{h}_{n+1}(\xi).$$

As a result,

$$\mathbb{E}\Big(\big(u(x) \diamond \xi\big)\mathrm{h}_0(\xi)\Big) = \mathbb{E}\big(u(x) \diamond \xi\big) = \big(\mathbb{E}u(x)\big)\big(\mathbb{E}\xi\big) = 0,$$

and, for $k \geq 1$,

$$\mathbb{E}\left(\left(u(x) \diamond \xi\right)\mathrm{h}_k(\xi)\right) = \sum_{n=0}^{\infty} u_n(x)\,\mathbb{E}\left(\mathrm{h}_{n+1}(\xi) \diamond \mathrm{h}_k(\xi)\right) = u_{k-1}(x) \tag{5.2.5}$$

Therefore,

$$\boldsymbol{\Delta} u_0(x) = f(x),\ x \in (0,1),\ u(0) = u(1) = 0 \tag{5.2.6}$$

and for $k \geq 1$,

$$\begin{aligned} &\boldsymbol{\Delta} u_k(x) + u_{k-1}(x) = 0,\ x \in (0,1) \\ &u_k(0) = u_k(1) = 0 \end{aligned} \tag{5.2.7}$$

System (5.2.6),(5.2.7) is often referred to as the (deterministic) `propagator` of Eq. (5.2.2). The propagator of Eq. (5.2.2) is a **lower-triangular system** and therefore can be solved sequentially, starting with Eq. (5.2.6). As soon as the propagator solved, the solution of the equation is obtained by the Cameron-Martin expansion

$$u(x) = \sum_{k=0}^{\infty} u_k(x)\, \mathrm{h}_k(\xi) \tag{5.2.8}$$

To summarize, the solution of our SPDE was given by: (a) Cameron-Martin expansion (5.2.8) and (b) the (deterministic) propagator (5.2.6)–(5.2.7).

Solution with such structure is usually referred to as the `polynomial chaos solution`. If, as in our example, the underlying random variables are Gaussian, then the solution is often referred to as `Wiener chaos solution`.

In this section we will develop a systematic approach to construction of Wiener chaos solutions to bilinear SPDEs driven by purely spatial Gaussian noise. In particular, we will investigate bilinear elliptic equations

$$\mathbf{A}u(x) + \delta_{\dot{W}(x)}\mathbf{M}u(x) = f(x) \tag{5.2.9}$$

for a wide range of operators **A** and **M**; the corresponding parabolic equations

$$\frac{\partial v(t,x)}{\partial t} = \mathbf{A}v(t,x) + \delta_{\dot{W}(x)}\mathbf{M}v(t,x) - f(x) \tag{5.2.10}$$

are the subject of the following section.

Purely spatial white noise is an important type of stationary perturbations. So far we have discussed elliptic equations only with additive random forcing (Sect. 4.2). However, the methodology developed in Sect. 4.2 is not suitable for elliptic SPDEs with multiplicative random forcing as in examples above.

We saw in Sect. 4.4 that, in the case of parabolic equations, the crucial assumption on the operators **A** and **M** was

$$\text{The operator } \mathbf{A} - \frac{1}{2}\mathbf{M}\mathbf{M}^{\star} \text{ is elliptic.} \tag{5.2.11}$$

This assumption ensured square integrability of the solution.

In this and the following sections we will study Wiener chaos solutions for important classes of SPDEs that are not covered by assumption (5.2.11).

Let us start with a simple example illustrating that the solution of a stationary stochastic equation is not square integrable.

Example 5.2.2 Consider a algebraic equation

$$v = 1 + v \diamond \xi. \tag{5.2.12}$$

It is easy to see that $\{v_n = \mathbb{E}\,(v\mathrm{h}_n(\xi))\,,\ n \geq 0\}$ solve the following system

$$v_0 = 1,\ v_n = I_{n=0} + \sqrt{n}v_{n-1},\ n \geq 1$$

Then $v_n = \sqrt{n!}$ and $v = 1 + \sum_{n=1}^{\infty} \sqrt{n!}\, \mathrm{h}_n(\xi)$, Therefore,

$$\mathbb{E}v^2 = 1 + \sum_{n \geq 1} n! = \infty.$$

Exercise 5.2.3 (B) Consider

$$u = 1 + u \diamond \xi + u \diamond \left(\xi^2 - 1\right) \tag{5.2.13}$$

and recall that $\xi^2 - 1 = H_2(\xi)$.

(a) Find the equations for u_n.
(b) Confirm that the u_n *is the n-th Fibonacci number.*
(c) Conclude that $Eu^2 = \infty$.

Hint: The asymptotic of Fibonacci numbers is given by

$$u_n \sim (1 + \sqrt{5})^n / 2^n \text{ as } n \to \infty. \tag{5.2.14}$$

Example 5.2.2 and Exercise 5.2.3 indicate that one should not expect a solution of a stationary equation to have finite variance. Nevertheless, Wiener chaos solutions can be defined in spaces larger then those we have considered before, and do not require assumption (5.2.11).

In the next section we will establish existence and uniqueness of Wiener Chaos solutions for stationary (elliptic) equations of the type (5.2.9).

5.2.2 Solving Stationary SPDEs by Weighted Wiener Chaos

In Sect. 4.2 we discussed a general elliptic SPDE of the form

$$\mathbf{A}u\,(x) = \dot{W}\,(x)\,,\ x \in G, \tag{5.2.15}$$

where $\mathbf{A}$ is an elliptic operator, $\dot{W}\,(x)$ is Gaussian white noise, and G is a bounded domain in $\mathbb{R}^{\mathrm{d}}$. Popular examples of elliptic SPDEs discussed below include: a random Poisson equation

$$\boldsymbol{\Delta}u\,(x) = \dot{W}\,(x)\,,\ x \in G,$$

and the Euclidean free field equation

$$\sqrt{1 - \boldsymbol{\Delta}}u\,(x) = \dot{W}\,(x)\,,\ x \in G,$$

etc. (see Sect. 4.2.1).

Now we will demonstrate that the Wiener chaos methodology is a powerful tool for analysis of stochastic elliptic PDEs with multiplicative noise.

In this section we will deal with equation

$$\mathbf{A}u + \delta_{\mathfrak{B}}(\mathbf{M}u) = f \tag{5.2.16}$$

in the normal triple (V, H, V') of Hilbert spaces (see Definition 3.1.5 on page 78).

Everywhere in this section $\mathbf{A} : V \to V'$ and $\mathbf{M} : V \to V' \otimes H$ are bounded linear operators. The white noise $\mathfrak{B}$ will be identified with it's chaos expansion (5.1.7)

$$\mathfrak{B} = \sum_{k \geq 1} \mathfrak{u}_k \, \xi_k.$$

Exercise 5.2.4 (C) Show that Eq. (5.2.16) can be rewritten in the form

$$\mathbf{A}u + \sum_{k \geq 1} \mathbf{M}_k u \diamond \xi_k = f, \tag{5.2.17}$$

where

$$\mathbf{M}_k u = \sum_{\alpha \in \mathcal{J}^1} u_{k,\alpha} \otimes \mathfrak{u}_k \xi_\alpha \tag{5.2.18}$$

Example 5.2.2 and Exercise 5.2.3 indicate that one should not expect a solution of equation (5.2.16) to have a finite second moment. However, returning to Example 5.2.2 one could see that the "weighted" norm

$$\mathbb{E}\, \|v\|_R^2 = \sum_{n \geq 1} r_n^2 v_n^2$$

with a sequence of weights r_n such that $\sum_{n \geq 1} r_n^2 n! < \infty$ is appropriate for the solution of equation (5.2.5). Therefore, one could expect that a solution of equation (5.2.16) has a finite "weighted" second moment.

With this in mind, we will now introduce a version of Wiener chaos expansion suitable for functions [of Gaussian random variables] with infinite variance.

Let $\mathcal{R}$ be a linear operator on $L_2(\Xi)$ defined by $\mathcal{R}\xi_\alpha = r_\alpha \xi_\alpha$ for every $\alpha \in \mathcal{J}^1$, where the *weights* $\{r_\alpha,\ \alpha \in \mathcal{J}^1\}$ are positive numbers. By Theorem 5.1.12, $\mathcal{R}$ is bounded if the weights r_α are uniformly bounded from above: $r_\alpha < C$ for all $\alpha \in \mathcal{J}^1$, with C independent of α.

Exercise 5.2.5 (C) Prove that if $\mathcal{R}$ is bounded linear operator on $L_2(\Xi)$, then r_α are uniformly bounded from above and the inverse operator $\mathcal{R}^{-1}$ is defined by $\mathcal{R}^{-1}\xi_\alpha = r_\alpha^{-1}\xi_\alpha$.

Let H be a Hilbert space and $(\cdot,\cdot)_H$ and $\|\cdot\|_H$ denote the inner product and the norm in H.

Exercise 5.2.6 (C) Prove that the operator $\mathcal{R}$ can be extended to an operator on $L_2(\Xi;H)$ by defining $\mathcal{R}f$ as the unique element of $L_2(\Xi;H)$ so that, for all $g \in L_2(\Xi;H)$,

$$\mathbb{E}(\mathcal{R}f,g)_H = \sum_{\alpha\in\mathcal{J}^1} r_\alpha \mathbb{E}\big((f,g)_H\xi_\alpha\big).$$

Let us denote by $\mathcal{R}L_2(\Xi;H)$ the closure of $L_2(\Xi;H)$ with respect to the norm

$$\|f\|^2_{\mathcal{R}L_2(\Xi;H)} := \|\mathcal{R}f\|^2_{L_2(\Xi;H)}.$$

It is readily seen that the elements of $\mathcal{R}L_2(\Xi;H)$ can be identified with a formal series $\sum_{\alpha\in\mathcal{J}^1} f_\alpha\xi_\alpha$, where $f_\alpha \in H$ and $\sum_{\alpha\in\mathcal{J}^1}\|f_\alpha\|^2_H r^2_\alpha < \infty$.

In the future, the argument H will be omitted if $H = \mathbb{R}$.

Next, we define the space $\mathcal{R}^{-1}L_2(\Xi;H)$ as the dual of $\mathcal{R}L_2\,(\Xi;H)$ relative to the inner product in the space $L_2(\mathbb{R};H)$:

$$\mathcal{R}^{-1}L_2(\Xi;H) = \left\{g \in L_2(\Xi;H) : \mathcal{R}^{-1}g \in L_2(\Xi;H)\right\}.$$

The duality for $f \in \mathcal{R}L_2(\Xi;H)$ and $g \in \mathcal{R}^{-1}L_2(\Xi)$ is defined by

$$\langle\!\langle f,g\rangle\!\rangle := \mathbb{E}\big((\mathcal{R}f)(\mathcal{R}^{-1}g)\big) \in H. \tag{5.2.19}$$

In what follows, the operator $\mathcal{R}$ will often be identified with the corresponding collection $(r_\alpha,\ \boldsymbol{\alpha} \in \mathcal{J}^1)$. Note that if $u \in \mathcal{R}_1L_2(\Xi;H)$ and $v \in \mathcal{R}_2L_2(\Xi;H)$, then both u and v belong to $\mathcal{R}L_2(\mathbb{F};H)$, where $r_\alpha = \min(r_{1,\alpha}, r_{2,\alpha})$.

Exercise 5.2.7 (C) Let $\{q_k,\ k \geq 1\}$ be a sequence of real numbers. What else needs to be assumed about this sequence to ensure that the numbers

$$r^2_\alpha = \prod_{k=1}^{\infty} q_k^{\alpha_k}$$

define a proper $\mathcal{R}L_2(\Xi;H)$ space?

The following notation will be often used in this section:

$$(2\mathbb{N})^{\gamma\alpha} = \prod_{k\geq 1}(2k)^{\gamma\alpha_k} \tag{5.2.20}$$

The following lemma, due to Zhang (see [78, 234]) is an important technical fact.

Lemma 5.2.8 *The sum*

$$\sum_{\alpha\in\mathcal{J}}(2\mathbb{N})^{-\gamma\alpha}<\infty$$

if and only if $\gamma>1$

Proof Note that by the formula for the sum of geometric progression

$$\begin{aligned}\sum_{\alpha\in\mathcal{J}}\prod_{i\geq 1}(2i)^{-\gamma\alpha_i}&=\prod_{i\geq 1}\sum_{n\geq 0}((2i)^{-\gamma})^n\\&=\prod_{i\geq 1}(1-(2i)^{-\gamma})^{-1}.\end{aligned}\tag{5.2.21}$$

The infinite product on the right hand of (5.2.21) converges if and only if

$$\sum_{i=1}^{\infty}(2i)^{-\gamma}<\infty,$$

that is, if and only if $\gamma>1$.

Let

$$r_\alpha^2=(\alpha!)^\rho(2\mathbb{N})^{\ell\alpha},\ \rho\leq 0,\ \ell\leq 0\tag{5.2.22}$$

The weights given by (5.2.22) define the class of spaces $\mathcal{R}L_2(\mathbb{F};H)$ often called the Hida-Kondratiev spaces and denoted by $(\mathcal{S})_{\rho,\ell}(H)$.

Exercise 5.2.9 (B) Prove that the solution of equation (5.2.12) belongs to Hida-Kondratiev's space $(\mathcal{S})_{-1,\ell}(\mathbb{R})$ for any $\ell<0$. [Keep in mind that $\boldsymbol{\alpha}=n$.]

Exercise 5.2.10 (A) Specify ρ,ℓ such that the solution of equation (5.2.13) $u\in(\mathcal{S})_{\rho,\ell}(\mathbb{R})$.

Next, we extend the Wick product to weighted spaces $\mathcal{R}L_2$.

Suppose that $f\in\mathcal{R}L_2(\varXi;H)$, where H is a Hilbert space, and $\eta\in\mathcal{R}L_2(\varXi;\mathbb{R})$. Then, the Wick product $f\diamond\eta$ is defined by

$$f\diamond\eta=\sum_{\alpha,\beta}f_\alpha\eta_\beta\xi_\alpha\diamond\xi_\beta.\tag{5.2.23}$$

Proposition 5.2.11 *If* $f\in\mathcal{R}L_2(\mathbb{F};X)$ *and* $\eta\in\mathcal{R}L_2(\varXi;\mathbb{R})$, *then* $f\diamond\eta$ *is an element of* $\bar{\mathcal{R}}L_2(\mathbb{F};X)$ *for a suitable operator* $\bar{\mathcal{R}}$.

Proof It follows from (5.2.23) that $f \diamond \eta = \sum_{\alpha\in\mathcal{J}^1} F_\alpha \xi_\alpha$ and

$$F_\alpha = \sum_{\beta,\gamma\in\mathcal{J}^1:\beta+\gamma=\alpha} \sqrt{\left(\frac{\alpha!}{\beta!\gamma!}\right)} f_\beta \eta_\gamma.$$

Therefore, each F_α is an element of X, because, for every $\boldsymbol{\alpha} \in \mathcal{J}^1$, there are only finitely many multi-indices β, $\boldsymbol{\gamma}$ satisfying $\boldsymbol{\beta} + \boldsymbol{\gamma} = \boldsymbol{\alpha}$. By Lemma 5.2.8,

$$\sum_{\alpha\in\mathcal{J}^1} (2\mathbb{N})^{q\alpha} < \infty \text{ if and only if } q < -1. \tag{5.2.24}$$

Therefore, $f \diamond \eta \in \bar{\mathcal{R}}L_2(\mathbb{F}; X)$, where the operator $\bar{\mathcal{R}}$ can be defined using the weights $\bar{r}_\alpha^2 = (2\mathbb{N})^{-2a}/(1 + \|F_\alpha\|_X^2)$.

Next, we extend the divergence operator $\delta_{\mathfrak{B}}$ to weighted spaces.

Definition 5.2.12 For $f \in \mathcal{R}L_2(\Xi; X\otimes H)$, we define $\delta_{\mathfrak{B}}(f)$ as the element of $\mathcal{R}L_2(\Xi; X)$ such that

$$\langle\langle \delta_{\mathfrak{B}}(f), \varphi\rangle\rangle = \mathbb{E}(\mathcal{R}f, \mathcal{R}^{-1}\mathbb{D}_{\mathfrak{B}}\varphi)_H \tag{5.2.25}$$

for every φ satisfying $\varphi \in \mathcal{R}^{-1}L_2(\Xi)$ and $\mathbb{D}_{\mathfrak{B}}\varphi \in \mathcal{R}^{-1}L_2(\Xi; H)$.

We can now generalize the relation (5.1.61) on page 249 connecting the Wick product with the divergence operator.

Theorem 5.2.13 *If f is an element of $\mathcal{R}L_2(\mathbb{F}; X\otimes H)$ so that $f = \sum_{k\geq 1} f_k \otimes \mathfrak{u}_k$, with $f_k = \sum_{\alpha\in\mathcal{J}^1} f_{k,\alpha}\, \xi_\alpha \in \mathcal{R}L_2(\mathbb{F}; X)$, then*

$$\delta_{\mathfrak{B}}(f) = \sum_{k\geq 1} f_k \diamond \xi_k, \tag{5.2.26}$$

and

$$(\delta_{\mathfrak{B}}(f))_\alpha = \sum_{k\geq 1} \sqrt{\alpha_k} f_{k,\alpha-\epsilon_k}. \tag{5.2.27}$$

Proof By linearity and (5.2.23),

$$\delta_{\mathfrak{B}}(f) = \sum_{k\geq 1}\sum_{\alpha\in\mathcal{J}^1} \delta_{\mathfrak{B}}(\xi_\alpha f_{k,\alpha} \otimes \mathfrak{u}_k) = \sum_{k\geq 1}\sum_{\alpha\in\mathcal{J}^1} f_{k,\alpha}\xi_\alpha \diamond \xi_k = \sum_{k\geq 1} f_k \diamond \xi_k,$$

which is (5.2.26). On the other hand, by (5.1.21),

$$\delta_{\mathfrak{B}}(f) = \sum_{k\ge 1}\sum_{\alpha\in\mathcal{J}^1} f_{k,\alpha}\sqrt{\alpha_k+1}\,\xi_{\alpha+\epsilon_k} = \sum_{k\ge 1}\sum_{\alpha\in\mathcal{J}^1} f_{k,\alpha-\epsilon_k}\sqrt{\alpha_k}\,\xi_\alpha,$$

and (5.2.27) follows.

Now we are in a position to construct a solution to Eq. (5.2.1).

5.2.2.1 Existence and Uniqueness of Solutions

We start with a rigorous definition of the solution to Eq. (5.2.1).

Definition 5.2.14 The solution of equation (5.2.1) with $f \in \mathcal{R}L_2(\Xi; V')$, is a random element $u \in \mathcal{R}L_2(\mathbb{F}; V)$ so that the equality

$$\langle\!\langle \mathbf{A}u, \varphi\rangle\!\rangle + \langle\!\langle \delta_{\mathfrak{B}}(\mathbf{M}u), \varphi\rangle\!\rangle = \langle\!\langle f, \varphi\rangle\!\rangle \tag{5.2.28}$$

holds in V' for every φ satisfying $\varphi \in \mathcal{R}^{-1}L_2(\Xi)$ and $\mathbb{D}\varphi \in \mathcal{R}^{-1}L_2(\Xi;\ H)$.

Taking $\varphi = \xi_\alpha$ in (5.2.28) and using relation (5.2.25) we conclude that Eq. (5.2.28) leads to the following system of equations for $u_\alpha = \mathbb{E}\,(u\xi_\alpha)$:

$$\begin{cases} \mathbf{A}u_{(0)} = \mathbb{E}f, \\ \mathbf{A}u_\alpha + \sum_{k\ge 1}\sqrt{\alpha_k}\,\mathbf{M}_k u_{\alpha-\epsilon_k} = f_\alpha \text{ for } |\boldsymbol{\alpha}| > 0. \end{cases} \tag{5.2.29}$$

In the future we will refer to system (5.2.29) as the `propagator`.

Exercise 5.2.15 (B) Derive the propagators for Eqs. (5.2.2)–(5.2.4).

The following theorem establishes equivalence of Eq. (5.2.1) and the propagator.

Theorem 5.2.16 *Let $u = \sum_{\alpha\in\mathcal{J}^1} u_\alpha\xi_\alpha$ be an element of $\mathcal{R}L_2(\Xi; V)$. Then u is a solution of equation (5.2.1) if and only if the non-random coefficients u_α have the following properties:*

1. *every u_α is an element of H,*
2. *the system of equalities (5.2.29) holds in V' for all $\boldsymbol{\alpha} \in \mathcal{J}^1$.*

Proof Let u be a solution of

$$\mathbf{A}u + \delta_{\mathfrak{B}}(\mathbf{M}u) = f$$

in $\mathcal{R}L_2(\Xi; V)$. Taking $\varphi = \xi_\alpha$ in (5.2.28) and using relation (5.2.27)

$$(\delta_{\mathfrak{B}}(f))_\alpha = \sum_{k\ge 1}\sqrt{\alpha_k} f_{k,\alpha-\epsilon_k},$$

we obtain Eq. (5.2.29). By Theorem 4.1.18 on page 161, $u_{\alpha} \in V$ for all $\boldsymbol{\alpha}$.

Conversely, let $\{u_{\alpha}, \boldsymbol{\alpha} \in \mathcal{J}\}$ be a collection of functions from V satisfying (5.2.29). Set $u = \sum_{\alpha \in \mathcal{J}^1} u_{\alpha}\xi_{\alpha}$. Then, by Theorem 5.2.13, the equality

$$\langle\langle u, \xi_{\alpha}\rangle\rangle + \langle\langle \mathbf{A}u + \delta_{\mathfrak{B}}(\mathbf{M}u), \xi_{\alpha}\rangle\rangle = \langle\langle f, \xi_{\alpha}\rangle\rangle$$

holds in V'. By linearity, we conclude that, for any $\varphi \in \mathcal{R}^{-1}L_2(\mathbb{F})$ such that $\mathbb{D}\varphi \in \mathcal{R}^{-1}L_2(\mathbb{F};H)$, the equality

$$\langle\langle u, \varphi\rangle\rangle + \langle\langle \mathbf{A}u + \delta_{\mathfrak{B}}(\mathbf{M}u), \varphi\rangle\rangle = \langle\langle f, \varphi\rangle\rangle$$

holds in V' as well.

The system of equations (5.2.29) is lower-triangular and can be solved by induction on $|\boldsymbol{\alpha}|$. Together with Theorem 5.2.16, this leads to the main result about existence and uniqueness of solution of (5.2.1).

Theorem 5.2.17 *Consider Eq.* (5.2.1) *in which* $f \in \bar{\mathcal{R}}L_2(\mathbb{F}; V')$ *for some* $\bar{\mathcal{R}}$. *Assume that the deterministic equation* $\mathbf{A}U = F$ *is uniquely solvable in the normal triple* (V, H, V'), *that is, for every* $F \in V'$, *there exists a unique solution* $U = \mathbf{A}^{-1}F \in V$ *and* $\|U\|_V \leq C_A\|F\|_{V'}$. *Assume also that each* $\mathbf{M}_k$ *is a bounded linear operator from* V *to* V' *so that, for all* $v \in V$,

$$\|\mathbf{A}^{-1}\mathbf{M}_k v\|_V \leq C_k\|v\|_V, \tag{5.2.30}$$

with C_k *independent of* v.

Then there exists an operator $\mathcal{R}$ *and a unique solution* $u \in \mathcal{R}L_2(\mathbb{F}; V)$ *of (5.2.1).*

Proof By assumption,

$$u_{\alpha} = \mathbf{A}^{-1}f_{\alpha} - \sum_k \sqrt{\alpha_k}\mathbf{A}^{-1}\mathbf{M}_k u_{\alpha-\epsilon(k)}$$

is the unique solution of (5.2.29) and $u_{\alpha} \in V$. It remains to take

$$r_{\alpha}^2 = \frac{(2\mathbb{N})^{-2\alpha}}{1 + \|u_{\alpha}\|_V^2}.$$

Remark 5.2.18 The assumption of the theorem about solvability of the deterministic equation holds if the operator $\mathbf{A}$ satisfies $\langle \mathbf{A}v, v\rangle \geq \kappa\|v\|_V^2$ for every $v \in V$, with $\kappa > 0$ independent of v.

If f is non-random, then there is a reasonably manageable closed-form expression for u_{α}, and a more explicit choice of the weights r_{α}. To state the corresponding result, introduce the notations

$$\delta_{\mathbf{B}}^{(0)}(\eta) = \eta, \;\; \delta_{\mathbf{B}}^{(n)}(\eta) = \delta_{\mathfrak{B}}(\mathbf{B}\delta_{\mathbf{B}}^{(n-1)}(\eta)), \;\; \eta \in \mathcal{R}L_2(\mathbb{F}; V),$$

where $\mathbf{B}$ is a bounded linear operator from V to $V \otimes H$.

Theorem 5.2.19 *Under the assumptions of Theorem 5.2.17, if f is non-random, then the following holds:*

1. *the coefficient u_{α}, corresponding to the multi-index α with $|\alpha| = n \geq 1$ and the characteristic set $K_{\alpha} = \{k_1, \ldots, k_n\}$, is given by*

$$u_{\alpha} = \frac{1}{\sqrt{\alpha!}} \sum_{\sigma \in \mathcal{P}_n} \mathbf{B}_{k_{\sigma(n)}} \cdots \mathbf{B}_{k_{\sigma(1)}} u_{(\mathbf{0})}, \tag{5.2.31}$$

where

- *$\mathcal{P}_n$ is the permutation group of the set $(1, \ldots, n)$;*
- *$\mathbf{B}_k = -\mathbf{A}^{-1}\mathbf{M}_k$;*
- *$u_{(\mathbf{0})} = \mathbf{A}^{-1} f$.*

2. *the operator $\mathcal{R}$ can be defined by the weights*

$$r_{\alpha} = \frac{q^{\alpha}}{\sqrt{|\alpha|!}}, \text{ where } q^{\alpha} = \prod_{k=1}^{\infty} q_k^{\alpha_k}, \tag{5.2.32}$$

where the numbers q_k, $k \geq 1$, are chosen so that $\sum_{k \geq 1} q_k^2 C_k^2 < 1$, and C_k are defined in (5.2.30).

3. *With r_{α} and q_k defined by (5.2.32),*

$$\sum_{|\alpha|=n} q^{\alpha} u_{\alpha} \xi_{\alpha} = \delta_{\overline{\mathbf{B}}}^{(n)}(\mathbf{A}^{-1} f), \tag{5.2.33}$$

where $\overline{\mathbf{B}} = -(q_1 \mathbf{A}^{-1}\mathbf{M}_1, q_2 \mathbf{A}^{-1}\mathbf{M}_2, \ldots)$, and

$$\mathcal{R}u = \mathbf{A}^{-1} f + \sum_{n \geq 1} \frac{1}{\sqrt{n!}} \delta_{\overline{\mathbf{B}}}^{(n)}(\mathbf{A}^{-1} f). \tag{5.2.34}$$

Proof Define $\widetilde{u}_{\alpha} = \sqrt{\alpha!}\, u_{\alpha}$. If f is non-random [so that $f_{(\mathbf{0})} = f$ and $f_{\alpha} = 0$, $|\alpha| > 0$], then $\widetilde{u}_{(\mathbf{0})} = \mathbf{A}^{-1} f$ and, for $|\alpha| \geq 1$,

$$\mathbf{A}\widetilde{u}_{\alpha} + \sum_{k \geq 1} \alpha_k \mathbf{M}_k \widetilde{u}_{\alpha - \epsilon_k} = 0,$$

or

$$\widetilde{u}_{\alpha} = \sum_{k \geq 1} \alpha_k \mathbf{B}_k \widetilde{u}_{\alpha - \epsilon_k} = \sum_{k \in K_{\alpha}} \mathbf{B}_k \widetilde{u}_{\alpha - \epsilon_k},$$

where $K_{\boldsymbol{\alpha}} = \{k_1, \ldots, k_n\}$ is the characteristic set of $\boldsymbol{\alpha}$ and $n = |\boldsymbol{\alpha}|$. By induction on n,

$$\widetilde{u}_{\boldsymbol{\alpha}} = \sum_{\sigma \in \mathcal{P}_n} \mathbf{B}_{k_{\sigma(n)}} \cdots \mathbf{B}_{k_{\sigma(1)}} u_{(\mathbf{0})},$$

and (5.2.31) follows.

Next, define

$$U_n = \sum_{|\boldsymbol{\alpha}|=n} q^{\boldsymbol{\alpha}} u_{\boldsymbol{\alpha}} \xi_{\boldsymbol{\alpha}}, \; n \geq 0.$$

Let us first show that, for each $n \geq 1$, $U_n \in L_2(\Xi; V)$. By (5.2.31) we have

$$\|u_{\boldsymbol{\alpha}}\|_V^2 \leq C_A^2 \frac{(|\boldsymbol{\alpha}|!)^2}{\boldsymbol{\alpha}!} \|f\|_{V'}^2 \prod_{k \geq 1} C_k^{\alpha_k}. \tag{5.2.35}$$

By direct computation,

$$\begin{aligned}\sum_{|\boldsymbol{\alpha}|=n} q^{2a} \|u_{\boldsymbol{\alpha}}\|_V^2 &\leq C_A^2 \|f\|_{V'}^2 \, n! \sum_{|\boldsymbol{\alpha}|=n} \left(\frac{n!}{\boldsymbol{\alpha}!} \prod_{k \geq 1} (C_k q_k)^{2\alpha_k} \right) \\ &= C_A^2 \|f\|_{V'}^2 \, n! \left(\sum_{k \geq 1} C_k^2 q_k^2 \right)^n < \infty,\end{aligned}$$

because of the selection of q_k, and so $U_n \in L_2(\Xi; V)$. If the weights $r_{\boldsymbol{\alpha}}$ are defined by (5.2.32), then

$$\sum_{\boldsymbol{\alpha} \in \mathcal{J}^1} r_{\boldsymbol{\alpha}}^2 \|u\|_V^2 = \sum_{n \geq 0} \sum_{|\boldsymbol{\alpha}|=n} r_{\boldsymbol{\alpha}}^2 \|u\|_V^2 \leq C_A^2 \|f\|_{V'}^2 \sum_{n \geq 0} \left(\sum_{k \geq 1} C_k^2 q_k^2 \right)^n < \infty,$$

because of the assumption $\sum_{k \geq 1} C_k^2 q_k^2 < 1$.

Since (5.2.34) follows directly from (5.2.33), it remains to establish (5.2.33), that is,

$$U_n = \delta_{\overline{\mathbf{B}}}(U_{n-1}), \; n \geq 1. \tag{5.2.36}$$

For $n = 1$ we have

$$U_1 = \sum_{k \geq 1} q_k u_{\boldsymbol{\epsilon}_k} \xi_k = \sum_{k \geq 1} \overline{\mathbf{B}}_k u_{(\mathbf{0})} \xi_k = \delta_{\overline{\mathbf{B}}}(U_0),$$

where the last equality follows from (5.2.26). More generally, for $n > 1$ we have by definition of U_n that

$$(U_n)_\alpha = \begin{cases} q^\alpha u_\alpha, & \text{if } |\alpha| = n, \\ 0, & \text{otherwise.} \end{cases}$$

From the equation

$$q^\alpha \mathbf{A} u_\alpha + \sum_{k\geq 1} q_k \sqrt{\alpha_k}\, \mathbf{M}_k q^{\alpha-\epsilon_k} u_{\alpha-\epsilon_k} = 0$$

we find

$$\begin{aligned}(U_n)_\alpha &= \begin{cases} \sum_{k\geq 1} \sqrt{\alpha_k}\, q_k \mathbf{B}_k q^{\alpha-\epsilon_k} u_{\alpha-\epsilon_k}, & \text{if } |\alpha| = n, \\ 0, & \text{otherwise.} \end{cases} \\ &= \sum_{k\geq 1} \sqrt{\alpha_k}\, \overline{\mathbf{B}}_k (U_{n-1})_{\alpha-\epsilon_k},\end{aligned}$$

and then (5.2.36) follows from (5.2.27). Theorem 5.2.19 is proved.

Here is another result about solvability of (5.2.17), this time with random f. We use the space $(\mathcal{S})_{\rho,q}$, defined by the weights (5.2.22). We will also use a certain inequality for multi-indices.

Exercise 5.2.20 (C) Let $\boldsymbol{x} = (x_1, x_2, \ldots)$ be a sequence of positive numbers such that

$$\sum_{j=1}^{\infty} x_j < \infty.$$

Verify that, for $n = 1, 2, 3, \ldots$,

$$\left(\sum_{j=1}^{\infty} x_j\right)^n = \sum_{|\alpha|=n} \frac{n!}{\alpha!} \prod_j x_j^{\alpha_j}. \tag{5.2.37}$$

Then take $x_j = 1/q_j$ such that $\sum_j (1/q_j) \leq 1$ and deduce a very useful inequality:

$$\alpha! \leq |\alpha|! \leq \left(\prod_j q_j^{\alpha_j}\right) \alpha!. \tag{5.2.38}$$

In particular, with $q_j = 2j^2$,

$$\alpha! \le |\alpha|! \le (2\mathbb{N})^{2\alpha}\alpha!. \tag{5.2.39}$$

[Clearly, $|\alpha|! \ge \alpha!$ because $|\alpha|!/\alpha! \ge 1$ is a multinomial coefficient.]

Theorem 5.2.21 *In addition to the assumptions of Theorem 5.2.17, let $C_A \le 1$ and $C_k \le 1$ for all k. If $f \in (\mathcal{S})_{-1,-\ell}(V')$ for some $\ell > 1$, then there exists a unique solution $u \in (\mathcal{S})_{-1,-\ell-4}(V)$ of (5.2.17) and*

$$\|u\|_{(\mathcal{S})_{-1,-\ell-4}(V)} \le C(\ell)\|f\|_{(\mathcal{S})_{-1,-\ell}(V')}. \tag{5.2.40}$$

Proof Denote by $u(g;\boldsymbol{\gamma})$, $\boldsymbol{\gamma} \in \mathcal{J}^1$, $g \in V'$, the solution of (5.2.17) with $f_\alpha = gI_{(\alpha=\gamma)}$, and define $\bar{u}_\alpha = (\alpha!)^{-1/2}u_\alpha$. Clearly, $u_\alpha(g,\boldsymbol{\gamma}) = 0$ if $|\alpha| < |\boldsymbol{\gamma}|$ and so

$$\sum_{\alpha\in\mathcal{J}^1} \|u_\alpha(f_\gamma;\boldsymbol{\gamma})\|_V^2 r_\alpha^2 = \sum_{\alpha\in\mathcal{J}^1} \|u_{a+\gamma}(f_\gamma;\boldsymbol{\gamma})\|_V^2 r_{a+\gamma}^2. \tag{5.2.41}$$

It follows from (5.2.29) that

$$\bar{u}_{\alpha+\gamma}(f_\gamma;\boldsymbol{\gamma}) = \bar{u}_\alpha\big(f_\gamma(\boldsymbol{\gamma}!)^{-1/2};(\mathbf{0})\big). \tag{5.2.42}$$

Now we use (5.2.35) to conclude that

$$\|\bar{u}_{a+\gamma}(f_\gamma;\boldsymbol{\gamma})\|_V \le \frac{|\alpha|!}{\sqrt{\alpha!\boldsymbol{\gamma}!}}\|f\|_{V'}. \tag{5.2.43}$$

Coming back to (5.2.41) with $r_\alpha^2 = (a!)^{-1}(2\mathbb{N})^{(-\ell-4)a}$ and using (5.2.39), we find:

$$\|u(f_\gamma;\boldsymbol{\gamma})\|_{(\mathcal{S})_{-1,-\ell-4}(V)} \le C(\ell)(2\mathbb{N})^{-2\gamma}\frac{\|f_\gamma\|_{V'}}{(2\mathbb{N})^{(\ell/2)\gamma}\sqrt{\boldsymbol{\gamma}!}},$$

where

$$C(\ell) = \left(\sum_{\alpha\in\mathcal{J}^1}\left(\frac{|\alpha|!}{\alpha!}\right)^2(2\mathbb{N})^{(-\ell-4)a}\right)^{1/2};$$

(5.2.24) and (5.2.39) imply $C(\ell) < \infty$. Then (5.2.40) follows by the triangle inequality after summing over all $\boldsymbol{\gamma}$ and using the Cauchy-Schwartz inequality.

Remark 5.2.22 Example 5.2.2, in which $f \in (\mathcal{S})_{0,0}(\mathbb{R})$ and $u \in (\mathcal{S})_{-1,q}(\mathbb{R})$, $q < 0$, shows that, while the results of Theorem 5.2.21 are not sharp, a bound of the type $\|u\|_{(\mathcal{S})_{\rho,q}(V)} \le C\|f\|_{(\mathcal{S})_{\rho,\ell}(V')}$ is, in general, impossible if $\rho > -1$ or $q \ge \ell$.

5.3 Elements of Malliavin Calculus for Brownian Motion

5.3.1 Cameron-Martin Basis for Scalar Brownian Motion

To begin, let us limit the discussion to a one-dimensional Brownian motion $w(t)$ on the probability space $\mathbb{F} = (\Omega, \mathcal{F}, \mathbb{P})$. Recall (see Definition 3.2.32 on page 114) that the standard (one-dimensional) Brownian motion $w = w(t)$, $t \in [0, T]$, is a real-valued Gaussian process such that $\mathbb{E}w(t) = 0$ and $\mathbb{E}w(t)\,w(s) = \min(t, s)$.

Let $\mathcal{F}_t^w$ be the $\mathbb{P}$-completed σ-algebra generated by $\{w(s),\ s \le t\}$.

Definition 5.3.1 The space $L_2(w) = L_2\left(\Omega, \left(\mathcal{F}_t^w\right)_{t\le T}, \mathbb{P}\right)$ will be referred to as `Wiener chaos space for Brownian motion` w.

Let $T > 0$ be non-random and let $\{\mathfrak{m}_i, i \ge 1\}$ be an orthonormal basis in $L_2((0,T))$. Define

$$\xi_i = \int_0^T \mathfrak{m}_i(s)\,dw(s).$$

For simplicity, and with little loss of generality, it will be assumed in the future that, for each i, $\sup_{0\le s\le T} |\mathfrak{m}_i(s)| < \infty$.

It follows that

- Each ξ_i is standard Gaussian and $\mathcal{F}_T^w$-measurable;
- For $i \ne j$,

$$\mathbb{E}\xi_i(w)\,\xi_j(w) = \int_0^T \mathfrak{m}_i(s)\,\mathfrak{m}_j(s)\,ds = 0,$$

  so that $\{\xi_i, i \ge 1\}$ are independent;
- The Brownian motion $w(t)$ admits the following Wiener chaos expansion (cf. (3.2.27) on page 116):

$$w(t) = \sum_{k\ge 1} M_k(t)\,\xi_k, \tag{5.3.1}$$

  where $M_k(t) = \int_0^t \mathfrak{m}_k(s)\,ds$.

Definition 5.3.2 Denote by $L_2(w)$ the collection of $\mathcal{F}_T^w$-measurable square integrable random variables.

Next, we apply the Cameron-Martin theorem (Theorem 5.1.12) to functions of the Brownian motion $w(t)$. The difference between the constructions here and in the previous section is that the Cameron-Martin basis $\{\xi_\alpha(w),\ \boldsymbol{\alpha} \in \mathcal{J}^1\}$, associated with the Brownian motion $w(t)$ is, in a sense, more "focused" than the general version based on an *arbitrary* system of independent standard Gaussian random variables $\{\xi_k, k \geq 1\}$ (as in (5.1.12) and Lemma 5.1.9), and, as a consequence, has a number of additional interesting and useful properties.

As before, the elements of the Cameron-Martin basis generated by the sequence $\{\xi_k,\ k \geq 1\}$ are

$$\xi_\alpha = \prod_{k\geq 1} \mathrm{H}_{\alpha_k}(\xi_k(w))/\sqrt{\alpha_k!}. \tag{5.3.2}$$

Let $\mathbf{z} = \{z_k,\ k \geq 1\}$ be a sequence of real numbers such that

$$\sum z_k^2 < \infty. \tag{5.3.3}$$

Write

$$\mathcal{E}(\mathbf{z}, t, w) = \exp\left(\sum_{k=1}^{\infty} z_k \int_0^t \mathfrak{m}_k(s)dw(s) - \frac{1}{2}\sum_{k=1}^{\infty} z_k^2 \int_0^t \mathfrak{m}_k^2(s)ds\right),\quad 0 \leq t \leq T. \tag{5.3.4}$$

In particular, $\mathbb{E}\mathcal{E}(\mathbf{z}, t, w) = 1,\ 0 \leq t \leq T$, and

$$\mathcal{E}(\mathbf{z}, T, w) = \exp\left\{\sum_{k=1}^{\infty} z_k\xi_k - \frac{1}{2}\sum_{k=1}^{\infty} z_k^2\right\}. \tag{5.3.5}$$

is a special version of the stochastic exponent (5.1.16).

By the Itô formula, the function $\mathcal{E}(\mathbf{z}, t, w)$ solves the following stochastic differential equation

$$d\mathcal{E}(\mathbf{z}, t, w) = \mathcal{E}(\mathbf{z}, t, w)\sum_{k=1}^{\infty} z_k \mathfrak{m}_k(t)dw(t), \tag{5.3.6}$$

which, in particular, implies that the process $\mathcal{E}(\mathbf{z}, t, w)$ is a martingale:

$$\mathbb{E}\Big(\mathcal{E}(\mathbf{z}, T, w)|\mathcal{F}_t^w\Big) = \mathcal{E}(\mathbf{z}, t, w). \tag{5.3.7}$$

At this point, it might appear to the reader that there is very little new substance in the Cameron-Martin basis associated with Brownian motion, as compared to the original basis generated by an arbitrary system of independent standard Gaussian random variables. Below, we will see that the "Brownian" Cameron-Martin basis is more nuanced then the time independent one.

For $\boldsymbol{\alpha} \in \mathcal{J}^1$, define

$$\xi_\alpha(t) = \frac{1}{\sqrt{\alpha!}}\frac{\partial^{|\alpha|}\mathcal{E}(z, t, w)}{\partial \mathbf{z}^\alpha}\Big|_{z=0}. \tag{5.3.8}$$

In particular,

$$\xi_\alpha(T) = \xi_\alpha.$$

It follows from (5.3.7) that, with probability 1,

$$\xi_\alpha(t) = \mathbb{E}\left[\xi_\alpha | \mathcal{F}_t^w\right]. \tag{5.3.9}$$

Theorem 5.3.3 *Let $f = f(t)$ be a process such that $f(t) \in L_2(w)$. and $f(t)$ is $\mathcal{F}_t^w$-measurable for every t. Then*

$$f(t) = \sum_{\alpha \in \mathcal{J}^1} f_\alpha(t)\, \xi_\alpha = \sum_{\alpha \in \mathcal{J}^1} f_\alpha(t)\, \xi_\alpha(t)$$

with probability 1.

Proof By Theorem 5.1.12, with $\xi_\alpha = \xi_\alpha(T)$,

$$f(t) = \sum_{\alpha \in \mathcal{J}^1} f_\alpha(t)\, \xi_\alpha(T),$$

where $f_\alpha(t) = \mathbb{E}\big(f(t)\xi_\alpha\big)$ is non-random. Since $f(t)$ is $\mathcal{F}_t^w$-measurable, we find using (5.3.9) that

$$f(t) = \mathbb{E}\big(f(t)|\mathcal{F}_t^w\big) = \sum_{\alpha \in \mathcal{J}^1} f_\alpha(t)\, \mathbb{E}\big(\xi_\alpha|\mathcal{F}_t^w\big)$$

$$= \sum_{\alpha \in \mathcal{J}^1} f_\alpha(t)\, \xi_\alpha(t).$$

Exercise 5.3.4 (A) Let us consider the equation

$$du(t,x) = a^2 u_{xx}(t,x)\, dt + \sigma u_x(t,x)\, dw(t), \; u(0,x) = \phi(x) \in L_2(\mathbb{R}). \tag{5.3.10}$$

Assume that $2a^2 - \sigma^2 \geq 0$ and $\|\phi\|^2_{L_2(\mathbb{R})} < \infty$. We know (cf. page 53) that the solution to this equation is square integrable and $\mathcal{F}_t^w$-adapted.

Find the system of equations for deterministic coefficients $u_\alpha(t,x) = \mathbb{E}\left[u(t,x)\, \xi_\alpha(t)\right]$ in the Cameron-Martin expansion

$$u(t,x) = \sum_{\alpha \in \mathcal{J}^1} u_\alpha(t,x)\, \xi_\alpha(t) \tag{5.3.11}$$

of the solution of equation (5.3.10).

Our next objective is to represent ξ_α as multiple stochastic integrals with respect to the Brownian motion w.

Exercise 5.3.5 (C) Let $T_n = \{t_1, \ldots, t_n : 0 \le t_1 \le \cdots \le t_n \le T\}$ and functions $g_i(t_1, t_2, \ldots, t_n) \in L_2(T_n)$, $i = 1, 2$. Define

$$I_n(g_i) = \int_0^T \int_0^{t_{n-1}} \cdots \int_0^{t_1} g_i(t_0, \ldots, t_{n-1}) dw(t_0) \ldots dw(t_{n-1}).$$

Prove that

$$\mathbb{E}\left(I_m(g_i) I_k(g_j)\right) = \begin{cases} (g_i, g_j)_{L_2(T_n)} & \text{if } m = k, \\ 0 & \text{if } m \ne k. \end{cases}$$

Definition 5.3.6

(a) A function $f(x_1, \ldots, x_n)$ is called `symmetric` if, for all permutations $(x_{\sigma_1}, \ldots, x_{\sigma_n})$ of $(x_1, \ldots, x_n)$,

$$f(x_1, \ldots x_n) = f(x_{\sigma_1}, \ldots, x_{\sigma_n}). \tag{5.3.12}$$

(b) A `symmetrization` of a function $f = f(x_1, \ldots, x_n)$ is

$$\tilde{f}(x_1, \ldots, x_n) = \frac{1}{n!} \sum_{\sigma} f(x_{\sigma_1}, \ldots, x_{\sigma_n}), \tag{5.3.13}$$

where σ is running over all permutations of the set $(1, \ldots, n)$.

Exercise 5.3.7 (C) Verify that, if function $f(t_1 \ldots, t_n) \in L_2(T_n)$ is symmetric, then

$$\int_{T_n} f^2(t_1, \ldots, t_n) dt_1 \ldots dt_n = \frac{1}{n!} \|f\|^2_{L_2([0,T]^n)}.$$

Now we can establish the multiple integral representation of ξ_{α} when $|\boldsymbol{\alpha}| > 0$ [recall that $\xi_{(\mathbf{0})} = 1$.]

Theorem 5.3.8

(a) If $|\boldsymbol{\alpha}| > 0$, *then*

$$\xi_{\alpha}(t) = \int_0^t \sum_{k=1}^{\infty} \sqrt{\alpha_k} \xi_{\alpha - \epsilon(k)}(s) \mathfrak{m}_k(s) dw(s). \tag{5.3.14}$$

(b) If $|\boldsymbol{\alpha}| = n \ge 1$, *then*

$$\xi_{\alpha}(T) = \frac{\sqrt{n!}}{\sqrt{\boldsymbol{\alpha}!}} \int_0^T \cdots \int_0^{t_2} \tilde{\mathfrak{m}}_{\alpha}(t_1, \ldots, t_n) dw(t_1) \ldots dw(t_n), \tag{5.3.15}$$

where

$$\mathfrak{m}_\alpha(t_1,\ldots,t_n)=\prod_{i=1}^{n}\mathfrak{m}_{j_i}(t_i),$$

$\{j_1,\ldots,j_n\}$ *is the characteristic set of* $\boldsymbol{\alpha}$, *and* $\tilde{\mathfrak{m}}_\alpha$ *is the symmetrization of* $\mathfrak{m}_\alpha$.

Proof By (5.3.4),

$$\frac{\partial}{\partial \mathbf{z}^{\epsilon}(i)}\mathcal{E}(\mathbf{z},t,w)\Big|_{z=0}=\int_0^t \mathfrak{m}_i(s)\,dw(s).$$

By (5.3.6),

$$\begin{aligned} d\left(\frac{\partial^{|\alpha|}}{\partial z^{\alpha}}\mathcal{E}(\mathbf{z},t,w)\right) &= \sum_{k=1}^{\infty}\left(\frac{\partial^{|\alpha|}}{\partial \mathbf{z}^{\alpha}}\mathcal{E}(\mathbf{z},t,w)\right)\mathfrak{m}_k(t)\,z_k dw(t) \\ &+\sum_{k=1}^{\infty}\left(\mathfrak{m}_k(t)\frac{\partial^{|\alpha|-1}}{\partial \mathbf{z}^{\alpha-\epsilon(k)}}\mathcal{E}(\mathbf{z},t,w)\right)dw(t). \end{aligned} \tag{5.3.16}$$

Integrating in time and setting $\mathbf{z}=0$ yields

$$\frac{\partial^{|\alpha|}}{\partial z^{\alpha}}\mathcal{E}(\mathbf{z},t,w)\Big|_{\mathbf{z}=0}=\int_0^t\sum_{k=1}^{\infty}\left(\frac{\partial^{|\alpha|-1}}{\partial \mathbf{z}^{\alpha-\epsilon(k)}}\mathcal{E}(\mathbf{z},s,w)\right)\Big|_{\mathbf{z}=0}\mathfrak{m}_k(s)\,dw(s), \tag{5.3.17}$$

leading to (5.3.14). After that, (5.3.15) follows after a repeated application of (5.3.14).

The linear subspace of $L_2(w)$ generated by ξ_α, $|\alpha|=n$, is called the n-th Wiener chaos. The multiple stochastic integral in the right-hand side of (5.3.15) is usually referred to as multiple Wiener-Itô integral. Integrals of this type have been introduced and studied extensively by Itô in [90].

5.3.1.1 Cameron-Martin Basis for Brownian Motion in a Hilbert Space

Let $\mathbb{F}=(\Omega,\mathcal{F},\{\mathcal{F}_t\}_{0\le t\le T},\mathbb{P})$ be a stochastic basis with the usual assumptions and let Y be a separable Hilbert space with inner product $(\cdot,\cdot)_Y$ and an orthonormal basis $\{y_k,\ k\ge 1\}$. Also, let us fix an orthonormal basis $\{\mathfrak{m}_i,\ i\ge 1\}$ in $L_2((0,T),\mathbb{R})$, so that each $\mathfrak{m}_i$ belongs to $L_\infty((0,T))$.

Exercise 5.3.9 (C) Prove that $\{\mathfrak{u}_{i,k}=\mathfrak{m}_i y_k,\ i,k\ge 1\}$ is an orthonormal basis in the Hilbert space $L_2((0,T);Y)$.

On $\mathbb{F}$ and Y, consider a cylindrical Brownian motion W, that is, a family of continuous $\mathcal{F}_t$-adapted Gaussian martingales $W_y(t)$, $y\in Y$, so that $W_y(0)=0$ and

$E(W_{y_1}(t)W_{y_2}(s)) = \min(t,s)(y_1, y_2)_Y$. In particular,

$$w_k(t) = W_{y_k}(t),\ k \geq 1,\ t \geq 0, \tag{5.3.18}$$

are independent standard Brownian motions on $\mathbb{F}$. The σ-algebra generated by $\{w_k(s),\ s \leq t\}$ will be denoted by $\mathcal{F}_t^{w_k}$. The sigma-algebra generated by the full collection $\{w_k(s),\ k \geq 1,\ 0 \leq s \leq t\}$ will be denoted by $\mathcal{F}_t^W$.

We write

$$W(t) = \sum_{k\geq 1} y_k w_k(t) \tag{5.3.19}$$

in the sense that

$$W_y(t) = \sum_{k\geq 1} (y, y_k)_Y\, w_k(t); \tag{5.3.20}$$

cf. (3.2.31) on page 117. Similarly,

$$\dot{W}(t) = \sum_{k\geq 1} y_k \dot{w}_k(t), \tag{5.3.21}$$

is the white noise associated with W

Next, we construct the Cameron-Martin basis associated with the cylindrical Brownian motion (5.3.20). Define

$$\xi_{i,k} = \int_0^T \mathfrak{m}_i(s)dw_k(s) = \int_0^T \mathfrak{m}_i(s)dw_k(s). \tag{5.3.22}$$

Exercise 5.3.10 (C)

(a) Prove that $\{\xi_{i,k},\ i,k \geq 1\}$ are independent standard Gaussian random variables.
(b) Confirm that

$$\mathbb{E}\left(\xi_{i,k}|\mathcal{F}_t^W\right) = \int_0^t \mathfrak{m}_i(s)dw_k(s). \tag{5.3.23}$$

Definition 5.3.11 Let $\mathcal{J}^2$ be the collection of multi-indices $\boldsymbol{\alpha}$ with $\boldsymbol{\alpha} = (\alpha_{i,k})_{i,k\geq 1}$ so that each $\alpha_{i,k}$ is a non-negative integer and $|\boldsymbol{\alpha}| := \sum_{i,k\geq 1} \alpha_{i,k} < \infty$. For $\boldsymbol{\alpha}, \boldsymbol{\beta} \in \mathcal{J}^2$, define

$$\boldsymbol{\alpha} + \boldsymbol{\beta} = (\alpha_{i,k} + \beta_{i,k})_{i,k\geq 1} \text{ and } \quad \boldsymbol{\alpha}! = \prod_{i,k\geq 1} \alpha_{i,k}!.$$

The multi-index consisting of all zeroes will be denoted by $(\mathbf{0})$.

The entry $\alpha_{i,k}$ of a multi-index $\boldsymbol{\alpha} \in \mathcal{J}^2$ is similar to a matrix entry, with i representing the row and k representing the column.

For $\boldsymbol{\alpha} \in \mathcal{J}^2$, define

$$|\boldsymbol{\alpha}| = \sum_{i,k} \alpha_{i,k}, \ \boldsymbol{\alpha}! = \prod_{i,k} \alpha_{i,k}!,$$

and

$$\xi_{\boldsymbol{\alpha}} = \frac{1}{\sqrt{\boldsymbol{\alpha}!}} \prod_{i,k} \mathrm{H}_{\alpha_{i,k}}(\xi_{i,k}), \tag{5.3.24}$$

where H_n is n^{th} Hermite polynomial and $\xi_{i,k}$ is defined by (5.3.22)

Example 5.3.12 Let

$$\boldsymbol{\alpha} = \begin{pmatrix} 0\,1\,0\,3\,0\,0 \cdots \\ 2\,0\,0\,0\,4\,0 \cdots \\ 0\,0\,0\,0\,0\,0 \cdots \\ \vdots\,\vdots\,\vdots\,\vdots\,\vdots\,\vdots\, \ldots \end{pmatrix} \tag{5.3.25}$$

with only four non-zero entries $\alpha_{1,2} = 1$; $\alpha_{1,4} = 3$; $\alpha_{2,1} = 2$; $\alpha_{2,5} = 4$, then

$$\xi_{\boldsymbol{\alpha}} = \frac{\mathrm{H}_1\,(\xi_{1,2})}{\sqrt{1!}} \cdot \frac{\mathrm{H}_3(\xi_{1,4})}{\sqrt{3!}} \cdot \frac{\mathrm{H}_2(\xi_{2,1})}{\sqrt{2!}} \cdot \frac{\mathrm{H}_4(\xi_{2,5})}{\sqrt{4!}}.$$

For every multi-index $\boldsymbol{\alpha} \in \mathcal{J}^2$ with $|\boldsymbol{\alpha}| = n$, we also define the `characteristic set` $K_{\boldsymbol{\alpha}}$ of $\boldsymbol{\alpha}$ in a way similar to $\boldsymbol{\alpha} \in \mathcal{J}^1$:

$$K_{\boldsymbol{\alpha}} = \{(i_1^{\alpha}, k_1^{\alpha}), \ldots, (i_n^{\alpha}, k_n^{\alpha})\}, \tag{5.3.26}$$

$i_1^{\alpha} \leq i_2^{\alpha} \leq \ldots \leq i_n^{\alpha}$, and if $i_j^{\alpha} = i_{j+1}^{\alpha}$, then $k_j^{\alpha} \leq k_{j+1}^{\alpha}$. The first pair $(i_1^{\alpha}, k_1^{\alpha})$ in $K_{\boldsymbol{\alpha}}$ is the position numbers of the first nonzero element of α. The second pair is the same as the first if the first nonzero element of α is greater than one; otherwise, the second pair is the position numbers of the second nonzero element of α and so on. As a result, if $\alpha_{i,k} > 0$, then exactly $\alpha_{i,k}$ pairs in $K_{\boldsymbol{\alpha}}$ are equal to (i, k). For example, if

$$\boldsymbol{\alpha} = \begin{pmatrix} 0\,1\,0\,2\,3\,0\,0 \cdots \\ 1\,2\,0\,0\,0\,1\,0 \cdots \\ 0\,0\,0\,0\,0\,0\,0 \cdots \\ \vdots\,\vdots\,\vdots\,\vdots\,\vdots\,\vdots\,\vdots\, \ldots \end{pmatrix}$$

with nonzero elements

$$\alpha_{1,2} = \alpha_{2,1} = \alpha_{2,6} = 1,\ \alpha_2^2 = \alpha_{1,4} = 2,\ \alpha_{1,5} = 3,$$

then the characteristic set is

$$K_{\boldsymbol{\alpha}} = \{(1,2), (1,4), (1,4), (1,5), (1,5), (1,5), (2,1), (2,2), (2,2), (2,6)\}.$$

Remark 5.3.13 Here and below we use the same notation $\boldsymbol{\alpha}, \boldsymbol{\beta}, \boldsymbol{\gamma}$, etc. [that is, a bold-face lower-case letter from the beginning of the Greek alphabet] for matrix-valued multi-indices, as in (5.3.25), and vector-valued multi-indices, as in Definition 5.1.8. However, the meaning of the notation will always be clear from the context.

Let $\mathbf{z} = \{z_{i,k}, i \geq 1, k \geq 1\}$ be a sequence of real numbers such that $\sum_{i,k} |z_{i,k}|^2 < \infty$. The stochastic exponent $\mathcal{E}(z, t, \mathbb{W})$ associated with $W(t)$ is

$$\mathcal{E}(\mathbf{z}, t, \mathbb{W}) = \exp\left(\sum_{i,k=1}^{\infty} z_{i,k} \int_0^t \mathfrak{m}_i(s) dw_k(s) - \frac{1}{2} \sum_{i,k=1}^{\infty} |z_{i,k}|^2 \int_0^t \mathfrak{m}_i^2(s) ds\right) \tag{5.3.27}$$

$$= \prod_{k=1}^{\infty} \exp\left(\sum_{i=1}^{\infty} z_{i,k} \int_0^t \mathfrak{m}_i(s) dw_k(s) - \frac{1}{2} \sum_{i=1}^{\infty} |z_{i,k}|^2 \int_0^t \mathfrak{m}_i^2(s) ds\right)$$

Remark 5.3.14 If $h \in L_2((0,T); Y)$ and

$$\mathcal{E}_t(h) = \exp\left(\int_0^t (h(s), dW(s))_Y - \frac{1}{2} \int_0^t \|h(t)\|_Y^2 dt\right), \tag{5.3.28}$$

then, by the Itô formula,

$$d\mathcal{E}_t(h) = \mathcal{E}_t(h)(h(t), dW(t))_Y. \tag{5.3.29}$$

This is a alternative representation of the stochastic exponential, with the correspondence given by

$$h(t) = \sum_{i,k} z_{i,k} \mathfrak{m}_i(t) y_k.$$

Similarly to (5.3.8), for $\boldsymbol{\alpha} \in \mathcal{J}^2$ we have

$$\xi_{\boldsymbol{\alpha}}(t) = \frac{1}{\sqrt{\boldsymbol{\alpha}!}} \frac{\partial^{|\boldsymbol{\alpha}|} \mathcal{E}(\mathbf{z}, t, \mathbb{W})}{\partial \mathbf{z}^{\boldsymbol{\alpha}}}\Big|_{\mathbf{z}=0}, \tag{5.3.30}$$

where

$$\frac{\partial^{|\alpha|}}{\partial^{\alpha}\mathbf{z}} = \prod_{i,k\geq 1} \frac{\partial^{\alpha_{i,k}}}{\partial z_{i,k}^{\alpha_{i,k}}},$$

and then

$$\xi_{\alpha}(t) = \int_0^t \sum_{i,k=1}^{\infty} \sqrt{\alpha_{i,k}}\xi_{\alpha-\epsilon(i,k)}(s)\,\mathfrak{m}_i(s)\,dw_k(s). \tag{5.3.31}$$

Due to independence of the Brownian motions $w_k(t)$, the elements ξ_α of the Cameron-Martin basis associated with W, are the products of the elements of the Cameron-Martin bases associated with each Brownian motion $w_k(t)$. More specifically,

$$\xi_{\alpha} = \prod_{k,i\geq 1} \frac{\mathrm{H}_{\alpha_{i,k}}(\xi_{i,k})}{\sqrt{\alpha_{i,k}!}} = \prod_{k\geq 1} \xi_{\alpha[k]} \tag{5.3.32}$$

where $\boldsymbol{\alpha}[k] = (\alpha_{1,k}, \alpha_{2,k}, \ldots) \in \mathcal{J}^1$ and

$$\xi_{\alpha[k]} = \frac{\sqrt{n!}}{\sqrt{\alpha[k]!}} \int_0^T \cdots \int_0^{t_2} \tilde{\mathfrak{m}}_{\alpha[k]}(t_1, \ldots, t_n) dw_k(t_1) \ldots dw_k(t_n), \tag{5.3.33}$$

where

$$\tilde{\mathfrak{m}}_{\alpha[k]}(t_1, \ldots, t_n) = \frac{1}{n!} \sum_{\sigma} \mathfrak{m}_{j_1}(t_{\sigma(1)}) \cdots \mathfrak{m}_{j_n}(t_{\sigma(n)})$$

and $\{j_1, \ldots, j_n\}$ is the characteristic set of $\boldsymbol{\alpha}[k]$.

In particular, for every k and $\boldsymbol{\alpha}$, $\xi_{\alpha[k]}(t)$ is a continuous square integrable martingale with respect to the filtration $\{\mathcal{F}_t^{w_k}\}_{t\geq 0}$ generated by the Brownian motion $w_k(t)$).

5.3.2 *The Malliavin Derivative and Its Adjoint*

In this section, we construct the Malliavin derivative and an analog of the Itô stochastic integral for *generalized random processes*.

By Definition 5.1.15 the action of Malliavin derivative $\mathbb{D}_{\xi_k}$ on the element Cameron-Martin basis $\{\xi_\alpha,\ \boldsymbol{\alpha} \in \mathcal{J}^1\}$ is defined by

$$\mathbb{D}_{\xi_k}(\xi_\alpha) = \sqrt{\alpha_k}\, \xi_{\alpha-\epsilon(k)}, \tag{5.3.34}$$

where $\boldsymbol{\epsilon}(k)$ is a multi-index of length 1 with the only positive component at the kth entry. Recall that the definition of $\mathcal{J}^1$ and $\mathcal{J}^2$ were introduced in Definitions 5.1.8 and 5.3.11, respectively.

A natural extension of the Malliavin derivative to the Cameron-Martin basis $\{\xi_\alpha,\ \boldsymbol{\alpha} \in \mathcal{J}^2\}$ is given by the formula

$$\mathbb{D}_{\xi_k}(\xi_\alpha) = \sqrt{\alpha_{i,k}}\, \xi_{\alpha-\epsilon(i,k)}, \tag{5.3.35}$$

where

$$\left(\epsilon\,(i,k)\right)_{j,\ell} = \begin{cases} 1 \text{ if } i=j, k=\ell \\ 0 \quad \text{otherwise} \end{cases} \tag{5.3.36}$$

is the $\mathcal{J}^2$-analog of $\boldsymbol{\epsilon}(k)$.

By linearity, we extend (5.3.35) to the definition of the Malliavin derivative

$$\mathbb{D}_{\mathfrak{B}}\xi_\alpha = \sum_{k\geq 1} \mathbb{D}_{\xi_k}(\xi_\alpha)\,\mathfrak{u}_k = \sum_{i,k\geq 1} \sqrt{\alpha_{i,k}}\xi_{\alpha-\epsilon(i,k)}\mathfrak{u}_{i,k}, \tag{5.3.37}$$

where $\mathfrak{B} = \sum_{i,k\geq 1} \mathfrak{u}_{i,k}\,\xi_{i,k}$, and $(\mathfrak{u}_{i,k} = \ i,k \geq 1)$ is a complete orthonormal basis in the Hilbert space $\mathcal{U}^2 := \mathcal{U}_1 \otimes \mathcal{U}_2$.

Exercise 5.3.15 (C) Prove that

$$\mathbb{E}\,\|\mathbb{D}_{\mathfrak{B}}\xi_\alpha\|^2_{\mathcal{U}^2} = \sum_{i,k\geq 1} \alpha_{i,k} = |\alpha|\,.$$

The following simple example will be instrumental in what follows.

Example 5.3.16 Let Y be a separable Hilbert space with a fixed orthonormal bases $\{y_k,\ k \geq 1\}$ and $\{\mathfrak{m}_i, i \geq 1\}$ be an orthonormal basis in $L_2[0,T]$. In this setting, the Malliavin derivative $\mathbb{D}_{\mathfrak{B}}\xi_\alpha$ is

$$(\mathbb{D}_{\mathfrak{B}}\xi_\alpha)(t) = \sum_{i,k} \sqrt{\alpha_{i,k}}\xi_{\alpha-\epsilon(i,k)}\,(t)\,\mathfrak{m}_i(t) y_k. \tag{5.3.38}$$

Lemma 5.3.17 *In the setting of Example 5.3.16,*

$$\mathbb{E}\,\|\mathbb{D}_{\mathfrak{B}}\xi_\alpha\,(t)\|^2_Y = \sum_k \sum_i \alpha_{i,k}\mathfrak{m}_i^2\,(t)\,. \tag{5.3.39}$$

Proof It follows from (5.3.38) that for $t \le T$

$$\mathbb{E}\left\|\mathbb{D}_{\mathfrak{B}}\xi_\alpha(t)\right\|_Y^2 = \mathbb{E}\left\|\sum_k\left(\sum_i\sqrt{\alpha_{i,k}}\xi_{\alpha-\epsilon(i,k)}(t)\,\mathfrak{m}_i(t)\right)y_k\right\|_Y^2$$

$$= \mathbb{E}\sum_k\left(\sum_i\sqrt{\alpha_{i,k}}\xi_{\alpha-\epsilon(i,k)}(t)\,\mathfrak{m}_i(t)\right)^2$$

$$= \mathbb{E}\sum_i\sqrt{\alpha_{i,k}}\xi_{\alpha-\epsilon(i,k)}(t)\,\mathfrak{m}_i(t)\sum_j\sqrt{\alpha_{j,k}}\xi_{\alpha-\epsilon(j,k)}(t)\,\mathfrak{m}_j(t) \tag{5.3.40}$$

$$= \sum_k\left(\sum_i\sqrt{\alpha_{i,k}}m_i(t)\delta(i,j)\sum_j\sqrt{\alpha_{j,k}}(t)\,\mathfrak{m}_j(t)\right)$$

$$= \sum_k\sum_i\alpha_{i,k}\mathfrak{m}_i^2(t).$$

Exercise 5.3.18 Confirm the equality

$$\mathbb{E}\left[\xi_{\alpha-\epsilon(i,k)}\xi_{\alpha-\epsilon(j,r)}(t)\right]\begin{cases}1 \text{ if } i=j,\ k=r\\ 0 \quad \text{otherwise,}\end{cases} \tag{5.3.41}$$

used in (5.3.40).

An immediate consequence of (5.3.39) is

$$\int_0^T \mathbb{E}\left\|\mathbb{D}_{\mathfrak{B}}\xi_\alpha(t)\right\|_Y^2 dt = \sum_{i,k}\alpha_{i,k} = |\alpha|. \tag{5.3.42}$$

Our next task is to extend the definition of Malliavin derivative to a more general random variable

$$v = \sum_{\alpha\in\mathcal{J}^2} v_\alpha\xi_\alpha$$

under the standard assumption $\sum_{\alpha\in\mathcal{J}^2} v_\alpha^2 < \infty$.

Since $(\mathbb{D}_{\mathfrak{B}}\xi_\alpha)(t) = \sum_{i,k}\sqrt{\alpha_{i,k}}\xi_{\alpha-\epsilon(i,k)}(t)\,\mathfrak{m}_i(t)y_k$, we have that

$$\mathbb{D}_{\mathfrak{B}}\left(\sum_{\alpha\in\mathcal{J}^2} v_\alpha\xi_\alpha\right)(t) = \sum_{\alpha\in\mathcal{J}^2} v_\alpha\sum_{i,k}\sqrt{\alpha_{i,k}}\xi_{\alpha-\epsilon(i,k)}(t)\,\mathfrak{m}_i(t)y_k.$$

Then

$$\int_0^T \mathbb{E} \left\| \mathbb{D}_{\mathfrak{B}} \sum_{\alpha \in \mathcal{J}^2} v_\alpha \xi_\alpha \right\|^2 dt =$$

$$\sum_{\alpha,\beta \in \mathcal{J}^2} v_\alpha v_\beta \tag{5.3.43}$$

$$\int_0^T \mathbb{E} \left(\sum_{i,k} \sqrt{\alpha_{i,k}} \xi_{\alpha - \epsilon(i,k)} (t)\, \mathfrak{m}_i(t) y_k, \sum_{j,n} \sqrt{\beta_{j,n}} \xi_{\beta - \epsilon(j,n)} (t)\, \mathfrak{m}_j(t) y_n \right)_Y dt.$$

By orthogonality,

$$\int_0^T \mathbb{E} \left(\sum_{i,k} \sqrt{\alpha_{i,k}} \xi_{\alpha - \epsilon(i,k)} (t)\, \mathfrak{m}_i(t) y_k, \sum_{j,n} \sqrt{\beta_{j,n}} \xi_{\beta - \epsilon(j,n)} (t)\, \mathfrak{m}_j(t) y_n \right)_Y dt$$

$$= \begin{cases} |\boldsymbol{\alpha}| & \text{if } \boldsymbol{\alpha} = \boldsymbol{\beta}, \\ 0 & \text{otherwise}, \end{cases} \tag{5.3.44}$$

Therefore, it follows from (5.3.43) and (5.3.44)

$$\int_0^T \mathbb{E} \left\| \mathbb{D}_{\mathfrak{B}} \left(\sum_{\alpha \in \mathcal{J}^2} v_\alpha \xi_\alpha \right) (t) \right\|^2 = \sum_{\alpha \in \mathcal{J}^2} |\boldsymbol{\alpha}| v_{\boldsymbol{\alpha}}^2, \tag{5.3.45}$$

which can be infinite.

Formula (5.3.45) implies that the Malliavin derivative $\mathbb{D}_{\mathfrak{B}}$ of the process $v(t) = \sum_{\alpha \in \mathcal{J}^2} v_\alpha (t)\, \xi_\alpha (t)$ with the standard assumption $\sum_{\alpha \in \mathcal{J}^2} v_{\boldsymbol{\alpha}}^2 (t) < \infty$ is not necessarily square integrable. The formula also implies that if we assume that

$$\sum_{\alpha \in \mathcal{J}} |\boldsymbol{\alpha}| v_{\boldsymbol{\alpha}}^2 < \infty,$$

then

$$\mathbb{E} \int_0^T \left| \mathbb{D}_{\mathfrak{B}} \sum_{\alpha \in \mathcal{J}^2} v_\alpha (t)\, \xi_\alpha (t) \right|^2 dt < \infty.$$

Let us now introduce the Hilbert space

$$L_2^1(\mathbb{W}) = \Big\{u \in L_2(\mathbb{W}) : \sum_{\alpha\in\mathcal{J}} |\alpha| u_\alpha^2 < \infty\Big\}. \tag{5.3.46}$$

Exercise 5.3.19 (C) Prove that the Malliavin derivative $\mathbb{D}$ extends to a continuous linear operator from $L_2^1(\mathbb{W})$ to $L_2(\mathbb{W})$.

For the sake of completeness and to justify further definitions, let us establish the standard connection between the Malliavin derivative and the stochastic Itô integral.

If u is an $\mathcal{F}_t^W$-adapted process from $L_2(\mathbb{W}; L_2((0,T);Y))$, then $u(t) = \sum_{k\geq 1} u_k(t)y_k$, where the random variable $u_k(t)$ is $\mathcal{F}_t^W$-measurable for each t and k, and

$$\sum_{k\geq 1}\int_0^T \mathbb{E}|u_k(t)|^2 dt < \infty.$$

We define the stochastic Itô integral

$$U(t) = \int_0^t (u(s), dW(s))_Y = \sum_{k\geq 1}\int_0^t u_k(s)dw_k(s). \tag{5.3.47}$$

Note that $U(t)$ is real-valued (as opposed to Y-valued) and is $\mathcal{F}_t^W$-measurable, and $\mathbb{E}|U(t)|^2 = \sum_{k\geq 1}\int_0^t E|u_k(s)|^2 ds$.

The next result establishes a connection between the Malliavin derivative and the stochastic Itô integral.

Lemma 5.3.20 *Suppose that u is an $\mathcal{F}_t^W$-adapted process from $L_2(\mathbb{W}; L_2((0,T); Y))$, and define the process U according to (5.3.47). Then, for every $0 < t \leq T$ and $\alpha \in J$,*

$$\mathbb{E}(U(t)\xi_\alpha) = \mathbb{E}\int_0^t (u(s), (\mathbb{D}_{\mathfrak{B}}\xi_\alpha)(s))_Y ds. \tag{5.3.48}$$

Proof Define $\xi_\alpha(t) = \mathbb{E}(\xi_\alpha|\mathcal{F}_t^W)$. By (5.3.31),

$$d\xi_\alpha(t) = \sum_{i,k}\sqrt{\alpha_{i,k}}\,\xi_{\alpha-\epsilon(i,k)}(t)\mathfrak{m}_i(t)dw_k(t). \tag{5.3.49}$$

Due to $\mathcal{F}_t^W$-measurability of $u_k(t)$, we have

$$u_{k,\alpha}(t) = \mathbb{E}\Big(u_k(t)\mathbb{E}(\xi_\alpha|\mathcal{F}_t^W)\Big) = \mathbb{E}(u_k(t)\xi_\alpha(t)). \tag{5.3.50}$$

The definition of U implies $dU(t) = \sum_{k\geq 1} u_k(t)dw_k(t)$, so that, by (5.3.49), (5.3.50), and the Itô formula,

$$U_\alpha(t) = \mathbb{E}(U(t)\xi_\alpha) = \int_0^t \sum_{i,k} \sqrt{\alpha_{i,k}} u_{k,\alpha-\epsilon(i,k)}(s)\mathfrak{m}_i(s)ds. \tag{5.3.51}$$

Together with (5.3.38), the last equality implies (5.3.48). Lemma 5.3.20 is proved.

Note that the coefficients $u_{k,\alpha}$ of $u \in L_2(W; L_2((0,T); H))$ belong to $L_2((0,T))$. We therefore define $u_{k,\alpha,i} = \int_0^T u_{k,\alpha}(t)\mathfrak{m}_i(t)dt$. Then, by (5.3.51),

$$U_\alpha(T) = \sum_{i,k} \sqrt{\alpha_{i,k}} u_{k,\alpha-\epsilon(i,k),i}. \tag{5.3.52}$$

Since $U(T) = \sum_{\alpha\in\mathcal{J}} U_\alpha(T)\xi_\alpha$, we shift the summation index in (5.3.52) and conclude that

$$U(T) = \sum_{\alpha\in\mathcal{J}^2} \sum_{i,k} \sqrt{\alpha_{i,k}+1}\, u_{k,\alpha,i}\xi_{\alpha+\epsilon(i,k)}. \tag{5.3.53}$$

As a result, $U(T) = \delta_{\mathfrak{B}}(u)$, where $\delta_{\mathfrak{B}}$ is the adjoint of the Malliavin derivative, also known as the Skorokhod integral; see [175] for details.

Lemma 5.3.20 suggests the following definition. For an $\mathcal{F}_t^W$-adapted process v from $L_2(\mathbb{W}; L_2((0,T)))$, let $\mathbb{D}_k^* u$ be the $\mathcal{F}_t^W$-adapted process from $L_2(\mathbb{W}; L_2((0,T)))$ so that

$$(\mathbb{D}_k^* u)_\alpha(t) = \int_0^t \sum_i \sqrt{\alpha_{i,k}} v_{\alpha-\epsilon(i,k)}(s)\mathfrak{m}_i(s)ds. \tag{5.3.54}$$

If $u = \sum_k u_k y_k \in L_2(\mathbb{W}; L_2((0,T); Y))$ is $\mathcal{F}_t^W$-adapted, then u is in the domain of the operator $\delta_{\mathfrak{B}}$ and $\delta_{\mathfrak{B}}(u) = \sum_{k\geq 1}(\mathbb{D}_k^* u_k)(t)$.

Next, we introduce `generalized random processes` and extend the operators $\mathbb{D}_k^*$ to such processes. Let X be a Banach space with norm $\|\cdot\|_X$.

Definition 5.3.21

(1) The space $D(L_2(\mathbb{W}))$ of test functions is the collection of elements from $L_2(\mathbb{W})$ that can be written in the form

$$v = \sum_{\alpha\in\mathcal{J}_v} v_\alpha \xi_\alpha$$

for some $v_\alpha \in \mathbb{R}$ and a finite subset J_v of $\mathcal{J}^2$.

(2) A sequence v_n converges to v in $D(L_2(\mathbb{W}))$ if and only if $J_{v_n} \subseteq J_v$ for all n and $\lim_{n\to\infty} |v_{n,\alpha} - v_\alpha| = 0$ for all α.

Definition 5.3.22 For a linear topological space X define the space $D'(L_2(\mathbb{W});X)$ of X-valued generalized random elements as the collection of continuous linear maps from the linear topological space $D(L_2(\mathbb{W}))$ to X. The elements of $D'(L_2(\mathbb{W});L_1((0,T);X))$ are called X-valued `generalized random processes`.

The element u of $D'(L_2(\mathbb{W});X)$ can be identified with a formal Cameron-Martin-Fourier series

$$u = \sum_{\alpha\in\mathcal{J}^2} u_\alpha \xi_\alpha,$$

and $u_\alpha \in X$ are then called generalized Fourier coefficients of u. For such a series and for $v \in D(L_2(\mathbb{W}))$, we have

$$u(v) = \sum_{\alpha\in\mathcal{J}_v} v_\alpha u_\alpha.$$

Conversely, for $u \in D'(L_2(\mathbb{W});X)$, we define the formal Cameron-Martin-Fourier series of u by setting $u_\alpha = u(\xi_\alpha)$. If $u \in L_2(\mathbb{W})$, then $u \in D'(L_2(\mathbb{W}))$ and $u(v) = \mathbb{E}(uv)$.

By Definition 5.3.22, a sequence $\{u_n,\ n \geq 1\}$ converges to u in $D'(L_2(\mathbb{W});X)$ if and only if $u_n(v)$ converges to $u(v)$ in the topology of X for every $v \in D(L_2(\mathbb{W}))$. In terms of generalized Fourier coefficients, this is equivalent to $\lim_{n\to\infty} u_{n,\alpha} = u_\alpha$ in the topology of X for every $\alpha \in J$.

The construction of the space $D'(L_2(\mathbb{W});X)$ can be extended to Hilbert spaces other than $L_2(\mathbb{W})$. Let H be a real separable Hilbert space with an orthonormal basis $\{e_k,\ k \geq 1\}$. Define the space

$$D(H) = \left\{ v \in H : v = \sum_{k\in\mathcal{J}_v} v_k e_k,\ v_k \in \mathbb{R},\ \mathcal{J}_v\ -\ \text{a finite subset of } \{1,2,\ldots\} \right\}.$$

By definition, v_n converges to v in $D(H)$ as $n \to \infty$ if and only if $J_{v_n} \subseteq J_v$ for all n and $\lim_{n\to\infty} |v_{n,k} - v_k| = 0$ for all k.

For a linear topological space X, $D'(H;X)$ is the space of continuous linear maps from $D(H)$ to X. An element g of $D'(H;X)$ can be identified with a formal series $\sum_{k\geq 1} g_k \otimes e_k$ so that $g_k = g(e_k) \in X$ and, for $v \in D(H)$, $g(v) = \sum_{k\in\mathcal{J}^2_v} g_k v_k$. If $X = \mathbb{R}$ and $\sum_{k\geq 1} g_k^2 < \infty$, then $g = \sum_{k\geq 1} g_k e_k \in H$ and $g(v) = (g, v)_H$, the inner product in H. The space X is naturally imbedded into $D'(H;X)$: if $u \in X$, then $\sum_{k\geq 1} u \otimes e_k \in D'(H;X)$.

A sequence $g_n = \sum_{k\geq 1} g_{n,k} \otimes e_k,\ n \geq 1$, converges to $g = \sum_{k\geq 1} g_k \otimes e_k$ in $D'(H;X)$ if and only if, for every $k \geq 1$, $\lim_{n\to\infty} g_{n,k} = g_k$ in the topology of X.

A collection $\{L_k,\ k \geq 1\}$ of linear operators from X_1 to X_2 naturally defines a linear operator L from $D'(H;X_1)$ to $D'(H;X_2)$:

$$L\left(\sum_{k\geq 1} g_k \otimes e_k\right) = \sum_{k\geq 1} L_k(g_k) \otimes e_k.$$

Similarly, a linear operator $L : D'(H;X_1) \to D'(H;X_2)$ can be identified with a collection $\{L_k,\ k \geq 1\}$ of linear operators from X_1 to X_2 by setting $L_k(u) = L(u \otimes e_k)$. We will see later that introduction of spaces $D'(H;X)$ and the corresponding operators makes it possible to study stochastic equations without worrying about square integrability of the solution.

Definition 5.3.23

(a) If u is an X-valued generalized random process, then $\mathbb{D}_k^* u$ is the X-valued generalized random process so that

$$(\mathbb{D}_k^* u)_{\alpha}(t) = \sum_i \int_0^t u_{\alpha-\epsilon(i,k)}(s)\sqrt{\alpha_{i,k}}m_i(s)ds. \tag{5.3.55}$$

(b) If $g \in D'\Big(Y; D'\left(L_2(\mathbb{W}); L_1((0,T);X)\right)\Big)$, with $g = \sum_{k\geq 1} g_k \otimes y_k$, $g_k \in D'(L_2(\mathbb{W}); L_1((0,T);X))$, then $\mathbb{D}^* g$ is the X-valued generalized random process with

$$(\mathbb{D}^* g)_{\alpha}(t) = \sum_k (\mathbb{D}_k^* g_k)_{\alpha}(t) = \sum_{i,k} \int_0^t g_{k,\alpha-\epsilon(i,k)}(s)\sqrt{\alpha_{i,k}}\mathfrak{m}_i(s)ds. \tag{5.3.56}$$

Using (5.3.38), we get a generalization of equality (5.3.48):

$$(\mathbb{D}^* g)_{\alpha}(t) = \int_0^t g\Big(\mathbb{D}\xi_{\alpha}\Big)(s)ds. \tag{5.3.57}$$

Indeed, by linearity,

$$g_k\Big(\sqrt{\alpha_{i,k}}\mathfrak{m}_i(s)\xi_{\alpha-\epsilon(i,k)}\Big)(s) = \sqrt{\alpha_{i,k}}\mathfrak{m}_i(s)g_{k,\alpha-\epsilon(i,k)}(s).$$

Theorem 5.3.24 *If $T < \infty$, then $\mathbb{D}_k^*$ and $\mathbb{D}^*$ are continuous linear operators.*

Proof It is enough to show that, if $u, u_n \in D'\left(L_2(\mathcal{F}_T^W); L_1((0,T);X)\right)$ and $\lim_{n\to\infty} \|u_{\alpha} - u_{n,\alpha}\|_{L_1((0,T);X)} = 0$ for every $\alpha \in J$, then, for every $k \geq 1$ and $\alpha \in J$,

$$\lim_{n\to\infty} \|(D_k^* u)_{\alpha} - (D_k^* u_n)_{\alpha}\|_{L_1((0,T);X)} = 0.$$

Using (5.3.55), we find

$$\|(\mathbb{D}_k^* u)_\alpha - (\mathbb{D}_k^* u_n)_\alpha\|_X(t)$$
$$\leq \sum_i \int_0^T \sqrt{\alpha_{i,k}}\|u_{\alpha-\epsilon(i,k)} - u_{n,\alpha-\epsilon(i,k)}\|_X(s)|\mathfrak{m}_i(s)|ds.$$

Note that the sum contains finitely many terms. By assumption, $|\mathfrak{m}_i(t)| \leq C_i$, and so

$$\|(\mathbb{D}_k^* u)_\alpha - (\mathbb{D}_k^* u_n)_\alpha\|_{L_1((0,T);X)}$$
$$\leq C(\alpha)\sum_i \sqrt{\alpha_{i,k}}\|u_{\alpha-\epsilon(i,k)} - u_{n,\alpha-\epsilon(i,k)}\|_{L_1((0,T);X)}. \tag{5.3.58}$$

Theorem 5.3.24 is proved.

5.4 Wiener Chaos Solutions for Parabolic SPDEs

5.4.1 The Propagator

In this section we build on the ideas from [153] to introduce the Wiener Chaos solution and the corresponding propagator for a general stochastic evolution equation. The notations from Sects. 5.1 and 5.3.2 will remain in force. It will be convenient to interpret the cylindrical Brownian motion W as a collection $\{w_k,\ k \geq 1\}$ of independent standard Wiener processes. As before, $T \in (0, \infty)$ is fixed and non-random. Introduce the following objects:

- The Banach spaces A, X, and U so that $U \subseteq X$.
- Linear operators

$$\mathcal{A} : L_1((0,T);A) \to L_1((0,T);X) \text{ and}$$
$$\mathcal{M}_k : L_1((0,T);A) \to L_1((0,T);X).$$

- Generalized random processes $f \in D'(L_2(\mathbb{W}); L_1((0,T);X))$ and $g_k \in D'(L_2(\mathbb{W}); L_1((0,T);X))$.
- The initial condition $u_0 \in D'(L_2(\mathbb{W}); U)$.

Consider the deterministic equation

$$v(t) = v_0 + \int_0^t (\mathcal{A}v)(s)ds + \int_0^t \varphi(s)ds, \tag{5.4.1}$$

where $v_0 \in U$ and $\varphi \in L_1((0,T);X)$.

Definition 5.4.1 A function v is called a $w(A, X)$ solution of (5.4.1) if and only if $v \in L_1((0,T);A)$ and equality (5.4.1) holds in the space $L_1((0,T);A)$.

Definition 5.4.2 An A-valued generalized random process u is called a $w(A, X)$ Wiener Chaos solution of the stochastic differential equation

$$du(t) = (\mathcal{A}u(t) + f(t))dt + (\mathcal{M}_k u(t) + g_k(t))dw_k(t),\ 0 < t \le T,\ u|_{t=0} = u_0, \tag{5.4.2}$$

if and only if the equality

$$u(t) = u_0 + \int_0^t (\mathcal{A}u + f)(s)ds + \sum_{k\ge 1} (\mathbb{D}_k^*(\mathcal{M}_k u + g_k))(t) \tag{5.4.3}$$

holds in $D'(L_2(\mathbb{W}); L_1((0,T);X))$.

Sometimes, to stress the dependence of the Wiener Chaos solution on the terminal time T, the notation $w_T(A, X)$ will be used.

By (5.3.56), equality (5.4.3) means that, for every $\boldsymbol{\alpha} \in \mathcal{J}^2$, the generalized Fourier coefficient $u_{\boldsymbol{\alpha}}$ of u satisfies

$$u_{\boldsymbol{\alpha}}(t) = u_{0,\alpha} + \int_0^t (\mathcal{A}u + f)_{\boldsymbol{\alpha}}(s)ds + \int_0^t \sum_{i,k} \sqrt{\alpha_{i,k}}(\mathcal{M}_k u + g_k)_{\boldsymbol{\alpha}-\boldsymbol{\epsilon}(i,k)}(s)\mathfrak{m}_i(s)ds. \tag{5.4.4}$$

Definition 5.4.3 System (5.4.4) is called the propagator for Eq. (5.4.2).

The propagator is a lower triangular system. Indeed, if $\boldsymbol{\alpha} = (\mathbf{0})$, that is, $|\boldsymbol{\alpha}| = 0$, then the corresponding equation in (5.4.4) becomes

$$u_{(\mathbf{0})}(t) = u_{0,(\mathbf{0})} + \int_0^t (\mathcal{A}u_{(\mathbf{0})}(s) + f_{(\mathbf{0})}(s))ds. \tag{5.4.5}$$

If $\boldsymbol{\alpha} = \boldsymbol{\epsilon}(j,\ell)$, that is, $\alpha_{j,\ell} = 1$ for some fixed j and ℓ and $\alpha_{i,k} = 0$ for all other $i, k \ge 1$, then the corresponding equation in (5.4.4) becomes

$$\begin{aligned} u_{\boldsymbol{\epsilon}(j,\ell)}(t) &= u_{0,\boldsymbol{\epsilon}(j,\ell)} + \int_0^t (\mathcal{A}u_{\boldsymbol{\epsilon}(j,\ell)}(s) + f_{\boldsymbol{\epsilon}(j,\ell)}(s))ds \\ &+ \int_0^t (\mathcal{M}_k u_{(\mathbf{0})}(s) + g_{\ell,(\mathbf{0})}(s))\mathfrak{m}_j(s)ds. \end{aligned} \tag{5.4.6}$$

Continuing in this way, we conclude that (5.4.4) can be solved by induction on $|\boldsymbol{\alpha}|$ as long as the corresponding deterministic equation (5.4.1) is solvable. The precise result is as follows.

Theorem 5.4.4 *If, for every $v_0 \in U$ and $\varphi \in L_1((0,T);X)$, Eq. (5.4.1) has a unique $w(A, X)$ solution $v(t) = V(t, v_0, \varphi)$, then Eq. (5.4.2) has a unique $w(A, X)$ Wiener*

Chaos solution so that

$$
\begin{aligned}
u_\alpha(t) &= V(t, u_{0,\alpha}, f_\alpha) + \sum_{i,k} \sqrt{\alpha_{i,k}}\, V(t, 0, \mathfrak{m}_i \mathcal{M}_k u_{\alpha-\epsilon(i,k)}) \\
&\quad + \sum_{i,k} \sqrt{\alpha_{i,k}}\, V(t, 0, \mathfrak{m}_i g_{k,\alpha-\epsilon(i,k)}).
\end{aligned}
\tag{5.4.7}
$$

Proof Using the assumptions of the theorem and linearity, we conclude that (5.4.7) is the unique solution of (5.4.4).

If the functions f, g, u_0 are non-random, we can derive a more explicit formula for u_α using the characteristic set K_α of the multi-index $\boldsymbol{\alpha}$; see (5.3.26) on page 271.

Theorem 5.4.5 *Assume that*

1. *for every $v_0 \in U$ and $\varphi \in L_1((0, T); X)$, Eq. (5.4.1) has a unique $w(A, X)$ solution $v(t) = V(t, v_0, \varphi)$,*
2. *the input data in (5.4.4) satisfy $g_k = 0$ and $f_\alpha = u_{0,\alpha} = 0$ if $|\alpha| > 0$.*

Let $u_{(0)}(t) = V(t, u_0, 0)$ be the solution of (5.4.4) for $|\alpha| = 0$. For $\boldsymbol{\alpha} \in \mathcal{J}^2$ with $|\boldsymbol{\alpha}| = n \geq 1$ and the characteristic set K_α, define functions $F^n = F^n(t; \boldsymbol{\alpha})$ by induction as follows:

$$
\begin{aligned}
F^1(t; \boldsymbol{\alpha}) &= V(t, 0, \mathfrak{m}_i \mathcal{M}_k u_{(0)}) \text{ if } K_\alpha = \{(i, k)\}; \\
F^n(t; \boldsymbol{\alpha}) &= \sum_{j=1}^{n} V(t, 0, \mathfrak{m}_{i_j} \mathcal{M}_{k_j} F^{n-1}(\cdot; \boldsymbol{\alpha} - \boldsymbol{\epsilon}(i_j, k_j))) \\
&\qquad \text{if } K_\alpha = \{(i_1, k_1), \ldots, (i_n, k_n)\}.
\end{aligned}
\tag{5.4.8}
$$

Then

$$
u_\alpha(t) = \frac{1}{\sqrt{\boldsymbol{\alpha}!}} F^n(t; \boldsymbol{\alpha}). \tag{5.4.9}
$$

Proof If $|\boldsymbol{\alpha}| = 1$, then representation (5.4.9) follows from (5.4.6). For $|\boldsymbol{\alpha}| > 1$, observe that

- If $\bar{u}_\alpha(t) = \sqrt{\boldsymbol{\alpha}!}\, u_\alpha$ and $|\boldsymbol{\alpha}| \geq 1$, then (5.4.4) implies

$$
\bar{u}(t) = \int_0^t \mathcal{A}\bar{u}_\alpha(s)ds + \sum_{i,k} \int_0^t \alpha_{i,k} \mathfrak{m}_i(s) \mathcal{M}_k \bar{u}_{\alpha-\epsilon(i,k)}(s)ds.
$$

- If $K_\alpha = \{(i_1, k_1), \ldots, (i_n, k_n)\}$, then, for every $j = 1, \ldots, n$, the characteristic set $K_{\alpha-\epsilon(i_j,k_j)}$ of $\boldsymbol{\alpha} - \boldsymbol{\epsilon}(i_j, k_j)$ is obtained from K_α by removing the pair (i_j, k_j).

- By the definition of the characteristic set,

$$\sum_{i,k} \alpha_{i,k}\mathfrak{m}_i(s)\mathcal{M}_k\bar{u}_{\alpha-\epsilon(i,k)}(s) = \sum_{j=1}^{n} \mathfrak{m}_{i_j}(s)\mathcal{M}_{k_j}\bar{u}_{\alpha-\epsilon(i_j,k_j)}(s).$$

As a result, representation (5.4.9) follows by induction on $|\alpha|$ using (5.4.7): if $|\alpha| = n > 1$, then

$$\begin{aligned}\bar{u}_\alpha(t) &= \sum_{j=1}^{n} V(t, 0, \mathfrak{m}_{i_j}\mathcal{M}_{k_j}\bar{u}_{\alpha-\epsilon(i_j,k_j)}) \\ &= \sum_{j=1}^{n} V(t, 0, \mathfrak{m}_{i_j}\mathcal{M}_{k_j}F^{(n-1)}(\cdot;\alpha - \epsilon(i_j, k_j)) = F^n(t;\alpha).\end{aligned} \quad (5.4.10)$$

Theorem 5.4.5 is proved.

Corollary 5.4.6 *Assume that the operator $\mathcal{A}$ is a generator of a strongly continuous semi-group $\Phi = \Phi_{t,s}$, $t \geq s \geq 0$, in some Hilbert space H so that $A \subset H$, each M_k is a bounded operator from A to H, and the solution $V(t, 0, \varphi)$ of Eq. (5.4.1) is written as*

$$V(t, 0, \varphi) = \int_0^T \Phi_{t,s}\varphi(s)ds, \quad \varphi \in L_2((0, T); H)). \quad (5.4.11)$$

Denote by $\mathcal{P}^n$ the permutation group of $\{1, \ldots, n\}$. If $u_{(\mathbf{0})} \in L_2((0, T); H))$, then, for α with $|\alpha| = n > 1$ and the characteristic set $K_\alpha = \{(i_1, k_1), \ldots, (i_n, k_n)\}$, representation (5.4.9) becomes

$$u_\alpha(t) = \frac{1}{\sqrt{\alpha!}} \sum_{\sigma\in\mathcal{P}^n} \int_0^t \int_0^{s_n} \cdots \int_0^{s_2}$$
$$\Phi_{t,s_n}\mathcal{M}_{k_{\sigma(n)}} \cdots \Phi_{s_2,s_1}\mathcal{M}_{k_{\sigma(1)}}u_{(\mathbf{0})}(s_1)\mathfrak{m}_{i_{\sigma(n)}}(s_n)\cdots\mathfrak{m}_{i_{\sigma(1)}}(s_1)ds_1\ldots ds_n. \quad (5.4.12)$$

Also,

$$\sum_{|\alpha|=n} u_\alpha(t)\xi_\alpha = \sum_{k_1,\ldots,k_n\geq 1} \int_0^t \int_0^{s_n} \cdots \int_0^{s_2}$$
$$\Phi_{t,s_n}\mathcal{M}_{k_n} \cdots \Phi_{s_2,s_1}\left(\mathcal{M}_{k_1}u_{(\mathbf{0})} + g_{k_1}(s_1)\right) dw_{k_1}(s_1)\cdots dw_{k_n}(s_n), \ n \geq 1, \quad (5.4.13)$$

and, for every Hilbert space X, the following energy equality holds:

$$\sum_{|\alpha|=n} \|u_\alpha(t)\|_X^2 = \sum_{k_1,\dots,k_n=1}^{\infty} \int_0^t \int_0^{s_n} \cdots \int_0^{s_2} \|\Phi_{t,s_n}\mathcal{M}_{k_n} \cdots \Phi_{s_2,s_1}\mathcal{M}_{k_1} u_{(\mathbf{0})}(s_1)\|_X^2 ds_1 \dots ds_n; \tag{5.4.14}$$

both sides in the last equality can be infinite. For $n = 1$*, formulas (5.4.12) and (5.4.14) become*

$$u_{(ik)}(t) = \int_0^t \Phi_{t,s}\mathcal{M}_k u_{(\mathbf{0})}(s)\, \mathfrak{m}_i(s) ds; \tag{5.4.15}$$

$$\sum_{|\alpha|=1} \|u_\alpha(t)\|_X^2 = \sum_{k=1}^{\infty} \int_0^t \|\Phi_{t,s}\mathcal{M}_k u_{(\mathbf{0})}(s)\|_X^2 ds. \tag{5.4.16}$$

Proof Using the semi-group representation (5.4.11), we conclude that (5.4.12) is just an expanded version of (5.4.9).

Since $\{\mathfrak{m}_i,\ i \geq 1\}$ is an orthonormal basis in $L_2(0,T)$, equality (5.4.16) follows from (5.4.15) and the Parseval identity. Similarly, equality (5.4.14) will follow from (5.4.12) after an application of an appropriate Parseval's identity.

To carry out the necessary arguments when $|\boldsymbol{\alpha}| > 1$, denote by $\mathcal{J}^1$ the collection of one-dimensional multi-indices $\boldsymbol{\beta} = (\beta_1, \beta_2, \dots)$ so that each β_i is a non-negative integer and $|\boldsymbol{\beta}| = \sum_{i\geq 1} \beta_i < \infty$. Given a $\boldsymbol{\beta} \in \mathcal{J}^1$ with $|\boldsymbol{\beta}| = n$, we define $K_{\boldsymbol{\beta}} = \{i_1, \dots, i_n\}$, the characteristic set of $\boldsymbol{\beta}$ and the function

$$E_{\boldsymbol{\beta}}(s_1, \dots, s_n) = \frac{1}{\sqrt{\boldsymbol{\beta}!\, n!}} \sum_{\sigma \in \mathcal{P}^n} \mathfrak{m}_{i_1}(s_{\sigma(1)}) \cdots \mathfrak{m}_{i_n}(s_{\sigma(n)}). \tag{5.4.17}$$

By construction, the collection $\{E_{\boldsymbol{\beta}}, \boldsymbol{\beta} \in \mathcal{J}^1, |\boldsymbol{\beta}| = n\}$ is an orthonormal basis in the sub-space of symmetric functions in $L_2((0,T)^n; X)$.

Next, we re-write (5.4.12) in a symmetrized form. To make the notations shorter, denote by $s^{(n)}$ the ordered set $(s_1, \dots, s_n)$ and write $ds^n = ds_1 \dots ds_n$. Fix $t \in (0, T]$ and the set $k^{(n)} = \{k_1, \dots, k_n\}$ of the second components of the characteristic set $K_{\boldsymbol{\alpha}}$. Define the symmetric function

$$G(t, k^{(n)}; s^{(n)}) = \frac{1}{\sqrt{n!}} \sum_{\sigma \in \mathcal{P}^n} \Phi_{t,s_{\sigma(n)}}\mathcal{M}_{k_n} \cdots \Phi_{s_{\sigma(2)},s_{\sigma(1)}}\mathcal{M}_{k_1} u_{(\mathbf{0})}(s_{\sigma(1)}) 1_{s_{\sigma(1)}<\cdots<s_{\sigma(n)}<t}(s^{(n)}). \tag{5.4.18}$$

Then (5.4.12) becomes

$$u_\alpha(t) = \int_{[0,T]^n} G(t, k^{(n)}; s^{(n)}) E_{\beta(\alpha)}(s^{(n)}) ds^n, \tag{5.4.19}$$

where the multi-indices $\boldsymbol{\alpha}$ and $\boldsymbol{\beta}(\boldsymbol{\alpha})$ are related via their characteristic sets: if

$$K_\alpha = \{(i_1, k_1), \ldots, (i_n, k_n)\},$$

then

$$K_{\beta(\alpha)} = \{i_1, \ldots, i_n\}.$$

Equality (5.4.19) means that, for fixed $k^{(n)}$, the function u_α is a Fourier coefficient of the symmetric function $G(t, k^{(n)}; s^{(n)})$ in the space $L_2((0,T)^n; X)$. Parseval's identity and summation over all possible $k^{(n)}$ yield

$$\sum_{|\alpha|=n} \|u_\alpha(t)\|_X^2 = \frac{1}{n!} \sum_{k_1,\ldots,k_n=1}^{\infty} \int_{[0,T]^n} \|G(t, k^{(n)}; s^{(n)})\|_X^2 ds^n,$$

which, due to (5.4.18), is the same as (5.4.14).

To prove equality (5.4.13), relating the Cameron-Martin and multiple Itô integral expansions of the solution, we use the following result [90, Theorem 3.1]:

$$\xi_\alpha = \frac{1}{\sqrt{\alpha!}} \int_0^T \int_0^{s_n} \cdots \int_0^{s_2} E_{\beta(\alpha)}(s^{(n)}) dw_{k_1}(s_1) \cdots dw_{k_n}(s_n);$$

see also [175, pp. 12–13]. Since the collection of all E_β is an orthonormal basis, equality (5.4.13) follows from (5.4.19) after summation over all $k_1, \ldots, k_n$.

Corollary 5.4.6 is proved.

We now present several examples to illustrate the general results.

Example 5.4.7 Consider the following equation:

$$\begin{aligned} du(t,x) &= (au_{xx}(t,x) + f(t,x))dt + (\sigma u_x(t,x) + g(t,x))dw(t), \\ &\quad t > 0,\ x \in \mathbb{R}, \end{aligned} \tag{5.4.20}$$

where $a > 0$, $\sigma \in \mathbb{R}$, $f \in L_2((0,T); H_2^{-1}(\mathbb{R}))$, $g \in L_2((0,T); L_2(\mathbb{R}))$, and $u|_{t=0} = u_0 \in L_2(\mathbb{R})$. By Theorem 4.4.3 on page 199, if $\sigma^2 < 2a$, then Eq. (5.4.20) has a unique square-integrable solution $u \in L_2\big(\mathbb{W}; L_2((0,T); H_2^1(\mathbb{R}))\big)$.

By $\mathcal{F}_t^W$-measurability of $u(t)$, we have

$$\mathbb{E}(u(t)\xi_\alpha) = \mathbb{E}(u(t)\mathbb{E}(\xi_\alpha|\mathcal{F}_t^W)).$$

Using the relation (5.3.49) and the Itô formula, we find that u_α satisfy

$$du_\alpha = a(u_\alpha)_{xx}dt + \sum_i \sqrt{\alpha_i}\sigma(u_{\alpha-\epsilon(i)})_x \mathfrak{m}_i(t)dt,$$

which is precisely the propagator for Eq. (5.4.20). In other words, if $2a > \sigma^2$, then the square-integrable solution of (5.4.20) coincides with the Wiener Chaos solution.

On the other hand, the heat equation

$$v(t,x) = v_0(x) + \int_0^t v_{xx}(s,x)ds + \int_0^t \varphi(s,x)ds, \ v_0 \in L_2(\mathbb{R})$$

with $\varphi \in L_2((0,T); H_2^{-1}(\mathbb{R}))$ has a unique $w(H_2^1(\mathbb{R}), H_2^{-1}(\mathbb{R}))$ solution. Therefore, by Theorem 5.4.4, the unique $w(H_2^1(\mathbb{R}), H_2^{-1}(\mathbb{R}))$ Wiener Chaos solution of (5.4.20) exists for all $\sigma \in \mathbb{R}$.

In the next example, the equation, although not parabolic, can be solved explicitly.

Example 5.4.8 Consider the following equation:

$$du(t,x) = u_x(t,x)dw(t), \ t > 0, \ x \in \mathbb{R}; \ u(0,x) = x. \tag{5.4.21}$$

Clearly, $u(t,x) = x + w(t)$ satisfies (5.4.21).

To find the Wiener Chaos solution of (5.4.21), note that, with one-dimensional Wiener process, $\alpha_{i,k} = \alpha_i$, and the propagator in this case becomes

$$u_\alpha(t,x) = x\mathbf{1}_{|\alpha|=0} + \int_0^t \sum_i \sqrt{\alpha_i}(u_{\alpha-\epsilon(i)}(s,x))_x \mathfrak{m}_i(s)ds.$$

Then $u_\alpha = 0$ if $|\alpha| > 1$, and

$$u(t,x) = x + \sum_{i\geq 1} \xi_i \int_0^t \mathfrak{m}_i(s)ds = x + w(t). \tag{5.4.22}$$

Even though Theorem 5.4.4 does not apply, the above arguments show that $u(t,x) = x + w(t)$ is the unique $w(A,X)$ Wiener Chaos solution of (5.4.21) for suitable spaces A and X, for example,

$$X = \left\{ f : \int_{\mathbb{R}} (1+x^2)^{-2} f^2(x)dx < \infty \right\} \text{ and } A = \{f : f, f' \in X\}.$$

Section 5.6.4 provides a more detailed analysis of Eq. (5.4.21).

If Eq. (5.4.2) is anticipating, that is, the initial condition is not deterministic/$\mathcal{F}_0$-measurable and/or the free terms f, g are not $\mathcal{F}_t^W$-adapted, then the Wiener Chaos solution generalizes the Skorokhod integral interpretation of the equation.

Example 5.4.9 Consider the equation

$$du(t,x) = \frac{1}{2}u_{xx}(t,x)dt + u_x(t,x)dw(t),\ t \in (0,T],\ x \in \mathbb{R}, \tag{5.4.23}$$

with initial condition $u(0,x) = x^2 w(T)$. Since $w(T) = \sqrt{T}\xi_1$, we find

$$(u_\alpha)_t(t,x) = \frac{1}{2}(u_\alpha)_{xx}(t,x) + \sum_i \sqrt{\alpha_i}\mathfrak{m}_i(t)(u_{\alpha-\epsilon(i)})_x(t,x) \tag{5.4.24}$$

with initial condition $u_\alpha(0,x) = \sqrt{T}x^2\mathbf{1}_{|\alpha|=1,\alpha_1=1}$. By Theorem 5.4.4, there exists a unique $w(A,X)$ Wiener Chaos solution of (5.4.23) for suitable spaces A and X. For example, we can take

$$X = \left\{f : \int_{\mathbb{R}}(1+x^2)^{-8}f^2(x)dx < \infty\right\}\ \text{and}\ A = \{f : f, f', f'' \in X\}.$$

System (5.4.24) can be solved explicitly. Indeed, $u_\alpha \equiv 0$ if $|\alpha| = 0$ or $|\alpha| > 3$ or if $\alpha_1 = 0$. Otherwise, writing $M_i(t) = \int_0^t \mathfrak{m}_i(s)ds$, we find:

$$\begin{aligned}
u_\alpha(t,x) &= (t+x^2)\sqrt{T},\ \text{ if } |\alpha| = 1,\ \alpha_1 = 1;\\
u_\alpha(t,x) &= 4\,xt,\ \text{ if } |\alpha| = 2,\ \alpha_1 = 2;\\
u_\alpha(t,x) &= 2\sqrt{T}\,xM_i(t),\ \text{ if } |\alpha| = 2,\ \alpha_1 = \alpha_i = 1,\ 1 < i;\\
u_\alpha(t,x) &= \frac{6}{\sqrt{T}}t^2,\ \text{ if } |\alpha| = 3,\ \alpha_1 = 3;\\
u_\alpha(t,x) &= 4\sqrt{T}\,M_1(t)M_i(t),\ \text{ if } |\alpha| = 3,\ \alpha_1 = 2,\ \alpha_i = 1,\ 1 < i;\\
u_\alpha(t,x) &= 2\sqrt{T}\,M_i^2(t),\ \text{ if } |\alpha| = 3,\ \alpha_1 = 1,\ \alpha_i = 2,\ 1 < i;\\
u_\alpha(t,x) &= 2\sqrt{T}\,M_i(t)M_j(t),\ \text{ if } |\alpha| = 3,\ \alpha_1 = \alpha_i = \alpha_j = 1,\ 1 < i < j.
\end{aligned}$$

Then, after straightforward but long calculations, we conclude that

$$u(t,x) = \sum_{\alpha\in\mathcal{J}} u_\alpha\xi_\alpha = w(T)w^2(t) - 2tw(t) + 2(w(T)w(t) - t)x + x^2 w(T) \tag{5.4.25}$$

is the Wiener Chaos solution of (5.4.23). It can be verified using the properties of the Skorokhod integral [175] that the function u defined by (5.4.25) satisfies

$$u(t,x) = x^2 w(T) + \frac{1}{2}\int_0^t u_{xx}(s,x)ds + \int_0^t u_x(s,x)dw(s),\ t \in [0,T],\ x \in \mathbb{R},$$

where the stochastic integral is in the sense of Skorokhod.

5.4.2 Special Weights and The S-Transform

The space $D'(L_2(\mathbb{W}); X)$ is too big to provide any reasonably useful information about the Wiener Chaos solution. Introduction of Wiener chaos spaces with special weights makes it possible to resolve this difficulty.

As before, let $\Xi = \{\xi_\alpha,\ \alpha \in \mathcal{J}^2\}$ be the Cameron-Martin basis in $L_2(\mathbb{W})$, and $D(L_2(\mathbb{W}); X)$, the collection of finite linear combinations of ξ_α with coefficients in a Banach space X.

Definition 5.4.10 Given a collection $\{r_\alpha,\ \alpha \in \mathcal{J}^2\}$ of positive numbers, the space $\mathcal{R}L_2(W; X)$ is the closure of $D(L_2(\mathbb{W}); X)$ with respect to the norm

$$\|v\|^2_{\mathcal{R}L_2(\mathbb{W};X)} := \sum_{\alpha\in\mathcal{J}^2} r_\alpha^2 \|v_\alpha\|_X^2.$$

The operator $\mathcal{R}$ defined by $(\mathcal{R}v)_\alpha := r_\alpha v_\alpha$ is a linear homeomorphism from $\mathcal{R}L_2(\mathbb{W}; X)$ to $L_2(\mathbb{W}; X)$.

There are several special choices of the weight sequence $\mathcal{R} = \{r_\alpha,\ \alpha \in \mathcal{J}^2\}$ and special notations for the corresponding weighted Wiener chaos spaces.

- If $Q = \{q_1, q_2, \ldots\}$ is a sequence of positive numbers, define

$$q^\alpha = \prod_{i,k} q_k^{\alpha_{i,k}}.$$

  The operator $\mathcal{R}$, corresponding to $r_\alpha = q^\alpha$, is denotes by $\mathcal{Q}$. The space $\mathcal{Q}L_2(\mathbb{W}; X)$ is denoted by $L_{2,Q}(\mathbb{W}; X)$ and is called a Q-weighted Wiener chaos space. The significance of this choice of weights will be explained shortly (see, in particular, Proposition 5.4.13).
- If

$$r_\alpha^2 = (\alpha!)^\rho \prod_{i,k} (2ik)^{\gamma\alpha_{i,k}},\ \rho, \gamma \in \mathbb{R},$$

  then the corresponding space $\mathcal{R}L_2(\mathbb{W}; X)$ is denoted by $(\mathcal{S})_{\rho,\gamma}(X)$. As always, the argument X will be omitted if $X = \mathbb{R}$. Note the analogy with (5.2.22) on page 257.

The structure of weights in the spaces $L_{2,Q}$ and $(\mathcal{S})_{\rho,\gamma}$ is different, and in general these two classes of spaces are not related. There exist generalized random elements that belong to some $L_{2,Q}(\mathbb{W}; X)$, but do not belong to any $(\mathcal{S})_{\rho,\gamma}(X)$. For example, $u = \sum_{k\geq1} e^{k^2}\xi_{1,k}$ belongs to $L_{2,Q}(\mathbb{W})$ with $q_k = e^{-2k^2}$, but to no $(\mathcal{S})_{\rho,\gamma}$, because the series $\sum_{k\geq1} e^{2k^2}(k!)^\rho(2k)^\gamma$ diverges for every $\rho, \gamma \in \mathbb{R}$. Similarly, there exist generalized random elements that belong to some $(\mathcal{S})_{\rho,\gamma}(X)$, but to no $L_{2,Q}(\mathbb{W}; X)$. For example, $u = \sum_{n\geq1} \sqrt{n!}\xi_{(n)}$, where (n) is the multi-index with $\alpha_{1,1} = n$ and

$\alpha_{i,k} = 0$ elsewhere, belongs to $(\mathcal{S})_{-1,-1}$, but does not belong to any $L_{2,Q}(\mathbb{W})$, because the series $\sum_{n\geq 1} q^n n!$ diverges for every $q > 0$.

The next result is the space-time analog of Proposition 2.3.3 in [78].

Proposition 5.4.11 *The sum*

$$\sum_{\alpha\in\mathcal{J}}\prod_{i,k\geq 1}(2ik)^{-\gamma\alpha_{i,k}}$$

converges if and only if $\gamma > 1$.

Proof Note that

$$\sum_{\alpha\in\mathcal{J}}\prod_{i,k\geq 1}(2ik)^{-\gamma\alpha_{i,k}} = \prod_{i,k\geq 1}\left(\sum_{n\geq 0}((2ik)^{-\gamma})^n\right) = \prod_{i,k}\frac{1}{(1-(2ik)^{-\gamma})},\ \gamma > 0 \tag{5.4.26}$$

The infinite product on the right of (5.4.26) converges if and only if each of the sums $\sum_{i\geq 1} i^{-\gamma}$, $\sum_{k\geq 1} k^{-\gamma}$ converges, that is, if an only if $\gamma > 1$.

Corollary 5.4.12 *For every $u \in D'(\mathbb{W}; X)$, there exists an operator $\mathcal{R}$ so that $\mathcal{R}u \in L_2(\mathbb{W}; X)$.*

Proof Define

$$r_\alpha^2 = \frac{1}{1+\|u_\alpha\|_X^2}\prod_{i,k\geq 1}(2ik)^{-2\alpha_{i,k}}.$$

Then

$$\begin{aligned}\|\mathcal{R}u\|^2_{L_2(\mathbb{W};X)} &= \sum_{\alpha\in\mathcal{J}^2}\frac{\|u_\alpha\|_X^2}{1+\|u_\alpha\|_X^2}\prod_{i,k\geq 1}(2ik)^{-2\alpha_{i,k}}\\ &\leq \sum_{\alpha\in\mathcal{J}^2}\prod_{i,k\geq 1}(2ik)^{-2\alpha_{i,k}} < \infty.\end{aligned}$$

The importance of the operator $\mathcal{Q}$ in the study of stochastic equations is due to the fact that the operator $\mathcal{R}$ maps a Wiener Chaos solution to a Wiener Chaos solution if and only $\mathcal{R} = \mathcal{Q}$ for some sequence $\mathcal{Q}$. Indeed, direct calculations show that the functions $u_\alpha, \alpha \in \mathcal{J}^2$, satisfy the propagator (5.4.4) if and only if $v_\alpha = (\mathcal{R}u)_\alpha$ satisfy

$$\begin{aligned}v_\alpha(t) &= (\mathcal{R}u_0)_\alpha + \int_0^t(\mathcal{A}v + \mathcal{R}f)_\alpha(s)ds\\ &+ \int_0^t\sum_{i,k}\sqrt{\alpha_{i,k}}\frac{r_\alpha}{r_{\alpha-\epsilon(i,k)}}(\mathcal{M}_k\mathcal{R}u + \mathcal{R}g_k)_{\alpha-\epsilon(i,k)}(s)\mathfrak{m}_i(s)ds.\end{aligned} \tag{5.4.27}$$

Therefore, the operator $\mathcal{R}$ preserves the structure of the propagator if and only if

$$\frac{r_\alpha}{r_{\alpha-\epsilon(i,k)}} = q_k,$$

that is, $r_\alpha = q^\alpha$ for some sequence Q.

Below is the summary of the main properties of the operator $\mathcal{Q}$.

Proposition 5.4.13

1. *If $q_k \le q < 1$ for all $k \ge 1$, then $L_{2,Q}(\mathbb{W}) \subset (\mathcal{S})_{0,-\gamma}$ for some $\gamma > 0$.*
2. *If $q_k \ge q > 1$ for all k, then $L_{2,Q}(\mathbb{W}) \subset L_2^n(\mathbb{W})$ for all $n \ge 1$, that is, the elements of $L_{2,Q}(\mathbb{W})$ are infinitely differentiable in the Malliavin sense.*
3. *If $u \in L_{2,Q}(\mathbb{W}; X)$ with generalized Fourier coefficients u_α satisfying the propagator (5.4.4), and $v = \mathcal{Q}u$, then the corresponding system for the generalized Fourier coefficients of v is*

$$\begin{aligned} v_\alpha(t) &= (\mathcal{Q}u_{0,\alpha} + \int_0^t (\mathcal{A}v + \mathcal{Q}f)_\alpha(s)ds \\ &+ \int_0^t \sum_{i,k} \sqrt{\alpha_{i,k}}(\mathcal{M}_k v + \mathcal{Q}g_k)_{\alpha-\epsilon(i,k)}(s)ds. \end{aligned} \tag{5.4.28}$$

4. *The function u is a Wiener Chaos solution of*

$$u(t) = u_0 + \int_0^t (\mathcal{A}u(s) + f(s))dt + \int_0^t (\mathcal{M}u(s) + g(s), dW(s))_Y \tag{5.4.29}$$

if and only if $v = \mathcal{Q}u$ is a Wiener Chaos solution of

$$v(t) = (\mathcal{Q}u)_0 + \int_0^t (\mathcal{A}v(s) + \mathcal{Q}f(s))dt + \int_0^t (\mathcal{M}v(s) + \mathcal{Q}g(s), dW^{\mathcal{Q}}(s))_Y, \tag{5.4.30}$$

where, for $h \in Y$, $W_h^{Q}(t) = \sum_{k\ge1}(h, y_k)_Y q_k w_k(t)$.

The following examples demonstrate how the operator $\mathcal{Q}$ helps with the analysis of various stochastic evolution equations.

Example 5.4.14 Consider the $w(H_2^1(\mathbb{R}), H_2^{-1}(\mathbb{R}))$ Wiener Chaos solution u of the equation

$$du(t,x) = (au_{xx}(t,x) + f(t,x))dt + \sigma u_x(t,x)dw(t),\ 0 < t \le T,\ x \in \mathbb{R}, \tag{5.4.31}$$

with $f \in L_2(\Omega \times (0,T); H_2^{-1}(\mathbb{R}))$, $g \in L_2(\Omega \times (0,T); L_2(\mathbb{R}))$, and $u|_{t=0} = u_0 \in L_2(\mathbb{R})$. Assume that $\sigma > 0$ and define the sequence Q so that $q_k = q$ for all $k \ge 1$

and $q < \sqrt{2a}/\sigma$. By Theorem 4.4.3, equation

$$dv = (av_{xx} + f)dt + (q\sigma u_x + g)dw$$

with $v|_{t=0} = u_0$, has a unique square-integrable solution

$$v \in L_2\left(\mathbb{W}; L_2((0,T); H_2^1(\mathbb{R}))\right) \bigcap L_2\left(\mathbb{W}; \mathcal{C}((0,T); L_2(\mathbb{R}))\right).$$

By Proposition 5.4.13, the $w(H_2^1(\mathbb{R}), H_2^{-1}(\mathbb{R}))$ Wiener Chaos solution u of equation (5.4.31) satisfies $u = \mathcal{Q}^{-1}v$ and

$$u \in L_{2,Q}\left(\mathbb{W}; L_2((0,T); H_2^1(\mathbb{R}))\right) \bigcap L_{2,Q}\left(\mathbb{W}; \mathcal{C}((0,T); L_2(\mathbb{R}))\right).$$

Note that if Eq. (5.4.31) is strongly parabolic, that is, $2a > \sigma^2$, then the weight q can be taken bigger than one, and, according to the first statement of Proposition 5.4.13, regularity of the solution is better than the one guaranteed by Theorem 4.4.3.

Example 5.4.15 The Wiener Chaos solutions can be constructed for stochastic ordinary differential equations. Consider, for example,

$$u(t) = 1 + \int_0^t \sum_{k\geq 1} u(s)dw_k(s), \tag{5.4.32}$$

which clearly does not have a square-integrable solution. On the other hand, the unique $w(\mathbb{R}, \mathbb{R})$ Wiener Chaos solution of this equation belongs to $L_{2,Q}(\mathbb{W}; L_2((0,T))$ for every sequence Q satisfying $\sum_k q_k^2 < \infty$. Indeed, for (5.4.32), Eq. (5.4.30) becomes

$$v(t) = 1 + \int_0^t \sum_k v(s)q_k dw_k(s).$$

If $\sum_k q_k^2 < \infty$, then the square-integrable solution of this equation exists and belongs to $L_2(\mathbb{W}; L_2((0,T)))$.

There exist equations for which the Wiener Chaos solution does not belong to any weighted Wiener chaos space $L_{2,Q}$. An example is given below in Sect. 5.6.4.

The `S-transform`, which is yet another variation on the theme of the Fourier transform, is necessary to introduce yet another notion of the solution (white noise solution). To define the S-transform, consider yet another version of the stochastic exponential:

$$\mathcal{E}(h) = \exp\left(\int_0^T (h(t), dW(t))_Y - \frac{1}{2}\int_0^T \|h(t)\|_Y^2 dt\right).$$

Lemma 5.4.16 *If $h \in D(L_2((0,T);Y))$, then*

- $\mathcal{E}(h) \in L_{2,Q}(\mathbb{W})$ *for every sequence* Q.
- $\mathcal{E}(h) \in (\mathcal{S})_{\rho,\gamma}$ *for* $0 \le \rho < 1$ *and* $\gamma \ge 0$.
- $\mathcal{E}(h) \in (\mathcal{S})_{1,\gamma}$, $\gamma \ge 0$, *as long as* $\|h\|^2_{L_2((0,T);Y)}$ *is sufficiently small.*

Proof Recall that, if $h \in D(L_2((0,T);Y))$, then $h(t) = \sum_{i,k\in I_h} h_{i,k}\mathfrak{m}_i(t)y_k$, where I_h is a finite set. Direct computations show that

$$\mathcal{E}(h) = \prod_{i,k}\left(\sum_{n\ge 0}\frac{H_n(\xi_{i,k})}{n!}(h_{i,k})^n\right) = \sum_{\alpha\in\mathcal{J}^2}\frac{h^\alpha}{\sqrt{\alpha!}}\xi_\alpha,$$

where $h^\alpha = \prod_{i,k} h_{i,k}^{\alpha_{i,k}}$. In particular,

$$(\mathcal{E}(h))_\alpha = \frac{h^\alpha}{\sqrt{\alpha!}}. \tag{5.4.33}$$

Consequently, for every sequence Q of positive numbers,

$$\|\mathcal{E}(h)\|^2_{L_{2,Q}(\mathbb{W})} = \exp\left(\sum_{i,k\in I_h} h^2_{i,k}q^2_k\right) < \infty. \tag{5.4.34}$$

Similarly, for $0 \le \rho < 1$ and $\gamma \ge 0$,

$$\|\mathcal{E}(h)\|^2_{(\mathcal{S})_{\rho,\gamma}} = \sum_{\alpha\in\mathcal{J}}\prod_{i,k}\frac{((2ik)^\gamma h_{i,k})^{2\alpha_{i,k}}}{(\alpha_{i,k}!)^{1-\rho}} = \prod_{i,k\in I_h}\left(\sum_{n\ge 0}\frac{((2ik)^\gamma h_{i,k})^{2n}}{(n!)^{1-\rho}}\right) < \infty, \tag{5.4.35}$$

and, for $\rho = 1$,

$$\|\mathcal{E}(h)\|^2_{(\mathcal{S})_{1,\gamma}} = \sum_{\alpha\in\mathcal{J}}\prod_{i,k}((2ik)^\gamma h_{i,k})^{2\alpha_{i,k}} = \prod_{i,k\in I_h}\left(\sum_{n\ge 0}((2ik)^\gamma h_{i,k})^{2n}\right) < \infty, \tag{5.4.36}$$

if $2\left(\max_{(m,n)\in J_h}(mn)^\gamma\right)\sum_{i,k} h^2_{i,k} < 1$. Lemma 5.4.16 is proved.

Remark 5.4.17 It is well-known (see, for example, [139, Proof of Theorem 5.5]) that the family $\{\mathcal{E}(h),\ h \in D(L_2((0,T);Y))\}$ is dense in $L_2(\mathbb{W})$ and consequently in every $L_{2,Q}(\mathbb{W})$ and every $(\mathcal{S})_{\rho,\gamma}$, $-1 < \rho \le 1$, $\gamma \in \mathbb{R}$.

Definition 5.4.18 If $u \in L_{2,Q}(\mathbb{W};X)$ for some Q, or if $u \in \bigcup_{q\geq 0}(\mathcal{S})_{-\rho,-\gamma}(X)$, $0 \leq \rho \leq 1$, then the deterministic function

$$S[u](h) = \sum_{\alpha\in\mathcal{J}} \frac{u_\alpha h^\alpha}{\sqrt{\alpha!}} \in X \tag{5.4.37}$$

is called the S-transform of u. Similarly, for $g \in D'\left(Y; L_{2,Q}(\mathbb{W};X)\right)$ the S-transform $S[g](h) \in D'(Y;X)$ is defined by setting $(S[g](h))_k = S[g_k](h)$.

Note that if $u \in L_2(\mathbb{W};X)$, then $S[u](h) = \mathbb{E}(u\mathcal{E}(h))$. If u belongs to $L_{2,Q}(\mathbb{W};X)$ or to $\bigcup_{q\geq 0}(\mathcal{S})_{-\rho,-\gamma}(X)$, $0 \leq \rho < 1$, then $S[u](h)$ is defined for all $h \in D\left(L_2((0,T);Y)\right)$. If $u \in \bigcup_{\gamma\geq 0}(\mathcal{S})_{-1,-\gamma}(X)$, then $S[u](h)$ is defined only for h sufficiently close to zero.

By Remark 5.4.17, an element u from $L_{2,Q}(\mathbb{W};X)$ or $\bigcup_{\gamma\geq 0}(\mathcal{S})_{-\rho,-\gamma}(X)$, $0 \leq \rho < 1$, is uniquely determined by the collection of deterministic functions $S[u](h)$, $h \in D\left(L_2((0,T);Y)\right)$. Since $\mathcal{E}(h) > 0$ for all $h \in D\left(L_2((0,T);Y)\right)$, Remark 5.4.17 also suggests the following definition.

Definition 5.4.19 An element u from $L_{2,Q}(\mathbb{W})$ or $\bigcup_{\gamma\geq 0}(\mathcal{S})_{-\rho,-\gamma}$, $0 \leq \rho < 1$ is called non-negative ($u \geq 0$) if and only if $S[u](h) \geq 0$ for all $h \in D\left(L_2((0,T);Y)\right)$.

The definition of the operator $\mathcal{Q}$ and Definition 5.4.19 imply the following result.

Proposition 5.4.20 *A generalized random element u from $L_{2,Q}(\mathbb{W})$ is non-negative if and only if $\mathcal{Q}u \geq 0$.*

For example, the solution of equation (5.4.32) is non-negative because

$$\mathcal{Q}u(t) = \exp\left(\sum_{k\geq 1}\left(q_k w_k(t) - (q_k^2/2)\right)\right).$$

We conclude this section with one technical remark.

Definition 5.4.18 expresses the S-transform in terms of the generalized Fourier coefficients. The following results makes it possible to recover generalized Fourier coefficients from the corresponding S-transform.

Proposition 5.4.21 *If u belongs to some $L_{2,Q}(\mathbb{W};X)$ or $\bigcup_{\gamma\geq 0}(\mathcal{S})_{-\rho,-\gamma}(X)$, $0 \leq \rho \leq 1$, then*

$$u_\alpha = \frac{1}{\sqrt{\alpha!}}\left(\prod_{i,k}\frac{\partial^{\alpha_{i,k}}S[u](h)}{\partial h_{i,k}^{\alpha_{i,k}}}\right)\Bigg|_{h=0}. \tag{5.4.38}$$

Proof For each $\boldsymbol{\alpha} \in \mathcal{J}^2$ with K non-zero entries, equality (5.4.37) and Lemma 5.4.16 imply that the function $S[u](h)$, as a function of K variables $h_{i,k}$, is analytic in some neighborhood of zero. Then (5.4.38) follows after differentiation of the series (5.4.37).

5.5 Further Properties of the Wiener Chaos Solutions

5.5.1 *White Noise and Square-Integrable Solutions*

Using notations and assumptions from Sect. 5.4.1, consider the linear evolution equation

$$du(t) = (\mathcal{A}(t)u(t) + f(t))dt + \big(\mathcal{M}(t)u(t) + g(t), dW(t)\big)_Y,$$
$$0 < t \le T,\ u|_{t=0} = u_0. \tag{5.5.1}$$

The objective of this section is to study how the Wiener Chaos compares with the square-integrable and white noise solutions.

To make the presentation shorter, call an X-valued generalized random element `S-admissible` if and only if it belongs to $L_{2,Q}(\mathbb{F}^W; X)$ for some Q or to $(\mathcal{S})_{\rho,\gamma}(X)$ for some $\rho \in [-1, 1]$ and $\gamma \in \mathbb{R}$. It was shown in Sect. 5.4.2 that, for every S-admissible u, the S-transform $S[u](h)$ is defined when $h = \sum_{i,k} h_{i,k}\mathfrak{m}_i y_k \in D(L_2((0,T); Y))$ and is an analytic function of $h_{i,k}$ in some neighborhood of $h = 0$.

The next result describes the S-transform of the Wiener Chaos solution.

Theorem 5.5.1 *Assume that*

1. *There exists a unique w(A, X) Wiener Chaos solution u of (5.5.1) and u is S-admissible;*
2. *For each* $t \in [0,T]$, *the linear operators* $\mathcal{A}(t), \mathcal{M}_k(t)$ *are bounded from A to X;*
3. *The generalized random elements* u_0, f, g_k *are S-admissible.*

Then, for every $h \in D(L_2((0,T); Y))$ *with* $\|h\|^2_{L_2((0,T);Y)}$ *sufficiently small, the function* $v = S[u](h)$ *is a w(A, X) solution of the deterministic equation*

$$v(t) = S[u_0](h) + \int_0^t \Big(\mathcal{A}v + S[f](h) + (\mathcal{M}_k v + S[g_k](h))h_k\Big)(s)ds. \tag{5.5.2}$$

Proof By assumption, $S[u](h)$ exists for suitable functions h. Then the S-transformed equation (5.5.2) follows from the definition of the S-transform (5.4.37) and the propagator equation (5.4.4) satisfied by the generalized Fourier coefficients of u. Indeed, continuity of operator $\mathcal{A}(t)$ implies

$$S[\mathcal{A}u](h) = \sum_\alpha \frac{h^\alpha}{\sqrt{\alpha!}}\mathcal{A}u_\alpha = \mathcal{A}\sum_\alpha \frac{h^\alpha}{\sqrt{\alpha!}}u_\alpha = \mathcal{A}(S[u](h)).$$

Similarly,

$$\sum_{\alpha} \frac{h^{\alpha}}{\sqrt{\alpha!}} \sum_{i,k} \sqrt{\alpha_{i,k}} \mathcal{M}_k u_{\alpha-\epsilon(i,k)} \mathfrak{m}_i = \sum_{\alpha} \sum_{i,k} \frac{h^{\alpha-\epsilon(i,k)}}{\sqrt{\alpha - \epsilon(i,k)!}} \mathcal{M}_k u_{\alpha-\epsilon(i,k)} \mathfrak{m}_i h_{i,k}$$
$$= \sum_{i,k} \left(\sum_{\alpha} \frac{h^{\alpha}}{\sqrt{\alpha}} \mathcal{M}_k u_{\alpha} \right) \mathfrak{m}_i h_{i,k}$$
$$= \mathcal{M}_k(S[u](h))h_k.$$

Computations for the other terms are similar. Theorem 5.5.1 is proved.

Remark 5.5.2 If $h \in D(L_2((0,T);Y))$ and

$$\mathcal{E}_t(h) = \exp\left(\int_0^t (h(s), dW(s))_Y - \frac{1}{2} \int_0^t \|h(t)\|_Y^2 dt \right), \tag{5.5.3}$$

then, by the Itô formula,

$$d\mathcal{E}_t(h) = \mathcal{E}_t(h)(h(t), dW(t))_Y. \tag{5.5.4}$$

If u_0 is deterministic, f and g_k are $\mathcal{F}_t^W$-adapted, and u is a square-integrable solution of (5.5.1), then equality (5.5.2) is obtained by multiplying Eqs. (5.5.4) and (5.5.1) according to the Itô formula and taking the expectation.

A partial converse of Theorem 5.5.1 is that, under some regularity conditions, the Wiener Chaos solution can be recovered from the solution of the S-transformed equation (5.5.2).

Theorem 5.5.3 *Assume that the linear operators $\mathcal{A}(t)$, $\mathcal{M}_k(t)$, $t \in [0,T]$, are bounded from A to X, the input data u_0, f, g_k are S-admissible, and, for every $h \in D(L_2((0,T);Y))$ with $\|h\|^2_{L_2((0,T);Y)}$ sufficiently small, there exists a $w(A,X)$ solution $v = v(t;h)$ of Eq. (5.5.2). Writing*

$$h = \sum_{i,k} h_{i,k} \mathfrak{m}_i y_k,$$

we consider v as a function of the variables $h_{i,k}$. Assume that all the derivatives of v at the point $h = 0$ exist, and, for $\boldsymbol{\alpha} \in \mathcal{J}^2$, define

$$u_{\alpha}(t) = \frac{1}{\sqrt{\alpha!}} \left(\prod_{i,k} \frac{\partial^{\alpha_{i,k}} v(t;h)}{\partial h_{i,k}^{\alpha_{i,k}}} \right) \Bigg|_{h=0}. \tag{5.5.5}$$

Then the generalized random process $u(t) = \sum_{\alpha \in \mathcal{J}^2} u_{\alpha}(t)\xi_{\alpha}$ is a $w(A,X)$ Wiener Chaos solution of (5.5.1).

Proof Differentiation of (5.5.2) and application of Proposition 5.4.21 show that the functions u_α satisfy the propagator (5.4.4).

Definition 5.5.4 A white noise solution of equation (5.5.1) is an S-admissible process u such that $S[u]$ satisfies (5.5.2).

Remark 5.5.5 The central part in the construction of the white noise solution of (5.5.1) is proving that the solution of (5.5.2) is an S-transform of a suitable generalized random process. For many particular cases of Eq. (5.5.1), the corresponding analysis is carried out in [76, 78, 162, 189]. The consequence of Theorems 5.5.1 and 5.5.3 is that a white noise solution of (5.5.1), if exists, must coincide with the Wiener Chaos solution.

The next theorem establishes the connection between the Wiener Chaos solution and the square-integrable solution. Recall that the square-integrable solution of (5.5.1) was introduced in Definition 4.4.1 on page 199. Accordingly, the notations from Sect. 4.4 will be used.

Theorem 5.5.6 *Let (V, H, V') be a normal triple of Hilbert spaces. Take deterministic functions u_0, f, and g_k such that*

$$u_0 \in H,\ f \in L_2\big((0,T); V'\big),\ \sum_k \|g_k\|^2_{L_2((0,T);H)} < \infty.$$

Then

1. *An $\mathcal{F}_t^W$-adapted square-integrable solution of (5.5.1) is also a Wiener Chaos solution.*
2. *If u is a $w(V, V')$ Wiener Chaos solution of (5.5.1) and*

$$\sum_{\alpha\in\mathcal{J}^2}\left(\int_0^T \|u_\alpha(t)\|_V^2 dt + \sup_{0\le t\le T}\|u_\alpha(t)\|_H^2\right) < \infty, \tag{5.5.6}$$

then u is an $\mathcal{F}_t^W$-adapted square-integrable solution of (5.5.1).

Proof

(1) If $u = u(t)$ is an $\mathcal{F}_t^W$-adapted square-integrable solution, then

$$u_\alpha(t) = \mathbb{E}(u(t)\xi_\alpha) = \mathbb{E}\left(u(t)\mathbb{E}(\xi_\alpha|\mathcal{F}_t^W)\right) = \mathbb{E}(u(t)\xi_\alpha(t)).$$

Then the propagator (5.4.4) for u_α follows after applying the Itô formula to the product $u(t)\xi_\alpha(t)$ and using (5.3.49).

(2) Assumption (5.5.6) implies

$$u \in L_2(\Omega\times(0,T); V)\bigcap L_2(\Omega; \mathcal{C}((0,T); H)).$$

Then, by Theorem 5.5.1, for every $\varphi \in V$ and $h \in D((0,T);Y)$, the S-transform u_h of u satisfies

$$\begin{aligned}(u_h(t),\varphi)_H &= (u_0,\varphi)_H + \int_0^t \langle \mathcal{A}u_h(s),\varphi\rangle ds + \int_0^t \langle f(s),\varphi\rangle ds\\ &+ \sum_{\alpha\in\mathcal{J}^2} \frac{h^\alpha}{\alpha!}\sum_{i,k}\int_0^t \sqrt{\alpha_{i,k}}\mathfrak{m}_i(s)\Big((\mathcal{M}_k u_{\alpha-\epsilon(i,k)}(s),\varphi)_H\\ &+ (g_k(s),\varphi)_H \mathbf{1}_{|\alpha|=1}\Big)ds.\end{aligned}$$

If $I(t) = \int_0^t (M_k u(s),\varphi)_H dw_k(s)$, then

$$\mathbb{E}(I(t)\xi_\alpha(t)) = \int_0^t \sum_{i,k}\sqrt{\alpha_{i,k}}\mathfrak{m}_i(s)(\mathcal{M}_k u_{\alpha-\epsilon(i,k)}(s),\varphi)_H ds. \tag{5.5.7}$$

Similarly,

$$\mathbb{E}\left(\xi_\alpha(t)\int_0^t (g_k(s),\varphi)_H dw_k(s)\right) = \sum_{i,k}\int_0^t \sqrt{\alpha_{i,k}}\mathfrak{m}_i(s)(g_k(s),\varphi)_H \mathbf{1}_{|\alpha|=1} ds.$$

Therefore,

$$\begin{aligned}&\sum_{\alpha\in\mathcal{J}^2}\frac{h^\alpha}{\alpha!}\sum_{i,k}\int_0^t \sqrt{\alpha_{i,k}}\mathfrak{m}_i(s)(\mathcal{M}_k u_{\alpha-\epsilon(i,k)}(s),\varphi)_H ds\\ &= \mathbb{E}\left(\mathcal{E}(h)\int_0^t \Big((\mathcal{M}_k u(s),\varphi)_H + (g_k(s),\varphi)_H\Big)dw_k(s)\right).\end{aligned}$$

As a result,

$$\begin{aligned}\mathbb{E}\Big(\mathcal{E}(h)(u(t),\varphi)_H\Big) &= \mathbb{E}\Big(\mathcal{E}(h)(u_0,\varphi)_H\Big)\\ &+ \mathbb{E}\left(\mathcal{E}(h)\int_0^t \langle \mathcal{A}u(s),\varphi\rangle ds\right) + \mathbb{E}\left(\mathcal{E}(h)\int_0^t \langle f(s),\varphi\rangle ds\right)\\ &+ \mathbb{E}\left(\mathcal{E}(h)\int_0^t \Big((\mathcal{M}_k u(s),\varphi)_H + (g_k(s),\varphi)_H\Big)dw_k(s)\right).\end{aligned} \tag{5.5.8}$$

Equality (5.5.8) and Remark 5.4.17 imply that (4.4.3) on page 199 holds. Theorem 5.5.11 is proved.

5.5.2 Additional Regularity

Let $\mathbb{F} = (\Omega, F, \{F_t\}_{t\geq 0}, P)$ be a stochastic basis with the usual assumptions and $w_k = w_k(t),\ k \geq 1,\ t \geq 0$, a collection of standard Wiener processes on $\mathbb{F}$. Let (V, H, V') be a normal triple of Hilbert spaces and $A(t) : V \to V'$, $M_k(t) : V \to H$, linear bounded operators; $t \in [0, T]$.

In this section we study the linear equation

$$u(t) = u_0 + \int_0^t (\mathcal{A}u(s) + f(s))ds + \int_0^t (\mathcal{M}_k u(s) + g_k(s))dw_k,\ 0 \leq t \leq T, \quad (5.5.9)$$

under the following assumptions:

A1 There exist positive numbers C_1 and δ so that

$$\langle \mathcal{A}(t)v, v\rangle + \delta \|v\|_V^2 \leq C_1 \|v\|_H^2,\ v \in V,\ t \in [0, T]. \quad (5.5.10)$$

A2 There exists a real number C_2 so that

$$2\langle \mathcal{A}(t)v, v\rangle + \sum_{k\geq 1} \|\mathcal{M}_k(t)v\|_H^2 \leq C_2 \|v\|_H^2,\ v \in V,\ t \in [0, T]. \quad (5.5.11)$$

A3 The initial condition u_0 is non-random and belongs to H; the process $f = f(t)$ is deterministic and $\int_0^T \|f(t)\|_{V'}^2 dt < \infty$; each $g_k = g_k(t)$ is a deterministic processes and $\sum_{k\geq 1} \int_0^T \|g_k(t)\|_H^2 dt < \infty$.

Note that condition (5.5.11) is weaker than standard stochastic parabolicity condition (4.4.4) on page 199. Traditional analysis of Eq. (5.5.9) under (5.5.11) requires additional regularity assumptions on the input data and additional Hilbert space constructions beyond the normal triple: see Theorem 4.4.12 on page 208. The Wiener chaos approach provides new existence and regularity results for Eq. (5.5.9). A different version of the following theorem is in [152].

Theorem 5.5.7 *Under assumptions **A1–A3**, for every $T > 0$, Eq. (5.5.9) has a unique $w(V, V')$ Wiener Chaos solution. This solution $u = u(t)$ has the following properties:*

1. There exists a weight sequence Q so that

$$u \in L_{2,Q}(\mathbb{W}; L_2((0, T); V)) \bigcap L_{2,Q}(\mathbb{W}; C((0, T); H)).$$

2. For every $0 \leq t \leq T$, $u(t) \in L_2(\Omega; H)$ and

$$\mathbb{E}\|u(t)\|_H^2 \leq 3e^{C_2 t} \left(\|u_0\|_H^2 + C_f \int_0^t \|f(s)\|_{V'}^2 ds + \sum_{k\geq 1} \int_0^t \|g_k(s)\|_H^2 ds \right), \quad (5.5.12)$$

where the number C_2 is from (5.5.11) and the positive number C_f depends only on δ and C_1 from (5.5.10).

3. *For every $0 \le t \le T$,*

$$u(t) = u_{(0)} + \sum_{n\geq 1} \sum_{k_1,\ldots,k_n\geq 1} \int_0^t \int_0^{s_n} \ldots \int_0^{s_2}$$

$$\Phi_{t,s_n} \mathcal{M}_{k_n} \cdots \Phi_{s_2,s_1} \left(\mathcal{M}_{k_1} u_{(0)} + g_{k_1}(s_1)\right) dw_{k_1}(s_1) \cdots dw_{k_n}(s_n), \tag{5.5.13}$$

where $\Phi_{t,s}$ is the semi-group of the operator $\mathcal{A}$.

Proof Assumption **A2** and the properties of the normal triple imply that there exists a positive number C^* so that

$$\sum_{k\geq 1} \|\mathcal{M}_k(t)v\|_H^2 \le C^* \|v\|_V^2, \ v \in V, \ t \in [0, T]. \tag{5.5.14}$$

Define the sequence Q by

$$q_k = \left(\frac{\mu\delta}{C^*}\right)^{1/2} := q, \ k \geq 1, \tag{5.5.15}$$

where $\mu \in (0, 2)$ and δ is from Assumption **A1**. Then, by Assumption **A2**,

$$2\langle \mathcal{A}v, v\rangle + \sum_{k\geq 1} q^2 \|\mathcal{M}_k v\|_H^2 \le -(2-\mu)\delta \|v\|_V^2 + C_1 \|v\|_H^2. \tag{5.5.16}$$

It follows from Theorem 4.4.3 that equation

$$v(t) = u_0 + \int_0^t (\mathcal{A}v + f)(s)ds + \sum_{k\geq 1} \int_0^t q(\mathcal{M}_k v + g_k)(s)dw_k(s) \tag{5.5.17}$$

has a unique solution

$$v \in L_2(\mathbb{W}; L_2((0, T); V)) \bigcap L_2(\mathbb{W}; \mathcal{C}((0, T); H)).$$

Comparison of the propagators for Eqs. (5.5.9) and (5.5.17) shows that $u = Q^{-1}v$ is the unique $w(V, V')$ solution of (5.5.9) and

$$u \in L_{2,Q}(\mathbb{W}; L_2((0, T); V)) \bigcap L_{2,Q}(\mathbb{W}; \mathcal{C}((0, T); H)). \tag{5.5.18}$$

If $C^* < 2\delta$, then Eq. (5.5.9) is strongly parabolic and $q > 1$ is an admissible choice of the weight. As a result, for strongly parabolic equations, the result (5.5.18) is stronger than the conclusion of Theorem 4.4.3.

The proof of (5.5.12) is based on the analysis of the propagator

$$u_\alpha(t) = u_0\mathbf{1}_{|\alpha|=0} + \int_0^t \Big(\mathcal{A}u_\alpha(s) + f(s)\mathbf{1}_{|\alpha|=0}\Big)ds$$
$$+ \int_0^t \sum_{i,k} \sqrt{\alpha_{i,k}}(\mathcal{M}_k u_{\alpha-\epsilon(i,k)}(s) + g_k(s)\mathbf{1}_{|\alpha|=1})\mathfrak{m}_i(s)ds. \tag{5.5.19}$$

We consider three particular cases: (1) $f = g_k = 0$ (the homogeneous equation); (2) $u_0 = g_k = 0$; (3) $u_0 = f = 0$. The general case will then follow by linearity and the triangle inequality.

Denote by $(\Phi_{t,s},\ t \geq s \geq 0)$ the (two-parameter) semi-group generated by the operator $\mathcal{A}(t)$; $\Phi_t := \Phi_{t,0}$. One of the consequence of Theorem 4.4.3 is that, under Assumption **A1**, this semi-group exists and is strongly continuous in H.

Consider the homogeneous equation: $f = g_k = 0$. By Corollary 5.4.6,

$$\sum_{|\alpha|=n} \|u_\alpha(t)\|_H^2 = \sum_{k_1,\ldots,k_n\geq 1} \int_0^t \int_0^{s_n} \cdots \int_0^{s_2} \|\Phi_{t,s_n}\mathcal{M}_{k_n}\cdots\Phi_{s_2,s_1}\mathcal{M}_{k_1}\Phi_{s_1}u_0\|_H^2 ds^n, \tag{5.5.20}$$

where $ds^n = ds_1\ldots ds_n$. Define $F_n(t) = \sum_{|\alpha|=n}\|u_\alpha(t)\|_H^2$, $n \geq 0$. Direct application of (5.5.11) shows that

$$\frac{d}{dt}F_0(t) \leq C_2F_0(t) - \sum_{k\geq 1}\|\mathcal{M}_k\Phi_t u_0\|_H^2. \tag{5.5.21}$$

For $n \geq 1$, equality (5.5.20) implies

$$\frac{d}{dt}F_n(t) = \sum_{k_1,\ldots,k_n\geq 1} \int_0^t \int_0^{s_{n-1}} \cdots \int_0^{s_2} \|\mathcal{M}_{k_n}\Phi_{t,s_{n-1}}\cdots\mathcal{M}_{k_1}\Phi_{s_1}u_0\|_H^2 ds^{n-1}$$
$$+ \sum_{k_1,\ldots,k_n\geq 1} \int_0^t \int_0^{s_n} \cdots \int_0^{s_2} \langle\mathcal{A}\Phi_{t,s_n}\mathcal{M}_{k_n}\ldots\Phi_{s_1}u_0, \Phi_{t,s_n}\mathcal{M}_{k_n}\ldots\Phi_{s_1}u_0\rangle ds^n. \tag{5.5.22}$$

By (5.5.11),

$$\sum_{k_1,\ldots,k_n\geq 1} \int_0^t \int_0^{s_n} \cdots \int_0^{s_2} \langle\mathcal{A}\Phi_{t,s_n}\mathcal{M}_{k_n}\ldots\Phi_{s_1}u_0, \Phi_{t,s_n}\mathcal{M}_{k_n}\ldots\Phi_{s_1}u_0\rangle ds^n$$
$$\leq - \sum_{k_1,\ldots,k_{n+1}\geq 1} \int_0^t \int_0^{s_n} \cdots \int_0^{s_2} \|\mathcal{M}_{k_{n+1}}\Phi_{t,s_n}\mathcal{M}_{k_n}\ldots\mathcal{M}_{k_1}\Phi_{s_1}u_0\|_H^2 ds^n \tag{5.5.23}$$
$$+ C_2 \sum_{k_1,\ldots,k_n\geq 1} \int_0^t \int_0^{s_n} \cdots \int_0^{s_2} \|\Phi_{t,s_n}\mathcal{M}_{k_n}\ldots\mathcal{M}_{k_1}\Phi_{s_1}u_0\|_H^2 ds^n.$$

As a result, for $n \geq 1$,

$$\begin{aligned} &\frac{d}{dt}F_n(t) \leq C_2F_n(t) \\ &+ \sum_{k_1,\dots,k_n\geq 1} \int_0^t \int_0^{s_{n-1}} \dots \int_0^{s_2} \|\mathcal{M}_{k_n}\Phi_{t,s_{n-1}}\mathcal{M}_{k_{n-1}}\dots\mathcal{M}_{k_1}\Phi_{s_1}u_0\|_H^2 ds^{n-1} \\ &- \sum_{k_1,\dots,k_{n+1}\geq 1} \int_0^t \int_0^{s_n} \dots \int_0^{s_2} \|\mathcal{M}_{k_{n+1}}\Phi_{t,s_n}\mathcal{M}_{k_n}\dots\mathcal{M}_{k_1}\Phi_{s_1}u_0\|_H^2 ds^n. \end{aligned} \tag{5.5.24}$$

Consequently,

$$\frac{d}{dt}\sum_{n=0}^{N}\sum_{|\alpha|=n}\|u_\alpha(t)\|_H^2 \leq C_2\sum_{n=0}^{N}\sum_{|\alpha|=n}\|u_\alpha(t)\|_H^2, \tag{5.5.25}$$

so that, by the Gronwall inequality,

$$\sum_{n=0}^{N}\sum_{|\alpha|=n}\|u_\alpha(t)\|_H^2 \leq e^{C_2t}\|u_0\|_H^2 \tag{5.5.26}$$

or

$$\mathbb{E}\|u(t)\|_H^2 \leq e^{C_2t}\|u_0\|_H^2. \tag{5.5.27}$$

Next, let us assume that $u_0 = g_k = 0$. Then the propagator (5.5.19) becomes

$$u_\alpha(t) = \int_0^t (\mathcal{A}u_\alpha(s) + f(s)\mathbf{1}_{|\alpha|=0})ds + \int_0^t \sum_{i,k}\sqrt{\alpha_{i,k}}\mathcal{M}_k u_{\alpha-\epsilon(i,k)}(s)\mathfrak{m}_i(s)ds. \tag{5.5.28}$$

If $\alpha = (\mathbf{0})$, then

$$\begin{aligned} \|u_{(\mathbf{0})}(t)\|_H^2 &= 2\int_0^t \langle \mathcal{A}u_{(\mathbf{0})}(s), u_{(\mathbf{0})}(s)\rangle ds + 2\int_0^t \langle f(s), u_{(\mathbf{0})}(s)\rangle ds \\ &\leq C_2\int_0^t \|u_{(\mathbf{0})}(s)\|_H^2 ds - \int_0^t \sum_{k\geq 1}\|\mathcal{M}_k u_{(\mathbf{0})}(s)\|_H^2 ds + C_f\int_0^t \|f(s)\|_{V'}^2 ds. \end{aligned}$$

By Corollary 5.4.6,

$$\sum_{|\alpha|=n}\|u_\alpha(t)\|_H^2 = \sum_{k_1,\dots,k_n\geq 1}\int_0^t\int_0^{s_n}\dots\int_0^{s_2}\|\Phi_{t,s_n}\mathcal{M}_{k_n}\dots\mathcal{M}_{k_1}u_{(\mathbf{0})}(s_1)\|_H^2 ds^n \tag{5.5.29}$$

for $n \geq 1$. Then, repeating the calculations (5.5.22)–(5.5.24), we conclude that

$$\sum_{n=1}^{N}\sum_{|\alpha|=n}\|u_\alpha(t)\|_H^2 \leq C_f\int_0^t \|f(s)\|_{V'}^2 ds + C_2\int_0^t \sum_{n=1}^{N}\sum_{|\alpha|=n}\|u_\alpha(s)\|_H^2 ds, \tag{5.5.30}$$

and, by the Gronwall inequality,

$$\mathbb{E}\|u(t)\|_H^2 \leq C_f e^{C_2 t}\int_0^t \|f(s)\|_{V'}^2 ds. \tag{5.5.31}$$

Finally, let us assume that $u_0 = f = 0$. Then the propagator (5.5.19) becomes

$$\begin{aligned} u_\alpha(t) &= \int_0^t \mathcal{A}u_\alpha(s)ds \\ &+ \int_0^t \left(\sum_{i,k}\sqrt{\alpha_{i,k}}\mathcal{M}_k u_{\alpha-\epsilon(i,k)}(s) + g_k(s)\mathbf{1}_{|\alpha|=1}\right)\mathfrak{m}_i(s)ds. \end{aligned} \tag{5.5.32}$$

Even though $u_{(\mathbf{0})}(t) = 0$, we have

$$u_{\epsilon(i,k)} = \int_0^t \Phi_{t,s} g_k(s)\mathfrak{m}_i(s)ds, \tag{5.5.33}$$

and then the arguments from the proof of Corollary 5.4.6 apply, resulting in

$$\sum_{|\alpha|=n}\|u_\alpha(t)\|_H^2 = \sum_{k_1,\ldots,k_n\geq 1}\int_0^t\int_0^{s_n}\ldots\int_0^{s_2}\|\Phi_{t,s_n}\mathcal{M}_{k_n}\ldots\Phi_{s_2,s_1}g_{k_1}(s_1)\|_H^2 ds^n$$

for $n \geq 1$. Note that

$$\sum_{|\alpha|=1}\|u_\alpha(t)\|_H^2 = \sum_{k\geq 1}\int_0^t \|g_k(s)\|_H^2 ds + 2\sum_{k\geq 1}\int_0^t \langle \mathcal{A}\Phi_{t,s}g_k(s), \Phi_{t,s}g_k(s)\rangle ds.$$

Then, repeating the calculations (5.5.22)–(5.5.24), we conclude that

$$\sum_{n=1}^{N}\sum_{|\alpha|=n}\|u_\alpha(t)\|_H^2 \leq \sum_{k\geq 1}\int_0^t \|g_k(s)\|_H^2 ds + C_2\int_0^t \sum_{n=1}^{N}\sum_{|\alpha|=n}\|u_\alpha(s)\|_H^2 ds, \tag{5.5.34}$$

and, by the Gronwall inequality,

$$\mathbb{E}\|u(t)\|_H^2 \leq e^{C_2 t}\sum_{k\geq 1}\int_0^t \|g_k(s)\|_H^2 ds. \tag{5.5.35}$$

To derive (5.5.12), it remains to combine (5.5.27), (5.5.31), and (5.5.35) with the inequality $(a+b+c)^2 \le 3(a^2+b^2+c^2)$.

Representation (5.5.13) of the Wiener chaos solution as a sum of iterated Itô integrals now follows from Corollary 5.4.6. Theorem 5.5.7 is proved.

Corollary 5.5.8 *If* $\sum_{\alpha\in\mathcal{J}^2}\int_0^T \|u_\alpha(s)\|_V^2 ds < \infty$, *then* $\sum_{\alpha\in\mathcal{J}^2}\sup_{0\le t\le T}\|u_\alpha(t)\|_H^2 < \infty$.

Proof The proof of Theorem 5.5.7 shows that it is enough to consider the homogeneous equation. Then by inequalities (5.5.23)–(5.5.24),

$$\begin{aligned}&\sum_{\ell=n+1}^{n_1}\sum_{|\alpha|=\ell}\|u_\alpha(t)\|_H^2 = \sum_{\ell=n+1}^{n_1} F_\ell(t)\\ &\le e^{C_2T}\sum_{k_1,\dots,k_{n+1}\ge 1}\int_0^T\int_0^t\int_0^{s_n}\cdots\int_0^{s_2}\|\mathcal{M}_{k_{n+1}}\Phi_{t,s_n}\mathcal{M}_{k_n}\dots\Phi_{s_1}u_0\|_H^2 ds^n dt.\end{aligned}\tag{5.5.36}$$

By Corollary 5.4.6,

$$\begin{aligned}&\int_0^T\|u_\alpha(s)\|_V^2 ds\\ &=\sum_{n\ge 1}\sum_{k_1,\dots,k_n\ge 1}\int_0^T\int_0^t\int_0^{s_n}\cdots\int_0^{s_2}\|\mathcal{M}_{k_n}\Phi_{t,s_n}\mathcal{M}_{k_n}\dots\Phi_{s_1}u_0\|_V^2 ds^n dt < \infty.\end{aligned}\tag{5.5.37}$$

As a result, (5.5.14) and (5.5.37) imply

$$\lim_{n\to\infty}\int_0^T\int_0^t\int_0^{s_n}\cdots\int_0^{s_2}\|\mathcal{M}_{k_{n+1}}\Phi_{t,s_n}\mathcal{M}_{k_n}\dots\mathcal{M}_{k_1}\Phi_{s_1}u_0\|_H^2 ds^n dt = 0,$$

which, by (5.5.36), implies uniform, with respect to t, convergence of the series $\sum_{\alpha\in\mathcal{J}^2}\|u_\alpha(t)\|_H^2$. Corollary 5.5.8 is proved.

Corollary 5.5.9 *Let* $a_{ij}, b_i, c, \sigma_{ik}, \nu_k$ *be deterministic measurable functions of* (t,x) *so that*

$$|a_{ij}(t,x)|+|b_i(t,x)|+|c(t,x)|+|\sigma_{ik}(t,x)|+|\nu_k(t,x)|\le K,$$

$i,j=1,\dots,\mathrm{d},\ k\ge 1,\ x\in\mathbb{R}^{\mathrm{d}},\ 0\le t\le T;$

$$\left(a_{ij}(t,x)-\frac{1}{2}\sigma_{ik}(t,x)\sigma_{jk}(t,x)\right)y_iy_j\ge 0,$$

$x, y \in \mathbb{R}^d$, $0 \le t \le T$; and

$$\sum_{k\ge1} |\nu_k(t,x)|^2 \le C_\nu < \infty,$$

$x \in \mathbb{R}^d$, $0 \le t \le T$. Consider the equation

$$du = (D_i(a_{ij}D_ju) + b_iD_iu + c\,u + f)dt + (\sigma_{ik}D_iu + \nu_ku + g_k)dw_k. \qquad (5.5.38)$$

In (5.5.38), we use the summation convention (summation over the repeated indices is not shown but is assumed). Assume that the input data satisfy $u_0 \in L_2(\mathbb{R}^d)$, $f \in L_2((0,T); H_2^{-1}(\mathbb{R}^d))$, $\sum_{k\ge1} \|g_k\|^2_{L_2((0,T)\times\mathbb{R}^d)} < \infty$, and there exists an $\varepsilon > 0$ so that

$$a_{ij}(t,x)y_iy_j \ge \varepsilon|y|^2,\ x, y \in \mathbb{R}^d,\ 0 \le t \le T.$$

Then there exists a unique Wiener Chaos solution $u = u(t,x)$ of (5.5.38). The solution has the following regularity:

$$u(t,\cdot) \in L_2(\mathbb{W}; L_2(\mathbb{R}^d)),\ 0 \le t \le T, \qquad (5.5.39)$$

and

$$\mathbb{E}\|u\|^2_{L_2(\mathbb{R}^d)}(t) \le C^*\Big(\|u_0\|^2_{L_2(\mathbb{R}^d)} + \|f\|^2_{L_2((0,T);H_2^{-1}(\mathbb{R}^d))} + \sum_{k\ge1} \|g_k\|^2_{L_2((0,T)\times\mathbb{R}^d)}\Big), \qquad (5.5.40)$$

where the positive number C^* depends only on C_ν, K, T, and ε.

If $f = g_k = 0$ and the equation is fully degenerate, that is, $2\langle A(t)v, v\rangle + \sum_{k\ge1} \|M_k(t)v\|^2_H = 0$, $t \in [0,T]$, then it is natural to expect conservation of energy. Once again, analysis of (5.5.23)–(5.5.24) shows that equality

$$\mathbb{E}\|u(t)\|^2_H = \|u_0\|^2_H$$

holds if and only if

$$\lim_{n\to\infty} \int_0^T \int_0^t \int_0^{s_n} \cdots \int_0^{s_2} \|\mathcal{M}_{k_{n+1}} \Phi_{t,s_n} \mathcal{M}_{k_n} \ldots \mathcal{M}_{k_1} \Phi_{s_1} u_0\|^2_H ds^n dt = 0.$$

The proof of Corollary 5.5.8 shows that a sufficient condition for the conservation of energy in a fully degenerate homogeneous equation is $\mathbb{E}\int_0^T \|u(t)\|^2_V dt < \infty$.

One of applications of the Wiener Chaos solution is new numerical methods for solving the evolution equations. Indeed, an approximation of the solution is obtained by truncating the sum $\sum_{\alpha\in\mathcal{J}^2} u_\alpha(t)\xi_\alpha$. For the Zakai filtering equation, these numerical methods were studied in [147, 148, 153]; see also Sect. 5.6.1 below. The main question in the analysis is the rate of convergence, in n, of the series $\sum_{n\geq 1}\sum_{|\alpha|=n}\|u(t)\|_H^2$. In general, this convergence can be arbitrarily slow. For example, consider the equation

$$du = \frac{1}{2}u_{xx}dt + u_x dw(t),\ t>0,\ x\in\mathbb{R},$$

in the normal triple $(H_2^1(\mathbb{R}), L_2(\mathbb{R}), H_2^{-1}(\mathbb{R}))$, with initial condition $u|_{t=0} = u_0 \in L_2(\mathbb{R})$. It follows from (5.5.20) that

$$F_n(t) = \sum_{|\alpha|=n}\|u\|^2_{L_2(\mathbb{R})}(t) = \frac{t^n}{n!}\int_{\mathbb{R}}|y|^{2n}e^{-y^2t}|\hat{u}_0|^2dy,$$

where $\hat{u}_0$ is the Fourier transform of u_0. If

$$|\hat{u}_0(y)|^2 = \frac{1}{(1+|y|^2)^{\gamma}},\ \gamma > 1/2,$$

then the rate of decay of $F_n(t)$ is close to $n^{-(1+2\gamma)/2}$. Note that, in this example, $E\|u\|^2_{L_2(\mathbb{R})}(t) = \|u_0\|^2_{L_2(\mathbb{R})}$.

An exponential convergence rate that is uniform in $\|u_0\|_H^2$ is achieved under strong parabolicity condition (4.4.4). An even faster factorial rate is achieved when the operators $\mathcal{M}_k$ are bounded on H.

Theorem 5.5.10 *Assume that there exist a positive number ε and a real number C_0 so that*

$$2\langle\mathcal{A}(t)v, v\rangle + \sum_{k\geq 1}\|\mathcal{M}_k(t)v\|_H^2 + \varepsilon\|v\|_V^2 \leq C_0\|v\|_H^2,\ t\in[0,T],\ v\in V.$$

Then there exists a positive number b so that, for all $t\in[0,T]$,

$$\sum_{|\alpha|=n}\|u_\alpha(t)\|_H^2 \leq \frac{\|u_0\|_H^2}{(1+b)^n}. \tag{5.5.41}$$

If, in addition, $\sum_{k\geq 1}\|\mathcal{M}_k(t)\varphi\|_H^2 \leq C_3\|\varphi\|_H^2$, then

$$\sum_{|\alpha|=n}\|u_\alpha(t)\|_H^2 \leq \frac{(C_3t)^n}{n!}e^{C_1t}\|u_0\|_H^2. \tag{5.5.42}$$

Proof If C^* is from (5.5.14) and $b = \varepsilon/C^*$, then the operators $\sqrt{1+b}\mathcal{M}_k$ satisfy

$$2\langle\mathcal{A}(t)v, v\rangle + (1+b)\sum_{k\geq 1}\|\mathcal{M}_k(t)\|_H^2 \leq C_0\|v\|_H^2.$$

By Theorem 5.5.7,

$$(1+b)^n \sum_{k_1,\ldots,k_n\geq 1} \int_0^t \int_0^{s_n} \cdots \int_0^{s_2} \|\Phi_{t,s_n}\mathcal{M}_{k_n}\ldots\mathcal{M}_{k_1}\Phi_{s_1}u_0\|_H^2 ds^n \leq \|u_0\|_H^2,$$

and (5.5.41) follows.

To establish (5.5.42), note that, by (5.5.10),

$$\|\Phi_t f\|_H^2 \leq e^{C_1 t}\|f\|_H^2,$$

and therefore the result follows from (5.5.20). Theorem 5.5.10 is proved.

The Wiener Chaos solution of (5.5.9) is not, in general, a solution of the equation in the sense of Definition 4.4.1. Indeed, if $u \notin L_2(\Omega \times (0,T); V)$, then the expressions $\langle Au(s), \varphi\rangle$ and $(M_k u(s), \varphi)_H$ are not defined. On the other hand, if there is a possibility to move the operators A and M from the solution process u to the test function φ, then Eq. (5.5.9) admits a natural analog of the standard variations formulation.

Theorem 5.5.11 *In addition to **A1–A3**, assume that there exist operators $\mathcal{A}^*(t)$, $\mathcal{M}_k^*(t)$ and a dense subset V_0 of the space V so that*

1. $\mathcal{A}^*(t)(V_0) \subseteq H$, $\mathcal{M}_k^*(t)(V_0) \subseteq H$, $t \in [0,T]$;
2. *for every* $v \in V$, $\varphi \in V_0$, *and* $t \in [0,T]$, $\langle \mathcal{A}(t)v, \varphi\rangle = (v, \mathcal{A}^*(t)\varphi)_H$, $(\mathcal{M}_k(t)v, \varphi)_H = (v, \mathcal{M}_k^*(t)\varphi)_H$.

If $u = u(t)$ is the Wiener Chaos solution of (5.5.9), then, for every $\varphi \in V_0$ and every $t \in [0,T]$, the equality

$$\begin{aligned}(u(t),\varphi)_H = (u_0,\varphi)_H &+ \int_0^t (u(s),\mathcal{A}^*(s)\varphi)_H ds + \int_0^t \langle f(s),\varphi\rangle ds \\ &+ \sum_{k\geq 1}\int_0^t (u(s),\mathcal{M}_k^*(s)\varphi)_H dw_k(s) + \sum_{k\geq 1}\int_0^t (g_k(s),\varphi)_H dw_k(s)\end{aligned} \tag{5.5.43}$$

holds in $L_2(\mathbb{W})$.

Proof The arguments are identical to the proof of Theorem 5.5.6(2).

As was mentioned earlier, the Wiener Chaos solution can be constructed for anticipating equations, that is, equations with $\mathcal{F}_T^W$-measurable (rather than $\mathcal{F}_t^W$-adapted) input data. With obvious modifications, inequality (5.5.12) holds if each of the input functions u_0, f, and g_k in (5.5.9) is a finite linear combination of the basis elements ξ_α. The following example demonstrates that inequality (5.5.12) is impossible for general anticipating equation.

Example 5.5.12 Let $u = u(t,x)$ be a Wiener Chaos solution of an ordinary differential equation

$$du = udw(t), 0 < t \leq 1, \tag{5.5.44}$$

with $u_0 = \sum_{\alpha \in \mathcal{J}} a_\alpha \xi_\alpha$. For $n \geq 0$, denote by (n) the multi-index with $\alpha_1 = n$ and $\alpha_i = 0$, $i \geq 2$, and assume that $a_{(n)} > 0$, $n \geq 0$. Then

$$\mathbb{E}u^2(1) \geq C \sum_{n \geq 0} e^{\sqrt{n}} a_{(n)}^2. \tag{5.5.45}$$

Indeed, the first column of propagator for $\boldsymbol{\alpha} = (n)$ is $u_{(0)}(t) = a_{(0)}$ and

$$u_{(n)}(t) = a_{(n)} + \sqrt{n} \int_0^t u_{(n-1)}(s)ds,$$

so that

$$u_{(n)}(t) = \sum_{k=0}^{n} \frac{\sqrt{n!}}{\sqrt{(n-k)!k!}} \frac{a_{(n-k)}}{\sqrt{k!}} t^k.$$

Then

$$u_{(n)}^2(1) \geq \sum_{k=0}^{n} \binom{n}{k} \frac{a_{(n-k)}^2}{k!}$$

and

$$\sum_{n \geq 0} u_{(n)}^2(1) \geq \sum_{n \geq 0} \left(\sum_{k \geq 0} \frac{1}{k!} \binom{n+k}{n} \right) a_{(n)}^2.$$

Since

$$\sum_{k \geq 0} \frac{1}{k!} \binom{n+k}{n} \geq \sum_{k \geq 0} \frac{n^k}{(k!)^2} \geq C e^{\sqrt{n}},$$

the result follows.

The consequence of Example 5.5.12 is that it is possible, in (5.5.9), to have $u_0 \in L_2^n(\mathbb{W}; H)$ for every n, and still get $E\|u(t)\|_H^2 = +\infty$ for all $t > 0$. More generally, the solution operator for (5.5.9) is not bounded on any $L_{2,Q}$ or $(\mathcal{S})_{-\rho,-\gamma}$. On the other hand, the following result holds.

Theorem 5.5.13 *In addition to Assumptions **A1**, **A2**, let u_0 be an element of $D'(\mathbb{W}; H)$, f, an element of $D'(\mathbb{W}; L_2((0,T), V'))$, and each g_k, an element of*

$D'(\mathbb{W}; L_2((0,T),H))$. Then the Wiener Chaos solution of equation (5.5.9) satisfies

$$\sqrt{\sum_{\alpha\in\mathcal{J}^2}\frac{\|u_\alpha(t)\|_H^2}{\alpha!}} \le C\sum_{\alpha\in\mathcal{J}^2}\frac{1}{\sqrt{\alpha!}}\left(\|u_{0,\alpha}\|_H + \left(\int_0^t \|f_\alpha(s)\|_{V'}^2 ds\right)^{1/2} + \left(\sum_{k\ge 1}\int_0^t \|g_{k,\alpha}(s)\|_H^2 ds\right)^{1/2}\right), \tag{5.5.46}$$

where $C > 0$ depends only on T and the numbers δ, C_1, and C_2 from (5.5.10) and (5.5.11).

Proof To simplify the presentation, assume that $f = g_k = 0$. For fixed $\gamma \in \mathcal{J}^2$, denote by $u(t;\varphi;\gamma)$ the Wiener Chaos solution of Eq. (5.5.9) with initial condition $u(0;\varphi;\gamma) = \varphi\xi_\gamma$. The structure of the propagator implies the following relation:

$$\frac{u_{\alpha+\gamma}(t;\varphi;\gamma)}{\sqrt{(\alpha+\gamma)!}} = \frac{u_\alpha\left(t;\frac{\varphi}{\sqrt{\gamma!}};(0)\right)}{\sqrt{\alpha!}}. \tag{5.5.47}$$

Clearly, $u_\alpha(t;\varphi;\gamma) = 0$ if $|\alpha| < |\gamma|$. By definition,

$$\|v(t)\|_{(\mathcal{S})_{-1,0}(H)}^2 = \sum_{\alpha\in\mathcal{J}^2}\frac{\|v_\alpha(t)\|_H^2}{\alpha!},$$

and then, by linearity and triangle inequality,

$$\|u(t)\|_{(\mathcal{S})_{-1,0}(H)} \le \sum_{\gamma\in\mathcal{J}^2}\|u(t;u_{0,\gamma};\gamma)\|_{(\mathcal{S})_{-1,0}(H)}.$$

We also have by (5.5.47) and Theorem 5.5.7

$$\|u(t;u_{0,\gamma};\gamma)\|_{(\mathcal{S})_{-1,0}(H)}^2 = \left\|u\left(t;\frac{u_{0,\gamma}}{\sqrt{\gamma!}};(0)\right)\right\|_{(\mathcal{S})_{-1,0}(H)}^2$$
$$\le \mathbb{E}\left\|u\left(t;\frac{u_{0,\gamma}}{\sqrt{\gamma!}};(0)\right)\right\|_H^2 \le e^{C_2 t}\frac{\|u_{0,\gamma}\|_H^2}{\gamma!}.$$

Inequality (5.5.46) then follows. Theorem 5.5.13 is proved.

Remark 5.5.14 Using Proposition 5.4.11 and the Cauchy-Schwartz inequality, (5.5.46) can be re-written in a slightly weaker form to reveal continuity of the solution operator for Eq. (5.5.9) from $(\mathcal{S})_{-1,\gamma}$ to $(\mathcal{S})_{-1,0}$ for every $\gamma > 1$:

$$\|u(t)\|^2_{(\mathcal{S})_{-1,0}(H)} \leq C\left(\|u_0\|^2_{(\mathcal{S})_{-1,\gamma}(H)} + \int_0^t \|f(s)\|^2_{(\mathcal{S})_{-1,\gamma}(V')} ds \right.$$
$$\left. + \sum_{k\geq 1} \int_0^t \|g_k(s)\|^2_{(\mathcal{S})_{-1,\gamma}(H)} ds \right).$$

5.5.3 *Probabilistic Representation*

The general discussion so far has been dealing with the abstract evolution equation

$$du = (\mathcal{A}u + f)dt + \sum_{k\geq 1} (\mathcal{M}_k u + g_k) dw_k.$$

By further specifying the operators $\mathcal{A}$ and $\mathcal{M}_k$, as well as the input data u_0, f, and g_k, it is possible to get additional information about the Wiener Chaos solution of the equation.

Definition 5.5.15 For $r \in \mathbb{R}$, the space $L_{2,(r)} = L_{2,(r)}(\mathbb{R}^d)$ is the collection of real-valued measurable functions so that $f \in L_{2,(r)}$ if and only if $\int_{\mathbb{R}^d} |f(x)|^2 (1+|x|^2)^r dx < \infty$. The space $H^1_{2,(r)} = H^1_{2,(r)}(\mathbb{R}^d)$ is the collection of real-valued measurable functions so that $f \in H^1_{2,(r)}$ if and only if f and all the first-order generalized derivatives $D_i f$ of f belong to $L_{2,(r)}$.

It is known, for example, from Theorem 3.4.7 in [199], that $L_{2,(r)}$ is a Hilbert space with norm

$$\|f\|^2_{0,(r)} = \int_{\mathbb{R}^d} |f(x)|^2 (1 + |x|^2)^r dx,$$

and $H^1_{2,(r)}$ is a Hilbert space with norm

$$\|f\|_{1,(r)} = \|f\|_{0,(r)} + \sum_{i=1}^{d} \|D_i f\|_{0,(r)}.$$

Denote by $H^{-1}_{2,(r)}$ the dual of $H^1_{2,(r)}$ with respect to the inner product in $L_{2,(r)}$. Then $(H^1_{2,(r)}, L_{2,(r)}, H^{-1}_{2,(r)})$ is a normal triple of Hilbert spaces.

Let $\mathbb{F} = (\Omega, \mathcal{F}, \{\mathcal{F}_t\}_{t\geq 0}, \mathbb{P})$ be a stochastic basis with the usual assumptions and $w_k = w_k(t),\ k \geq 1,\ t \geq 0$, a collection of standard Wiener processes on $\mathbb{F}$. Using the summation conventions, consider the linear equation

$$du = (a_{ij} D_i D_j u + b_i D_i u + cu + f)dt + (\sigma_{ik} D_i u + \nu_k u + g_k) dw_k \qquad (5.5.48)$$

under the following assumptions:

B0 All coefficients, free terms, and the initial condition are non-random.

B1 The functions $a_{ij} = a_{ij}(t,x)$ and their first-order derivatives with respect to x are uniformly bounded in (t,x), and the matrix (a_{ij}) is uniformly positive definite, that is, there exists a $\delta > 0$ so that, for all vectors $y \in \mathbb{R}^{\mathrm{d}}$ and all (t,x), $a_{ij}y_iy_j \geq \delta|y|^2$.

B2 The functions $b_i = b_i(t,x)$, $c = c(t,x)$, and $\nu_k = \nu_k(t,x)$ are measurable and bounded in (t,x).

B3 The functions $\sigma_{ik} = \sigma_{ik}(t,x)$ are continuous and bounded in (t,x).

B4 The functions $f = f(t,x)$ and $g_k = g_k(t,x)$ belong to $L_2((0,T);L_{2,(r)})$ for some $r \in \mathbb{R}$.

B5 The initial condition $u_0 = u_0(x)$ belongs to $L_{2,(r)}$.

Under Assumptions **B2–B4**, there exists a sequence $Q = \{q_k, k \geq 1\}$ of positive numbers with the following properties:

P1 The matrix A with $A_{ij} = a_{ij} - (1/2)\sum_{k\geq 1} q_k\sigma_{ik}\sigma_{jk}$ satisfies

$$A_{ij}(t,x)y_iy_j \geq 0,$$

$x, y \in \mathbb{R}^{\mathrm{d}}$, $0 \leq t \leq T$.

P2 There exists a number $C > 0$ so that

$$\sum_{k\geq 1}\left(\sup_{t,x}|q_k\nu_k(t,x)|^2 + \int_0^T \|q_kg_k\|^p_{0,(r)}(t)dt\right) \leq C.$$

For the matrix A and each t, x, we have $A_{ij}(t,x) = \tilde{\sigma}_{ik}(t,x)\tilde{\sigma}_{jk}(t,x)$, where the functions $\tilde{\sigma}_{ik}$ are bounded. This representation might not be unique; see, for example, [54, Theorem III.2.2] or [214, Lemma 5.2.1]. Given any such representation of A, consider the following backward Itô equation

$$X_{t,x,i}(s) = x_i + \int_s^t B_i(\tau, X_{t,x}(\tau))\,d\tau + \sum_{k\geq 1} q_k\sigma_{ik}(\tau, X_{t,x}(\tau))\,\overleftarrow{dw_k}(\tau)$$
$$+ \int_s^t \tilde{\sigma}_{ik}(\tau, X_{t,x}(\tau))\,\overleftarrow{d\tilde{w}_k}(\tau)\,;\ s \in (0,t),\ t \in (0,T],\ t - \text{fixed}, \tag{5.5.49}$$

where

$$B_i = b_i - \sum_{k\geq 1} q_k^2\sigma_{ik}\nu_k$$

and $\tilde{w}_k$, $k \geq 1$, are independent standard Wiener processes on $\mathbb{F}$ that are independent of w_k, $k \geq 1$. This equation might not have a strong solution, but does have weak, or martingale, solutions due to Assumptions B1–B3 and properties P1 and P2 of the sequence Q; this weak solution is unique in the sense of probability law [214, Theorem 7.2.1].

The following result is a variation of Theorem 4.1 in [152].

Theorem 5.5.16 *Under assumptions **B0–B5** Eq. (5.5.48) has a unique $w(H^1_{2,(r)}, H^{-1}_{2,(r)})$ Wiener Chaos solution. If Q is a sequence with properties **P1** and **P2**, then the solution of (5.5.48) belongs to*

$$L_{2,Q}\left(\mathbb{W}; L_2((0,T); H^1_{2,(r)})\right) \bigcap L_{2,Q}\left(\mathbb{W}; \mathcal{C}((0,T); L_{2,(r)})\right)$$

and has the following representation:

$$\begin{aligned} u(t,x) = \mathcal{Q}^{-1}\mathbb{E}\Bigg(& \int_0^t f(s, X_{t,x}(s))\gamma(t,s,x)ds \\ & + \sum_{k\geq 1}\int_0^t q_k g_k(s, X_{t,x}(s))\gamma(t,s,x)\overleftarrow{dw_k}(s) + u_0(X_{t,x}(0))\gamma(t,0,x)\Big|\mathcal{F}^W_t\Bigg), \\ t \leq T, & \end{aligned} \tag{5.5.50}$$

where $X_{t,x}(s)$ is a weak solution of (5.5.49), and

$$\begin{aligned} \gamma(t,s,x) = \exp\Bigg(& \int_s^t c(\tau, X_{t,x}(\tau))d\tau + \sum_{k\geq 1}\int_s^t q_k \nu_k(\tau, X_{t,x}(\tau))\overleftarrow{dw_k}(\tau) \\ & - \frac{1}{2}\int_s^t \sum_{k\geq 1} q_k^2|\nu_k(\tau, X_{t,x}(\tau))|^2 d\tau\Bigg). \end{aligned} \tag{5.5.51}$$

Proof It is enough to establish (5.5.50) when $t = T$. Keeping in mind the summation convention, consider the equation

$$dU = (a_{ij}D_iD_jU + b_iD_iU + cU + f)dt + \sum_{k\geq 1}(\sigma_{ik}D_iU + \nu_kU + g_k)q_kdw_k \tag{5.5.52}$$

with initial condition $U(0,x) = u_0(x)$. Applying Theorem 4.4.3 in the normal triple $(H^1_{2,(r)}, L_{2,(r)}, H^{-1}_{2,(r)})$, we conclude that there is a unique solution

$$U \in L_2\left(\mathbb{W}; L_2((0,T); H^1_{2,(r)})\right) \bigcap L_2\left(\mathbb{W}; \mathcal{C}((0,T); L_{2,(r)})\right)$$

of this equation. By Proposition 5.4.13, the process $u = \mathcal{Q}^{-1}U$ is the corresponding Wiener Chaos solution of (5.5.48). To establish representation (5.5.50), consider the S-transform U_h of U. According to Theorem 5.5.1, the function U_h is the unique

$w(H^1_{2,(r)}, H^{-1}_{2,(r)})$ solution of the equation

$$dU_h = (a_{ij}D_iD_jU_h + b_iD_iU_h + cU_h + f)dt + \sum_{k\geq 1}(\sigma_{ik}D_iU_h + \nu_kU_h + g_k)q_kh_kdt \tag{5.5.53}$$

with initial condition $U_h|_{t=0} = u_0$. We also define

$$\begin{aligned} Y(T,x) &= \int_0^T f(s, X_{T,x}(s))\gamma(T,s,x)ds \\ &+ \sum_{k\geq 1}\int_0^T g_k(s, X_{T,x}(s))\gamma(T,s)q_k\overleftarrow{dw_k}(s) + u_0(X_{T,x}(0))\gamma(T,0,x). \end{aligned} \tag{5.5.54}$$

By direct computation,

$$\mathbb{E}\left(\mathbb{E}\left(\mathcal{E}(h)Y(T,x)|\mathcal{F}^W_T\right)\right) = \mathbb{E}\left(\mathcal{E}(h)Y(T,x)\right) = \mathbb{E}'Y(T,x),$$

where $\mathbb{E}'$ is the expectation with respect to the measure $d\mathbb{P}'_T = \mathcal{E}(h)d\mathbb{P}_T$ and $\mathbb{P}_T$ is the restriction of $\mathbb{P}$ to $\mathcal{F}^W_T$.

To proceed, let us first assume that the input data u_0, f, and g_k are all smooth functions with compact support. Then, applying the Feynmann-Kac formula to the solution of equation (5.5.53) and using the Girsanov theorem (see, for example, Theorems 3.5.1 and 5.7.6 in [103]), we conclude that $U_h(T,x) = \mathcal{E}'Y(T,x)$ or

$$\mathbb{E}\left(\mathcal{E}(h)\mathbb{E}Y(t,x)|\mathcal{F}^W_T\right) = \mathbb{E}\left(\mathcal{E}\left(h\right)U(T,x)\right).$$

By Remark 5.4.17, the last equality implies $U(T,\cdot) = \mathcal{E}\left(Y(T,\cdot)|\mathcal{F}^W_T\right)$ as elements of $L_2\left(\Omega; L_{2,(r)}(\mathbb{R}^d)\right)$.

To remove the additional smoothness assumption on the input data, let u^n_0, f^n, and g^n_k be sequences of smooth compactly supported functions so that

$$\begin{aligned} \lim_{n\to\infty} \Bigg(&\|u_0 - u^n_0\|^2_{L_{2,(r)}(\mathbb{R}^d)} + \int_0^T \|f - f^n\|^2_{L_{2,(r)}(\mathbb{R}^d)}(t)dt \\ &+ \sum_{k\geq 1}\int_0^T q_k^2\|g_k - g^n_k\|^2_{L_{2,(r)}(\mathbb{R}^d)}(t)dt\Bigg) = 0. \end{aligned} \tag{5.5.55}$$

Denote by U^n and Y^n the corresponding objects defined by (5.5.52) and (5.5.54) respectively. By Theorem 5.5.7, we have

$$\lim_{n\to\infty}\mathbb{E}\|U - U^n\|^2_{L_{2,(r)}(\mathbb{R}^d)}(T) = 0. \tag{5.5.56}$$

To complete the proof, it remains to show that

$$\lim_{n\to\infty} \mathbb{E} \left\| \mathbb{E}\left(Y(T,\cdot) - Y^n(T,\cdot) \middle| \mathcal{F}_T^W\right) \right\|^2_{L_{2,(r)}(\mathbb{R}^d)} = 0. \tag{5.5.57}$$

To this end, introduce a new probability measure $\mathbb{P}_T''$ by

$$d\mathbb{P}_T'' = \tilde{\gamma}(T,s,x)d\mathbb{P}_T,$$

$$\tilde{\gamma}(T,s,x) = \exp\left(2\sum_{k\geq 1}\int_0^T \nu_k(s,X_{T,x}(s))q_k\overleftarrow{dw_k}(s)\right.$$
$$\left. - 2\int_0^T \sum_{k\geq 1} q_k^2|\nu_k(s,X_{T,x}(s))|^2 ds\right).$$

By Girsanov's theorem, Eq. (5.5.49) can be rewritten as

$$X_{T,x,i}(s) = x_i + \int_s^T \sum_{k\geq 1} \sigma_{ik}(\tau, X_{T,x}(\tau))\, h_k(\tau)\, q_k d\tau$$
$$+ \int_s^t (b_i + \sum_{k\geq 1} q_k^2 \sigma_{ik}\nu_k)(\tau, X_{T,x}(\tau))\, d\tau$$
$$+ \int_s^t \sum_{k\geq 1} q_k\sigma_{ik}(\tau, X_{T,x}(\tau))\, \overleftarrow{dw_k''}(\tau) + \int_s^t \tilde{\sigma}_{ik}(\tau, X_{T,x}(\tau))\, \overleftarrow{d\tilde{w}''_k}(\tau), \tag{5.5.58}$$

where w_k'' and $\tilde{w}''_k$ are independent Winer processes with respect to the measure P_T''. Denote by $p(s,y|x)$ the probability density function of $X_{T,x}(s)$ and write $\ell(x) = (1+|x|^2)^r$. It then follows by the Hölder and Jensen inequalities that

$$\begin{aligned}
&\mathbb{E}\left\|\mathbb{E}\left(\int_0^T \tilde{\gamma}(T,s,\cdot)(f-f^n)(s,X_{T,\cdot}(s))ds \middle| \mathcal{F}_T^W\right)\right\|^2_{L_{2,(r)}(\mathbb{R}^d)} \\
&\leq K_1 \int_{\mathbb{R}^d}\left(\int_0^T \mathbb{E}\left(\tilde{\gamma}(T,s,x)(f-f^n)^2(s,X_{T,x}(s))\right)ds\right)\ell(x)dx \\
&\leq K_2 \int_{\mathbb{R}^d}\left(\int_0^T \mathbb{E}''(f-f^n)^2(s,X_{T,x}(s))ds\right)\ell(x)dx \\
&= K_2 \int_{\mathbb{R}^d}\int_0^T\int_{\mathbb{R}^d}(f(s,y)-f^n(s,y))^2 p(s,y|x)dy\, ds\, \ell(x)dx,
\end{aligned} \tag{5.5.59}$$

where the number K_1 depends only on T, and the number K_2 depends only on T and $\sup_{(t,x)} |c(t,x)| + \sum_{k\geq 1} q_k^2 \sup_{(t,x)} |\nu_k(t,x)|^2$. Assumptions **B0–B3** imply that there exist positive numbers K_3 and K_4 so that

$$p(s,y|x) \leq \frac{K_3}{(T-s)^{d/2}} \exp\left(-K_4 \frac{|x-y|^2}{T-s}\right); \tag{5.5.60}$$

see, for example, [46]. As a result,

$$\int_{\mathbb{R}^d} p(s,y|x)\ell(x)dx \leq K_5 \ell(y),$$

and

$$\begin{aligned}&\int_{\mathbb{R}^d} \int_0^T \int_{\mathbb{R}^d} (f(s,y) - f^n(s,y))^2 p(s,y|x)dy\, ds\, \ell(x)dx \\ &\leq K_5 \int_0^T \|f - f^n\|^2_{L_{2,(r)}(\mathbb{R}^d)}(s)ds \to 0,\ n \to \infty,\end{aligned} \tag{5.5.61}$$

where the number K_5 depends only on K_3, K_4, T, and r.

Calculations similar to (5.5.59)–(5.5.61) show that

$$\begin{aligned}&\mathbb{E}\left\|\mathbb{E}\left(\gamma^2(T,0,\cdot)(u_0 - u_0^n)(X_{T,\cdot}(0))\Big|\mathcal{F}_T^W\right)\right\|^2_{L_{2,(r)}(\mathbb{R}^d)} \\ &+ \mathbb{E}\left\|\mathbb{E}\left(\int_0^T \sum_{k\geq 1}(g_k - g_k^n)(s, X_{T,\cdot}(s))\gamma(t,s,\cdot)q_k \overleftarrow{dw_k}(s)\Big|\mathcal{F}_T^W\right)\right\|^2_{L_{2,(r)}(\mathbb{R}^d)} \to 0\end{aligned} \tag{5.5.62}$$

as $n \to \infty$. Then convergence (5.5.57) follows, which, together with (5.5.56), implies that $U(T,\cdot) = \mathbb{E}\left(U^Q(T,\cdot)|\mathcal{F}_T^W\right)$ as elements of $L_2\left(\Omega; L_{2,(r)}(\mathbb{R}^d)\right)$. It remains to note that $u = \mathcal{Q}^{-1}U$. Theorem 5.5.16 is proved.

Given $f \in L_{2,(r)}$, we say that $f \geq 0$ if and only if

$$\int_{\mathbb{R}^d} f(x)\varphi(x)dx \geq 0$$

for every non-negative $\varphi \in C_0^\infty(\mathbb{R}^d)$. Then Theorem 5.5.16 implies the following result.

Corollary 5.5.17 *In addition to Assumptions* ***B0–B5****, let* $u_0 \geq 0, f \geq 0$, *and* $g_k = 0$ *for all* $k \geq 1$. *Then* $u \geq 0$.

Proof This follows from (5.5.50) and Proposition 5.4.20.

Example 5.5.18 (Krylov-Veretennikov Formula) Consider the equation

$$du = \left(a_{ij}D_iD_ju + b_iD_iu\right)dt + \sum_{k=1}^{\mathrm{d}} \sigma_{ik}D_iudw_k,\ u(0,x) = u_0(x). \tag{5.5.63}$$

Assume **B0–B5** and suppose that $a_{ij}(t,x) = \frac{1}{2}\sigma_{ik}(t,x)\sigma_{jk}(t,x)$. By Theorem 5.5.7, Eq. (5.5.63) has a unique Wiener chaos solution so that

$$\mathbb{E}\|u\|^2_{L_2(\mathbb{R}^{\mathrm{d}})}(t) \le C^*\|u_0\|^2_{L_2(\mathbb{R}^{\mathrm{d}})}$$

and

$$u(t,x) = \sum_{n=1}^{\infty}\sum_{|\alpha|=n} u_\alpha(t,x)\xi_\alpha = u_0(x) + \sum_{n=1}^{\infty}\sum_{k_1,\ldots,k_n=1}^{\mathrm{d}} \int_0^t\int_0^{s_n}\cdots\int_0^{s_2} \tag{5.5.64}$$

$$\Phi_{t,s_n}\sigma_{jk_n}D_j\cdots\Phi_{s_2,s_1}\sigma_{ik_1}D_i\Phi_{s_1,0}u_0(x)dw_{k_1}(s_1)\cdots dw_{k_n}(s_n),$$

where $\Phi_{t,s}$ is the semi-group generated by the operator $A = a_{ij}D_iD_ju + b_iD_iu$. On the other hand, in this case, Theorem 5.5.16 yields

$$u(t,x) = \mathbb{E}\left(u_0(X_{t,x}(0))\,|\mathcal{F}_t^W\right),$$

where $W = (w_1, \ldots, w_d)$ and

$$X_{t,x,i}(s) = x_i + \int_s^t b_i(\tau, X_{t,x}(\tau))\,d\tau + \sum_{k=1}^{\mathrm{d}}\sigma_{ik}(\tau, X_{t,x}(\tau))\,\overleftarrow{dw_k}(\tau) \tag{5.5.65}$$

$$s \in (0,t),\ t \in (0,T],\ t - \text{fixed}.$$

Thus, we have arrived at the `Krylov-Veretennikov formula` [123, Theorem 4]

$$\mathbb{E}\left(u_0(X_{t,x}(0))\,|\mathcal{F}_t^W\right) = u_0(x) + \sum_{n=1}^{\infty}\sum_{k_1,\ldots,k_n=1}^{\mathrm{d}}\int_0^t\int_0^{s_n}\cdots\int_0^{s_2}$$

$$\Phi_{t,s_n}\sigma_{jk_n}D_j\cdots\Phi_{s_2,s_1}\sigma_{ik_1}D_i\Phi_{s_1,0}u_0(x)dw_{k_1}(s_1)\cdots dw_{k_n}(s_n). \tag{5.5.66}$$

5.6 Examples

5.6.1 *Wiener Chaos and Nonlinear Filtering*

In this section, we discuss some applications of the Wiener Chaos expansion to numerical solution of the nonlinear filtering problem for diffusion processes; the presentation is essentially based on [153].

Let $(\Omega, \mathcal{F}, \mathbb{P})$ be a complete probability space with independent standard Wiener processes $W = W(t)$ and $V = V(t)$ of dimensions d_1 and r respectively. Let X_0 be a random variable independent of W and V. In the diffusion filtering model, the unobserved d-dimensional state (or signal) process $X = X(t)$ and the r-dimensional observation process $Y = Y(t)$ are defined by the stochastic ordinary differential equations

$$\begin{aligned} dX(t) &= b(X(t))dt + \sigma(X(t))dW(t) + \rho(X(t))dV(t), \\ dY(t) &= h(X(t))dt + dV(t),\ 0 < t \leq T; \\ X(0) &= X_0, \quad Y(0) = 0, \end{aligned} \tag{5.6.1}$$

where $b(x) \in \mathbb{R}^{\mathrm{d}}$, $\sigma(x) \in \mathbb{R}^{\mathrm{d}\times d_1}$, $\rho(x) \in \mathbb{R}^{\mathrm{d}\times r}$, $h(x) \in \mathbb{R}^r$.

Denote by $\mathcal{C}^n(\mathbb{R}^{\mathrm{d}})$ the Banach space of bounded, n times continuously differentiable functions on $\mathbb{R}^{\mathrm{d}}$ with finite norm

$$\|f\|_{\mathcal{C}^n(\mathbb{R}^{\mathrm{d}})} = \sup_{x\in\mathbb{R}^{\mathrm{d}}} |f(x)| + \max_{1\leq k\leq n} \sup_{x\in\mathbb{R}^{\mathrm{d}}} |D^k f(x)|.$$

Assumption R1 The the components of the functions σ and ρ are in $\mathcal{C}^2(\mathbb{R}^{\mathrm{d}})$, the components of the functions b are in $\mathcal{C}^1(\mathbb{R}^{\mathrm{d}})$, the components of the function h are bounded measurable, and the random variable X_0 has a density u_0.

Assumption R2 The matrix $\sigma\sigma^*$ is uniformly positive definite: there exists an $\varepsilon > 0$ so that

$$\sum_{i,j=1}^{\mathrm{d}} \sum_{k=1}^{d_1} \sigma_{ik}(x)\sigma_{jk}(x)y_i y_j \geq \varepsilon|y|^2,\ x, y \in \mathbb{R}^{\mathrm{d}}.$$

Under Assumption R1 system (5.6.1) has a unique strong solution [103, Theorems 5.2.5 and 5.2.9]. Extra smoothness of the coefficients in assumption R1 ensure the existence of a convenient representation of the optimal filter.

If $f = f(x)$ is a scalar measurable function on $\mathbb{R}^{\mathrm{d}}$ and $\sup_{0\leq t\leq T} \mathbb{E}|f(X(t))|^2 < \infty$, then the filtering problem for (5.6.1) is to find the best mean square estimate $\hat{f}_t$ of $f(X(t))$, $t \leq T$, given the observations $Y(s)$, $0 < s \leq t$.

Denote by $\mathcal{F}_t^Y$ the σ-algebra generated by $Y(s)$, $0 \le s \le t$. Then the properties of the conditional expectation imply that the solution of the filtering problem is

$$\hat{f}_t = \mathbb{E}\left(f(X(t))|\mathcal{F}_t^Y\right).$$

To derive an alternative representation of $\hat{f}_t$, some additional constructions will be necessary.

Define a new probability measure $\widetilde{\mathbb{P}}$ on $(\Omega, \mathcal{F})$ as follows: for $A \in \mathcal{F}$,

$$\widetilde{\mathbb{P}}(A) = \int_A Z_T^{-1} d\mathbb{P},$$

where

$$Z_t = \exp\left\{\int_0^t h^*(X(s))dY(s) - \frac{1}{2}\int_0^t |h(X(s))|^2 ds\right\}$$

(here and below, if $\zeta \in \mathbb{R}^k$, then ζ is a column vector, $\zeta^* = (\zeta_1, \ldots, \zeta_k)$, and $|\zeta|^2 = \zeta^*\zeta$). If the function h is bounded, then the measures $\mathbb{P}$ and $\widetilde{\mathbb{P}}$ are equivalent. The expectation with respect to the measure $\widetilde{\mathbb{P}}$ will be denoted by $\widetilde{\mathbb{E}}$.

The following properties of the measure $\widetilde{\mathbb{P}}$ are well known [101, 199]:

P1. Under the measure $\widetilde{\mathbb{P}}$, the distributions of the Wiener process W and the random variable X_0 are unchanged, the observation process Y is a standard Wiener process, and, for $0 < t \le T$, the state process X satisfies

$$dX(t) = b(X(t))dt + \sigma(X(t))dW(t) + \rho(X(t))\left(dY(t) - h(X(t))dt\right),$$
$$X(0) = X_0;$$

P2. Under the measure $\widetilde{\mathbb{P}}$, the Wiener processes W and Y and the random variable X_0 are jointly independent;

P3. The optimal filter $\hat{f}_t$ satisfies

$$\hat{f}_t = \frac{\widetilde{\mathbb{E}}\left[f(X(t))Z_t|\mathcal{F}_t^Y\right]}{\widetilde{\mathbb{E}}[Z_t|\mathcal{F}_t^Y]}. \quad (5.6.2)$$

Because of property **P2** of the measure $\widetilde{\mathbb{P}}$ the filtering problem will be studied on the probability space $(\Omega, \mathcal{F}, \widetilde{\mathbb{P}})$. In particular, we will consider the stochastic basis $\widetilde{\mathbb{F}} = \{\Omega, \mathcal{F}, \{\mathcal{F}_t^Y\}_{0\le t\le T}, \widetilde{\mathbb{P}}\}$ and the Wiener Chaos space $\widetilde{L}_2(Y)$ of $\mathcal{F}_T^Y$-measurable random variables η with $\widetilde{\mathbb{E}}|\eta|^2 < \infty$.

If the function h is bounded, then, by the Cauchy-Schwarz inequality,

$$\mathbb{E}|\eta| \le C(h,T)\sqrt{\widetilde{\mathbb{E}}|\eta|^2},\ \eta \in \widetilde{L}_2(Y). \quad (5.6.3)$$

Next, consider the partial differential operators

$$\mathcal{L}g(x) = \frac{1}{2}\sum_{i,j=1}^{\mathrm{d}} \big((\sigma(x)\sigma^*(x))_{ij} + (\rho(x)\rho^*(x))_{ij}\big)\frac{\partial^2 g(x)}{\partial x_i \partial x_j} + \sum_{i=1}^{\mathrm{d}} b_i(x)\frac{\partial g(x)}{\partial x_i};$$

$$\mathcal{M}_l g(x) = h_l(x)g(x) + \sum_{i=1}^{\mathrm{d}} \rho_{il}(x)\frac{\partial g(x)}{\partial x_i},\ l = 1,\ldots,r;$$

and their adjoints

$$\mathcal{L}^* g(x) = \frac{1}{2}\sum_{i,j=1}^{\mathrm{d}} \frac{\partial^2}{\partial x_i \partial x_j}\big((\sigma(x)\sigma^*(x))_{ij}g(x) + (\rho(x)\rho^*(x))_{ij}g(x)\big)$$
$$-\sum_{i=1}^{\mathrm{d}} \frac{\partial}{\partial x_i}\left(b_i(x)g(x)\right);$$

$$\mathcal{M}^*_l g(x) = h_l(x)g(x) - \sum_{i=1}^{\mathrm{d}} \frac{\partial}{\partial x_i}\left(\rho_{il}(x)g(x)\right),\ l = 1,\ldots,r.$$

Note that, under the assumptions R1 and R2, the operators $\mathcal{L}, \mathcal{L}^*$ are bounded from $H^1_2(\mathbb{R}^{\mathrm{d}})$ to $H^{-1}_2(\mathbb{R}^{\mathrm{d}})$, operators $\mathcal{M}, \mathcal{M}^*$ are bounded from $H^1_2(\mathbb{R}^{\mathrm{d}})$ to $L_2(\mathbb{R}^{\mathrm{d}})$, and

$$2\langle \mathcal{L}^* v, v\rangle + \sum_{l=1}^{r} \|\mathcal{M}^*_l v\|^2_{L_2(\mathbb{R}^{\mathrm{d}})} + \varepsilon\|v\|^2_{H^2_1(\mathbb{R}^{\mathrm{d}})} \le C\|v\|^2_{L_2(\mathbb{R}^{\mathrm{d}})},\ v \in H^1_2(\mathbb{R}^{\mathrm{d}}), \qquad (5.6.4)$$

where $\langle\cdot,\cdot\rangle$ is the duality between $H^1_2(\mathbb{R}^{\mathrm{d}})$ and $H^{-1}_2(\mathbb{R}^{\mathrm{d}})$. The following result is well known: see, for example, Theorem 4.4.27 on page 227 or [199, Theorem 6.2.1].

Proposition 5.6.1 *In addition to Assumptions R1 and R1 suppose that the initial density u_0 belongs to $L_2(\mathbb{R}^{\mathrm{d}})$. Then there exists a random field $u = u(t,x)$, $t \in [0,T]$, $x \in \mathbb{R}^{\mathrm{d}}$, with the following properties:*

1. $u \in \widetilde{L}_2(Y; L_2((0,T); H^1_2(\mathbb{R}^{\mathrm{d}}))) \cap \widetilde{L}_2(Y; C([0,T], L_2(\mathbb{R}^{\mathrm{d}})))$.
2. *The function $u(t,x)$ is a square-integrable solution of the stochastic partial differential equation*

$$\begin{aligned} du(t,x) &= \mathcal{L}^* u(t,x)dt + \sum_{l=1}^{r} \mathcal{M}^*_l u(t,x)dY_l(t),\ 0 < t \le T,\ x \in \mathbb{R}^{\mathrm{d}}; \\ u(0,x) &= u_0(x). \end{aligned} \qquad (5.6.5)$$

3. The equality

$$\widetilde{\mathbb{E}}\left[f(X(t))Z_t|\mathcal{F}_t^Y\right] = \int_{\mathbb{R}^{\mathrm{d}}} f(x)u(t,x)dx \tag{5.6.6}$$

holds for all bounded measurable functions f.

The random field $u = u(t,x)$ is called the unnormalized filtering density (UFD) and the random variable $\phi_t[f] = \widetilde{\mathbb{E}}\left[f(X(t))Z_t|\mathcal{F}_t^Y\right]$, the unnormalized optimal filter.

A number of authors studied the nonlinear filtering problem using the multiple Itô integral version of the Wiener chaos [17, 124, 180, 228, etc.]. In what follows, we construct approximations of u and $\phi_t[f]$ using the Cameron-Martin version.

By Theorem 5.5.6,

$$u(t,x) = \sum_{\alpha \in \mathcal{J}^2} u_\alpha(t,x)\xi_\alpha, \tag{5.6.7}$$

where

$$\xi_\alpha = \frac{1}{\sqrt{\alpha!}} \prod_{i,k} \mathrm{H}_{\alpha_{i,k}}(\xi_{i,k}),\ \xi_{i,k} = \int_0^T \mathfrak{m}_i(t)dY_k(t),\ k = 1,\dots,r; \tag{5.6.8}$$

as before, $\mathrm{H}_n(\cdot)$ is the Hermite polynomial (2.3.37) and $\mathfrak{m}_i \in L_\infty((0,T))$ is an orthonormal basis in $L_2((0,T))$. The functions u_α satisfy the corresponding propagator

$$\begin{aligned}\frac{\partial}{\partial t}u_\alpha(t,x) &= \mathcal{L}^* u_\alpha(t,x)\\ &\quad + \sum_{i,k} \sqrt{\alpha_{i,k}}\mathcal{M}_k^* u_{\alpha-\epsilon(i,k)}(t,x)\mathfrak{m}_i(t),\ 0 < t \le T,\ x \in \mathbb{R}^{\mathrm{d}};\\ u(0,x) &= u_0(x)I(|\alpha| = 0).\end{aligned} \tag{5.6.9}$$

Writing

$$f_\alpha(t) = \int_{\mathbb{R}^{\mathrm{d}}} f(x)u_\alpha(t,x)dx,$$

we also get a Wiener chaos expansion for the unnormalized optimal filter:

$$\phi_t[f] = \sum_{\alpha \in \mathcal{J}^2} f_\alpha(t)\xi_\alpha,\ t \in [0,T]. \tag{5.6.10}$$

For a positive integer N, define

$$u_N(t,x) = \sum_{|\alpha|\le N} u_\alpha(t,x)\xi_\alpha. \tag{5.6.11}$$

Theorem 5.6.2 *Under Assumptions R1 and R2, there exists a positive number ν, depending only on the functions h and ρ, so that*

$$\widetilde{\mathbb{E}}\|u - u_N\|^2_{L_2(\mathbb{R}^d)}(t) \le \frac{\|u_0\|^2_{L_2(\mathbb{R}^d)}}{\nu(1+\nu)^N},\ t \in [0,T]. \tag{5.6.12}$$

If, in addition, $\rho = 0$, then there exists a real number C, depending only on the functions b and σ, so that

$$\widetilde{\mathbb{E}}\|u - u_N\|^2_{L_2(\mathbb{R}^d)}(t) \le \frac{(4h_\infty t)^{N+1}}{(N+1)!}e^{Ct}\|u_0\|^2_{L_2(\mathbb{R}^d)},\ t \in [0,T], \tag{5.6.13}$$

where $h_\infty = \max_{k=1,\dots,r}\sup_x |h_k(x)|$.

For positive integers N, n, define a set of multi-indices

$$\mathcal{J}_N^n = \{\alpha = (\alpha_{i,k},\ k = 1,\dots,r,\ i = 1,\dots,n) : |\alpha| \le N\}.$$

and let

$$u_N^n(t,x) = \sum_{\alpha\in\mathcal{J}_N^n} u_\alpha(t,x)\xi_\alpha. \tag{5.6.14}$$

Unlike Theorem 5.6.2, to compute the approximation error in this case we need to choose special basis functions $\mathfrak{m}_k$—to do the error analysis for the Fourier approximation in time. We also need extra regularity of the coefficients in the state and observation equations—to have the semi-group generated by the operator L^* continuous not only in $L_2(\mathbb{R}^d)$ but also in $H_2^2(\mathbb{R}^d)$. The resulting error bound is presented below; the proof can be found in [153].

Theorem 5.6.3 *Assume that*

1. *The basis $\mathfrak{m}$ is the Fourier cosine basis*

$$\mathfrak{m}_1(s) = \frac{1}{\sqrt{T}};\ \mathfrak{m}_k(t) = \sqrt{\frac{2}{T}}\cos\left(\frac{\pi(k-1)t}{T}\right),\ k>1;\ 0\le t\le T, \tag{5.6.15}$$

2. *The components of the functions σ are in $C^4(\mathbb{R}^d)$, the components of the functions b are in $C^3(\mathbb{R})$, the components of the function h are in $C^2(\mathbb{R}^d)$; $\rho = 0$; $u_0 \in H_2^2(\mathbb{R}^d)$.*

Then there exist a positive number B_1 and a real number B_2, both depending only on the functions b and σ so that

$$\widetilde{\mathbb{E}}\|u - u_N^n\|^2_{L_2(\mathbb{R}^d)}(T) \le B_1 e^{B_2 T}\left(\frac{(4h_\infty T)^{N+1}}{(N+1)!}e^{Ct}\|u_0\|^2_{L_2(\mathbb{R}^d)} + \frac{T^3}{n}\|u_0\|^2_{H_2^2(\mathbb{R}^d)}\right), \tag{5.6.16}$$

where $h_\infty = \max_{k=1,\dots,r} \sup_x |h_k(x)|$.

5.6.2 Passive Scalar in a Gaussian Field

This section presents the results from [152] and [150] about the stochastic transport equation.

The following viscous transport equation is used to describe time evolution of a scalar quantity θ in a given velocity field v:

$$\dot{\theta}(t,x) = \nu \boldsymbol{\Delta}\theta(t,x) - \mathbf{v}(t,x)\cdot\nabla\theta(t,x) + f(t,x);\ x \in \mathbb{R}^{\mathrm{d}},\ \mathrm{d} > 1. \tag{5.6.17}$$

The scalar θ is called passive because it does not affect the velocity field $\mathbf{v}$.

We assume that $\mathbf{v} = \mathbf{v}(t,x) \in \mathbb{R}^{\mathrm{d}}$ with components $v^1,\dots,v^{\mathrm{d}}$ is an isotropic Gaussian vector field with zero mean and covariance

$$\mathbb{E}(v^i(t,x)v^j(s,y)) = \delta(t-s)C^{ij}(x-y),$$

where $C = (C^{ij}(x), i,j = 1,\dots,\mathrm{d})$ is a matrix-valued function so that $C(0)$ is a scalar matrix; with no loss of generality we will assume that $C(0) = I$, the identity matrix.

It is known from [130, Sect. 10.1] that, for an isotropic Gaussian vector field, the Fourier transform $\hat{C} = \hat{C}(z)$ of the function $C = C(x)$ is

$$\hat{C}(y) = \frac{A_0}{(1+|y|^2)^{(\mathrm{d}+\gamma)/2}}\left(a\frac{yy^*}{|y|^2} + \frac{b}{\mathrm{d}-1}\left(I - \frac{yy^*}{|y|^2}\right)\right), \tag{5.6.18}$$

where y^* is the row vector $(y_1,\dots,y_d)$, y is the corresponding column vector, $|y|^2 = y^*y$; $\gamma > 0$, $a \ge 0$, $b \ge 0$, $A_0 > 0$ are real numbers, I is the identity matrix. Similar to [130], we assume that $0 < \gamma < 2$. This range of values of γ corresponds to a turbulent velocity field v, also known as the generalized Kraichnan model [60]; the original Kraichnan model [116] corresponds to $a = 0$. For small x, the asymptotics of $C^{ij}(x)$ is $(\delta_{ij} - c^{ij}|x|^\gamma)$ [130, Sect. 10.2].

By direct computation (cf. [8]), the vector field $\mathbf{v} = (v^1,\dots,v^{\mathrm{d}})$ can be written as

$$v^i(t,x) = \sum_k \sigma_k^i(x)\dot{w}_k(t), \tag{5.6.19}$$

where $\{\sigma_k,\ k \geq 1\}$ is an orthonormal basis in the space H_C, the reproducing kernel Hilbert space corresponding to the kernel function C. It is known from [130] that H_C is all or part of the Sobolev space $H^{(\mathrm{d}+\gamma)/2}(\mathbb{R}^{\mathrm{d}};\mathbb{R}^{\mathrm{d}})$.

If $a > 0$ and $b > 0$, then the matrix $\hat{C}$ is invertible and

$$H_C = \left\{ f \in \mathbb{R}^{\mathrm{d}} : \int_{\mathbb{R}^{\mathrm{d}}} \hat{f}^*(y)\hat{C}^{-1}(y)\hat{f}(y)dy < \infty \right\} = H^{(\mathrm{d}+\gamma)/2}(\mathbb{R}^{\mathrm{d}};\mathbb{R}^{\mathrm{d}}),$$

because $\|\hat{C}(y)\| \sim (1+|y|^2)^{-(\mathrm{d}+\gamma)/2}$.

If $a > 0$ and $b = 0$, then

$$H_C = \left\{ f \in \mathbb{R}^{\mathrm{d}} : \int_{\mathbb{R}^{\mathrm{d}}} |\hat{f}(y)|^2(1+|y|^2)^{(\mathrm{d}+\gamma)/2}dy < \infty;\ yy^*\hat{f}(y) = |y|^2\hat{f}(y) \right\},$$

the subset of gradient fields in $H^{(\mathrm{d}+\gamma)/2}(\mathbb{R}^{\mathrm{d}};\mathbb{R}^{\mathrm{d}})$, that is, vector fields f for which $\hat{f}(y) = y\hat{F}(y)$ for some scalar $F \in H^{(\mathrm{d}+\gamma+2)/2}(\mathbb{R}^{\mathrm{d}})$.

If $a = 0$ and $b > 0$, then

$$H_C = \left\{ f \in \mathbb{R}^{\mathrm{d}} : \int_{\mathbb{R}^{\mathrm{d}}} |\hat{f}(y)|^2(1+|y|^2)^{(\mathrm{d}+\gamma)/2}dy < \infty;\ y^*\hat{f}(y) = 0 \right\},$$

the subset of divergence-free fields in $H^{(\mathrm{d}+\gamma)/2}(\mathbb{R}^{\mathrm{d}};\mathbb{R}^{\mathrm{d}})$.

By the embedding theorems, each σ_k^i is a bounded continuous function on $\mathbb{R}^{\mathrm{d}}$; in fact, every σ_k^i is Hölder continuous of order $\gamma/2$. In addition, being an element of the corresponding space H_C, each σ_k is a gradient field if $b = 0$ and is divergence free if $a = 0$.

Equation (5.6.17) becomes

$$d\theta(t,x) = (\nu\boldsymbol{\Delta}\theta(t,x) + f(t,x))dt - \sum_k \sigma_k(x)\cdot\nabla\theta(t,x)dw_k(t). \tag{5.6.20}$$

We summarize the above constructions in the following assumptions:

S1 There is a fixed stochastic basis $\mathbb{F} = (\Omega,\mathcal{F},\{\mathcal{F}_t\}_{t\geq 0},\mathbb{P})$ with the usual assumptions and $(w_k(t), k \geq 1, t \geq 0)$ is a collection of independent standard Wiener processes on $\mathbb{F}$.

S2 For each k, the vector field σ_k is an element of the Sobolev space $H_2^{(\mathrm{d}+\gamma)/2}(\mathbb{R}^{\mathrm{d}};\mathbb{R}^{\mathrm{d}}), 0 < \gamma < 2, d \geq 2$.

S3 For all x, y in $\mathbb{R}^{\mathrm{d}}$, $\sum_k \sigma_k^i(x)\sigma_k^j(y) = C^{ij}(x-y)$ so that the matrix-valued function $C = C(x)$ satisfies (5.6.18) and $C(0) = I$.

S4 The input data θ_0, f are deterministic and satisfy

$$\theta_0 \in L_2(\mathbb{R}^{\mathrm{d}}),\ f \in L_2((0,T);H_2^{-1}(\mathbb{R}^{\mathrm{d}}));$$

$\nu > 0$ is a real number.

Theorem 5.6.4 *Let Q be the sequence with $q_k = q$ for all $k \geq 1$, and*

$$q < \sqrt{2\nu}.$$

*Under assumptions **S1–S4**, there exits a unique $w(H_2^1(\mathbb{R}^d), H_2^{-1}(\mathbb{R}^d))$ Wiener Chaos solution of (5.6.20). This solution is an $\mathcal{F}_t^W$-adapted process and satisfies*

$$\|\theta\|^2_{L_{2,Q}(\mathbb{W};L_2((0,T);H_2^1(\mathbb{R}^d)))} + \|\theta\|^2_{L_{2,Q}(\mathbb{W};C((0,T);L_2(\mathbb{R}^d)))} \leq C(\nu, q, T)\left(\|\theta_0\|^2_{L_2(\mathbb{R}^d)} + \|f\|^2_{L_2((0,T);H_2^{-1}(\mathbb{R}^d))}\right).$$

Theorem 5.6.4 provides useful information about the solution of equation (5.6.17) for all values of $\nu > 0$. Indeed, if $\sqrt{2\nu} > 1$, then $q > 1$ is an admissible choice of the weights, and, by Proposition 5.4.13(1), the solution θ has Malliavin derivatives of every order. If $\sqrt{2\nu} \leq 1$, then Eq. (5.6.20) does not have a square-integrable solution.

Note that if $q = \sqrt{2\nu}$, then Eq. (5.6.17) can still be analyzed using Theorem 5.5.7 in the normal triple $(H_2^1(\mathbb{R}^d), L_2(\mathbb{R}^d), H_2^{-1}(\mathbb{R}^d))$.

If $\nu = 0$, Eq. (5.6.20) must be interpreted in the sense of Stratonovich:

$$du(t,x) = f(t,x)dt - \sigma_k(x) \cdot \nabla\theta(t,x) \circ dw_k(t). \tag{5.6.21}$$

To simplify the presentation, we assume that $f = 0$. If (5.6.18) holds with $a = 0$, then each σ_k is divergence free and (5.6.21) has an equivalent Itô form

$$d\theta(t,x) = \frac{1}{2}\boldsymbol{\Delta}\theta(t,x)dt - \sum_{i,k}\sigma_k^i(x)D_i\theta(t,x)dw_k(t). \tag{5.6.22}$$

Equation (5.6.22) is a model of non-viscous turbulent transport [44]. The propagator for (5.6.22) is

$$\frac{\partial}{\partial t}\theta_{\boldsymbol{\alpha}}(t,x) = \frac{1}{2}\boldsymbol{\Delta}\theta_{\boldsymbol{\alpha}}(t,x) - \sum_{i,k}\sqrt{\alpha_{i,k}}\sigma_k^i D_j\theta_{\boldsymbol{\alpha}-\epsilon(i,k)}(t,x)\mathfrak{m}_i(t),\ 0 < t \leq T, \tag{5.6.23}$$

with initial condition $\theta_{\boldsymbol{\alpha}}(0,x) = \theta_0(x)I(|\boldsymbol{\alpha}| = 0)$.

The following result about solvability of (5.6.22) is proved in [152] and, in a slightly weaker form, in [150].

Theorem 5.6.5 *In addition to **S1–S4**, assume that each σ_k is divergence free. Then there exits a unique $w(H_2^1(\mathbb{R}^d), H_2^{-1}(\mathbb{R}^d))$ Wiener Chaos solution $\theta = \theta(t,x)$ of*

(5.6.22). This solution has the following properties:

(A) For every $\varphi \in \mathcal{C}_0^\infty(\mathbb{R}^d)$ *and all* $t \in [0,T]$, *the equality*

$$(\theta,\varphi)(t) = (\theta_0,\varphi)+\frac{1}{2}\int_0^t (\theta, \Delta\varphi)(s)ds+\sum_{i,k}\int_0^t (\theta, \sigma_k^i D_i\varphi)dw_k(s) \qquad (5.6.24)$$

holds in $L_2(\mathcal{F}_t^W)$, *where* $(\cdot,\cdot)$ *is the inner product in* $L_2(\mathbb{R}^d)$.

(B) If $X = X_{t,x}$ *is a weak solution of*

$$X_{t,x} = x + \int_0^t \sigma_k(X_{s,x})\,dw_k(s), \qquad (5.6.25)$$

then, for each $t \in [0,T]$,

$$\theta(t,x) = \mathbb{E}\left(\theta_0(X_{t,x})\,|\mathcal{F}_t^W\right). \qquad (5.6.26)$$

(C) For $1 \le p < \infty$ *and* $r \in \mathbb{R}$, *define* $L_{p,(r)}(\mathbb{R}^d)$ *as the Banach space of measurable functions with norm*

$$\|f\|^p_{L_{p,(r)}(\mathbb{R}^d)} = \int_{\mathbb{R}^d} |f(x)|^p(1+|x|^2)^{pr/2}dx$$

is finite. Then there exits a number K depending only on p, r so that, for each $t > 0$,

$$\mathbb{E}\|\theta\|^p_{L_{p,(r)}(\mathbb{R}^d)}(t) \le e^{Kt}\|\theta_0\|^p_{L_{p,(r)}(\mathbb{R}^d)}. \qquad (5.6.27)$$

In particular, if $r = 0$, *then* $K = 0$.

It follows that, for all s,t and almost all x,y,

$$\mathbb{E}\theta(t,x) = \theta_{(\mathbf{0})}$$

and

$$\mathbb{E}\theta(t,x)\,\theta(s,y) = \sum_{\alpha\in\mathcal{J}^2}\theta_\alpha(t,x)\,\theta_\alpha(s,y).$$

If the initial condition θ_0 belongs to $L_2(\mathbb{R}^d)\cap L_p(\mathbb{R}^d)$ for $p \ge 3$, then, by (5.6.27), higher order moments of θ exist. To obtain the expressions of the higher-order moments in terms of the coefficients θ_α, we need some auxiliary constructions.

For $\boldsymbol{\alpha},\ \boldsymbol{\beta} \in \mathcal{J}^2$, define $\boldsymbol{\alpha}+\boldsymbol{\beta}$ as the multi-index with components $\alpha_{i,k}+\beta_{i,k}$. Similarly, we define the multi-indices $|\boldsymbol{\alpha}-\boldsymbol{\beta}|$ and $\boldsymbol{\alpha}\wedge\boldsymbol{\beta} = \min(\boldsymbol{\alpha},\boldsymbol{\beta})$. We write

$\boldsymbol{\beta} \le \alpha$ if and only if $\beta_{i,k} \le \alpha_{i,k}$ for all $i, k \ge 1$. If $\boldsymbol{\beta} \le \boldsymbol{\alpha}$, we define

$$\binom{\boldsymbol{\alpha}}{\boldsymbol{\beta}} := \prod_{i,k} \frac{\alpha_{i,k}!}{\beta_{i,k}!(\alpha_{i,k} - \beta_{i,k})!}.$$

Definition 5.6.6 We say that a triple of multi-indices $(\boldsymbol{\alpha}, \boldsymbol{\beta}, \boldsymbol{\gamma})$ is complete and write $(\boldsymbol{\alpha}, \boldsymbol{\beta}, \boldsymbol{\gamma}) \in \triangle$ if all the entries of the multi-index $\boldsymbol{\alpha} + \boldsymbol{\beta} + \boldsymbol{\gamma}$ are even numbers and $|\boldsymbol{\alpha} - \boldsymbol{\beta}| \le \boldsymbol{\gamma} \le \boldsymbol{\alpha} + \boldsymbol{\beta}$. For fixed $\boldsymbol{\alpha}, \boldsymbol{\beta} \in \mathcal{J}^2$, we write

$$\triangle(\boldsymbol{\alpha}) := \left\{\boldsymbol{\gamma}, \boldsymbol{\mu} \in \mathcal{J}^2 : (\boldsymbol{\alpha}, \boldsymbol{\gamma}, \boldsymbol{\mu}) \in \triangle\right\}$$

and

$$\triangle(\boldsymbol{\alpha}, \boldsymbol{\beta}) := \{\boldsymbol{\gamma} \in \mathcal{J}^2 : (\boldsymbol{\alpha}, \boldsymbol{\beta}, \boldsymbol{\gamma})) \in \triangle\}.$$

For $(\boldsymbol{\alpha}, \boldsymbol{\beta}, \boldsymbol{\gamma}) \in \triangle$, we define

$$\Psi(\boldsymbol{\alpha}, \boldsymbol{\beta}, \boldsymbol{\gamma}) := \frac{\sqrt{\boldsymbol{\alpha}!\boldsymbol{\beta}!\boldsymbol{\gamma}!}}{\left(\frac{\boldsymbol{\alpha} - \boldsymbol{\beta} + \boldsymbol{\gamma}}{2}\right)! \left(\frac{\boldsymbol{\beta} - \boldsymbol{\alpha} + \boldsymbol{\gamma}}{2}\right)! \left(\frac{\boldsymbol{\alpha} + \boldsymbol{\beta} - \boldsymbol{\gamma}}{2}\right)!}. \tag{5.6.28}$$

Note that the triple $(\boldsymbol{\alpha}, \boldsymbol{\beta}, \boldsymbol{\gamma})$ is complete if and only if every permutation of the triple $(\boldsymbol{\alpha}, \boldsymbol{\beta}, \boldsymbol{\gamma})$ is complete. Similarly, the value of $\Psi(\boldsymbol{\alpha}, \boldsymbol{\beta}, \boldsymbol{\gamma})$ is invariant under permutation of the arguments.

We also define

$$C(\boldsymbol{\gamma}, \boldsymbol{\beta}, \boldsymbol{\mu}) := \left[\binom{\boldsymbol{\gamma} + \boldsymbol{\beta} - 2\boldsymbol{\mu}}{\boldsymbol{\gamma} - \boldsymbol{\mu}} \binom{\boldsymbol{\gamma}}{\boldsymbol{\mu}} \binom{\boldsymbol{\beta}}{\boldsymbol{\mu}}\right]^{1/2}, \quad \boldsymbol{\mu} \le \boldsymbol{\gamma} \wedge \boldsymbol{\beta}. \tag{5.6.29}$$

It is readily checked that if f is a function on $\mathcal{J}^2$, then, for $\boldsymbol{\gamma}, \boldsymbol{\beta} \in \mathcal{J}^2$,

$$\sum_{\boldsymbol{\mu} \le \boldsymbol{\gamma} \wedge \boldsymbol{\beta}} C(\boldsymbol{\gamma}, \boldsymbol{\beta}, p) f(\boldsymbol{\gamma} + \boldsymbol{\beta} - 2\boldsymbol{\mu}) = \sum_{\boldsymbol{\mu} \in (\boldsymbol{\gamma}, \boldsymbol{\beta})} f(\boldsymbol{\mu})\, \Phi(\boldsymbol{\gamma}, \boldsymbol{\beta}, \boldsymbol{\mu}) \tag{5.6.30}$$

The next theorem presents the formulas for the third and fourth moments of the solution of equation (5.6.22) in terms of the coefficients $\theta_{\boldsymbol{\alpha}}$.

Theorem 5.6.7 *In addition to* **S1–S4**, *assume that each σ_k is divergence free and the initial condition θ_0 belongs to $L_4(\mathbb{R}^d)$. Then*

$$\mathbb{E}\theta(t, x)\theta(t', x')\,\theta(s, y) = \sum_{(\boldsymbol{\alpha}, \boldsymbol{\beta}, \boldsymbol{\gamma}) \in \triangle} \Psi(\boldsymbol{\alpha}, \boldsymbol{\beta}, \boldsymbol{\gamma})\, \theta_{\boldsymbol{\alpha}}(t, x)\theta_{\boldsymbol{\beta}}(t', x')\theta_{\boldsymbol{\gamma}}(s, y) \tag{5.6.31}$$

and

$$\mathbb{E}\theta(t,x)\theta(t',x')\theta(s,y)\,\theta\left(s',y'\right) \tag{5.6.32}$$
$$= \sum_{\rho\in\Delta(\alpha,\beta)\cap\Delta(\gamma,\kappa)} \Psi(\alpha,\beta,\rho)\,\Psi(\rho,\gamma,\kappa)\,\theta_\alpha(t,x)\,\theta_\beta(t',x')\theta_\gamma(s,y)\,\theta_\kappa\left(s',y'\right).$$

Proof It is known [160] that

$$\xi_\gamma\xi_\beta = \sum_{\mu\le\gamma\wedge\beta} C(\gamma,\beta,\mu)\,\xi_{\gamma+\beta-2\mu}. \tag{5.6.33}$$

Let us consider the triple product $\xi_\alpha\xi_\beta\xi_\gamma$. By (5.6.33),

$$\mathbb{E}\xi_\alpha\xi_\beta\xi_\gamma = \mathbb{E}\sum_{\mu\in\Delta(\alpha,\beta)} \xi_\gamma\xi_\mu\Psi(\alpha,\beta,\mu) = \begin{cases}\Psi(\alpha,\beta,\gamma), & (\alpha,\beta,\gamma)\in\Delta;\\ 0, & \text{otherwise.}\end{cases} \tag{5.6.34}$$

Equality (5.6.31) now follows.

To compute the fourth moment, note that

$$\begin{aligned}\xi_\alpha\xi_\beta\xi_\gamma &= \sum_{\mu\le\alpha\wedge\beta} C(\alpha,\beta,\mu)\,\xi_{\alpha+\beta-2\mu}\xi_\gamma\\ &= \sum_{\mu\le\alpha\wedge\beta} C(\alpha,\beta,\mu) \sum_{\rho\le(\alpha+\beta-2\mu)\wedge\gamma}\\ &\quad C(\alpha+\beta-2\mu,\gamma,\rho)\,\xi_{\alpha+\beta+\gamma-2\mu-2\rho}.\end{aligned} \tag{5.6.35}$$

Repeated applications of (5.6.30) yield

$$\begin{aligned}\xi_\alpha\xi_\beta\xi_\gamma &= \sum_{\mu\le\alpha\wedge\beta} C(\alpha,\beta,\mu) \sum_{\rho\in\Delta(\alpha+\beta-2\mu,\gamma)} \xi_\rho\Psi(\alpha+\beta-2\mu,\gamma,\rho)\\ &= \sum_{\mu\in\Delta(\alpha,\beta)}\sum_{\rho\in\Delta(\mu,\gamma)} \Psi(\alpha,\beta,\mu)\,\Psi(\mu,\gamma,\rho)\,\xi_\rho\end{aligned}$$

Thus,

$$\begin{aligned}\mathbb{E}\xi_\alpha\xi_\beta\xi_\gamma\xi_\kappa &= \sum_{\mu\in\Delta(\alpha,\beta)}\sum_{\rho\in\Delta(\mu,\gamma)} \Psi(\alpha,\beta,\mu)\,\Psi(\mu,\gamma,\rho)\,\mathbf{1}_{\{\mu=\kappa\}}\\ &= \sum_{\rho\in\Delta(\alpha,\beta)\cap\Delta(\gamma,\kappa)} \Psi(\alpha,\beta,\rho)\,\Psi(\rho,\gamma,\kappa).\end{aligned}$$

Equality (5.6.32) now follows.

In the same way, one can get formulas for fifth- and higher-order moments.

Remark 5.6.8 Expressions (5.6.31) and (5.6.32) do not depend on the structure of Eq. (5.6.22) and can be used to compute the third and fourth moments of any random field with a known Cameron-Martin expansion. The interested reader should keep in mind that the formulas for the moments of orders higher then two should be interpreted with care. In fact, they represent the pseudo-moments (for detail see [164]).

We now return to the analysis of the passive scalar equation (5.6.20). By reducing the smoothness assumptions on σ_k, it is possible to consider velocity fields $\mathbf{v}$ that are more turbulent than in the Kraichnan model, for example,

$$v^i(t,x) = \sum_{k\geq 1} \sigma_k^i(x)\dot{w}_k(t), \tag{5.6.36}$$

where $\{\sigma_k,\ k \geq 1\}$ is an orthonormal basis in $L_2(\mathbb{R}^{\mathrm{d}};\mathbb{R}^{\mathrm{d}})$. With v as in (5.6.36), the passive scalar equation (5.6.20) becomes

$$\dot{\theta}(t,x) = \nu \boldsymbol{\Delta}\theta(t,x) + f(t,x) - \nabla\theta(t,x)\cdot\dot{W}(t,x), \tag{5.6.37}$$

where $\dot{W} = \dot{W}(t,x)$ is a d-dimensional space-time white noise and the Itô stochastic differential is used. Previously, such equations have been studied using white noise approach in the space of Hida distributions [37, 189]. A summary of the related results can be found in [78, Sect. 4.3].

The Q-weighted Wiener chaos spaces allow us to state a result that is fully analogous to Theorem 5.6.4. The proof is derived from Theorem 5.5.7; see [152] for details.

Theorem 5.6.9 *Suppose that $\nu > 0$ is a real number, each $|\sigma_k^i(x)|$ is a bounded measurable function, and the input data are deterministic and satisfy $u_0 \in L_2(\mathbb{R}^{\mathrm{d}})$, $f \in L_2\left((0,T); H_2^{-1}(\mathbb{R}^{\mathrm{d}})\right)$.*

Fix $\varepsilon > 0$ and let $Q = \{q_k,\ k \geq 1\}$ be a sequence such that

$$2\nu|y|^2 - \sum_{k\geq 1} q_k^2\sigma_k^i(x)\sigma_k^j(x)y_iy_j \geq \varepsilon|y|^2,\ x,y \in \mathbb{R}^{\mathrm{d}}.$$

Then, for every $T > 0$, there exits a unique $w(H_2^1(\mathbb{R}^{\mathrm{d}}), H_2^{-1}(\mathbb{R}^{\mathrm{d}}))$ Wiener Chaos solution θ of equation

$$d\theta(t,x) = (\nu\boldsymbol{\Delta}\theta(t,x) + f(t,x))dt - \sigma_k(x)\cdot\nabla\theta(t,x)dw_k(t), \tag{5.6.38}$$

The solution is an $\mathcal{F}_t$-adapted process and satisfies

$$\begin{aligned}&\|\theta\|^2_{L_{2,Q}(\mathbb{W};L_2((0,T);H_2^1(\mathbb{R}^{\mathrm{d}})))} + \|\theta\|^2_{L_{2,Q}(\mathbb{W};C((0,T);L_2(\mathbb{R}^{\mathrm{d}})))}\\ &\leq C(\nu,q,T)\left(\|\theta_0\|^2_{L_2(\mathbb{R}^{\mathrm{d}})} + \|f\|^2_{L_2((0,T);H_2^{-1}(\mathbb{R}^{\mathrm{d}}))}\right).\end{aligned}$$

If $\max_i \sup_x |\sigma_k^i(x)| \le C_k$, $k \ge 1$, then a possible choice of Q is

$$q_k = (\delta\nu)^{1/2}/(\mathrm{d}2^k C_k),\ 0 < \delta < 2.$$

If $\sigma_k^i(x)\sigma_k^j(x) \le C_\sigma < +\infty$, $i, j = 1, \ldots, \mathrm{d}$, $x \in \mathbb{R}^{\mathrm{d}}$, then a possible choice of Q is

$$q_k = q = \varepsilon\,(2\nu/(C_\sigma \mathrm{d}))^{1/2},\ 0 < \varepsilon < 1.$$

5.6.3 Stochastic Navier-Stokes Equations

In this section, we review the main facts about the stochastic Navier-Stokes equations and indicate how the Wiener Chaos approach can be used in the study of non-linear equations. Most of the results of this section come from the papers [164] and [165].

A priori, it is not clear in what sense the motion described by Kraichnan's velocity (see Sect. 5.6.2) might fit into the paradigm of Newtonian mechanics. Accordingly, relating the Kraichnan velocity field $\mathbf{v}$ to classic fluid mechanics naturally leads to the question whether we can compensate $\mathbf{v}(t,x)$ by a field $\mathbf{u}(t,x)$ that is more regular with respect to the time variable, so that there is a balance of momentum for the resulting field $\mathbf{U}(t,x) = \mathbf{u}(t,x) + \mathbf{v}(t,x)$ or, equivalently, that the motion of a fluid particle in the velocity field $\mathbf{U}(t,x)$ satisfies the Second Law of Newton.

A positive answer to this question is given in [164], where it is shown that the equation for the smooth component $\mathbf{u} = (u^1, \ldots, u^{\mathrm{d}})$ of the velocity is given by

$$\begin{cases} du^i = [\nu \boldsymbol{\Delta} u^i - u^j D_j u^i - D_i P + f_i]dt \\ + \left(g_k^i - D_i \tilde{P}_k - D_j \sigma_k^j u^i\right) dw_k,\ i = 1, \ldots, \mathrm{d},\ 0 < t \le T; \\ \operatorname{div} \mathbf{u} = 0,\ \mathbf{u}(0,x) = \mathbf{u}_0(x). \end{cases} \tag{5.6.39}$$

In (5.6.39), w_k, $k \ge 1$, are independent standard Wiener processes on a stochastic basis $\mathbb{F}$, the functions σ_k^j are given by (5.6.19), the known functions $f = (f^1, \ldots, f^{\mathrm{d}})$, $g_k = (g_k^i)$, $i = i, \ldots, \mathrm{d}$, $k \ge 1$, are, respectively, the drift and the diffusion components of the free force, and the unknown functions P, $\tilde{P}_k$ are the drift and diffusion components of the pressure.

Remark 5.6.10 It is useful to study Eq. (5.6.39) for more general coefficients σ_k^j. So, in the future, σ_k^j are not necessarily the same as in Sect. 5.6.2.

We make the following assumptions:

NS1 The functions $\sigma_k^i = \sigma_k^i(t,x)$ are deterministic and measurable,

$$\sum_{k\geq 1}\left(\sum_{i=1}^{\mathrm{d}} |\sigma_k^i(t,x)|^2 + |D_i\sigma_k^i(t,x)|^2\right) \leq K,$$

and there exists $\varepsilon > 0$ so that, for all $y \in \mathbb{R}^{\mathrm{d}}$,

$$\nu|y|^2 - \frac{1}{2}\sigma_k^i(t,x)\sigma_k^j(t,x)y_iy_j \geq \varepsilon|y|^2,$$

$t \in [0,T]$, $x \in \mathbb{R}^{\mathrm{d}}$.

NS2 The functions f^i, g_k^i are non-random and

$$\sum_{i=1}^{\mathrm{d}}\left(\|f^i\|^2_{L_2((0,T);H_2^{-1}(\mathbb{R}^{\mathrm{d}}))} + \sum_{k\geq 1}\|g_k^i\|^2_{L_2((0,T);L_2(\mathbb{R}^{\mathrm{d}}))}\right) < \infty.$$

Remark 5.6.11 In **NS1**, the derivatives $D_i\sigma_k^i$ are understood as Schwartz distributions, but it is assumed that $\mathrm{div}(\sigma) := \sum_{i=1}^{\mathrm{d}} D_i\sigma^i$ is a bounded ℓ_2-valued function. Obviously, the latter assumption holds in the important case when $\sum_{i=1}^{\mathrm{d}} D_i\sigma^i = 0$.

Our next step is to use the divergence-free property of u to eliminate the pressure P and $\tilde{P}$ from Eq. (5.6.39). For that, we need the decomposition of $L_2(\mathbb{R}^{\mathrm{d}};\mathbb{R}^{\mathrm{d}})$ into potential and solenoidal components.

Write $\mathfrak{S}(L_2(\mathbb{R}^{\mathrm{d}};\mathbb{R}^{\mathrm{d}})) = \{\mathbf{v} \in L_2(\mathbb{R}^{\mathrm{d}};\mathbb{R}^{\mathrm{d}}) : \mathrm{div}(\mathbf{v}) = 0\}$. It is known (see e.g. [107]) that

$$L_2(\mathbb{R}^{\mathrm{d}};\mathbb{R}^{\mathrm{d}}) = \mathfrak{G}(L_2(\mathbb{R}^{\mathrm{d}};\mathbb{R}^{\mathrm{d}})) \oplus \mathfrak{S}(L_2(\mathbb{R}^{\mathrm{d}};\mathbb{R}^{\mathrm{d}})),$$

where $\mathfrak{G}(L_2(\mathbb{R}^{\mathrm{d}};\mathbb{R}^{\mathrm{d}}))$ is the orthogonal complement of $\mathfrak{S}(L_2(\mathbb{R}^{\mathrm{d}};\mathbb{R}^{\mathrm{d}}))$.

The functions $\mathfrak{G}(\mathbf{v})$ and $\mathfrak{S}(\mathbf{v})$ can be defined for $\mathbf{v}$ from any Sobolev space $H_2^\gamma(\mathbb{R}^{\mathrm{d}};\mathbb{R}^{\mathrm{d}})$ and are usually referred to as the potential and the divergence free (or solenoidal), projections, respectively, of the vector field $\mathbf{v}$.

Now let $\mathbf{u}$ be a solution of equation (5.6.39). Since $\mathrm{div}(\mathbf{u}) = 0$, we have

$$D_i(\nu\boldsymbol{\Delta}u^i - u^jD_ju^i - D_iP + f^i) = 0;\ D_i(\sigma_k^jD_ju^ju^i + g_k^i - D_i\tilde{P}_k) = 0,\ k \geq 1.$$

As a result,

$$D_iP = \mathfrak{G}(\nu\boldsymbol{\Delta}u^i - u^jD_ju^i + f^i);\ D_i\tilde{P}_k = \mathfrak{G}(\sigma_k^jD_ju^i + g_k^i),\ i = 1,\ldots,\mathrm{d},\ k \geq 1.$$

So, instead of Eq. (5.6.39), we can and will consider its equivalent form for the unknown vector $\mathbf{u} = (u^1, \dots, u^{\mathrm{d}})$:

$$d\mathbf{u} = \mathfrak{S}(\nu \boldsymbol{\Delta} \mathbf{u} - u^j D_j \mathbf{u} + f)dt + \mathfrak{S}(\sigma_k^j D_j \mathbf{u} + g_k)dw_k,\ 0 < t \le T, \tag{5.6.40}$$

with initial condition $\mathbf{u}|_{t=0} = \mathbf{u}_0$.

Definition 5.6.12 An $\mathcal{F}_t$-adapted random process $\mathbf{u}$ from the space $L_2(\Omega \times [0,T]; H_2^1(\mathbb{R}^{\mathrm{d}}; \mathbb{R}^{\mathrm{d}}))$ is called a solution of equation (5.6.40) if

1. With probability one, the process $\mathbf{u}$ is weakly continuous in $L_2(\mathbb{R}^{\mathrm{d}}; \mathbb{R}^{\mathrm{d}})$.
2. For every $\varphi \in \mathcal{C}_0^{\infty}(\mathbb{R}^{\mathrm{d}}, \mathbb{R}^{\mathrm{d}})$, with div $\varphi = 0$ there exists a measurable set $\Omega' \subset \Omega$ so that, for all $t \in [0,T]$, the equality

$$\begin{aligned}(u^i, \varphi^i)(t) = (u_0^i, \varphi^i) + \int_0^t \big((\nu D_j u^i, D_j \varphi^i)(s) + \langle f^i, \varphi^i \rangle(s)\big)ds \\ \int_0^t (\sigma_k^j D_j u^i + g^i, \varphi^i) dw_k(s)\end{aligned} \tag{5.6.41}$$

holds on Ω'. In (5.6.41), $(\cdot,\cdot)$ is the inner product in $L_2(\mathbb{R}^{\mathrm{d}})$ and $\langle \cdot, \cdot, \rangle$ is the duality between $H_2^1(\mathbb{R}^{\mathrm{d}})$ and $H_2^{-1}(\mathbb{R}^{\mathrm{d}})$.

The following existence and uniqueness result is proved in [165].

Theorem 5.6.13 *In addition to NS1 and NS2, assume that the initial condition* $\mathbf{u}_0$ *is non-random and belongs to* $L_2(\mathbb{R}^{\mathrm{d}}; \mathbb{R}^{\mathrm{d}})$. *Then there exist a stochastic basis* $\mathbb{F} = (\Omega, \mathcal{F}, \{\mathcal{F}_t\}_{t\ge 0}, \mathbb{P})$ *with the usual assumptions, a collection* $\{w_k, k \ge 1\}$ *of independent standard Wiener processes on* $\mathbb{F}$, *and a process* $\mathbf{u}$ *so that* $\mathbf{u}$ *is a solution of (5.6.40) and*

$$\mathbb{E}\left(\sup_{s \le T} \|\mathbf{u}(s)\|^2_{L_2(\mathbb{R}^{\mathrm{d}};\mathbb{R}^{\mathrm{d}})} + \int_0^T \|\nabla \mathbf{u}(s)\|^2_{L_2(\mathbb{R}^{\mathrm{d}};\mathbb{R}^{\mathrm{d}})}\, ds\right) < \infty.$$

If, in addition, $\mathrm{d} = 2$, *then the solution of (5.6.40) exists on any prescribed stochastic basis, is strongly continuous in t, is* $\mathcal{F}_t^W$*-adapted, and is unique, both path-wise and in distribution.*

When $\mathrm{d} \ge 3$, existence of a strong solution as well as uniqueness (strong or weak) for Eq. (5.6.40) are important open problems.

By the Cameron-Martin theorem,

$$\mathbf{u}(t,x) = \sum_{\alpha \in \mathcal{J}^2} \mathbf{u}_\alpha(t,x) \xi_\alpha.$$

If the solution of (5.6.40) is $\mathcal{F}_t^W$-adapted, then, using the Itô formula together with relation (5.3.49) for the time evolution of $\mathbb{E}(\xi_\alpha | \mathcal{F}_t^W)$ and relation (5.6.33) for the product of two elements of the Cameron-Martin basis, we can derive the propagator system for coefficients $\mathbf{u}_\alpha$ [165, Theorem 3.2]:

Theorem 5.6.14 *In addition to* ***NS1*** *and* ***NS2****, assume that* $\mathbf{u}_0 \in L_2(\mathbb{R}^{\mathrm{d}};\mathbb{R}^{\mathrm{d}})$ *and Eq. (5.6.40) has an* $\mathcal{F}_t^W$*-adapted solution* $\mathbf{u}$ *so that*

$$\sup_{t\le T}\mathbb{E}\|\mathbf{u}\|^2_{L_2(\mathbb{R}^{\mathrm{d}};\mathbb{R}^{\mathrm{d}})}(t) < \infty. \tag{5.6.42}$$

Then

$$\mathbf{u}(t,x) = \sum_{\alpha\in\mathcal{J}^2}\mathbf{u}_\alpha(t,x)\,\xi_\alpha, \tag{5.6.43}$$

and the Fourier coefficients $\mathbf{u}_\alpha(t,x)$ *are* $L_2(\mathbb{R}^{\mathrm{d}};\mathbb{R}^{\mathrm{d}})$*-valued weakly continuous functions so that*

$$\sup_{t\le T}\sum_{\alpha\in\mathcal{J}^2}\|\mathbf{u}_\alpha\|^2_{L_2(\mathbb{R}^{\mathrm{d}};\mathbb{R}^{\mathrm{d}})}(t) + \int_0^T\sum_{\alpha\in\mathcal{J}^2}\|\nabla\mathbf{u}_\alpha\|^2_{L_2(\mathbb{R}^{\mathrm{d}};\mathbb{R}^{\mathrm{d}\times\mathrm{d}})}(t)\,dt < \infty. \tag{5.6.44}$$

The functions $u_\alpha(t,x)$, $\alpha\in\mathcal{J}^2$*, satisfy the (nonlinear) propagator*

$$\begin{aligned}\frac{\partial}{\partial t}\mathbf{u}_\alpha &= \mathfrak{S}\Big(\boldsymbol{\Delta}\mathbf{u}_\alpha - \sum_{\gamma,\beta\in\boldsymbol{\Delta}(\alpha)}\Psi(\alpha,\beta,\gamma)\left(\mathbf{u}_\gamma,\nabla\mathbf{u}_\beta\right) + \mathbf{1}_{|\alpha|=0}f \\ &\quad + \sum_{j,k}\sqrt{\alpha_{j,k}}\left(\left(\sigma^k,\nabla\right)\mathbf{u}_{\alpha-\epsilon(j,k)} + \mathbf{1}_{|\alpha|=1}g^k\right)\mathfrak{m}_j(t)\Big),\ 0<t\le T; \qquad (5.6.45)\\ \mathbf{u}_\alpha|_{t=0} &= \mathbf{u}_0\mathbf{1}_{|\alpha|=0};\end{aligned}$$

recall that the numbers $\Psi(\alpha,\beta,\gamma)$ *are defined in (5.6.28).*

One of the questions in the theory of the Navier-Stokes equations is computation of the mean value $\bar{\mathbf{u}} = \mathbb{E}\mathbf{u}$ of the solution. The traditional approach relies on the Reynolds equation for the mean

$$\partial_t\bar{\mathbf{u}} - \nu\boldsymbol{\Delta}\bar{\mathbf{u}} + \overline{(\mathbf{u},\nabla)\ \mathbf{u}} = 0, \tag{5.6.46}$$

which is not really an equation with respect to $\bar{\mathbf{u}}$. Decoupling (5.6.46) has been an area of active research: Reynolds approximations, coupled equations for the moments, Gaussian closures, and so on (see e.g. [169, 221] and the references therein).

Another way to compute $\bar{\mathbf{u}}(t,x)$ is to find the probability distribution of $\mathbf{u}(t,x)$ using the infinite-dimensional Kolmogorov equation associated with (5.6.40). The complexity of this Kolmogorov equation is prohibitive for any realistic application, at least for now.

The propagator provides a third way: expressing the mean and other statistical moments of $\mathbf{u}$ in terms of $\mathbf{u}_\alpha$. Indeed, by Cameron-Martin Theorem,

$$\mathbb{E}\mathbf{u}(t,x) = \mathbf{u}_{(0)}(t,x),$$
$$\mathbb{E}u^i(t,x)u^j(s,y) = \sum_{\alpha\in\mathcal{J}^2} u^i_\alpha(t,x)u^j_\alpha(s,y)$$

If exist, the third- and fourth-order moments can be computed using (5.6.31) and (5.6.32).

The next theorem, proved in [165], shows that the existence of a solution of the propagator (5.6.45) is not only necessary but, to some extent, sufficient for the global existence of a probabilistically strong solution of the stochastic Navier-Stokes equation (5.6.40).

Theorem 5.6.15 *Let NS1 and NS2 hold and* $\mathbf{u}_0 \in L_2(\mathbb{R}^d;\mathbb{R}^d)$. *Assume that the propagator (5.6.45) has a solution* $\{\mathbf{u}_\alpha(t,x),\ \alpha\in\mathcal{J}^2\}$ *on the interval* $(0,T]$ *so that, for every* α, *the process* $\mathbf{u}_\alpha$ *is weakly continuous in* $L_2(\mathbb{R}^d;\mathbb{R}^d)$ *and the inequality*

$$\sup_{t\le T}\sum_{\alpha\in\mathcal{J}^2}\|\mathbf{u}_\alpha\|^2_{L_2(\mathbb{R}^d;\mathbb{R}^d)}(t) + \int_0^T\sum_{\alpha\in\mathcal{J}^2}\|\nabla\mathbf{u}_\alpha\|^2_{L_2(\mathbb{R}^d;\mathbb{R}^{d\times d})}(t)\,dt < \infty \tag{5.6.47}$$

holds. If the process

$$\bar{\mathbf{U}}(t,x) := \sum_{\alpha\in\mathcal{J}^2}\mathbf{u}_\alpha(t,x)\,\xi_\alpha \tag{5.6.48}$$

is $\mathcal{F}^W_t$*-adapted, then it is a solution of (5.6.40).*

The process $\bar{\mathbf{U}}$ *satisfies*

$$\mathbb{E}\left(\sup_{s\le T}\|\bar{\mathbf{U}}(s)\|^2_{L_2(\mathbb{R}^d;\mathbb{R}^d)} + \int_0^T\|\nabla\bar{\mathbf{U}}(s)\|^2_{L_2(\mathbb{R}^d;\mathbb{R}^{d\times d})}\,ds\right) < \infty$$

and, for every $\mathbf{v}\in L_2(\mathbb{R}^d;\mathbb{R}^d)$, $t\mapsto\mathbb{E}\left(\bar{\mathbf{U}},\mathbf{v}\right)_{\mathbb{R}^d}(t)$ *is a continuous function of* t.

Since $\bar{\mathbf{U}}$ is constructed on a prescribed stochastic basis and over a prescribed time interval $[0,T]$, this solution of (5.6.40) is strong in the probabilistic sense and is global in time. Being true in any space dimension d, Theorem 5.6.15 suggests another possible way to study Eq. (5.6.40) when $\mathrm{d}\ge 3$. Unlike the propagator for the linear equation, the system (5.6.45) is not lower-triangular and not solvable by induction, so that analysis of (5.6.45) is an open problem.

5.6.4 First-Order Itô Equations

The objective of this section is to study equation

$$du(t,x) = u_x(t,x)dw(t),\ t > 0,\ x \in \mathbb{R}, \tag{5.6.49}$$

and its analog for $x \in \mathbb{R}^d$.

Equation (5.6.49) was first encountered in Example 5.4.8 on page 287; see also [62]. With a non-random initial condition $u(0,x) = \varphi(x)$, direct computations show that, if exists, the Fourier transform $\hat{u} = \hat{u}(t,y)$ of the solution must satisfy

$$d\hat{u}(t,y) = \sqrt{-1}y\hat{u}(t,y)dw(t),\ \text{or } \hat{u}(t,y) = \hat{\varphi}(y)e^{\sqrt{-1}yw(t)+\frac{1}{2}y^2t}. \tag{5.6.50}$$

The last equality shows that the properties of the solution essentially depend on the initial condition, and, in general, the solution is not in $L_2(W)$.

The S-transformed equation, $v_t = h(t)v_x$, has a unique solution

$$v(t,x) = \varphi\left(x + \int_0^t h(s)ds\right),\ h(t) = \sum_{i=1}^N h_i m_i(t).$$

The results of Sect. 5.4.2 imply that a white noise solution of the equation can exist only if φ is a real analytic function. On the other hand, if φ is infinitely differentiable, then, by Theorem 5.5.3, the Wiener Chaos solution exists and can be recovered from v.

Theorem 5.6.16 *Assume that the initial condition φ belongs to the Schwarz space $\mathcal{S} = \mathcal{S}(\mathbb{R})$ of tempered distributions. Then there exists a generalized random process $u = u(t,x)$, $t \geq 0$, $x \in \mathbb{R}$, so that, for every $\gamma \in \mathbb{R}$ and $T > 0$, the process u is the unique $w(H_2^\gamma(\mathbb{R}), H_2^{\gamma-1}(\mathbb{R}))$ Wiener Chaos solution of equation (5.6.49).*

Proof The propagator for (5.6.49) is

$$u_\alpha(t,x) = \varphi(x)\mathbf{1}_{|\alpha|=0} + \int_0^t \sum_i \sqrt{\alpha_i}(u_{\alpha-\epsilon(i)}(s,x))_x \mathfrak{m}_i(s)ds. \tag{5.6.51}$$

Even though Theorem 5.4.4 is not applicable, the system can be solved by induction if φ is sufficiently smooth. Denote by $C_\varphi(k)$, $k \geq 0$, the square of the $L_2(\mathbb{R})$ norm of the kth derivative of φ:

$$C_\varphi(k) = \int_{-\infty}^{+\infty} |\varphi^{(k)}(x)|^2 dx. \tag{5.6.52}$$

By Corollary 5.4.6, for every $k \geq 0$ and $n \geq 0$,

$$\sum_{|\alpha|=k} \|(u_\alpha^{(n)})_x\|^2_{L_2(\mathbb{R})}(t) = \frac{t^k C_\varphi(n+k)}{k!}. \tag{5.6.53}$$

The statement of the theorem now follows.

Remark 5.6.17 Once interpreted in a suitable sense, the Wiener Chaos solution of (5.6.49) is $\mathcal{F}_t^W$-adapted and does not depend on the choice of the Cameron-Martin basis in $L_2(W)$. Indeed, choose the wight sequence so that

$$r_\alpha^2 = \frac{1}{1 + C_\varphi(|\alpha|)}.$$

By (5.6.53), we have $u \in \mathcal{R}L_2(W; L_2(\mathbb{R}))$.

Next, define

$$\psi_N(x) = \frac{1}{\pi}\frac{\sin(Nx)}{x}.$$

Direct computations show that the Fourier transform of ψ_N is supported in $[-N, N]$ and $\int_{\mathbb{R}} \psi_N(x)dx = 1$. Consider Eq. (5.6.49) with initial condition

$$\varphi_N(x) = \int_{\mathbb{R}} \varphi(x-y)\psi_N(y)dy.$$

By (5.6.50), this equation has a unique solution u_N so that $u_N(t, \cdot) \in L_2(W; H_2^\gamma(\mathbb{R}))$, $t \geq 0$, $\gamma \in \mathbb{R}$. Relation (5.6.53) and the definition of u_N imply

$$\lim_{N\to\infty} \sum_{|\alpha|=k} \|u_\alpha - u_{N,\alpha}\|^2_{L_2(\mathbb{R})}(t) = 0,\ t \geq 0,\ k \geq 0,$$

so that, by the Lebesgue dominated convergence theorem,

$$\lim_{N\to\infty} \|u - u_N\|^2_{\mathcal{R}L_2(W;L_2(\mathbb{R}))}(t) = 0,\ t \geq 0.$$

In other words, the solution of the propagator (5.6.51) corresponding to any basis $\mathfrak{m}$ in $L_2((0,T))$ is a limit in $\mathcal{R}L_2(W; L_2(\mathbb{R}))$ of the sequence $\{u_N,\ N \geq 1\}$ of $\mathcal{F}_t^W$-adapted processes.

The properties of the Wiener Chaos solution of (5.6.49) depend on the growth rate of the numbers $C_\varphi(n)$. In particular,

- If $C_\varphi(n) \leq C^n (n!)^\gamma,\ C > 0,\ 0 \leq \gamma < 1$, then $u \in L_2\left(W; L_2((0,T); H_2^n(\mathbb{R}))\right)$ for all $T > 0$ and every $n \geq 0$.
- If $C_\varphi(n) \leq C^n n!,\ C > 0$, then

  - for every $n \geq 0$, there is a $T > 0$ so that $u \in L_2\big(W; L_2((0,T); H_2^n(\mathbb{R}))\big)$. In other words, the square-integrable solution exists only for sufficiently small T.
  - for every $n \geq 0$ and every $T > 0$, there exists a number $\delta \in (0,1)$ so that $u \in L_{2,Q}\big(W; L_2((0,T); H_2^n(\mathbb{R}))\big)$ with $Q = (\delta, \delta, \delta, \ldots)$.
- If the numbers $C_\varphi(n)$ grow as $C^n (n!)^{1+\rho}$, $\rho \geq 0$, then, for every $T > 0$, there exists a number $\gamma > 0$ so that $u \in (\mathcal{S})_{-\rho,-\gamma}\big(L_2(W); L_2((0,T); H_2^n(\mathbb{R}))\big)$. If $\rho > 0$, then this solution does not belong to any $L_{2,Q}\big(W; L_2((0,T); H_2^n(\mathbb{R}))\big)$. If $\rho > 1$, then this solution does not have an S-transform.
- If the numbers $C_\varphi(n)$ grow faster than $C^n(n!)^b$ for any $b, C > 0$, then the Wiener Chaos solution of (5.6.49) does not belong to any $(\mathcal{S})_{-\rho,-\gamma}\big(L_2((0,T); H_2^n(\mathbb{R}))\big)$, $\rho,\ \gamma > 0$, or $L_{2,Q}\big(W; L_2((0,T); H_2^n(\mathbb{R}))\big)$.

To construct a function φ with the required rate of growth of $C_\varphi(n)$, consider

$$\varphi(x) = \int_0^\infty \cos(xy) e^{-g(y)} dy,$$

where g is a suitable positive, unbounded, even function. Note that, up to a multiplicative constant, the Fourier transform of φ is $e^{-g(y)}$, and so $C_\varphi(n)$ grows with n as $\int_0^{+\infty} |y|^{2n} e^{-2g(y)} dy$.

A more general first-order equation can be considered:

$$du(t,x) = \sum_{i,k} \sigma_{ik}(t,x) D_i u(t,x) dw_k(t),\ t > 0,\ x \in \mathbb{R}^{\mathrm{d}}. \tag{5.6.54}$$

Theorem 5.6.18 *Assume that, in Eq. (5.6.54), the initial condition $u(0,x)$ belongs to $\mathcal{S}(\mathbb{R}^{\mathrm{d}})$ and each σ_{ik} is infinitely differentiable with respect to x so that $\sup_{(t,x)} |D^n \sigma_{ik}(t,x)| \leq C_{ik}(n)$, $n \geq 0$. Then there exists a generalized random process $u = u(t,x)$, $t \geq 0$, $x \in \mathbb{R}^{\mathrm{d}}$, so that, for every $\gamma \in \mathbb{R}$ and $T > 0$, the process u is the unique $w(H_2^\gamma(\mathbb{R}^{\mathrm{d}}), H_2^{\gamma-1}(\mathbb{R}^{\mathrm{d}}))$ Wiener Chaos solution of equation (5.6.49).*

Proof The arguments are identical to the proof of Theorem 5.6.16.

Note that the S-transformed equation (5.6.54) is $v_t = h_k \sigma_{ik} D_i v$ and has a unique solution if each σ_{ik} is a Lipschitz continuous function of x. Still, without additional smoothness, it is impossible to relate this solution to any generalized random process.

5.6.5 Problems

Problems 5.6.1 and 5.6.2 suggest further investigation of the transport equation $u_t = u_x dw(t)$ and its higher-dimensional version. Problem 5.6.3 extends the same ideas to parabolic equations. Problem 5.6.4 encourages the reader to learn more about the Skorokhod integral. Problem 5.6.5 presents an easy example of getting additional regularity for parabolic equations using chaos expansion.

Problem 5.6.1 A classical example of a infinitely differentiable non-analytic function is

$$\varphi(x) = \begin{cases} e^{-1/x^2}, & \text{if } x \neq 0, \\ 0, \text{ if } x = 0. \end{cases}$$

In this case, $\varphi^{(n)}(0) = 0$ for all n, so $\varphi(x) \neq \sum_{n\geq 0} \varphi^{(n)}(0)x^n/n!$, $x \neq 0$.

(a) What happens to the derivatives of φ near $x = 0$? [Apparently they must grow pretty fast, to prevent the remainder in the Taylor formula from going to zero].
(b) Investigate the chaos solution of $u_t = u_x\dot{w}(t)$ with initial conditions $u(0,x) = 1 - \varphi(x)$ and $u(0,x) = \varphi(x)$.

Problem 5.6.2

(a) Construct a closed-form solution for the equation

$$du = u_x dw(t), \ u(0,x) = x^n, \ n > 1.$$

(b) Construct a chaos solution for the equation

$$du = \sum_{i,k} \sigma_{i,k}(t,x) u_{x_k} dw_i(t), \ x \in \mathbb{R}^d,$$

when $u(0,x)$ is an element of $\mathcal{S}(\mathbb{R}^d)$, each $\sigma_{i,k}(t,x)$ is infinitely differentiable in x, and $\sup_{t,x} |D_x^n \sigma_{i,k}(t,x)| = C_{i,k}(n) < \infty$.

Problem 5.6.3 By analogy with equation $u_t = u_x\dot{w}$, construct the chaos solution of

$$\dot{u} = u_{xx} + u^{(n)}(x)\dot{w}(t), \ t > 0, \ x \in \mathbb{R},$$

given a smooth, but not necessarily analytic, initial condition $u(0,x) = \varphi(x)$.

Problem 5.6.4 Given real numbers $a \neq 0, \sigma \neq 0$ and a positive integer n, construct a closed-form solution of the Skorokhod integral equation

$$u(t,x) = x^n w(T) + a\int_0^t a u_{xx}(s,x)ds + \sigma \int_0^t u_x(s,x)\delta w(s), \ 0 < t < T, \ x \in \mathbb{R}.$$

What conditions do you need to impose on a and σ (i) to be able to find the solution; (ii) to make the computations easier?

Problem 5.6.5 Consider a normal triple (V, H, V') and the stochastic parabolic equation

$$\dot{u} = \mathcal{A}u + \mathcal{M}u\dot{w}$$

with a standard Brownian motion $w = w(t)$, bounded linear operators $\mathcal{A} : V \to V'$ and $\mathcal{M} : V \to H$, and a non-random initial condition $u_0 \in H$, and assume that the stochastic parabolicity condition is satisfied:

$$2[\mathcal{A}v, v] + \|\mathcal{M}v\|_H^2 + \delta\|v\|_V^2 \le C_0\|v\|_H^2.$$

Denote by $\boldsymbol{e}$ the sequence with $e_k = 1, k \ge 1$.

Show that $u \in (\mathcal{S})_{0,q\boldsymbol{e}}\Big(W; L_2\big((0,T); V\big)\Big)$ for some $q > 1$.

[For example, you can take $q = 1 + \varepsilon$, if $\varepsilon\|\mathrm{M}v\|_H^2 < \delta\|v\|_V^2$].

5.7 Distribution Free Stochastic Analysis

5.7.1 Distribution Free Polynomial Chaos

So far in this book we have studied linear stochastic PDEs driven by a sequence of independent standard Gaussian random variables

$$\varXi := \{\xi_k,\ 1 \le k \le N\}. \tag{5.7.1}$$

The σ-algebra generated by $\varXi$ and completed by the sets of probability zero will be denoted $\sigma(\varXi)$.

In this section we will introduce and investigate an extension of this setting to an arbitrary sequence

$$\varPhi := \{\phi_k,\ 1 \le k \le M\} \tag{5.7.2}$$

of square integrable and uncorrelated random variables on the probability space $(\varOmega, \mathcal{F}, \mathbb{P})$.

Similarly to $\sigma(\varXi)$, the σ-sigma algebra generated by $\varPhi$ will be denoted by $\sigma(\varPhi)$. As usual, $\sigma(\varPhi)$ is assumed to be completed by the sets of probability zero.

We will assume that the upper bounds N in (5.7.1) and M in (5.7.2) can be either finite or infinite.

The most important feature of the sequence (5.7.2) is that the distribution functions of the random variables ϕ_k are not specified beyond some general assumptions. This property of system $\varPhi$ is often called distribution free (and abbreviated as DF).

Definition 5.7.1 Let $\sigma(\Phi)$ be the σ-algebra generated by $\{\phi_k,\ k \geq 1\}$ and completed by sets of probability zero.

Suppose that $g(\Phi) = g(\phi_1, \phi_2, \ldots)$ is a $\sigma(\Phi)$-measurable function such that $\mathbb{E}\,|g(\Phi)|^2 < \infty$. Clearly, the information regarding $g(\Phi)$ appears to be very limited. In construct, in the Gaussian setting $(\Phi = \Xi)$ one can construct the Cameron-Martin expansion for a random function $f(\Xi)$ with finite variance. For example, it follows from the Cameron-Martin Theorem (see (5.1.12)) that

$$f(\Xi) = \sum_{\alpha \in \mathcal{J}^1} f_\alpha \xi_\alpha, \tag{5.7.3}$$

where

$$\xi_\alpha = \prod_{k \geq 1} \mathrm{H}_{\alpha_k}(\xi_k)/\sqrt{\alpha!} \text{ and } f_\alpha = \mathbb{E}(f\xi_\alpha).$$

As we have already seen in this book, the application of Cameron-Martin Theorem to SPDEs driven by Gaussian noise/forcing (e.g. SPDE (5.2.10)) leads to construction of low-triangular system of deterministic equations. In particular, a solution of the linear parabolic SPDE

$$du(t,x) = (au_{xx}(t,x) + f(t,x))dt + (\sigma u_x(t,x) + g(t,x))dw(t),\ t > 0,\ x \in \mathbb{R} \tag{5.7.4}$$

is given by the Cameron-Martin formula

$$u(t,x) = \sum_{\alpha \in \mathcal{J}^1} u_\alpha(t,x)\,\xi_\alpha, \tag{5.7.5}$$

where

$$du_\alpha(t,x) = a\Big(u_\alpha(t,x)\Big)_{xx} dt + \sum_{i \geq 1} \sqrt{\alpha_i}\sigma\Big(u_{\alpha-\epsilon(i)}(t,x)\Big)_x \mathfrak{m}_i(t)dt, \tag{5.7.6}$$

and $\{\mathfrak{m}_i(t), i \geq 1\}$ is an orthonormal basis in $L_2[0,T]$. Obviously, (5.7.6) is a lower triangular system of deterministic PDEs (often referred to as the propagator).

An important question is whether one can generalize the Cameron-Martin formula and the related lower-triangular propagator to distribution free ($\sigma(\Phi)$-measurable) linear SPDEs.

The answer is positive and will be established below. To demonstrate this, we shall introduce some fundamental notions and concepts of stochastic calculus in the distribution free setting. It is convenient to start with a σ-finite complete measure space $(U, \mathcal{U}, \mu)$ and the related Hilbert space $\mathbf{H} = L_2(U, \mathcal{U}, \mu)$. In addition we shall introduce a complete probability space $(\Omega, \mathcal{F}, \mathbb{P})$.

Now we shall introduce a general definition of driving cylindrical noise.

Definition 5.7.2 A continuous linear functional Φ from $\mathbf{H}$ to $L_2(\Omega, \mathcal{F}, \mathbb{P})$ such that

$$\mathbb{E}[\Phi(f)^2] = \|f\|_{\mathbf{H}}^2 \text{ for all } f \in \mathbf{H} \tag{5.7.7}$$

is called a driving cylindrical random field on U.

Without loss of generality we assume that

$$\mathbb{E}\Phi(f) = 0 \text{ for all } f \in \mathbf{H}. \tag{5.7.8}$$

Example 5.7.3 Let $f \in L_2(0,T)$, $w(t)$ is a standard Brownian motion and $\Phi(f) = \int_0^T f(s)\,dw(s)$. Then $\mathbb{E}\Phi(f) = 0$ and

$$\mathbb{E}[\Phi(f)^2] = \int_0^T |f(s)|^2\,ds < \infty$$

for all $f \in L_2(0,T)$.

Exercise 5.7.4 (B) Prove that Φ is an isometric embedding of $\mathbf{H}$ into $L_2(\Omega, \mathcal{F}, \mathbb{P})$.

Let us fix an integer N_0 such that $2 \le N_0$ and (possibly) $N_0 \to \infty$. If $f \in \mathbf{H}$ and $\{\mathfrak{m}_k, k < N_0\}$ is a complete orthonormal system in $\mathbf{H}$ then

$$\Phi(f) = \sum_{k \ge 1} f_k \phi_k = \sum_{1 \le k < N_0} f_k \phi_k \text{ in } L_2(\Omega, \mathcal{F}, \mathbb{P}), \tag{5.7.9}$$

where $\phi_k = \Phi(\mathfrak{m}_k)$ and $f_k = \int_U f \mathfrak{m}_k d\mu$ (note that $f = \sum_k f_k \mathfrak{m}_k$ in $\mathbf{H}$).

Exercise 5.7.5 (C) Prove that

$$\mathbf{E}[\Phi(f)\,\Phi(g)] = \int_U f(u)\,g(u)\,d\mu(u),\ f, g \in \mathbf{H}. \tag{5.7.10}$$

In the distribution free setting, it is natural to construct the white noise $\dot{\Phi}$ as a formal series:

$$\dot{\Phi}(u) = \sum_k \mathfrak{m}_k(u)\,\phi_k = \sum_{1 \le k < N_0} \mathfrak{m}_k(u)\,\phi_k,\ \phi_k \in \Phi, u \in U \tag{5.7.11}$$

where ϕ_k are uncorrelated random variables, $E\phi_k = 0$, $E|\phi_k|^2 = 1$ and $\{\mathfrak{m}_k, k \ge 1\}$ is an orthogonal basis in some Hilbert space $\mathbf{H}$.

Exercise 5.7.6 (C) Let $f \in \mathbf{H}$. Show that

$$\int_U \dot{\Phi}(u) f(u)\,d\mu(u) = \sum_k \phi_k f_k \tag{5.7.12}$$

The next example motivates a construction of stochastic integral with respect to the process $\Phi_t = \Phi\left(\mathbf{1}_{[0,t]}\right)$.

Example 5.7.7 Let ϕ_k be a sequence independent random variables with zero mean and unit variance and such that, for each ϕ_k, the moment generating function exists in a neighborhood of zero. Let $(U, \mathcal{U}, \mu)$ be a σ-finite measure space. Let $\{\mathfrak{m}_k, k < N_0\}$ be a CONS in $H = L_2(U, \mathcal{U}, \mu)$. We define $\Phi : L_2(U, \mathcal{U}, \mu) \to L_2(\Omega, \mathbb{P})$ by

$$\Phi(f) = \sum_{k<N_0} f_k \xi_k, \; f = \sum_{k<N_0} f_k \mathfrak{m}_k \in \mathbf{H}. \tag{5.7.13}$$

The Φ-white noise is

$$\dot{\Phi}(u) = \sum_k \mathfrak{m}_k(u) \phi_k.$$

Now let $U = L_2[0,T]$ and $\{\mathfrak{m}_k, \; k \geq 1\}$ is a CONS on $L_2([0,T])$. Then we can regard Φ in (5.7.13) as a stochastic process:

$$\Phi_t = \Phi\left(\mathbf{1}_{[0,t]}\right) = \sum_k \phi_k \int_0^t \mathfrak{m}_k(s)\, ds, \; 0 \leq t \leq T. \tag{5.7.14}$$

This process has uncorrelated increments, $\mathbb{E}\left(\Phi_t^2\right) = t, \mathbb{E}\left(\Phi_t \Phi s\right) = t \wedge s, \mathbb{E}\left(\Phi_t\right) = 0, t \geq 0$. For a continuous deterministic function $f = f(t)$ on $[0,T]$, define

$$\Phi(f) = \lim_n \sum_{i=1}^n f(t_i)\left[\Phi_{t_{i+1}} - \Phi_{t_i}\right] \text{ in } L_2(\Omega, \mathbb{P}), \tag{5.7.15}$$

where $t_i = t_i^n, \; 0 \leq i \leq n$, is a partition of $[0,T]$ into n disjoint subintervals whose maximal size converges to zero as $n \to \infty$.

Exercise 5.7.8 (B) Confirm that (5.7.15) and (5.7.13) define the same random variable $\Phi(f)$.

Our next task will be to construct a complete orthogonal system $\{\phi_{\boldsymbol{\alpha}}, \; \boldsymbol{\alpha} \in \mathcal{J}^1\}$ in $L_2(\Omega, \Phi, \mathbb{P})$, where $\mathcal{J}^1$ is the set of multiindices $\boldsymbol{\alpha} = (\alpha_1, \alpha_2, \ldots)$ and $|\boldsymbol{\alpha}| = \sum_k \alpha_k$. The Cameron-Martin basis $\{\xi_{\boldsymbol{\alpha}}, \boldsymbol{\alpha} \in \mathcal{J}^1\}$ for Gaussian random fields is a well studied particular case.

We make the following assumption about the random field Φ.

B1. Let $\phi_k = \Phi(\mathfrak{m}_k), k \geq 1$. For each random vector $(\phi_{i_1}, \ldots, \phi_{i_n}), n \geq 1$, the moment generating function

$$M_{i_1 \ldots i_n}(t) = M_{i_1 \ldots i_n}(t_1, \ldots, t_n) = \mathbf{E} \exp\{t_1 \phi_{i_1} + \ldots t_n \phi_{i_n}\}$$

exists for all $t = (t_1, \ldots, t_n)$ in some neighborhood of $0 \in \mathbb{R}^n$.

For $\boldsymbol{\alpha} = (\alpha_k) \in \mathcal{J}^1$, we denote

$$\phi^{\alpha} = \prod_k \phi_k^{\alpha_k}, \ \phi^{(0)} = 1, \tag{5.7.16}$$

where $\phi_k = \Phi(\mathfrak{m}_k), k \geq 1$.

Exponential integrability assumption **B1** implies that ϕ_k have all the moments (see [226, Theorem 5a on page 57]). Writing $\mathcal{F}^0 = \sigma(\phi_k, k \geq 1)$, we also conclude that every $f \in L_2(\Omega, \mathcal{F}^0, \mathbb{P})$ can be approximated in $L_2(\Omega, \mathcal{F}^0, \mathbb{P})$ by a sequence of polynomials in $\{\phi^{\alpha}, \alpha \in \mathcal{J}^1\}$.

In addition, we will need the following assumption:

B2. There exists an orthogonalization $\{\tilde{K}_{\alpha}, \alpha \in \mathcal{J}^1\}$ of the system $O = \{\phi^{\alpha}, \boldsymbol{\alpha} \in \mathcal{J}^1\}$ such that for each n, $\{\tilde{K}_{\alpha}, |\boldsymbol{\alpha}| \leq n\}$ spans the same linear subspace H_n as $\{\phi^{\alpha}, |\boldsymbol{\alpha}| \leq n\}$ and, for $|\boldsymbol{\alpha}| = n + 1$,

$$\tilde{K}_{\alpha} = \phi^{\alpha} - \text{projection}_{H_n} \phi^{\alpha}.$$

Assume that **B1** and **B2** hold. Set $\phi_{\alpha} = c_{\alpha} \tilde{K}_{\alpha}$ and choose the normalization c_{α} so that $\mathbb{E}\phi_{\alpha}^2 = \boldsymbol{\alpha}!$ if $\tilde{K}_{\alpha}$ is not identically zero, and $\phi_{\alpha} = 0$ if $\tilde{K}_{\alpha} \equiv 0$. Define the set of multi-indices $\bar{\mathcal{J}} = \{\boldsymbol{\alpha} \in \mathcal{J}^1 : \phi_{\alpha} \neq 0\}$.

Exercise 5.7.9 (A) Confirm that $\mathcal{F}^0 = \sigma(\Phi(f) : f \in \mathbf{H})$.

Remark 5.7.10

(a) If every $\phi_k = \Phi(\mathfrak{m}_k)$ is bounded, then **B1** obviously holds.
(b) The Hilbert space $\mathbf{H} = L_2(U, \mathcal{U}, \mu)$ can be finite dimensional. In particular, if $U = \{1\}$ and μ is the Dirac measure δ_1, then $\mathbf{H}$ is one-dimensional and $m_1 = 1$. In this case, the Φ-noise is the random variable ξ.

Proposition 5.7.11 *Assume that* ***B1, B2*** *hold. Then* $\{\phi_{\alpha}, \boldsymbol{\alpha} \in \mathcal{J}^1\}$ *is a complete orthogonal system of* $L_2(\Omega, \mathcal{F}^0, \mathbb{P})$*: for each* $\eta \in L_2(\Omega, \mathcal{F}^0, \mathbb{P})$,

$$\eta = \sum_{\alpha} \frac{\eta_{\alpha}}{\boldsymbol{\alpha}!} \phi_{\alpha}, \ \sum_{\alpha} \eta_{\alpha}^2 = \mathbb{E}\eta^2,$$

where

$$\eta_{\alpha} = \mathbb{E}[\eta_{\alpha} \phi_{\alpha}]$$

Proof Orthogonality is by construction; completeness follows from the exponential integrability assumption **B1**.

Example 5.7.12 For a single random variable ξ with $\mathbb{E}\xi = 0$, condition **B1** holds if the moment generating function of ξ exists in a neighborhood of the origin, and then **B2** holds as well. With $\boldsymbol{\alpha} = (n, 0, \ldots) := (n)$, we conclude that $\phi_{(n)}$ is equal, up to a multiple, to the corresponding orthogonal polynomial generated by the distribution of ξ [the result of applying the Gramm-Schmidt procedure to the collection $\{1, x, x^2, \ldots\}$ under the inner product $(f,g) = \mathbb{E}f(\xi)g(\xi)$]. In particular, $\phi_{(0)} = 1$ and $\phi_{(1)} = c_1\xi$, and , if ξ is standard normal, then $\phi_{(n)}$ are Hermite polynomials (2.3.37).

Exercise 5.7.13 (B) Prove that if ξ is discrete with m distinct values, then $\phi_{(n)} = 0$ for $n > m$.

Example 5.7.14 Let $\xi = \eta - Np$, where η is a binomial random variable with parameters N, p. Then $\xi = \sum_{k=1}^{N} \bar{\eta}_k$, where η_k are iid Bernoulli with $\mathbb{P}(\eta_k = 1) = p$, and $\bar{\eta}_k = \eta_k - p$. Define the function

$$p(z) = \prod_{k=1}^{N} (1 + z\bar{\eta}_k), \; z \in \mathbb{R},$$

and then define the random variables ξ_n by the equality

$$p(z) = \sum_{k=0}^{N} \xi_k z^k.$$

In particular,

$$\xi_0 = 1, \; \xi_1 = \xi, 2\xi_2 = \xi^2 - (1-2p)\xi - Np(1-p).$$

By direct computation,

$$\mathbb{E}\xi_k\xi_\ell = 0, \; k \neq \ell,$$

and therefore

$$\phi_{(n)} = \frac{n!}{\mathbb{E}\xi_n^2}\,\xi_n\,\mathbf{1}_{n\leq N}.$$

Exercise 5.7.15 (A) By noticing that $\mathbb{E}\big(p(z)p(w)\big)$ is a function of the product zw only, confirm that $\mathbb{E}\xi_k\xi_\ell = 0, \; k \neq \ell$, and compute $\mathbb{E}\xi_k^2$.

Example 5.7.16 Let ϕ_k be iid with distribution $\mathbb{P}(\phi_k = 1) = P(\phi_k = -1) = 1/2$. In this case conditions **B1** and **B2** trivially hold. The process Φ_t in (5.7.14) is not Gaussian: for each $n > 1, \lambda \in \mathbb{R}$,

$$\mathbb{E}\exp\left\{\mathrm{i}\lambda\sum_{k=1}^{n}\int_0^t \mathfrak{m}_k(s)ds\phi_k\right\} = \prod_{k=1}^{n}\cos\left\{\lambda\int_0^t \mathfrak{m}_k(s)ds\right\},$$

where i is imaginary unit $(\mathrm{i}^2 = -1)$.

Exercise 5.7.17 (B) Using $\phi_k^2 = 1$ a.s., confirm that

$$\mathbb{E}\,(\Phi_t - \Phi_s)^4 \le C|t-s|^2,\ s,t \in [0,T], \tag{5.7.17}$$

and so Φ_t has a continuous in t modification.

Example 5.7.18 Let ϕ_k be iid uniform on $[-\sqrt{3}, \sqrt{3}]$ (to ensure $\mathbb{E}\phi_k = 0, \mathbb{E}\phi_k^2 = 1$. Once again, conditions **B1** and **B2** trivially hold, and Φ_t is a non-Gaussian with continuous paths. Also,

$$\mathbb{E}\exp\left\{ \mathrm{i}\lambda \sum_{k=1}^{n} \int_0^t \mathfrak{m}_k(s)ds\phi_k \right\} = \prod_{k=1}^{n} \frac{\sin\left(\lambda \int_0^t \mathfrak{m}_k(s)ds\right)}{\left(u \int_0^t \mathfrak{m}_k(s)ds\right)}.$$

The orthogonal basis $\{\phi_{\alpha},\ \alpha \in \mathcal{J}^1\}$ is constructed in the following exercise by re-scaling the Legendre polynomials

$$L_n(x) = \frac{(-1)^n}{2^n n!}\frac{d^n}{dx^n}(1-x^2)^n. \tag{5.7.18}$$

Exercise 5.7.19 (A)

(a) Verify that

$$\frac{1}{2\sqrt{3}}\int_{-\sqrt{3}}^{\sqrt{3}} L_n\left(\frac{x}{\sqrt{3}}\right)L_m\left(\frac{x}{\sqrt{3}}\right)dx = \frac{1}{2}\int_{-1}^{1} L_n(x)L_m(x)dx = \frac{1}{(2n+1)}\mathbf{1}_{n=m}.$$

(b) Define

$$B_0(x) = 1,\ B_n(x) = \sqrt{n!(2n+1)}\,L_n\left(x/\sqrt{3}\right).$$

Confirm that

$$\phi_{\alpha} = \prod_k B_{\alpha_k}(\xi_k).$$

Example 5.7.20 (Poisson Random Field) Let N be a Poisson random measure on U with μ as its Lévy measure and $\tilde{N} = N - \mu$. Using the notations

$$N(f) = \int_U f(u)\,N(du),\ \mu(f) = \int_U \phi(u)\mu(du),$$

we get an isometry

$$\mathbb{E}\left[\tilde{N}(f)^2\right] = \int_U f^2 d\mu,\ f \in L_2(U,\mu).$$

Let $\mathfrak{m}_k,\ k < N_0$, be a CONS in $L_2(U,\mu)$ such that, for every k,

$$\mathfrak{m}_k \in L_\infty(U,\mu)\bigcap L_1(U,\mu);$$

note that μ is not necessarily a probability measure, and it is possible to have $\mu(U) = +\infty$. In general, the random variables $\phi_k = \tilde{N}(\mathfrak{m}_k)$ are only uncorrelated, not independent.

Let $\mathcal{Z}$ be the set of all real-valued sequences $z = (z_k)$ such that only the finite number of z_k is not zero. For $z \in \mathcal{Z}$ set, $m = m_z = \sum_k z_k\mathfrak{m}_k$. Under the assumptions above, the moment generating function

$$\begin{aligned} M(z) &= \mathbb{E}\exp\{\tilde{N}(m_z)\} \\ &= \exp\left\{\int_U \left[e^{m_z(u)} - 1 - m_z(u)\right]\mu(du)\right\} \end{aligned}$$

exists and Assumption **B1** is satisfied.

For small z, let

$$p(z) = \exp\left\{\int_U \ln[1 + m_z(\upsilon)]N(d\upsilon) - \int_U m_z(\upsilon)\mu(d\upsilon)\right\}.$$

For $\boldsymbol{\alpha} \in \mathcal{J}^1$, we define the random variables $C_{\boldsymbol{\alpha}}$ as the coefficients of the Taylor expansion of $p(z)$ at zero:

$$p(z) = \sum_{\boldsymbol{\alpha}\in\mathcal{J}^1} \frac{C_{\boldsymbol{\alpha}}}{\boldsymbol{\alpha}!} z^{\boldsymbol{\alpha}}.$$

In particular,

$$C_{(\mathbf{0})} = 1,\ C_{\boldsymbol{\epsilon}(k)} = \tilde{N}(\mathfrak{m}_k),$$

and

$$\begin{aligned} C_{\boldsymbol{\epsilon}(k_1)+\boldsymbol{\epsilon}(k_2)} &= N(\mathfrak{m}_{k_1})N(\mathfrak{m}_{k_2}) - N(\mathfrak{m}_{k_1}\mathfrak{m}_{k_2}) - N(\mathfrak{m}_{k_1})\mu(\mathfrak{m}_{k_2}) \\ &\quad - N(\mathfrak{m}_{k_2})\mu(\mathfrak{m}_{k_1}) + \mu(\mathfrak{m}_{k_1})\mu(\mathfrak{m}_{k_2}). \end{aligned}$$

Using

$$\mathbb{E}p(z)\,p(z') = \exp\left\{\int m_z(u)\,m_{z'}(u)\,\mu(du)\right\},\ z, z' \in \mathcal{Z},$$

it is possible to show that

$$\mathbb{E}C_{\boldsymbol{\alpha}}C_{\boldsymbol{\beta}} = \boldsymbol{\alpha}!\mathbf{1}_{\boldsymbol{\alpha}=\boldsymbol{\beta}},$$

although, unlike the Gaussian case and Hermite polynomials, C_α is not a product of simple polynomials of independent Poisson random variables.

The driving Poisson noise on U is $\Phi(f) = \tilde{N}(f), f \in L_2(U, d\mu)$; it satisfies **B1**, **B2**, and $\phi_\alpha = C_\alpha$, $\alpha \in \mathcal{J}^1$.

Exercise 5.7.21 (A+) Let N be a Poisson process on $[0, T]$, so that μ is the Lebesgue measure on $[0, T]$ and $U = L_2[0, T]$. Let $\mathfrak{m}_k$ be the Haar basis in U, that is, each $\mathfrak{m}_k$ is piece-wise constant. Show that in this case each C_α is indeed a product of Poisson random variables. Are those random variables independent? [Yes; in fact, almost everything follows from the following three facts: (a) $N(\mathbf{1}_{(a,b)})$ is a Poisson random variable; (b) $N(\mathbf{1}_{(a,b)})$ and $N(\mathbf{1}_{(c,d)})$ are independent for non-overlapping interval; (c) The sum of independent Poisons is again Poisson.]

5.7.2 Distribution Free Malliavin Calculus

Let Y be a separable Hilbert space and let $D(Y)$ be the space of all random variables $v = \sum_\alpha v_\alpha \phi_\alpha$, where $v_\alpha \in Y$, only finite number of v_α are not zero, and $v_\alpha = 0$ if $\phi_\alpha = 0$ i.e. if $\alpha \notin \bar{\mathcal{J}}$.

Definition 5.7.22 A generalized D-random variable u with values in Y is a formal series $u = \sum_\alpha u_\alpha \phi_\alpha$, where $u_\alpha \in Y$ and $u_\alpha = 0$ if $\phi_\alpha = 0$.

Denote the vector space of all generalized D-random variables with values in Y by $D'(Y)$. If $Y = \mathbb{R}$ we write simply D and D'. The elements of D are the test random variables for $D'(Y)$. We define the action of a generalized random variable $u \in D'(Y)$ on the test random variable $v \in D$ by $\langle u, v\rangle = \sum_\alpha v_\alpha u_\alpha$. If $u \in D'(Y)$, $v \in D(Y)$, then $\langle u, v\rangle = \sum_\alpha (v_\alpha, u_\alpha)_Y$.

Definition 5.7.23 For a sequence $u^n \in D'$ and $u \in D'$, we say that $u^n \to u$, if for every $v \in D$, $\langle u, v^n\rangle \to \langle u, v\rangle$.

Definition 5.7.24 If $u = \sum_\alpha u_\alpha \phi_\alpha \in D'(Y)$, F is a vector space, and $f : E \to F$ is a linear map, then we define

$$f(u) = \sum_\alpha f(u_\alpha)\phi_\alpha \in D'(F).$$

Definition 5.7.25 A Y-valued generalized D-field on a measurable space $(B, \mathcal{B})$ is a $D'(Y)$-valued function on B such that for each $x \in B$,

$$u(x) = \sum_\alpha u_\alpha(x)\phi_\alpha \in D'(E),$$

where $u_\alpha(x)$ are deterministic measurable E-valued functions.

We denote the linear space of Y-valued generalized D-fields by $D'(B; Y)$. If B is a topological space and a generalized D-field $u(x)$ is continuous on B we write

$u \in \mathcal{C}D'(B;E)$; note that $u(x)$ is continuous if and only if every u_α is continuous. In particular, if $B = [0,T]$, we say $u(t)$ is a generalized D-process. If there is no room for confusion, we will often say D-process (D-random variable) instead of generalized D-process (generalized D-random variable).

Exercise 5.7.26 (C) Let

$$u^n = \sum_\alpha u^n_\alpha \phi_\alpha, \; u = \sum_\alpha u_\alpha \phi_\alpha$$

Confirm that $u^n \to u$ if and only if $u^n_\alpha \to u_\alpha$ as $n \to \infty$ for every $\boldsymbol{\alpha}$.

Let $(B, \mathcal{B}, \kappa)$ be a measure space and let E be a normed vector space. For $p > 0$ we denote

$$L_p(D'(B;E),\kappa) \\ = \{u(x) = \sum_\alpha u_\alpha(x)\phi_\alpha \in D'(B;E) : \int_B |u_\alpha(x)|^p_E d\kappa < \infty, \boldsymbol{\alpha} \in \mathcal{J}^1\}.$$

For $u(t) = \sum_\alpha u_\alpha(t)\phi_\alpha \in L_1(D'([0,T],E))$ we define

$$\int_0^t u(s)ds, \;\; 0 \le t \le T$$

in $D'([0,T];E)$ by

$$\int_0^t u(s)ds = \sum_\alpha \left(\int_0^t u_\alpha(s)ds \right) \phi_\alpha, \;\; 0 \le t \le T.$$

If $u(t) - \sum_\alpha u_\alpha(t)\phi_\alpha \in D'([0,T];E)$, then $u(t)$ is differentiable in t if and only if $u_\alpha(t)$ are differentiable in t. In that case,

$$\frac{d}{dt}u(t) = \dot{u}(t) = \sum_\alpha \dot{u}_\alpha(t)\phi_\alpha \in D'([0,T],E).$$

The next step is construction of distribution free multiple stochastic integrals. Now, the Hilbert space $\mathbf{H} = L_2(U, \mathcal{U}, \mu)$ should be replaced by the product space $\mathbf{H}^{\otimes n} = L_2\left(U^n, \mathcal{U}^{\otimes n}, \mu_n\right)$, where $U^n = U \times \ldots \times U$ [n times], $\mu_n = \mu^{\otimes n}$; $\mathbf{H}^{\hat{\otimes} 0} = \mathbb{R}$. Let $\mathbf{H}^{\hat{\otimes} n}$ be the symmetric part of $\mathbf{H}^{\otimes n}$: it is the set of all symmetric μ_n-square integrable functions on U^n. Let $\mathcal{P}_n$ be the group of permutations of $\{1,\ldots,n\}$, let $\{\mathfrak{m}_k, \; k \ge 1\}$ be an orthonormal basis in $\mathbf{H}$, and, for $\boldsymbol{\alpha} \in \mathcal{J}^1$ with $|\boldsymbol{\alpha}| = n$ and characteristic set $(j_1,\ldots,j_n)$,

$$E_\alpha = \sum_{\sigma \in \mathcal{P}_n} \mathfrak{m}_{\sigma(j_1)} \otimes \ldots \otimes \mathfrak{m}_{\sigma(j_n)}.$$

Exercise 5.7.27 (C)

(a) Prove that

$$e_\alpha = \frac{E_\alpha}{\sqrt{\alpha!\,|\alpha|!}}, \quad \alpha \in \mathcal{J}^1, \ |\alpha| = n,$$

is a CONS of the symmetric part of $\mathbf{H}^{\hat{\otimes}n}$.

(b) Is it true that $E_{\epsilon(k)} = e_{\epsilon(k)} = \mathfrak{m}_k$?

Now we construct multiple integrals with respect to a distribution free driving noise Φ (see Definition 5.7.2) on the symmetric part $\mathbf{H}^{\hat{\otimes}n}$.

Let $\mathcal{J}_n = \left\{\alpha \in \mathcal{J}^1 : |\alpha| = n, \phi_\alpha \neq 0\right\}$. For $\alpha \in \mathcal{J}_n$ we set

$$I_n\left(E_\alpha\right) = n!\phi_\alpha. \tag{5.7.19}$$

It follows from (5.7.19) that, for $v = \sum_{|\alpha|=n} v_\alpha E_\alpha$,

$$I_n\left(v\right) = \sum_{|\alpha|=n} v_\alpha I_n\left(E_\alpha\right) = \sum_{|\alpha|=n} v_\alpha \left|\alpha\right|!\phi_\alpha \tag{5.7.20}$$

and $I_0\left(c\right) = c, \ c \in \mathbb{R}$.

The definition of ϕ_α implies $\mathbb{E}\phi_\alpha = 0$ for $|\alpha| > 0$, and therefore

$$\mathbb{E}\left(I_n\left(v\right)\right) = n! \sum_{|\alpha|=n} v_\alpha\left(\mathbb{E}\phi_\alpha\right) = 0. \tag{5.7.21}$$

Exercise 5.7.28 (C) Confirm that

$$\mathbb{E}\left(I_n(v)\right)^2 = (n!)^2 \sum_{|\alpha|=n} \alpha! v_\alpha^2$$

Now we shall introduce an extension of $\mathbf{H}^{\hat{\otimes}n}$ to a general Hilbert space Y.

Definition 5.7.29 Let $\mathbf{H}^{\hat{\otimes}n}(Y)$ be the space of all Y-valued symmetric functions $v = \sum_{|\alpha|=n} v_\alpha E_\alpha$ on U^n such that

$$|v|^2_{\mathbf{H}^{\hat{\otimes}n}(Y)} = \int_{U^n} |v\left(r\right)|^2_Y \, d\mu_n = \sum_\alpha |v_\alpha|^2_Y \left|\alpha\right|!\alpha! < \infty. \tag{5.7.22}$$

Let $\mathbf{H}^{\hat{\otimes}n}_\pi(Y) \subseteq \mathbf{H}^{\hat{\otimes}n}(Y)$ be the linear subspace of $\mathbf{H}^{\hat{\otimes}n}(Y)$ spanned on $\{E_\alpha : \alpha \in \mathcal{J}_n\}$. Set $\mathbf{H}^{\hat{\otimes}n}_\pi(Y) = \{0\}$ if $\mathcal{J}_n = \emptyset$. For $v = \sum_{|\alpha|=n} v_\alpha E_\alpha \in \mathbf{H}^{\hat{\otimes}n}(Y)$, let $v^{(\pi_n)}$ be its projection onto $\mathbf{H}^{\hat{\otimes}n}_\pi(Y) \subseteq \mathbf{H}^{\hat{\otimes}n}(Y)$ spanned on $\{E_\alpha : \alpha \in \mathcal{J}_n\}$

($\mathbf{H}_{\pi}^{\hat{\otimes}n}(Y) = \{0\}$ if $\mathcal{J}_n = \emptyset$). We define the n-th multiple integral of v as

$$I_n(v) = I_n\left(v^{(\pi_n)}\right) = \sum_{|\alpha|=n} v_\alpha I_n(E_\alpha) = n! \sum_{|\alpha|=n} v_\alpha \phi_\alpha,$$

and $I_0(c) = c,\ c \in \mathbb{R}$.

Exercise 5.7.30 (B) By induction on n, show that that

$$\int_{U^n} \left(\sum_{|\alpha|=n} v_\alpha E_\alpha \right)^2 d\mu_n = n! \sum_{|\alpha|=n} \|v_\alpha\|_Y^2 \alpha!$$

and

$$\mathbb{E}\left[|I_n(v)|_Y^2\right] = n!^2 \sum_{\alpha \in \mathcal{J}_n} \|v_\alpha\|_Y^2 \alpha! = n! \left\| v^{(\pi_n)} \right\|^2_{\mathbf{H}^{\hat{\otimes}n}(Y)}.$$

Let $\mathcal{S}(Y)$ be the space of all finite linear combinations $\sum_k \frac{I_k(F_k)}{k!}$ with $F_k \in \mathbf{H}_{\pi}^{\hat{\otimes}k}(Y)$.

Definition 5.7.31 A generalized $\mathcal{S}$-random variable is a formal sum

$$u = \sum_k \frac{I_k(F_k)}{k!} \text{ with } F_k \in \mathbf{H}_{\pi}^{\hat{\otimes}k}(Y).$$

We denote the set of all generalized $\mathcal{S}$-random variables by $\mathcal{S}'(Y)$.

The action of $u \in \mathcal{S}'(Y)$ on $v \in \mathcal{S}(Y)$ is defined as

$$\langle u, v \rangle = \sum_k \int_{U^k} (u_k, v_k)_Y d\mu,$$

where $u = \sum_k I_k(u_k)/k!,\ v = \sum_k I_k(v_k)/k!$ with $u_k, v_k \in \mathbf{H}_{\pi}^{\hat{\otimes}k}(Y)$.

Definition 5.7.32 A generalized $\mathcal{S}$-field on a measurable space $(B, \mathcal{B})$ is a $\mathcal{S}'(Y)$-valued function on B such that for each $x \in B$,

$$u(x) = \sum_n \frac{I_n(F_n(x))}{n!} \in \mathcal{S}'(Y),$$

where $x \mapsto F_n(x) = F_n(x; v_1, \ldots, v_n)$ are deterministic measurable $\mathbf{H}^{\hat{\otimes}n}(Y)$-valued functions on B.

We denote the linear space of all such fields by $\mathcal{S}'(B; E)$. In particular, if $B = [0, T]$, we say $u(t)$ is a generalized $\mathcal{S}$-process. If a generalized $\mathcal{S}$-field $u(x)$

is continuous on B, then we write $u \in \mathcal{CS}'(B;E)$; note that $u(x)$ is continuous if and only every function F_n is continuous.

Exercise 5.7.33 (B) Verify the following statements:

(a) $D(Y) \subseteq \mathcal{S}((Y))$ and $\mathcal{S}'(Y) \subseteq D'(Y)$.
(b)

$$\mathcal{S}'(Y) = \left\{ u = \sum_{\alpha} u_\alpha \Phi_\alpha \in D'(Y) : \sum_{|\alpha|=n} |u_\alpha|_Y^2 \alpha! < \infty \ \forall n \geq 1 \right\}.$$

Here are some of the potentially useful equalities. If $u = \sum_\alpha u_\alpha \phi_\alpha$ with $\sum_{|\alpha|=n} |u_\alpha|_Y^2 \alpha! < \infty, n \geq 1$, then

$$u_n = \sum_{|\alpha|=n} u_\alpha E_\alpha \in \mathbf{H}_\pi^{\hat{\otimes} n}(Y)$$

and

$$u = \sum_\alpha u_\alpha \Phi_\alpha = \sum_{n=0}^{\infty} \sum_{|\alpha|=n} u_\alpha \Phi_\alpha = \sum_{n=0}^{\infty} \frac{I_n(u_n)}{n!} \in \mathcal{S}'(Y).$$

Exercise 5.7.34 (C) Show that, for $|\alpha| = n$,

$$u_\alpha = \frac{1}{\alpha! n!} \int_{U^n} u_n(\upsilon) E_\alpha(\upsilon)\, d\mu_n, \ n \geq 1.$$

For $n \geq 0$, let $\mathcal{E}\mathbf{H}^{\hat{\otimes} n}$ be the space of all finite linear combinations of E_α, $|\alpha| = n$. The following statement provides some insight about the transition from n-th multiple integral to an integral on U^{n+1}.

Proposition 5.7.35 *Let $f \in \mathcal{E}\mathbf{H}^{\hat{\otimes} n}, g \in \mathcal{E}\mathbf{H}, f \otimes g = f(z) g(\upsilon), z \in U^n, \upsilon \in U$, and let $\widetilde{f \otimes g}$ be the standard symmetrization of $f \otimes g$. Then*

$$I_{n+1}\left(\widetilde{f \otimes g}\right) = I_n(f) I_1(g) - \mathit{projection}_{H_n} \left[I_n(f) I_1(g)\right].$$

Proof Let

$$f = \sum_{|\alpha|=n} f_\alpha E_\alpha, \ g = \sum_{|\alpha|=1} g_\alpha E_\alpha.$$

Then

$$fg = \sum_{\alpha,\alpha'} f_\alpha g_{\alpha'} E_\alpha E_{\alpha'},$$

$$\widetilde{f \otimes g} = \sum_{\alpha,\alpha'} f_\alpha g_{\alpha'} \widetilde{E_\alpha E_{\alpha'}} = \frac{1}{n+1} \sum_{\alpha,\alpha'} f_\alpha g_{\alpha'} E_{\alpha+\alpha'}$$

and

$$I_{n+1}\left(\widetilde{f \otimes g}\right) = \frac{1}{n+1} \sum_{\alpha,\alpha'} f_\alpha g_{\alpha'} I_{n+1}(E_{\alpha+\alpha'}) = n! \sum_{\alpha,\alpha'} f_\alpha g_{\alpha'} \phi_{\alpha+\alpha'},$$

$$I_n(f) = n! \sum_\alpha f_\alpha \phi_\alpha, I_1(g) = \sum_{\alpha'} g_{\alpha'} \phi_{\alpha'}$$

Since

$$\phi_{\alpha+\alpha'} = \phi_\alpha \phi_{\alpha'} - \text{projection}_{H_n}[\phi_\alpha \phi_{\alpha'}],$$

it follows that

$$\begin{aligned} I_{n+1}\left(\widetilde{f \otimes g}\right) &= n! \sum_{\alpha,\alpha'} f_\alpha g_{\alpha'} [\phi_\alpha \phi_{\alpha'} - \text{projection}_{H_n}(\phi_\alpha \phi_{\alpha'})] \\ &= I_n(f) I_1(g) - \text{projection}_{H_n}[I_n(f) I_1(g)]. \end{aligned}$$

Exercise 5.7.36 (C) Show that, if $U = [0, T]$, $d\mu = dt$, and $\mathfrak{m}_1 = \mathbf{1}_{(a,b)}$, then, according to Proposition 5.7.35, the "measure" of the square $(a, b)^2$ is

$$I_2\left(\mathbf{1}_{(a,b)}^{\otimes 2}\right) = I_1\left(\mathbf{1}_{(a,b)}\right)^2 - \text{projection}_{H_1}\left[I_1\left(\mathbf{1}_{(a,b)}\right)^2\right].$$

Wick Product and Skorokhod Integral
We define the Wick product of ϕ_α and ϕ_β by

$$\phi_\alpha \diamond \phi_\beta = \phi_{\alpha+\beta}, \ 1 \diamond \phi_\alpha = \phi_\alpha, \ \boldsymbol{\alpha}, \boldsymbol{\beta} \in \mathcal{J}^1.$$

For a Hilbert space Y and $u = \sum_\alpha u_\alpha \phi_\alpha, \ v = \sum_\alpha v_\alpha \phi_\alpha \in D'(Y)$

$$u \diamond v = \sum_\alpha \sum_{\beta \le \alpha} (u_\beta, v_{\alpha-\beta})_Y \phi_\alpha.$$

For a generalized random field on $u = \sum_{\alpha} u_{\alpha}\phi_{\alpha} \in D'(\mathbf{H}(Y))$, we define the Skorokhod integrals

$$\delta_{\epsilon(k)}(u) = \sum_{\alpha} \phi_{\alpha+\epsilon(k)} \int_U u_{\alpha}(\upsilon)E_k(\upsilon)d\mu,$$

and

$$\begin{aligned}\delta(u) &= \sum_k \delta_{\epsilon(k)}(u) = \sum_{\alpha}\sum_k \phi_{\alpha+\epsilon(k)} \int_U u_{\alpha}(\upsilon)E_k(\upsilon)d\mu(\upsilon) \\ &= \sum_{|\alpha|\geq 1}\sum_{k:\epsilon(k)\leq\alpha} \phi_{\alpha} \int_U u_{\alpha-\epsilon(k)}(\upsilon)E_k(\upsilon)d\mu(\upsilon).\end{aligned}$$

For a deterministic $u \in \mathbf{H}(Y)$,

$$\begin{aligned}\delta_{\epsilon(k)}(u) &= \phi_{\varepsilon_k} \int_U u(\upsilon)\mathfrak{m}_k(\upsilon)d\mu(\upsilon), \\ \delta(u) &= \sum_k \phi_k \int_U u(\upsilon)\mathfrak{m}_k(\upsilon)d\mu(\upsilon) = \Phi(u).\end{aligned}$$

Now, we describe the Skorokhod integral δ in terms of the multiple integrals I_n. We show that δ maps $\mathcal{S}$ to $\mathcal{S}$ and maps $\mathcal{S}'$ to $\mathcal{S}'$.

Proposition 5.7.37 *Let*

$$u = \sum_{n=0}^{\infty} \frac{1}{n!} I_n(u_n) \in D'(L_2(U, d\mu)),$$

where

$$u_n = u_n(\upsilon; \upsilon_1, \ldots, \upsilon_n) = \sum_{|\alpha|=n} u_{\alpha}(\upsilon)E_{\alpha}(\upsilon_1, \ldots, \upsilon_n)$$

and

$$\sum_{|\alpha|=n} \alpha! \int_U |u_{\alpha}(\upsilon)|^2 d\mu(\upsilon) < \infty \ \forall n.$$

Then

$$\delta(u) = \sum_{n=0}^{\infty} \frac{1}{n!} I_{n+1}(\tilde{u}_n),$$

where $\tilde{u}_n$ is the standard symmetrization of u_n on U^{n+1}.

Proof According to Exercise 5.7.33, for $|\boldsymbol{\alpha}| = n$,

$$u_{\alpha}(\upsilon) = \frac{1}{\alpha!n!}\int_{U^n} u_n(\upsilon;\upsilon')E_{\alpha}(\upsilon')\,\mu_n(d\upsilon')$$

and

$$u(\upsilon) = \sum_{\alpha} u_{\alpha}(\upsilon)\phi_{\alpha}.$$

Note that

$$\int_{U\times U^n} |u_n(\upsilon;\upsilon')|^2 d\mu_{n+1} < \infty,$$

$$\begin{aligned}\int_U u_{\alpha}(\upsilon)E_k(\upsilon)d\mu &= \frac{1}{\alpha!n!}\int_{U^n}\int_U u_n(r;\upsilon')E_k(r)\mu(dr)\,E_{\alpha}(\upsilon')\,\mu_n(d\upsilon')\\ &= \frac{1}{\sqrt{\alpha!n!}}\int_{U^n}\int_U u_n(r;\upsilon')E_k(r)\mu(dr)\,e_{\alpha}(\upsilon')\,\mu_n(d\upsilon'),\end{aligned}$$

and

$$u_n(\upsilon,\upsilon') = \sum_{|\alpha|=n,k\geq 1} E_{\alpha}(\upsilon')E_k(\upsilon)\int_U u_{\alpha}(r)E_k(r)d\mu(r).$$

Therefore, the standard symmetrization of $u_n(\upsilon,\upsilon')$, with $\upsilon \in U$, $\upsilon' = (\upsilon_1,\ldots,\upsilon_n) \in U^n$, is

$$\tilde{u}_n(\upsilon,\upsilon') = \sum_{|\alpha|=n+1,k\geq 1} \frac{n!}{(n+1)!}E_{\alpha}(\upsilon,\upsilon')\int_U u_{\alpha-\epsilon(k)}(r)E_p(r)d\mu(r).$$

By the definition of the Skorokhod integral,

$$\begin{aligned}\delta(u) &= \sum_{\alpha}\sum_{k\geq 1}\left(\int_U u_{\alpha}(r)E_p(r)d\mu\right)\phi_{\alpha+\epsilon(k)}\\ &= \sum_{n=0}^{\infty}\sum_{|\alpha|=n}\sum_{k\geq 1}\left(\int_U u_{\alpha}(r)E_k(r)d\mu(r)\right)\frac{I_{n+1}(E_{\alpha+\epsilon(k)})}{(n+1)!}\end{aligned}$$

$$= \sum_{n=0}^{\infty} \sum_{|\alpha|=n+1} \sum_{k\geq 1} \left(\int_U u_{\alpha-\epsilon(k)}(r) E_k(r) d\mu(r) \right) \frac{I_{n+1}(E_\alpha)}{(n+1)!}$$

$$= \sum_{n=0}^{\infty} \frac{1}{n!} I_{n+1}(\tilde{u}_n),$$

and the statement follows.

Multiple Skorokhod Integrals

For symmetric $u = \sum_\alpha u_\alpha \phi_\alpha \in D'\left(\mathbf{H}^{\hat{\otimes} n}\right)$, with $u_\alpha \in \mathbf{H}^{\hat{\otimes} n}$, and $\gamma \in \mathcal{J}^1$, $|\gamma| = n$ we define

$$\delta_\gamma(u) = \sum_\alpha \phi_{\alpha+\gamma} \int_{U^n} u_\alpha \frac{E_\gamma}{\gamma!} d\mu_n, \text{ and then}$$

$$\delta^n(u) = \sum_{|\gamma|=n} \delta_\gamma(u) = \sum_\alpha \sum_{|\gamma|=n} \phi_{\alpha+\gamma} \int_{U^n} u_\alpha \frac{E_\gamma}{\gamma!} d\mu_n.$$

Let $\delta^0(u) = u_0$.

For a deterministic $u(x_1, \ldots, x_n)$ in $\mathbf{H}^{\hat{\otimes} n}$,

$$\delta^n(u) = I_n(u).$$

Indeed,

$$\delta^n(u) = \sum_{|\gamma|=n} \phi_\gamma \int_{U^n} u \frac{E_\gamma}{\gamma!} d\mu_n = \sum_{|\gamma|=n} \frac{I_n(E_\gamma)}{n!} \int_{U^n} u \frac{E_\gamma}{\gamma!} d\mu_n,$$

$$\delta^n(E_\gamma) = \phi_\gamma \int_{U^n} E_\gamma \frac{E_\gamma}{\gamma!} d\mu_n = n! \phi_\gamma = I_n(E_\gamma).$$

Exercise 5.7.38 (B) Confirm that, for $u = \sum_\alpha u_\alpha \phi_\alpha \in \mathcal{S}'\left(\mathbf{H}^{\hat{\otimes} n}\right)$, $n \geq 1$, (with $u_\alpha \in \mathbf{H}^{\hat{\otimes} n}$), $\delta^n(u) = \delta\left(\delta^{n-1}(\tilde{u}(\upsilon))\right)$, where

$$\tilde{u}(\upsilon) = \tilde{u}(\upsilon; \upsilon_1, \ldots, \upsilon_{n-1}) = u(\upsilon, \upsilon_1, \ldots, \upsilon_{n-1}), \ \upsilon, \upsilon_i \in U.$$

Exercise 5.7.39 (C) Confirm that, in the framework of a single random variable ξ (see Example 5.7.12),

$$\delta^k(\phi_n) = \phi_{n+k}, \ k \geq 1.$$

For $n \geq 0$, define $\mathcal{J}_n = \{\alpha \in \mathcal{J}^1 : \phi_\alpha \neq 0, |\alpha| = n\}$. Let $\mathbf{H}_\pi^{\hat{\otimes} n}$ be the subspace of $\mathbf{H}^{\hat{\otimes} n}$ spanned on E_α, $\alpha \in \mathcal{J}_n$. For $u \in \mathbf{H}^{\hat{\otimes} n}$, we denote by $u^{(\pi_n)}$ the orthogonal projection of u onto $\mathbf{H}_\pi^{\hat{\otimes} n}$.

Lemma 5.7.40 *For a deterministic* $u =\in \mathbf{H}^{\hat{\otimes} n}$,

$$\mathbb{E}\left[\delta^n (u)^2\right] = n! \int_{U^n} \left|u^{(\pi_n)}\right|^2 d\mu_n,$$

where $u^{(\pi_n)}$ *is the projection of* u *onto* $\mathbf{H}_\pi^{\hat{\otimes} n}$.

Proof By definition,

$$\delta^n (u) = \sum_{|\gamma|=n} \phi_\gamma \int_{U^n} u \frac{E_\gamma}{\gamma!} d\mu_n,$$

$$\mathbb{E}\left[\delta^n (u)^2\right] = \sum_{\gamma \in \mathcal{J}_n} \gamma \left(\int_{U^n} u \frac{E_\gamma}{\gamma!} d\mu_n\right)^2 = n! \int_{U^n} \left|u^{(\pi_n)}\right|^2 d\mu_n,$$

which completes the proof.

Exercise 5.7.41 (B) Show that it is possible to rewrite Proposition 5.7.11 using multiple integrals as follows: for each $\eta \in L_2\left(\Omega, \mathcal{F}^0, \mathbb{P}\right)$,

$$\eta = \sum_\alpha \eta_\alpha \phi_\alpha = \sum_{n=0}^{\infty} \frac{1}{n!} \sum_{|\alpha|=n} \eta_\alpha I_n (E_\alpha) = \sum_{n=0}^{\infty} \frac{1}{n!} I_n (\eta_n) = \sum_{n=0}^{\infty} \frac{1}{n!} \delta^n (\eta_n)$$

with

$$\eta_\alpha = \frac{\mathbb{E}\left[\eta \phi_\alpha\right]}{\alpha!} = \frac{1}{\alpha! n!} \int_{U^n} \eta_n(\upsilon) E_\alpha (\upsilon)\, d\mu_n$$

and

$$\eta_n = \eta_n (\upsilon) = \sum_{|\alpha|=n} \eta_\alpha E_\alpha (\upsilon).$$

5.7.2.1 The Malliavin Derivative

If $\phi_\alpha \neq 0$, we define

$$\mathbb{D}\phi_\alpha = \sum_{|\gamma|=1, \gamma \leq \alpha} \frac{\alpha!}{(\alpha - \gamma)!} \phi_{\alpha-\gamma} E_\gamma (\upsilon) = \sum_{\gamma} \sum_{|\gamma|=1, \gamma+\gamma=\alpha} \frac{(\gamma + \gamma)!}{\gamma!} E_\gamma (\upsilon) \phi_\gamma,$$

and $\mathbb{D}\phi_\alpha = 0$ if $\phi_\alpha = 0$.

For $u = \sum_\alpha u_\alpha \phi_\alpha \in D$, and with the convention $u_\alpha = 0$ if $\phi_\alpha = 0$, we get

$$\mathbb{D}_k u = \sum_{|\alpha|\geq 1} \alpha_k u_\alpha \phi_{\alpha(k)} \mathfrak{m}_k(\upsilon), \mathbb{D}_\gamma u = \sum_{\alpha \geq p} \frac{\alpha!}{(\alpha-\gamma)!} u_\alpha \phi_{\alpha-\gamma} E_\gamma(\upsilon), |\gamma| = 1,$$

$$\mathbb{D}u = \sum_{|\gamma|=1} \mathbb{D}_\gamma u = \sum_{\alpha \geq p} \sum_{|\gamma|=1} \frac{\alpha!}{(\alpha-\gamma)!} u_\alpha \phi_{\alpha-\gamma} E_\gamma(\upsilon)$$
$$= \sum_\alpha \sum_{|\gamma|=1} \frac{(\alpha+\gamma)!}{\alpha!} u_{\alpha+\gamma} E_\gamma(\upsilon) \phi_\alpha.$$

In a standard way we define the higher order Malliavin derivatives: for $u = \sum_\alpha u_\alpha \phi_\alpha \in D$,

$$\mathbb{D}^n_\gamma u = \sum_{\alpha \geq \gamma} \frac{\alpha!}{(\alpha-\gamma)!} u_\alpha \phi_{\alpha-\gamma} \frac{E_\gamma(\upsilon_1, \ldots, \upsilon_n)}{\gamma!}, \quad |\gamma| = n,$$

$$\mathbb{D}^n u = \sum_{|\gamma|=n} \sum_{\alpha \geq \gamma} \frac{\alpha!}{(\alpha-\gamma)!p!} u_\alpha \phi_{\alpha-\gamma} E_\gamma(\upsilon_1, \ldots, \upsilon_n)$$
$$= \sum_\alpha \sum_{|\gamma|=n} \frac{(\alpha+\gamma)!}{\alpha!\gamma!} u_{\alpha+\gamma} E_\gamma(\upsilon_1, \ldots, \upsilon_n)\phi_\alpha.$$

We will now compute the Malliavin derivative for multiple integrals.

Proposition 5.7.42 *Let* $u = \sum_{|\alpha|=n} u_\alpha E_\alpha \in \mathbf{H}_\pi^{\hat{\otimes} n}(Y)$ *has only finite number of* $u_\alpha \neq 0$. *Then*

$$\mathbb{D}I_n(u) = nI_{n-1}(u(\cdot, t)), \quad t \in U.$$

Proof By definition,

$$I_n(u) = \sum_{|\alpha|=n} u_\alpha I_n(E_\alpha) = n! \sum_{|\alpha|=n} u_\alpha \phi_\alpha \in D.$$

Since

$$u_\alpha = \frac{1}{\alpha!n!} \int_{U^n} u E_\alpha d\mu_n,$$

and, for $t, \upsilon_1, \ldots, \upsilon_n \in U$,

$$u(t, \upsilon_1, \ldots, \upsilon_{n-1}) = \sum_k E_k(t) \int_U u(t', \upsilon_1, \ldots, \upsilon_{n-1}) E_k(t') \, d\mu$$

is a finite sum,

$$\mathbb{D}u = \sum_{\alpha}\sum_{k}\frac{(\alpha+\epsilon(k))!}{\alpha!}u_{\alpha+\epsilon(k)}E_k(\upsilon)\phi_\alpha.$$

Next, we use the equality

$$\left\{(\alpha,\epsilon(k)) : |\alpha+\epsilon(k)| = n,\ k\geq 1\right\} = \left\{(\alpha,\epsilon(k)) : |\alpha| = n-1,\ k\geq 1\right\}$$

to find

$$\begin{aligned}
\mathbb{D}I_n(u) &= \sum_{k\geq 1, |\alpha+\epsilon(k)|=n}\frac{(\alpha+\gamma)!}{\alpha!}u_{\alpha+\epsilon(k)}n!E_k(t)\phi_\alpha \\
&= \sum_{|\alpha|=n-1}\sum_{k\geq 1}\frac{(\alpha+\gamma)!}{\alpha!}u_{\alpha+\epsilon(k)}n!E_k(t)\phi_\alpha \\
&= \sum_{|\alpha|=n-1}\sum_{k\geq 1}\frac{1}{\alpha!}\left(\int uE_{\alpha+\epsilon(k)}d\mu_n\right)E_\gamma(t)\phi_\alpha \\
&= \sum_{|\alpha|=n-1}\sum_{k\geq 1}E_k(t)\phi_\alpha n\int u\frac{E_\alpha}{\alpha!}E_k d\mu_n \\
&= \sum_{|\alpha|=n-1} n(n-1)!\left(\int u(t,\cdot)\frac{E_\alpha}{\alpha!(n-1)!}d\mu_{n-1}\right)\phi_\alpha \\
&= \sum_{|\alpha|=n-1} n(n-1)!u_\alpha(t,\cdot)\phi_\alpha = nI_{n-1}(u(\cdot,t)).
\end{aligned}$$

We used here that

$$\widetilde{E_\alpha E_k} = \frac{|\alpha|!}{|\alpha+\epsilon(k)|!}E_{\alpha+\epsilon(k)} = \frac{1}{n}E_{\alpha+\epsilon(k)},$$

where $\widetilde{f}$ is the symmetrization of f.

If $|\alpha| = n-1$ and $\phi_\alpha = 0$, then, for each $k\geq 1$,

$$0 = u_{\alpha+\epsilon(k)} = \frac{1}{(\alpha+\epsilon(k))!(n-1)!}\int_{U^n} uE_\alpha E_k d\mu_n$$

and

$$u(t,\cdot)_\alpha = \int_{U^{n-1}} u(t,\upsilon')E_\alpha(\upsilon')\,d\mu_{n-1} = 0.$$

Exercise 5.7.43 (C) Confirm that if $u \in \mathbf{H}_{\pi}^{\hat{\otimes}n}(Y)$, then $u(\cdot,t) \in \mathbf{H}_{\pi}^{\hat{\otimes}(n-1)}(Y)$ for μ-almost all $t \in U$.

As suggested by Proposition 5.7.42, for an arbitrary $v = \sum_{|\alpha|=n} v_\alpha E_\alpha \in \mathbf{H}_{\pi}^{\hat{\otimes}n}(Y)$, we define

$$\mathbb{D}I_n(v) = nI_{n-1}(v(y,\cdot)), \; y \in U,$$

$I_0(c) = c,\ c \in \mathbb{R}$. For $u = \sum_n \frac{I_n(u_n)}{n!} \in \mathcal{S}'(Y)$ we define

$$\mathbb{D}u(y) = \sum_n \frac{\mathbb{D}I_n(u_n)}{n!} = \sum_n \frac{I_{n-1}(u_n(y,\cdot))}{(n-1)!}.$$

Exercise 5.7.44 (C) Confirm that $\mathbb{D}$ maps $\mathcal{S}(\mathbb{R})$ to $\mathcal{S}(Y)$.

Exercise 5.7.45 (B) Show that, in the framework of a single random variable ξ (see Example 5.7.12),

$$\mathbb{D}^k\left(\xi^{\diamond n}\right) = \frac{n!}{(n-k)!}\xi^{\diamond(n-k)},\ k \ge 1,\ \text{ if } \xi^{\diamond n} \neq 0 \text{ and } \mathbb{D}^k\left(\xi^{\diamond n}\right) = 0 \text{ if } \xi^{\diamond n} = 0.$$

5.7.3 *Adapted Stochastic Processes*

In this subsection we assume that

$$U = [0,T] \times V,\ \mathcal{U} = \mathcal{B}([0,T]) \times \mathcal{V},\ d\mu = dtd\pi.$$

Let

$$u = \sum_n \frac{I_n(u_n)}{n!} \in \mathcal{S}'(Y), \tag{5.7.23}$$

with $u_n \in \mathbf{H}_{\pi}^{\hat{\otimes}n}(Y), n \ge 0$: $u_n = u_n(t_1, v_1, \ldots, t_n, v_n), (t_i, v_i) \in U, i = 1, \ldots, n$.

For $t \in [0,T]$, let $Q_t^n = ([0,t] \times V)^n$.

Definition 5.7.46 Let $t_0 \in [0,T]$. A random variable u defined by (5.7.23) is called $\mathcal{F}_{t_0}$-measurable if for each n, $\operatorname{supp}(u_n) \subseteq Q_{t_0}^n$, that is

$$u_n(t_1, v_1, \ldots, t_n, v_n) = 0,\ \mu_{n-1} - \text{a.e. if } t_i > t_0 \text{ for some } i$$

Proposition 5.7.47 *A random variable $u \in \mathcal{S}'(Y)$ defined by (5.7.23) is $\mathcal{F}_{t_0}$-measurable if and only if*

$$\mathbb{D}u(t,v) = 0 \text{ for all } t > t_0.$$

Proof For $u \in \mathcal{S}'(Y)$ defined by (5.7.23),

$$\mathbb{D}u = \sum_{n\geq 1} \frac{1}{(n-1)!} I_{n-1}\left(u_n(t, \upsilon, \cdot)\right), \ (t, \upsilon) \in U. \tag{5.7.24}$$

and

$$\mathbf{E}\left[\mathbb{D}u\,(t, \upsilon)^2\right] = \sum_n \frac{1}{(n-1)!} \mathbf{E}[I_{n-1}\left(u_n\,(t, \upsilon, \cdot)\right)^2]$$
$$= \sum_n \int_{U^{n-1}} u_n\,(t, \upsilon, \cdot)^2\, d\mu_{n-1} = 0$$

for $t > t_0$ iff $u_n(t, \upsilon, \cdot) = 0$ μ_{n-1}-a.e for $t > t_0$.

Next, we introduce the notion of an adapted random process. Consider $u \in \mathcal{S}'(U; Y)$, i.e,

$$u\,(t, \upsilon) = \sum_n \frac{I_n\left(u_n\,(t, \upsilon, \cdot)\right)}{n!} \tag{5.7.25}$$

with $u_n\,(t, \upsilon, \cdot) \in \mathbf{H}_\pi^{\hat{\otimes} n}(Y)$ for all $(t, \upsilon) \in U$:

$$u_n(t, \upsilon, \cdot) = \sum_{|\alpha|=n} u_\alpha(t, \upsilon) E_\alpha\,(\cdot)\,, (t, \upsilon) \in U. \tag{5.7.26}$$

Definition 5.7.48 A random field $u(t, \upsilon)$ on U defined by (5.7.25) is called adapted if supp $(u_n(t, \upsilon, \cdot)) \subseteq Q_t^n, \upsilon \in V$, for every $t \in [0, T]$.

A straightforward consequence of Proposition 5.7.47 is the following result.

Corollary 5.7.49 *A random field $u \in \mathcal{S}'(U; Y)$ is adapted if and only if, for each $t \in [0, T]$, the Malliavin derivative $\mathbb{D}u(t, \upsilon; s_1, \upsilon_1, \ldots, s_n, \upsilon_n) = 0, \upsilon \in V$, if $s_i > t$ for some i.*

Given a random field $u(t, \upsilon)$ on $U = [0, T] \times V$, consider its Skorokhod integral

$$\delta\,(u)_t = \delta\left(\mathbf{1}_{[0,t]}u\right), \ 0 \leq t \leq T.$$

Proposition 5.7.50 *Consider a random field $u \in \mathcal{S}'(\mathbf{H}(Y)\,; Y)$, that is, (5.7.25) holds and*

$$\int_{U^{n+1}} \|u_n\|_Y^2\, d\mu_{n+1} < \infty\ \forall n.$$

If u is adapted, then $\delta\,(u)_t$, $0 \leq t \leq T$, is adapted as well.

Proof Since

$$u(t,\upsilon) = \sum_n \frac{I_n(u_n(t,\upsilon,\cdot))}{n!}$$

is adapted, with u_n satisfying (5.7.26), we have supp $(u_n(t,\upsilon,\cdot)) \subseteq Q_t^n, \upsilon \in V$, for all n and $t \in [0,T]$, that is, $u_n(t,\upsilon) = u_n(t,\upsilon)\mathbf{1}_{Q_t^n}$. By Proposition 5.7.37,

$$\delta(u)_t = \delta\left(u\,\mathbf{1}_{[0,t]}\right) = \sum_{n=0}^{\infty} \frac{1}{n!} I_{n+1}(\widetilde{u_n\,\mathbf{1}_{[0,t]}}),$$

where $\widetilde{u_n\,\mathbf{1}_{[0,t]}}$ is the standard symmetrization of $u_n\,\mathbf{1}_{[0,t]} = u_n \mathbf{1}_{Q_t^{n+1}}$. Since its support is obviously a subset of Q_t^{n+1}, the statement follows.

5.7.3.1 Itô-Skorokhod Isometry

Now we estimate the L_2-norm of the Skorokhod integral.

Proposition 5.7.51 *Let* $u = u(\upsilon) = \sum_\alpha u_\alpha(\upsilon)\phi_\alpha \in L_2(D(U;Y), d\mu)$, *i.e.* $u_\alpha \in \mathbf{H}(Y)$:

$$\int |u_\alpha(\upsilon)|_Y^2 d\mu < \infty,\ \boldsymbol{\alpha} \in \mathcal{J}^1,$$

and only finitely many of u_α *are not zero. Then*

$$\mathbb{E}\left[|\delta(u)|_Y^2\right] \le \mathbb{E}\left[\int_U |u(\upsilon)|_Y^2\, d\mu\right] + \mathbb{E}\left[\int_{U^2} \left(\mathbb{D}u(\upsilon;\upsilon'), \mathbb{D}u(\upsilon';\upsilon)\right)_Y \mu(d\upsilon)\mu(d\upsilon')\right],$$

where

$$\mathbb{D}u(\upsilon;\upsilon') = \sum_\alpha \sum_{k\ge1} \frac{(\boldsymbol{\alpha}+\boldsymbol{\epsilon}(k))!}{\boldsymbol{\alpha}!} u_{\alpha+\epsilon(k)}(\upsilon) E_k(\upsilon')\,\phi_\alpha.$$

An equality holds if $\phi_\alpha \ne 0$ *for all* $\boldsymbol{\alpha}$.

Proof By definition,

$$\delta(u) = \sum_{|\alpha|\ge1} \phi_\alpha \sum_{k\ge1} \int_U u_{\alpha-p}(\upsilon)E_k(\upsilon)d\mu = \sum_\alpha \sum_{k\ge1} \phi_{\alpha+\epsilon(k)} \int_U u_\alpha(\upsilon)E_k(\upsilon)d\mu.$$

Hence

$$\mathbb{E}\left[\|\delta(u)\|_Y^2\right] = \sum_{|\alpha|\geq 1, \varphi_\alpha \neq 0} \alpha! \left\| \sum_{k\geq 1} \int_U u_{\alpha-\epsilon(k)}(x) E_k(x) d\mu \right\|_Y^2$$

$$\leq \sum_{k,k'\geq 1} \sum_{|\alpha|\geq 1} \alpha! \int_{U^2} \left(u_{\alpha-\epsilon(k)}(x), u_{\alpha-\epsilon(k')}(x') \right)_Y$$

$$\times E_k(x) E_{k'}(x') \mu(dx) \mu\left(dx'\right)$$

$$= \sum_{k=k'\geq 1} \left(\cdots\right) + \sum_{k\neq k'\geq 1} \left(\cdots\right) := A + B.$$

Now

$$A = \sum_{k\geq 1} \sum_{\alpha \geq \epsilon(k)} \alpha! \left\| \int_U u_{\alpha-\epsilon(k)}(x) E_k(x) d\mu \right\|_Y^2$$

$$= \sum_\gamma \sum_{k\geq 1} (\gamma + \epsilon(k))! \left\| \int_U u_\gamma(x) E_k(x) d\mu \right\|_Y^2.$$

Also,

$$B = \sum_{k\neq k'\geq 1} \sum_{\alpha \geq \epsilon(k)+\epsilon(k')} \alpha! \int_{U^2} \left(u_{\alpha-\epsilon(k)}(x), u_{\alpha-\epsilon(k')}(x') \right)_Y$$

$$\times E_k(x) E_{k'}(x') \mu(dx) \mu\left(dx'\right)$$

$$= \sum_{k\neq k'\geq 1} \sum_\beta (\beta + \epsilon(k) + \epsilon(k'))! \int_{U^2} (u_{\beta+\epsilon(k')}(x), u_{\beta+\epsilon(k)}(x'))_Y$$

$$\times E_k(x) E_{k'}(x') \mu(dx) \mu\left(dx'\right).$$

On the other hand,

$$\mathbb{E}\Big(\mathbb{D}u(x; x'), \mathbb{D}u(x'; x)\Big)_Y$$

$$= \sum_\alpha \alpha! \sum_{k,k'\geq 1} \frac{(\alpha + \epsilon(k))!}{\alpha!} \left(u_{\alpha+\epsilon(k)}(x), u_{\alpha+\epsilon(k')}(x') \right)_Y$$

$$\times \frac{(\alpha+\epsilon(k'))!}{\alpha!} E_k\left(x'\right) E_{k'}(x)$$

$$= \sum_{\alpha} \sum_{k=k'\geq 1} \left(\cdots\right) + \sum_{\alpha} \sum_{k\neq k'\geq 1} \left(\cdots\right) := C + D,$$

and

$$C = \sum_{k\geq 1} \sum_{\alpha\geq\epsilon(k)} \alpha! \frac{\alpha!}{(\alpha-\epsilon(k))!} \left(u_\alpha(x), u_\alpha(x')\right)_Y E_k(x) E_k(x'),$$

$$D = \sum_{\alpha} \alpha! \sum_{k\neq k'\geq 1}$$

$$\frac{(\alpha+\epsilon(k))!}{\alpha!} \left(u_{\alpha+\epsilon(k)}(x) E_k\left(x'\right), u_{\alpha+\epsilon(k')}(x') E_{k'}(x)\right)_Y \frac{(\alpha+\epsilon(k'))!}{\alpha!}.$$

The statement of the theorem follows after comparing $\int_{U^2} C d\mu_2$, $\int_{U^2} D d\mu_2$ and A, B, because

$$A = \int_{U^2} C d\mu_2 + \mathbb{E}\left[\int_U |u(\upsilon)|_Y^2 \, d\mu\right], \quad B = \int_{U^2} D d\mu_2. \tag{5.7.27}$$

Exercise 5.7.52 (C) Verify (5.7.27)

Corollary 5.7.53 *Let $u \in \mathcal{S}'(\mathbf{H}(Y), Y)$, i.e. (5.7.25) holds with*

$$\int_{U^{n+1}} |u_n|_Y^2 \, d\mu_{n+1} < \infty \ \forall n.$$

Then the statement of Proposition 5.7.51 holds for $\delta(u)$.

Proof It is enough to prove the statement for $u(\upsilon) = I_n(u_n(\upsilon))$, where

$$u_n(\upsilon, \cdot) = \sum_{|\alpha|=n} u_\alpha(\upsilon) E_\alpha(\cdot)$$

with a finite number of nonzero $u_\alpha \in \mathbf{H}(Y)$.

In this case,

$$u = n! \sum_{|\alpha|=n} u_\alpha(\upsilon) \phi_\alpha \in L_2(D(U; Y), d\mu)$$

and Proposition 5.7.51 applies. We obtain the general case by linearity.

Exercise 5.7.54 (B) Confirm that, in the framework of a single r.v. ξ (see Example 5.7.12) with all $\phi_n \neq 0$, for $u = \sum_n \phi_n$ we have

$$\mathbf{E}\left[\delta(u)^2\right] = \mathbf{E}\left[|u|^2\right] + \mathbf{E}\left[(\mathbb{D}u)^2\right].$$

For an adapted random field on $U = [0, T] \times V$, $d\mu = dtd\pi$, the Itô-type inequality holds as a direct consequence of Corollary 5.7.53.

Corollary 5.7.55 *Let* $\mathbf{H} = L_2([0, T] \times V, dtd\pi)$. *Assume* $u \in \mathcal{S}(\mathbf{H}(Y), Y)$ *is an adapted random field on* $U = [0, T] \times V$. *Then*

$$\mathbb{E}\left[|\delta(u)|_Y^2\right] \leq \mathbb{E}\left[\int_U |u(t, \upsilon)|_Y^2 \, d\mu\right].$$

Duality Between δ and $\mathbb{D}$

Proposition 5.7.56 *Let* $u = u(\upsilon) = \sum_\alpha u_\alpha(\upsilon)\phi_\alpha \in L_2(D(U), d\mu)$, *i.e.*

$$\int |u_\alpha(\upsilon)|^2 d\mu < \infty, \ \boldsymbol{\alpha} \in \mathcal{J}^1,$$

with a finite number of $u_\alpha \neq 0$. *Let* $\psi = \sum_\alpha \psi_\alpha \phi_\alpha \in D$. *Then*

$$\mathbb{E}[\delta(u)\psi] = \mathbb{E}\left[\int_U u(\upsilon)\mathbb{D}\psi(\upsilon)d\mu\right].$$

Proof By direct computation,

$$\begin{aligned}
&\mathbb{E}[\delta(u)\psi] \\
&= \sum_\alpha \psi_\alpha \alpha! \sum_{k\geq 1} \int_U u_{\alpha-\epsilon(k)}(\upsilon)E_k(\upsilon)d\mu \\
&= \sum_\gamma \sum_{k\geq 1} \psi_{\gamma+\epsilon(k)} \frac{(\gamma+\epsilon(k))!}{\gamma!}\gamma! \int_U u_\gamma(\upsilon)E_k(\upsilon)d\mu = \mathbb{E}\left[\int_U u(\upsilon)\mathbb{D}\psi(\upsilon)\,d\mu\right].
\end{aligned}$$

5.7.4 Stochastic Differential Equations

To simplify the presentation, we assume from now on that $\phi_\alpha \neq 0$ for all $\boldsymbol{\alpha} \in \mathcal{J}^1$.

5.7.4.1 Wick Exponential

We start with the definition of Wick exponential.

Let $f = \sum_k f_k \mathfrak{m}_k \in L_2(U, d\mu)$. Writing

$$f^{\alpha} = \prod_k f_k^{\alpha_k}$$

and $\Phi(f) = \sum_k f_k \phi_k$, the definition of the Wick product implies that, for $n \geq 1$,

$$\Phi(f)^{\diamond n} := \Phi(f) \diamond \cdots \diamond \Phi(f) \text{ [}n\text{ times]} = \sum_{|\alpha|=n} \frac{n!}{\alpha!} f^{\alpha} \phi_{\alpha}.$$

Note that

$$\mathbb{E}\left[\left(\frac{1}{n!}\phi(f)^{\diamond n}\right)^2\right] = \sum_{|\alpha|=n} \frac{f^{2\alpha}}{\alpha!} = \frac{1}{n!} \sum_{|\alpha|=n} \frac{n! f^{2\alpha}}{\alpha!} \tag{5.7.28}$$
$$= \frac{1}{n!}\left(\sum_i f_i^2\right)^n = \frac{1}{n!} \|f\|_{L_2(\mu)}^{2n} < \infty,$$

that is, $\Phi(f)^{\diamond n} \in L_2(\Omega)$.

Let $\mathcal{Z}$ be the set of all real sequences $z = (z_k,\ k \geq 1)$ such that every sequence in $\mathcal{Z}$ has only finitely many non-zero terms. The following result holds.

Proposition 5.7.57

a) Let $f = \sum_k f_k \mathfrak{m}_k \in L_2(\mu)$. *Then*

$$\Phi(f)^{\diamond n} = I_n\left(f^{\otimes n}\right)$$

and

$$\exp^{\diamond}\{\Phi(f)\} := \sum_{n=0}^{\infty} \frac{\Phi(f)^{\diamond n}}{n!} = \sum_{n=0}^{\infty} \sum_{|\alpha|=n} \frac{f^{\alpha}}{\alpha!}\phi_{\alpha} = \sum_{\alpha} \frac{f^{\alpha}}{\alpha!}\phi_{\alpha} \in L_2(\Omega)$$

with

$$f^{\alpha} = \frac{1}{n!}\int f^{\otimes n} E_{\alpha} d\mu_n.$$

Moreover,

$$\mathbb{E}\left[\left(\exp^{\diamond}\{\phi(f)\}\right)^2\right] = \exp\left\{|f|_{L_2(\mu)}^2\right\}. \tag{5.7.29}$$

b) Let $z = (z_k) \in \mathcal{Z}$. *Then* **P**-*a.s.*

$$p(z) = \exp^{\diamond}\left\{\Phi\left(\sum_k z_k \mathfrak{m}_k\right)\right\} = \sum_{\alpha} \frac{z^{\alpha}}{\alpha!}\phi_{\alpha}, \quad z = (z_k) \in \mathcal{Z},$$

is an analytic function of z *and*

$$\left.\frac{\partial^{|\alpha|} p(z)}{\partial z^{\alpha}}\right|_{z=0} = \phi_{\alpha}.$$

Proof

a) In terms of multiple integrals we have

$$\begin{aligned}\Phi(f)^{\diamond n} &= \sum_{|\alpha|=n} \frac{n! f^{\alpha}}{\alpha!}\phi_{\alpha} = \sum_{|\alpha|=n} \frac{f^{\alpha}}{\alpha!} I_n(E_{\alpha}) \\ &= \sum_{k_i \geq 1, \epsilon(k_1)+\ldots+\epsilon(k_n)=\alpha} \frac{n! f^{\alpha}}{\alpha!} I_n\left(\widetilde{E_{k_1} \ldots E_{k_n}}\right) \\ &= I_n\left(f^{\otimes n}\right) = I_n\left(\sum_{|\alpha|=n} \frac{f^{\alpha}}{\alpha!} E_{\alpha}\right),\end{aligned}$$

with

$$f^{\otimes n} = \sum_{|\alpha|=n} \frac{f^{\alpha}}{\alpha!} E_{\alpha}, \quad f^{\alpha} = \frac{1}{n!}\int f^{\otimes n} E_{\alpha} d\mu_n.$$

Moreover (see (5.7.28)),

$$\mathbb{E}\left(\sum_n \left(\frac{\Phi(f)^{\diamond n}}{n!}\right)^2\right) = \sum_n \frac{1}{n!} \|f\|^{2n}_{L_2(\mu)} = \exp\left\{\|f\|^2_{L_2(\mu)}\right\}.$$

b) Let $z = (z_k) \in \mathcal{Z}$. Then

$$\mathbb{E}\,|p(z)|^2 = \sum_{\alpha} \frac{\left|z^{2\alpha}\right|}{\alpha!} = \prod_k \sum_n \frac{|z_k|^{2n}}{n!} = \exp\left(\sum_k |z_k|^2\right),$$

that is, $p(z)$ is represented by a power series that, with probability one, converges for all $z \in \mathcal{Z}$.

In a time dependent case the following statement holds.

Corollary 5.7.58 *Let* $U = [0, T] \times V$, $d\mu = dtd\pi$. *Let* $G \in L_2([0, T] \times V, d\mu)$. *Consider* $M_t = \exp^{\diamond}\{\Phi(\mathbf{1}_{[s,t]}G)\}$, $0 \le s \le t \le T$. *Then*

$$M_t = \sum_{\alpha} \frac{H^{\alpha}(s,t)}{\alpha!}\phi_{\alpha} = \sum_{n=0}^{\infty} \frac{I_n(H_{(n)}(s,t))}{n!}, \; s \le t \le T,$$

with

$$H^{\alpha}(s,t) = \prod_k H_k^{\alpha_k}(s,t), \;\; H_k(s,t) = \int_s^t \left(\int G(r, \upsilon)\, \mathfrak{m}_k(r, \upsilon) d\pi \right) dr,$$

$$H_{(n)}(s,t) = \left(\mathbf{1}_{[s,t]}G\right)^{\otimes n}.$$

Also,

$$\frac{H^{\alpha}(s,t)}{\alpha!} = \frac{1}{|\alpha|!\alpha!} \int H_{(|\alpha|)} E_{\alpha} d\mu_{|\alpha|},$$

and the process M is adapted.

Exercise 5.7.59 (B) Prove Corollary 5.7.58.

5.7.4.2 Linear SDEs

Let $U = [0, T] \times V, d\mu = dtd\pi$. Let $w = \sum_{\alpha} w_{\alpha}\phi_{\alpha} \in D'$, $f = \sum_{\alpha} f_{\alpha}(t, \upsilon)\phi_{\alpha} \in L_2(D'(U), d\mu)$,

$$\dot{\Phi}(t, \upsilon) = \sum_k \mathfrak{m}_k(t, \upsilon)\phi_k.$$

For $G \in L_2(\mu)$, consider a non-homogeneous linear equation

$$\dot{u}(t) = \int [u(t)G(t, \upsilon) + f(t, \upsilon)] \diamond \phi(t, \upsilon)\pi(d\upsilon), u(0) = w. \tag{5.7.30}$$

Using the notation

$$\Phi(dt, d\upsilon) = \dot{\Phi}(t, \upsilon)d\mu(t, \upsilon),$$

we write (5.7.30) in a more compact form

$$u(t) = w + \int_0^t [u(s)G(s, \upsilon) + f(s, \upsilon)] \diamond \Phi(ds, d\upsilon), \;\; 0 \le t \le T. \tag{5.7.31}$$

We seek a solution to (5.7.31) in the form

$$u(t) = \sum_{\alpha} u_\alpha(t)\phi_\alpha, \; 0 \le t \le T. \tag{5.7.32}$$

Lemma 5.7.60 *Let* $w = \sum_\alpha w_\alpha \phi_\alpha \in D'$, $f = \sum_\alpha f_\alpha(t, \upsilon)\phi_\alpha \in L_2(D'(U), d\mu)$. *Then there is a unique solution to (5.7.31) in* $\mathcal{C}D'([0,T])$ *(Recall that* $\mathcal{C}D'([0,T])$ *is the class of all real-valued generalized processes* $u = \sum_\alpha u_\alpha(t)\phi_\alpha$ *on* $[0,T]$ *such that each* u_α *is continuous on* $[0,T]$*). The solution u given by (5.7.32) has the following coefficients:* $u_{(\mathbf{0})}(t) = w_{(\mathbf{0})}$,

$$u_\alpha(t) = w_\alpha + \sum_{k\ge1} \int_0^t \int [u_{\alpha-\epsilon(k)}(r)\, G(r,\upsilon) + f_{\alpha-\epsilon(k)}(r,\upsilon)] E_k(r,\upsilon) d\pi dr,$$
$$0 \le t \le T. \tag{5.7.33}$$

Proof Plugging the series (5.7.32) into (5.7.31), we immediately get the system (5.7.33). The system is lower triangular, and so we start with $u_{(\mathbf{0})}(t) = w_{(\mathbf{0})}$ to find a unique continuous $u_\alpha(t)$ for every $\boldsymbol{\alpha}$ with $|\boldsymbol{\alpha}| \ge 1$.

Let

$$H_k(t) = \int_0^t \int G(s,\upsilon)\, \mathfrak{m}_k(s,\upsilon) ds d\pi, \;\; H^\alpha(t) = \prod_k H_k^{\alpha_k}(t), \;\; \boldsymbol{\alpha} \in \mathcal{J}^1.$$

For $w = \sum_\alpha w_\alpha \phi_\alpha \in D'$, let

$$||w||^2 = \sum_\alpha |w_\alpha|^2 \boldsymbol{\alpha}! + \sup_t \sum_\alpha \boldsymbol{\alpha}! \left(\sum_{\beta \le \alpha} w_{\alpha-\beta} \frac{H^\beta(t)}{\boldsymbol{\beta}!} \right)^2$$

Lemma 5.7.61 *Let* $f = 0, w = \sum_\alpha w_\alpha \phi_\alpha \in D'$.

(i) The solution to (5.7.31) is

$$u(t) = w \diamond \exp^\diamond \{\Phi(\mathbf{1}_{[0,t]}G)\} = \sum_\alpha \sum_{\beta \le \alpha} w_{\alpha-\beta} \frac{H^\beta(r)}{\boldsymbol{\beta}!} \phi_\alpha, 0 \le r \le T. \tag{5.7.34}$$

(ii) If $w \in \mathcal{S}'(\mathbb{R})$, *then* $u \in \mathcal{C}\mathcal{S}'([0,T];\mathbb{R})$;

(iii)

$$\sup_t \mathbb{E}\left[u^2(t)\right] \le ||w||^2,$$

in particular, $u \in L_2(\Omega, \mathbb{P})$ *if* $||w|| < \infty$ *(see Example 5.7.62 below).*

(iv) If $w = w_{(\mathbf{0})}$ is non-random, then $u(t)$ is adapted and

$$\sup_t \mathbb{E}\left[u^2(t)\right] = w_0^2 \exp\left\{\int |G|^2\, d\mu\right\}.$$

Proof

(i) Let $M_t = \exp^{\diamond}\left\{\Phi\left(\mathbf{1}_{[0,t]}G\right)\right\}$. By Corollary 5.7.58,

$$\psi(r) := w \diamond M_r = \sum_{\alpha}\sum_{\beta\le\alpha} w_{\alpha-\beta}\frac{H^{\beta}(r)}{\beta!}\phi_\alpha,\ \ 0 \le r \le T.$$

We will show that ψ solves (5.7.31). Indeed,

$$\begin{aligned}
&\delta\left(\mathbf{1}_{[0,t]}\psi G\right)\\
&= \sum_{|\alpha|\ge 1}\sum_{k\ge 1}\int_0^t\int_V \sum_{\beta\le\alpha-\epsilon(k)} w_{\alpha-\epsilon(k)-\beta}\frac{H^{\beta}(r)}{\beta!}G(r,\upsilon)\,E_k(r,\upsilon)d\mu\phi_\alpha\\
&= \sum_{|\alpha|\ge 1}\sum_{k\ge 1}\int_U \sum_{\beta\le\alpha-\epsilon(k)}\\
&w_{\alpha-\epsilon(k)-\beta}\frac{\int\left(\mathbf{1}_{[0,r]}G\right)^{\otimes|\beta|}E_\beta d\mu_{|\beta|}}{|\beta|!\beta!}\mathbf{1}_{[0,t]}(r)G(r,\upsilon)\,E_p(r,\upsilon)d\mu\phi_\alpha\\
&= \sum_{|\alpha|\ge 1}\sum_{|p|=1}\int_{U^{|\gamma|}}\sum_{\gamma\le\alpha,|\gamma|\ge 1}\\
&w_{\alpha-\gamma}\left(\mathbf{1}_{[0,r]}G\right)^{\otimes|\gamma|-1}\mathbf{1}_{[0,t]}(r)G(r,\upsilon)\frac{E_\gamma}{(|\gamma|-1)!\gamma!}d\mu_{|\gamma|}\phi_\alpha\\
&= \sum_{|\alpha|\ge 1}\sum_{\gamma\le\alpha,|\gamma|\ge 1}\int w_{\alpha-\gamma}\left(\mathbf{1}_{[0,t]}G\right)^{\otimes|\gamma|}\frac{E_\gamma}{|\gamma|!\gamma!}d\mu_{|\gamma|}\phi_\alpha = \psi(t) - w,
\end{aligned}$$

and (5.7.31) holds.

(ii) follows from (5.7.33). The part (iii) is a direct consequence of (5.7.34) and the definition of $\|w\|$. Finally, (iv) follows from (5.7.34), Proposition 5.7.57 and Corollary 5.7.58.

Example 5.7.62 Let $F \in L_2(U, d\mu)$. Taking $w = \exp^{\diamond}\{\Phi(F)\}$ in (5.7.34), we see that the solution to (5.7.31)

$$\begin{aligned}
u(t) &= \exp^{\diamond}\{\phi(F)\}\diamond\exp^{\diamond}\left\{\phi\left(\mathbf{1}_{[0,t]}G\right)\right\}\\
&= \exp^{\diamond}\left\{\phi(F) + \phi\left(\mathbf{1}_{[0,t]}G\right)\right\}
\end{aligned}$$

is, in general, non-adapted, but, by Proposition 5.7.57, $\sup_t \mathbf{E}\left[u(t)^2\right] < \infty$.

For $s \le t$ and $f = \sum_\alpha f_\alpha(t, \upsilon)\,\phi_\alpha \in L_2(D'(U), d\mu)$, define

$$H_k(s,t) = \int \mathbf{1}_{[s,t]} G\mathfrak{m}_k d\mu,\quad H^\alpha(s,t) = \prod_k H_k^{\alpha_k} k(s,t),\quad \boldsymbol{\alpha} \in \mathcal{J}^1,$$

$$||f||^2_{0,T} = \mathbb{E}\int_U |f|^2\, d\mu,$$

$$||f||^2_{T} = \sum_\alpha \alpha! \int_U |f_\alpha|^2\, d\mu + \sup_t \sum_\alpha \alpha!$$

$$\left(\sum_{k\ge 1}\int_0^t\int_U \sum_{\beta+\epsilon(k)\le\alpha, |\beta|\le n} f_{\alpha-\epsilon(k)-\beta}(s,\upsilon)E_k(s,\upsilon)\frac{H^\beta(s,t)}{\beta!}d\mu\right)^2.$$

Proposition 5.7.63 *Let* $f = \sum_\alpha f_\alpha(t, \upsilon)\,\phi_\alpha \in L_2(D'(U), d\mu)$, $w = \sum_\alpha w_\alpha \phi_\alpha \in D'$. *Then*

(i) The unique solution to (5.7.31) in $CD'([0,T]; \mathbf{R})$ *is*

$$u(t) \tag{5.7.35}$$

$$= w \diamond \exp^\diamond\left\{\Phi\left(\chi_{[0,t]}G\right)\right\} + \int_0^t\int \exp^\diamond\left\{\Phi\left(\mathbf{1}_{[s,t]}G\right)\right\} \diamond f(s,\upsilon) \diamond \Phi(ds, d\upsilon)$$

$$= \sum_{|\alpha|\ge 1}\phi_\alpha \sum_{k\ge 1}\int_0^t\int_U \sum_{\beta+\epsilon(k)\le\alpha, |\beta|\le n} f_{\alpha-\epsilon(k)-\beta}(s,\upsilon)E_k(s,\upsilon)\frac{H^\beta(s,t)}{\beta!}d\mu$$

$$+ \sum_\alpha \phi_\alpha \sum_{\beta\le\alpha} w_{\alpha-\beta}\frac{H(r)^\beta}{\beta!}.$$

(ii) The solution is the limit of the Picard iterations $u^n(t)$*:* $u^0(t) = w + \int_0^t \int f(s,\upsilon) \diamond \Phi(ds, d\upsilon)$,

$$u^{n+1}(t) = w + \int_0^t\int \left[u^n(s)G(s,\upsilon) + f(s,\upsilon)\right] \diamond \Phi(ds, d\upsilon),\quad 0 \le t \le T. \tag{5.7.36}$$

That is,

$$u^n(t) = w \diamond \sum_{k=0}^n \frac{\phi\left(\mathbf{1}_{[0,t]}G\right)^{\diamond k}}{k!} + \int_0^t\int \sum_{k=0}^n \frac{\phi\left(\mathbf{1}_{[s,t]}G\right)^{\diamond k}}{k!} \diamond f(s,\upsilon) \diamond \Phi(ds, d\upsilon). \tag{5.7.37}$$

(iii) *If* $f \in \mathcal{S}'(\mathbf{H})$ *and* $w \in \mathcal{S}'$, *then* u^n, $u \in \mathcal{CS}'([0,T];\mathbb{R})$;

$$\sup_t \mathbb{E}\left[u^2(t)\right] \le 2\left(||w||^2 + ||f||_T^2\right).$$

(iv) *If* $w = w_0$ *is deterministic and* f *is adapted with* $\|f\|_{0,T}^2 < \infty$, *then* $u(t)$ *is adapted and*

$$\sup_t \mathbb{E}\left[u^2(t)\right] \le C\left(w_0^2 + \mathbb{E}\int_U |f|^2 d\mu\right).$$

Proof Because of Lemma 5.7.61 we assume $w = 0$.

(i) Let

$$\begin{aligned} l(s,\upsilon) &= f(s,\upsilon) \diamond \exp^{\diamond}\left\{\Phi\left(\mathbf{1}_{[s,r]}\right)G\right\} \\ &= \sum_{\alpha} \sum_{\beta\le\alpha, |\beta|\le n} f_{\alpha-\beta}(s,\upsilon)\frac{H^{\beta}(s,r)}{\beta!}, \quad 0 \le s \le r \le T \\ &= \sum_{\alpha}\phi_\alpha \sum_{\beta\le\alpha,|\beta|\le n} f_{\alpha-\beta}(s,\upsilon)\frac{1}{|\beta|!\beta!}\int\left(\mathbf{1}_{[s,t]}G\right)^{\otimes|\beta|}E_\beta d\mu_{|\beta|}. \end{aligned}$$

and, for $0 \le r \le T$, set

$$\begin{aligned} \psi(r) &= \int_0^r \int l(s,\upsilon) \diamond \phi\,(ds, d\upsilon) \\ &= \sum_{|\alpha|\ge 1}\phi_\alpha \sum_{k\ge 1,\ \epsilon(k)\le\alpha}\int \sum_{\beta+\epsilon(k)\le\alpha, |\beta|\le n} f_{\alpha-(\epsilon(k)+\beta)}(s,\upsilon)E_k(s,\upsilon)d\mu\frac{H^{\beta}(s,t)}{\beta!}. \end{aligned}$$

For $0 \le r \le T$ and $\upsilon = (s_1, \upsilon_1, \dots, s_k, \upsilon_k) \in U^k$, $k \ge 1$, define

$$\begin{aligned} \Psi(r,k,G,f) &= \Psi(r,k,G,f)(s_1,\upsilon_1,\dots,s_k,\upsilon_k) \\ &= \sum_{j=1}^{k} f(\hat{s},\upsilon_j) \prod_{i=1,i\ne j}^{k} \chi_{[\hat{s},r]}(s_i)\, G(s_i,\upsilon_i), \end{aligned}$$

where $\hat{s} = \min\{s_i, 1 \le i \le k\}$.

By Corollary 5.7.58,

$$\begin{aligned} &\psi(r) \\ &= \sum_{|\alpha|\ge 1, k\ge 1,\ \epsilon(k)\le\alpha} \phi_\alpha \int_0^r \int \sum_{\beta+\epsilon(k)\le\alpha} f_{\alpha-(\epsilon(k)+\beta)}(s,\upsilon)\frac{E_p(s,\upsilon)}{|\beta|!\beta!} \end{aligned}$$

$$\int \left(\mathbf{1}_{[s,r]}G\right)^{\otimes|\beta|} E_\beta d\mu_{|\beta|} d\mu$$

$$= \sum_\alpha \phi_\alpha \sum_{\alpha \geq \beta', |\beta'| \geq 1} \int_{U^{|\beta|+1}} \left(\mathbf{1}_{[s,r]}G\right)^{\otimes(|\beta'|-1)} f_{\alpha-\beta'}(s,\upsilon) \frac{E_{\beta'}}{(|\beta'|-1)!\beta'!} d\mu_{|\beta'|}$$

$$= \sum_\alpha \phi_\alpha \sum_{\beta' \leq \alpha, 1 \leq |\beta'|} \int_{U^{|\beta'|}} \Psi\left(r, |\beta'|, G, f_{\alpha-\beta'}\right) \frac{E_{\beta'}}{|\beta'|!\beta'!} d\mu_{|\beta'|}.$$

We will show that ψ solves (5.7.31). Indeed,

$$\int_0^t \psi(r)G(r,\upsilon) \diamond \Phi(dr, d\upsilon)$$

$$= \sum_{|\alpha|\geq 2} \phi_\alpha \sum_{k\geq 1} \sum_{\beta'+\epsilon(k)\leq\alpha, 1\leq|\beta'|} \int \chi_{[0,t]}(r)G(r,\upsilon)E_k(r,\upsilon) \times$$

$$\times \int_{U^{|\beta'|}} \Psi\left(r, |\beta'|, G, f_{\alpha-(\epsilon(k)+\beta')}\right) \frac{E_{\beta'}}{|\beta'|!\beta'!} d\mu_{|\beta'|} d\mu$$

$$= \sum_{|\alpha|\geq 2} \phi_\alpha \sum_{k\geq 1} \sum_{\gamma\leq\alpha, 2\leq|\gamma|} \int_{U^{|\gamma|}} \mathbf{1}_{[0,t]}(r)G(r,\upsilon)\Psi\left(r, |\gamma|-1, G, f_{\alpha-\gamma}\right)$$

$$\times \frac{E_\gamma}{(|\gamma|-1)!\gamma!} d\mu_{|\beta'|} d\mu$$

$$= \sum_{|\alpha|\geq 2} \phi_\alpha \sum_{\gamma=\leq\alpha, 2\leq|\gamma|} \int_{U^{|\gamma|}} \Phi\left(t, |\gamma|, G, f_{\alpha-\gamma}\right) \frac{E_\gamma}{|\gamma|!\gamma!} d\mu_{|\beta'|} d\mu$$

and, by construction,

$$\int_0^t \psi(r)G(r,\upsilon) \diamond \Phi(dr, d\upsilon) = \psi(t) - \int_0^t \int_V f(r) \diamond G(r,\upsilon)\, \Phi(dr, d\upsilon).$$

(ii) Consider $u^n(t)$ defined by (5.7.37) with $w = 0$. Then

$$u^n(t) = \sum_{|\alpha|\geq 1} \phi_\alpha \sum_{k\geq 1, \epsilon(k)\leq\alpha} \sum_{\beta+\epsilon(k)\leq\alpha, |\beta|\leq n}$$

$$\frac{H^\beta(s,t)}{\beta!} \int_V f_{\alpha-(\epsilon(k)+\beta)}(s,\upsilon)E_k(s,\upsilon)d\mu, \tag{5.7.38}$$

and, after repeating corresponding arguments from the proof of part (i), we see that for $0 \leq t \leq, T$

$$\int_0^t \int_V u^n(r)\, G(r,\upsilon) \diamond \Phi(dr, d\upsilon) = u^{n+1}(t) - \int_0^t \int_V f(s,\upsilon) \diamond \Phi(ds, d\upsilon).$$

If $f \in \mathcal{S}'(\mathbf{H})$, then $u^0 \in \mathcal{CS}'([0,T])$. If $u^n \in \mathcal{CS}'([0,T])$, then $u^n G \in \mathcal{S}'(\mathbf{H})$. By Proposition 5.7.37, $u^{n+1} \in \mathcal{CS}'(L_2[0,T])$ and the statement follows by comparing (5.7.38) and (5.7.35).

The part (iii) is a direct consequence of (5.7.35).

(iv) Since $\mathbb{E}\int_U |f|^2 d\mu < \infty$, it follows that $f \in \mathcal{S}'(\mathbf{H}, \mathbb{R})$, and, according to part (ii) and Proposition 5.7.50, all the iterations are adapted. Therefore Itô isometry holds. By definition,

$$\sup_t \mathbb{E}\left[u^0(t)\right]^2 \leq \mathbb{E}\int_U |f|^2 d\mu < \infty.$$

Assume $\sup_t \mathbb{E}\left[u^n(t)\right]^2 < \infty$. Using (5.7.36) and Itô isometry,

$$\mathbb{E}\left[u^{n+1}(t)\right]^2 \leq C\left[\int_0^t \int_V \mathbb{E}\left(u^n(s)\right)^2 G(s,\upsilon)^2 d\pi ds + \mathbb{E}\int_0^t \int_V |f|^2 d\pi ds\right],$$

and, by Gronwall's lemma, there is a constant C independent of n such that

$$\sup_{n,t} \mathbb{E}\left(u^n(t)\right)^2 \leq C\mathbb{E}\int_0^T \int_V |f|^2 d\pi ds.$$

Similarly, using Gronwall's lemma, we show that

$$\sum_n \sup_t \mathbb{E}\left([u^{n+1}(t) - u^n(t)]^2\right) < \infty.$$

This concludes the proof of Proposition 5.7.63.

In this section we extend the results on the linear SDE to a simple parabolic SPDE.

As in the previous section, let $U = [0,T] \times V$, $d\mu = dtd\pi$,

$$\Phi(dt, d\upsilon) = \dot{\Phi}(t,\upsilon)d\mu(t,\upsilon)$$

We denote $\mathbf{R}_T^{\mathrm{d}} = \mathbb{R}^{\mathrm{d}} \times [0,T]$ and suppose that the following measurable functions are given

$$a : \mathbb{R}^{\mathrm{d}} \to \mathbb{R}^{\mathrm{d}\times\mathrm{d}}, \quad b : \mathbb{R}^{\mathrm{d}} \to \mathbb{R}^{\mathrm{d}}.$$

The following is assumed.

A1. The functions a, b are infinitely differentiable and bounded with all derivatives, and the matrix $a = \left(a^{ij}(x)\right)$ is symmetric and non-degenerate: for all x

$$a^{ij}(x)y_i y_j \geq \delta\,|y|^2, y \in \mathbb{R}^{\mathrm{d}},$$

for some $\delta > 0$.

Let $H_2^s = H_2^s\left(\mathbb{R}^d\right)$, $s = 1, 2, \ldots$, be the Sobolev class of square-integrable functions ψ on $\mathbb{R}^d$ having generalized space derivatives up to order s with the finite norm

$$|\psi|_{s,2} = |\psi|_2 + |D_x^s \psi|_2 ,$$

where $|\psi|_2 = (\int_{\mathbb{R}^d} |\psi|^2 \, dx)^{1/2}$.

Let $G \in L_2([0,T] \times V, d\mu)$ with $d\mu = dtd\pi$. Define

$$L^{2,1} = L^{2,1}\left(\mathbb{R}^d \times [0,T] \times V, dxdtd\pi\right)$$

as the space of all measurable functions g on $\mathbb{R}^d \times [0,T] \times V$ such that

$$||g||_{1,2}^2 = \int_0^T \int_{\mathbb{R}^d} \int_V [|g(s,x,\upsilon)|^2 + |D_x g(s,x,\upsilon)|^2] dsdxd\pi < \infty.$$

Let $w = \sum_\alpha w_\alpha(x)\, \phi_\alpha \in D'\left(H_2^3(\mathbb{R}^d)\right)$ and $g = \sum_\alpha g_\alpha(x,s,\upsilon)\phi_\alpha \in D'\left(L^{2,1}\right)$. The main objective of this section is to study the equation for $u(t) = u(t,x)$,

$$\begin{aligned} du(t,x) &= \mathcal{L}u(t,x)dt \\ &\quad + \int_U (u(t,x)G(t,\upsilon) + g(t,x,\upsilon)) \diamond \Phi(dt, d\upsilon) \\ u(0,x) &= w(x), \end{aligned} \tag{5.7.39}$$

where $\mathcal{L}u = a^{ij}(x)u_{x_i x_j} + b^i(x)u_{x_i}$. An equivalent form of (5.7.39) is

$$\begin{aligned} u(t,x) &= w(x) + \int_0^t \mathcal{L}u(s,x)ds \\ &\quad + \int_0^t \int_U [u(s,x)G(s,\upsilon) + g(s,\upsilon)] \diamond \Phi(ds, d\upsilon), \end{aligned} \tag{5.7.40}$$

$0 \le t \le T$.

We will seek a solution to (5.7.40) in the form

$$u(t,x) = \sum_\alpha u_\alpha(t,x)\phi_\alpha \in CD'([0,T]; H_2^2). \tag{5.7.41}$$

We start our analysis of Eq. (5.7.40) by introducing the definition of a solution in the "weak sense". Recall that $CD'([0,T], H_2^2)$ is the class of all generalized processes $u = \sum_\alpha u_\alpha(t)\phi_\alpha$ on $[0,T]$ such that each u_α is continuous on $[0,T]$ with values in H_2^2.

Definition 5.7.64 We say that a generalized D-process

$$u(t) = \sum_{\alpha} u_\alpha(t)\phi_\alpha \in \mathcal{C}D'([0,T],H_2^2)$$

is a D-H_2^2 solution of equation (5.7.40) in $[0,T]$, if the equality (5.7.40) holds in $D(L_2(\mathbb{R}^d))$ for every $0 \le t \le T$

5.7.4.3 Linear Parabolic SPDEs

Lemma 5.7.65 *Assume* ***A1*** *holds and*

$$w = \sum_{\alpha} w_\alpha \phi_\alpha \in D'(H_2^3), \;\; g = \sum_{\alpha} g_\alpha \phi_\alpha \in D'\left(L^{2,1}\right).$$

Then there is a unique solution to (5.7.40) in $\mathcal{C}D'([0,T],H_2^2)$. *The coefficients* u_α *of the solution* u *given by (5.7.41) satisfy* $u_{(\mathbf{0})}(t) = w_{(\mathbf{0})}$,

$$\begin{cases} \partial_t u_\alpha = \mathcal{L}u_\alpha + \sum_k \int_V \mathfrak{m}_k(u_{\alpha(k)}G + g_{\alpha(k)})d\pi \\ u_\alpha|_{t=0} = w_\alpha. \end{cases} \tag{5.7.42}$$

Proof Plugging the series (5.7.41) into (5.7.40) we get (5.7.42): by definition, for $t \in [0,T]$,

$$\sum_{\alpha} u_\alpha \phi_\alpha = \sum_{\alpha} w_\alpha \phi_\alpha$$
$$+ \sum_{\alpha} \left(\int_0^t \mathcal{L}u_\alpha(x,s)ds\phi_\alpha + \sum_{\alpha}\sum_{k} \int_0^t \int_V \mathfrak{m}_k[u_{\alpha(k)}G + g_{\alpha(k)}]d\pi ds \right) \phi_\alpha.$$

Denote by $T_t h$ the solution of the problem

$$\begin{cases} \partial_t u = \mathcal{L}u, \quad 0 \le t \le T, \\ u(0,x) = h(x), x \in \mathbf{R}^d. \end{cases}$$

If **A1** holds, then [128]

$$|T_t h|^2_{L_2(\mathbf{R}^d)} \le e^{Ct}|h|^2_{L_2(\mathbf{R}^d)}, \tag{5.7.43}$$

Since (5.7.42) is a lower-triangular system, starting with $u_{(\mathbf{0})}(t) = T_t w_{(\mathbf{0})}$ we find unique continuous $u_{\boldsymbol{\alpha}}(t)$ for every $\boldsymbol{\alpha}$ with $|\boldsymbol{\alpha}| \geq 1$:

$$u_\alpha(t) = T_t w_\alpha + \sum_k \int_0^t \int_V [\mathfrak{m}_k(s,\upsilon)(T_{t-s}u_{\alpha(k)}(s)G(s,\upsilon) \qquad (5.7.44)$$

$$+ T_{t-s}g_{\alpha(k)}(s,\upsilon))]dsd\pi.$$

Next, we establish an equivalent mild formulation of the equation.

Lemma 5.7.66 *Assume **A1** holds and*

$$w = \sum_\alpha w_\alpha \phi_\alpha \in D'(H_2^3), \quad g = \sum_\alpha g_\alpha \phi_\alpha \in D'\left(L^{2,1}\right).$$

Then u is the unique solution to (5.7.40) in $CD'([0,T], H_2^2)$ if and only if it satisfies

$$u(t) = \int_0^t \int_U [T_{t-s}u(s)G(s,\upsilon) + T_{t-s}g(s,\upsilon)] \diamond \phi(ds, d\upsilon) \qquad (5.7.45)$$

$$+ T_t w,$$

$0 \leq t \leq T$.

Proof Since (5.7.44) holds, the statement is an immediate consequence of Lemma 5.7.65.

Next, we derive a closed-form expression and the norm bounds for the solution.

Proposition 5.7.67 *Let **A1** hold and*

$$w = \sum_\alpha w_\alpha(x)\phi_\alpha \in D'\left(H_2^3(\mathbb{R}^d)\right), \quad g = \sum_\alpha f_\alpha(x,s,\upsilon)\phi_\alpha \in D'\left(L^{2,1}\right).$$

(i) The unique solution to (5.7.40) is given by

$$u(t) = T_t w(x) \diamond \exp^\diamond\left\{\Phi\left(\mathbf{1}_{[0,t]}G\right)\right\}$$
$$+ \int_0^t \int_U \exp^\diamond\left\{\Phi\left(\mathbf{1}_{[s,t]}G\right)\right\} \diamond T_{t-s}g(s,x,\upsilon) \diamond \Phi(ds,d\upsilon)$$

and has the chaos expansion

$$u(t) = \sum_{|\alpha|\geq 1}\left(\sum_{k\geq 1}\int_0^t\int_U \sum_{\beta+\epsilon(k)\leq\alpha} T_{t-s}g_{\alpha-p-\beta}(s,\upsilon)E_k(s,\upsilon)\frac{H^\beta(s,t)}{\beta!}d\mu\right)\phi_\alpha$$
$$+ \sum_\alpha\left(\sum_{\beta\leq\alpha} T_t w_{\alpha-\beta}\frac{H^\beta(t)}{\beta!}\right)\phi_\alpha.$$

(ii) The solution is the limit of Picards iterations $u^n(t)$ with

$$u^0(t) = T_t w + \int_0^t \int T_{t-s} g(s, \upsilon) \diamond \Phi(ds, d\upsilon),$$

$$u^{n+1}(t) = T_t w + \int_0^t \int_U [T_{t-s} u^n(s) G(s, \upsilon) + T_{t-s} g(s, \upsilon)] \diamond \Phi(ds, d\upsilon),$$

$0 \le t \le T$. In particular, for $0 \le t \le T$,

$$u^n(t) = T_t w \diamond \sum_{k=0}^{n} \frac{\Phi\left(\mathbf{1}_{[0,t]} G\right)^{\diamond k}}{k!}$$

$$+ \int_0^t \int_U \sum_{k=0}^{n} \frac{\Phi\left(\mathbf{1}_{[s,t]} G\right) c^{\diamond k}}{k!} \diamond T_{t-s} g(s, \upsilon) \diamond \Phi(ds, d\upsilon).$$

If $g \in \mathcal{S}'(\mathbf{H})$ and $w \in \mathcal{S}'$, then $u^n, u \in C\mathcal{S}'([0, T])$;

(iii) If w is deterministic and g is adapted, then the solution u is adapted and

$$\sup_t \mathbb{E}\left[\|u(t)\|^2_{L_2(\mathbb{R}^d)}\right] \le C\mathbb{E}\left[\|w\|_{L_2(\mathbb{R}^d)} + \int_0^T \int_{\mathbb{R}^d} \int_U |g(x, s, \upsilon)|^2 \, dx \, d\pi \, ds\right].$$

Proof We repeat the main arguments of Proposition 5.7.63 (as in the case of linear SDE). The changes in the proof (i) are obvious. The proof of (ii)–(iii) is identical to the proof of (ii), (iv) in Proposition 5.7.63 with the use of (5.7.43) for the estimate of the iterations $L_2(\Omega, \mathbb{P})$-norm.

5.7.4.4 Stationary SPDEs

Let us consider a stationary (time independent) equation

$$\mathbf{A}u + \delta_\Phi(\mathbf{M}u) = g \tag{5.7.46}$$

where, as previously, the Φ-noise is a formal series $\dot{\Phi} = \sum_k \mathfrak{m}_k \phi_k$, $\{\mathfrak{m}_k\}$ is a CONS in a Hilbert space H, ϕ_k are independent random variables with zero mean and variance 1, and $g = \sum_\alpha g_\alpha \phi_\alpha$ is a free term.

We will consider Eq. (5.7.46) in a normal triple of Hilbert spaces (V, H, V') :

- $V \subset H \subset V'$ and the embeddings $V \subset H$ and $H \subset V'$ are dense and continuous;
- The space V' is dual to V relative to the inner product in H;
- There exists a constant $C > 0$ such that $|(u, \upsilon)_H| \le C \|u\|_V \|\upsilon\|_{V'}$ for all u and υ.

A typical example of a normal triple is the Sobolev spaces

$$\left(H^{r+\gamma}\left(\mathbb{R}^d\right), H^r\left(\mathbb{R}^d\right), H^{r-\gamma}\left(\mathbb{R}^d\right)\right) \text{ for } \gamma > 0.$$

Everywhere in this section it is assumed that $\mathbf{A} : V \to V'$ and $\mathbf{M} : V \to V' \otimes \ell_2$ are bounded linear operators.

As we already know, Eq. (5.7.46) can be rewritten in the form

$$\mathbf{A}u + \sum_{n\geq 1} \mathbf{M}_n u \diamond \phi_n = g, \tag{5.7.47}$$

where $u = \sum_\alpha u_\alpha \phi_\alpha$. Since

$$\mathbf{M}_n u = \sum_{\alpha \in \mathcal{J}^1} \mathbf{M}_n u_\alpha \phi_\alpha, \tag{5.7.48}$$

we get

$$\begin{aligned}
&\sum_{n\geq 1} \mathbf{M}_n u \diamond \phi_n \\
&= \sum_{\alpha\in\mathcal{J}^1}\sum_{n\geq 1} \mathbf{M}_n u_\alpha \phi_\alpha \diamond \phi_n + \sum_{\alpha\in\mathcal{J}^1}\sum_{n\geq 1} \mathbf{M}_n u_\alpha \phi_{\alpha+\epsilon(n)} \\
&= \sum_{n\geq 1}\sum_{\beta\in:|\beta|\geq 1} \mathbf{M}_n u_{\beta-\epsilon(n)} \phi_\beta .
\end{aligned}$$

Therefore, for $\boldsymbol{\alpha} \in \mathcal{J}^1$ such that $|\boldsymbol{\alpha}| > 0$, we have

$$\left(\sum_{n\geq 1} \mathbf{M}_n u \diamond \phi_n\right)_\alpha = \sum_{n\geq 1} \mathbf{M}_n u_{\alpha-\epsilon(n)}$$

The propagator describing the coefficients $(u_\alpha, \boldsymbol{\alpha} \in \mathcal{J}^1)$ for the solution of (5.7.47),

$$u = \sum_{\alpha\in\mathcal{J}^1} u_\alpha \phi_\alpha, \tag{5.7.49}$$

is

$$\begin{aligned}
&\mathbf{A}u_\alpha = \mathbb{E}g \ \text{ if } |\boldsymbol{\alpha}| = 0 \\
&\mathbf{A}u_\alpha + \textstyle\sum_{n\geq 1} \mathbf{M}_n u_{\alpha-\epsilon(n)} = g_\alpha \ \text{ if } |\boldsymbol{\alpha}| > 0.
\end{aligned} \tag{5.7.50}$$

The propagator (5.7.50) is a lower triangular system. Therefore, if $\mathbf{A}$ has an inverse $\mathbf{A}^{-1}$, then the propagator can be solved sequentially.

The next question to ask whether the solution (5.7.49) of Eq. (5.7.47) has a finite variance. The following example demonstrates that, in general the answer to this question is negative.

Example 5.7.68 Consider the following simple version of the equation

$$u = 1 + u \diamond \phi,$$

with $\mathbb{E}\phi = 0$, $\mathbb{E}\phi^2 = 1$. In this setting, $\mathcal{J}^1 = (0, 1, 2, ...)$ and consists of one-dimensional indices $\boldsymbol{\alpha} = (\mathbf{0}), (1), (2), \ldots$. Recall that, by our general convention,

$$\mathbb{E}\phi_{(n)}^2 = n!.$$

System (5.7.50) in this case becomes

$$u_{(0)} = 1, \; u_{(n)} = I_{n=0} + u_{n-1}, \; n \geq 1,$$

that is, $u_n = 1$ and $u = \sum_n \phi_{(n)}$ so that

$$\mathbb{E}u^2 = \sum_{n \geq 1} n! = \infty.$$

As a result, we are essentially forced to define the solution to Eq. (5.7.47) as a generalized D-random variable with values in V, such that (5.7.47) holds in $D(V')$.

5.7.4.5 Weighted Norms

Since the space D' is often too large to provide any useful information about the solution, a popular definition of solutions in the Gaussian and Poisson cases is based on rescaling/weighting of the coefficients $u_{\boldsymbol{\alpha}}$; cf. page 255, as well as [38, 151, 166, 177, etc.]. This technique is also works in the distribution-free setting.

Given a separable Hilbert space X and a sequence of positive numbers $\mathcal{R} = \{r_{\boldsymbol{\alpha}}, \; \boldsymbol{\alpha} \in \mathcal{J}^1\}$, we define the space $\mathcal{R}L_2\,(X)$ as the collection of formal series $f = \sum_{\boldsymbol{\alpha}} f_{\boldsymbol{\alpha}} \phi_{\boldsymbol{\alpha}}, \; f_{\boldsymbol{\alpha}} \in X$, such that

$$\|f\|_{\mathcal{R}L_2(X)}^2 = \sum_{\boldsymbol{\alpha}} \|f_{\boldsymbol{\alpha}}\|_X^2 \, r_{\boldsymbol{\alpha}}^2 < \infty. \tag{5.7.51}$$

If (5.7.51) holds, then $\sum_{\boldsymbol{\alpha}} r_{\boldsymbol{\alpha}} f_{\boldsymbol{\alpha}} \phi_{\boldsymbol{\alpha}} \in L_2\,(X)$.

Similarly, the sequence $\mathcal{R}^{-1} = \{r_{\boldsymbol{\alpha}}^{-1}, \; \boldsymbol{\alpha} \in \mathcal{J}^1\}$ defines the space $R^{-1}L_2\,(X)$, consisting of the formal series $f = \sum_{\boldsymbol{\alpha}} f_{\boldsymbol{\alpha}} \phi_{\boldsymbol{\alpha}}, \; f_{\boldsymbol{\alpha}} \in X$, such that

$$\sum_{\boldsymbol{\alpha}} \|f_{\boldsymbol{\alpha}}\|_X^2 \, r_{\boldsymbol{\alpha}}^{-2} < \infty$$

Important and popular examples of the space $RL_2(X)$ correspond to the following weights:

(a) $r_\alpha^2 = \prod_{k=1}^{\infty} q_k^{\alpha_k}$, where $\{q_k, k \geq 1\}$ is a non-increasing sequence of positive numbers;

(b) Kondratiev's spaces $(\mathcal{S})_{\rho,\ell}$:

$$r_\alpha^2 = (\alpha!)^\rho (2\mathbb{N})^{\ell\alpha}, \ \rho \leq 0, \ \ell \leq 0.$$

In particular, in the setting of Example 5.7.68, $\mathbb{E}u^2 = \|u\|^2_{(\mathcal{S})_{0,0}} = \infty$, but $\|u\|^2_{(\mathcal{S})_{\rho,\ell}} < \infty$ for suitable ρ and ℓ

Exercise 5.7.69 (A) Determine a range of values for ρ and ℓ such that the generalized process $u = \sum_n \phi_{(n)}$ in Example 5.7.68 belong to $(\mathcal{S})_{\rho,\ell}$. [Recall a similar exercise in the Gaussian case on page 257.]

5.7.4.6 Wick-Nonlinear SPDEs

To illustrate the general idea, consider the equation

$$\mathbf{A}u - u^{\diamond 3} + \sum_{n\geq 1} \mathbf{M}_n u \diamond \phi_n = f, \tag{5.7.52}$$

where $u^{\diamond 3} = u \diamond u \diamond u$, $f = \sum_\alpha f_\alpha \phi_\alpha$. As in the previous sections, we will look for a chaos solution of the form

$$u = \sum_{\alpha\in\mathcal{J}^1} u_\alpha \phi_\alpha.$$

The definition of the Wick product implies

$$u^{\diamond 3} = \sum_{\alpha,\beta,\gamma\in\mathcal{J}^1} u_\alpha u_\beta u_\gamma \, \phi_{\alpha+\beta+\gamma},$$

that is,

$$\left(u^{\diamond 3}\right)_\alpha = \sum_{\beta,\beta',\beta'':\beta+\beta'+\beta''=\alpha} u_\beta u_{\beta'} u_{\beta''}$$

Therefore, the propagator for Eq. (5.7.52) is

$$\mathbf{A}u_\alpha - \sum_{\beta,\beta',\beta'':\beta+\beta'+\beta''=\alpha} u_\beta u_{\beta'} u_{\beta''} + \sum_{n\geq 1} \mathbf{M}_n u_{\alpha-\epsilon(n)} = f_\alpha. \tag{5.7.53}$$

Similar to (5.7.50), system (5.7.53) is also lower triangular and can be solved sequentially, assuming that operator **A** has an appropriate inverse.

If the Wick cubic $u^{\diamond 3}$ is replaced by any Wick power or a polynomial then the related propagator remains lower triangular. The same ideas also apply to evolution equations.

Exercise 5.7.70 (B) Write the propagator for the following equations:

$$u_{xx} - u \diamond u_x + \sum_{n\geq 1} u \diamond \phi_n = f;$$

$$u_t = u_{xx} - u \diamond u_x + \sum_{n\geq 1} u_x \diamond \phi_n - f.$$

Chapter 6
Parameter Estimation for Diagonal SPDEs

6.1 Examples and General Ideas

6.1.1 An Oceanographic Model and Its Simplifications

Let $U = U(t,x)$ be the temperature of the top layer of a body of water such as lake, sea, or ocean. Various historical data provide information about the long-time average value $\bar{U}$ of U. The quantity of interest then becomes the fluctuation $u = U - \bar{U}$, and time evolution of u can be modeled by the following `heat balance equation` (see Frankignoul [53] or Piterbarg and Rozovskii [186]):

$$u_t = \kappa \boldsymbol{\Delta} u + \boldsymbol{v} \cdot \nabla u - \lambda u + \dot{W}^Q. \tag{6.1.1}$$

Here, $\kappa > 0$ and $\lambda > 0$ are physical parameters and the $\boldsymbol{v} = \boldsymbol{v}(t,x)$ is the velocity of the water on the surface. In the most complete model, $\boldsymbol{v}$ is the solution of the Navier-Stocks equations (see (1.2.18) on page 19). Note that both x and $\boldsymbol{v}$ are two-dimensional.

Practical applications of (6.1.1) for modeling and prediction require knowledge of the parameters κ and λ, and these parameters can only be estimated using the measurements of the temperature. In other words, we have an inverse problem: determine κ and λ given the solution of (6.1.1).

We will simplify the problem by making the following assumptions:

1. the parameters κ and λ are constant, that is, non-random and independent of t and x;
2. the surface is not moving, that is, $\boldsymbol{v} = \boldsymbol{0}$;
3. the shape G of the body of water is a bounded domain in $\mathbb{R}^2$ that is smooth or otherwise nice, e.g. a rectangle, and $u(t,x) = 0$ on the boundary of G;

S.V. Lototsky, B.L. Rozovsky, *Stochastic Partial Differential Equations*, Universitext, DOI 10.1007/978-3-319-58647-2_6

4. Eq. (6.1.1) with $\boldsymbol{v} = \mathbf{0}$ is `diagonal` (or `diagonalizable`), that is, the covariance operator Q of the noise has pure point spectrum and the eigenfunctions of Q are the same as the eigenfunctions of the Laplace operator $\boldsymbol{\Delta}$ in G with zero boundary conditions;
5. there is no initial fluctuation: $u(0, x) = 0$;
6. the solution $u = u(t, x)$ of (6.1.1) is a continuous function of (t, x) can be measured without errors for all $t \in [0, T]$ and $x \in G$.

Equation (6.1.1) becomes

$$u_t = \kappa \mathrm{A} u - \lambda u + \dot{W}^{\mathrm{Q}}, \; 0 < t \leq T, \; x \in G; \; u(t, 0) = 0, \tag{6.1.2}$$

where A is the Laplace operator in G with zero boundary conditions.

Exercise 6.1.1 (B) Find sufficient conditions on the operator Q so that the variational solution of (6.1.2) is a continuous function of (t, x).

Denote by $\mathfrak{h}_k = \mathfrak{h}_k(x), \; k \geq 1$, the eigenfunctions of A. Then

$$\mathrm{A}\mathfrak{h}_k = -\nu_k \, \mathfrak{h}_k, \; \nu_k > 0, \tag{6.1.3}$$

$$\mathrm{Q}\mathfrak{h}_k = q_k \, \mathfrak{h}_k, \; q_k > 0, \tag{6.1.4}$$

$$\dot{W}^{\mathrm{Q}}(t, x) = \sum_{k \geq 1} q_k \dot{w}_k(t) \mathfrak{h}_k(x), \tag{6.1.5}$$

and if a reasonable solution of (6.1.2) exists, it has a representation as a Fourier series in space

$$u(t, x) = \sum_{k \geq 1} u_k(t) \mathfrak{h}_k(x). \tag{6.1.6}$$

Substituting (6.1.3)–(6.1.6) into (6.1.2) leads to the equation for the corresponding Fourier coefficient u_k:

$$du_k(t) = -(\kappa \, \nu_k + \lambda) u_k(t) dt + q_k dw_k(t), \; u_k(0) = 0. \tag{6.1.7}$$

In (6.1.4), we assume that $q_k > 0$ for all k. If $q_n = 0$ for some n, then (6.1.7) implies that $u_n(t) = 0$ for all $t \geq 0$.

The corresponding inverse problem can now be stated as follows: estimate the numbers κ and λ given the observations $u_k(t), k = 1, \ldots, N, \; t \in [0, T]$. In this setting, we can actually forget about the original SPDE (6.1.2) and work directly with (6.1.7).

Estimation of the number q_k from the continuous time observations of u_k is (mathematically) easy.

Exercise 6.1.2 (B) Take a positive integer M and consider a uniform partition $0 = t_0 < t_{1,M} < \ldots < t_{M,M} = T$ of $[0, T]$ with step $\Delta t_M = T/M$: $t_{m,M} = m\Delta t_M$. Verify that

$$q_k^2 = \frac{1}{T}\lim_{M\to\infty}\sum_{m=1}^{M}\Big(u_k\big((m+1)\Delta t_M\big) - u_k\big(m\Delta t_M\big)\Big)^2. \tag{6.1.8}$$

The result also shows that the sign of q_k is not observable, that is, we cannot distinguish between $du = udt + dw(t)$ and $du = udt - dw(t)$ based on the observations $u(t), t \geq 0$.

Hint. The martingale component of u_k is $q_k w_k$.

While it is still not immediately clear how to estimate the numbers κ and λ, we did make some progress by reducing an SPDE setting to an SODE one: from (6.1.2) to (6.1.7). To proceed, we simplify the problem even further and consider an SPDE in one space variable with one unknown parameter θ and with space-time white noise $\dot{W}(t, x)$ as the driving force:

$$u_t(t, x) = \theta\, u_{xx}(t, x)dt + q\dot{W}(t, x),\ 0 < t \leq T,\ x \in (0, \pi), \tag{6.1.9}$$

with zero initial and boundary conditions. When $\theta = 1$ and $q = 1$, we investigate this equation on page 55. In particular, we know that the solution of (6.1.9) is a continuous function of t and x. The reason for introducing q will become clear later, when we study asymptotic behavior of the estimators of θ and use dimensional analysis to check some of the results.

Even though, by gradually moving from (6.1.2) to (6.1.9), we seemingly lost all connections to the original application, we will eventually learn how the ideas and methods we develop while working with the easy equation (6.1.9) can be applied to more complicated equations, including (6.1.2).

Similar to (6.1.2), Eq. (6.1.9) is diagonal, and the solution of (6.1.9) has the Fourier series representation

$$u(t, x) = \sum_{k\geq 1} u_k(t)\mathfrak{h}_k(x),$$

where

$$du_k(t) = -\theta k^2\, u_k(t)dt + q dw_k(t),\ 0 < t \leq T, \tag{6.1.10}$$

with initial condition $u_k(0) = 0$; see (2.3.16) on page 55.

By direct computation,

$$u_k(t) = q\int_0^t e^{-\theta k^2(t-s)}dw_k(s), \tag{6.1.11}$$

$$\mathbb{E}u_k^2(t) = q^2\int_0^t e^{-2\theta k^2(t-s)}ds = \frac{q^2(1 - e^{-2\theta k^2 t})}{2\theta k^2},$$

$$\int_0^T \mathbb{E}u_k^2(t)dt = \frac{q^2}{2\theta k^2}\int_0^T (1-e^{-2\theta k^2 t})dt$$
$$= \frac{q^2}{2\theta k^2}\left(T - \frac{1-e^{-2\theta k^2 T}}{2\theta k^2}\right). \tag{6.1.12}$$

Fix $k \geq 1$ and consider the process $u_k = u_k(t)$, $0 \leq t \leq T$. Multiplying both sides of (6.1.10) by $u_k(t)$ and integrating from 0 to T, we get

$$\int_0^T u_k(t)du_k(t) = -\theta k^2 \int_0^T u_k^2(t)dt + q\int_0^T u_k(t)dw_k(t). \tag{6.1.13}$$

Note that $\int_0^T u_k^2(t)dt > 0$ with probability one. Indeed, $\int_0^T u_k^2(t)dt = 0$ implies $u_k^2(t) = 0$ for almost all t, which, by (6.1.11) is a probability zero event. Then equality (6.1.13) suggests the following estimator $\widehat{\theta}^{(k)}(T)$ of θ:

$$\hat{\theta}^{(k)}(T) = -\frac{\int_0^T u_k(t)du_k(t)}{k^2\int_0^T u_k^2(t)dt}, \tag{6.1.14}$$

or, using the Itô formula,

$$\hat{\theta}^{(k)}(T) = \frac{u_k^2(T) - q^2 T}{2k^2\int_0^T u_k^2(t)dt}.$$

For now, we ignore that, while expression (6.1.14) can be both positive and negative, the number θ must be positive if Eq. (6.1.9) is to be well posed in $L_2((0,\pi))$.

Substituting (6.1.10) into (6.1.14), we get the expression for the estimation error:

$$\hat{\theta}^{(k)}(T) - \theta = -\frac{q\int_0^T u_k(t)dw_k(t)}{k^2\int_0^T u_k^2(t)dt}. \tag{6.1.15}$$

6.1.2 Long Time vs Large Space

In what follows, we will look at the estimation error (6.1.15) in two asymptotic regimes, $T \to +\infty$ for fixed k (long time) and $k \to +\infty$ for fixed T (large space). We will use the following notations:

1. $\mathcal{N}(m,\sigma^2)$, to denote a Gaussian random variable with mean m and variance σ^2;
2. $\zeta \overset{d}{=} \eta$, to indicate that random variables ζ and η have the same distribution;
3. $X \overset{\mathcal{L}}{=} Y$, to indicate that random processes X and Y have the same distribution in the space of continuous functions.

For example, if $w = w(t)$ is a standard Brownian motion, $c > 0$, and $w_c(t) = w(c^2t)$, then

$$w(t) \stackrel{d}{=} \mathcal{N}(0,t) \stackrel{d}{=} \sqrt{t}\mathcal{N}(0,1) \text{ for fixed } t, \tag{6.1.16}$$

$$w(c^2t) \stackrel{\mathcal{L}}{=} cw(t), \text{ that is, } w_c(\cdot) \stackrel{\mathcal{L}}{=} c\,w(\cdot) \text{ as processes.} \tag{6.1.17}$$

Is important to distinguish equality $X(t) \stackrel{d}{=} Y(t)$ for fixed t and equality $X(t) \stackrel{\mathcal{L}}{=} Y(t)$ as processes because $X(t) \stackrel{\mathcal{L}}{=} Y(t)$ implies

$$\int_0^T X(t)dt \stackrel{d}{=} \int_0^T Y(t)dt \tag{6.1.18}$$

and more generally, $F(X) \stackrel{d}{=} F(Y)$ for a continuous functional F on the space of continuous functions: as (6.1.16) suggests, having $X(t) \stackrel{d}{=} Y(t)$ for every t is not enough to claim equalities like (6.1.18).

Exercise 6.1.3 (C) Verify that

$$\int_0^1 w(t)dt \stackrel{d}{=} \mathcal{N}(0,1/3).$$

Let us pass to the limit $T \to +\infty$ in (6.1.15). It is well-known that $u_k(t)$ is an ergodic process with stationary distribution $\mathcal{N}\big(0, q^2/(2\theta k^2)\big)$, a time average converges, to the corresponding population average with respect to the stationary distribution:

$$\lim_{T\to\infty} \frac{1}{T}\int_0^T F\big(u_k(t)\big)dt = \mathbb{E}F(\zeta_k),\ \zeta_k \stackrel{d}{=} \mathcal{N}\big(0, q^2/(2\theta k^2)\big),$$

and the convergence is with probability one. In particular,

$$\lim_{T\to\infty} \frac{1}{T}\int_0^T u_k^2(t)dt = \mathbb{E}\zeta_k^2 = \frac{q^2}{2\theta k^2} \tag{6.1.19}$$

with probability one. On the other hand, by (6.1.12),

$$\mathbb{E}\left(\int_0^T u_k(t)dw_k(t)\right)^2 = \int_0^T \mathbb{E}u_k^2(t)dt \simeq \frac{q^2T}{2\theta k^2},\ T \to +\infty;$$

recall that $f(T) \simeq g(T)$ as $T \to +\infty$ means that $\lim_{T\to+\infty} f(T)/g(T) = 1$. Therefore, by the Chebychev inequality,

$$\lim_{T\to\infty} \frac{1}{T}\int_0^T u_k(t)dw_k(t) = 0 \tag{6.1.20}$$

in probability, and, after combining (6.1.15), (6.1.19), and (6.1.20),

$$\lim_{T\to\infty} \hat{\theta}^{(k)}(T) = \theta \tag{6.1.21}$$

in probability[1] for every $k \geq 1$. In other words, estimator (6.1.14) is consistent in the long-time asymptotic $T \to +\infty$.

The next question is the `rate of convergence` of $\hat{\theta}^{(k)}(T)$: we want to find an increasing non-random function $v = v(T)$ such that $\lim_{T\to+\infty} v(T) = +\infty$ and the limit $\lim_{T\to+\infty} v(T)\big(\hat{\theta}^{(k)}(T) - \theta\big)$ exists in distribution and is a non-degenerate random variable.

To answer this question, we need some basic facts related to convergence in distribution. The first is a version of the martingale central limit theorem.

Theorem 6.1.4 *If, for $t \geq 0$ and $\varepsilon > 0$, $X_\varepsilon = X_\varepsilon(t)$ and $X = X(t)$ are real-valued, continuous square-integrable martingales such that X is a Gaussian process, $X_\varepsilon(0) = X(0) = 0$, and, for some $t_0 > 0$,*

$$\lim_{\varepsilon\to 0} \langle X_\varepsilon\rangle(t_0) = \langle X\rangle(t_0) \tag{6.1.22}$$

in probability, then $\lim_{\varepsilon\to 0} X_\varepsilon(t_0) \stackrel{d}{=} X(t_0)$.

Proof Here is an outline; for complete technical details, see Jacod and Shiryaev [93, Theorem VIII.4.17] or Liptser and Shiryaev [138, Theorem 5.5.4(II)].

First of all, recall that if $\zeta \stackrel{d}{=} \mathcal{N}(0, \sigma^2)$, then the characteristic function of ζ is $\mathbb{E}e^{\mathrm{i}\lambda\zeta} = e^{-\sigma^2\lambda^2/2}$. On the other hand, if M is a continuous square-integrable martingale, then, by the Itô formula, the process

$$\mathcal{G}_M^\lambda(t) = \exp\left(\mathrm{i}\lambda M(t) + \frac{1}{2}\lambda^2\langle M\rangle(t)\right)$$

is a (local) martingale for every real λ. If X is a Gaussian martingale with $X(0) = 0$, then $X(t) \stackrel{d}{=} \mathcal{N}\big(0, \mathbb{E}\langle X\rangle(t)\big)$, and the equalities

$$1 = \mathbb{E}\mathcal{G}_X^\lambda(t),\ 1 = \mathbb{E}\exp\left(\mathrm{i}\lambda X(t) + \frac{1}{2}\lambda^2\mathbb{E}\langle X\rangle(t)\right),$$

being true for all λ and t, suggest that $\langle X\rangle$ is non-random:

$$\langle X\rangle(t) = \mathbb{E}\langle X\rangle(t).$$

[1] With some extra work [140, Theorem 17.4], one can show that convergence in (6.1.20) and (6.1.21) is with probability one.

This is indeed the case; see [93, Sect. II.4(d)] for details. To complete the proof of the theorem, it remains to pass to the limit $\varepsilon \to 0$ in the equality

$$1 = \mathbb{E}\exp\left(\mathrm{i}\lambda X_\varepsilon(t) + \frac{1}{2}\lambda^2\langle X_\varepsilon\rangle(t)\right)$$

and to conclude that

$$\lim_{\varepsilon\to 0}\mathbb{E}\exp\left(\mathrm{i}\lambda X_\varepsilon(t)\right) = e^{-(\lambda^2/2)\langle X\rangle(t)} = \mathbb{E}\exp\left(\mathrm{i}\lambda X(t)\right).$$

Note that passing to the limit in the expectation is not a problem because, for every $\lambda \in \mathbb{R}$, the function $f(x) = e^{\mathrm{i}\lambda x}$ is uniformly bounded: $|f(x)| \le 1$.
This concludes the proof of Theorem 6.1.4.

The idea behind the proof of Theorem 6.1.4 is similar to the idea behind the proof of the `Lévy characterization of the Brownian motion`.

Theorem 6.1.5 (Lévy's Characterization of the Brownian Motion) *If $X = X(t)$ is a continuous square integrable martingale, $X(0) = 0$, and $\langle X\rangle(t) = t$, then X is a standard Brownian motion.*

Exercise 6.1.6 (A)

(a) Prove Theorem 6.1.5.
Hint. Use the Itô formula to show that

$$\mathbb{E}\left(e^{\mathrm{i}\lambda(X(t)-X(s))}\Big|\mathcal{F}_s\right) = e^{-\lambda^2(t-s)/2}.$$

(b) Why is continuity of X a necessary assumption?
Hint. Let X be a Poisson process with unit intensity. What compensates $\big(X(t)-t\big)^2$ to a martingale? At the very least confirm that $\mathbb{E}\big(X(t)-t\big)^2 = t$.

Next, we recall the following classical fact.

Theorem 6.1.7 *If $\zeta_\varepsilon, \eta_\varepsilon$, $\varepsilon > 0$, are random variables such that*

$$\lim_{\varepsilon\to 0}\zeta_\varepsilon \stackrel{d}{=} \zeta$$

and $\lim_{\varepsilon\to 0}\eta_\varepsilon = c$ in probability, where ζ is a random variable and c is a non-random number, then

$$\lim_{\varepsilon\to 0}(\zeta_\varepsilon + \eta_\varepsilon) \stackrel{d}{=} \zeta + c, \quad \lim_{\varepsilon\to 0}\zeta_\varepsilon\eta_\varepsilon \stackrel{d}{=} \zeta\, c.$$

Theorem 6.1.7 is known as the Slutsky theorem, after the Ukranian statistician EVGENY "EUGEN" EVGENIEVICH SLUTSKY (1880–1948), who published it in 1925. The proof is by the usual ϵ–δ argument.

Now we will establish the rate of convergence of $\hat{\theta}^{(k)}(T)$ to θ as $T \to +\infty$, and also identify the limit distribution.

Proposition 6.1.8 *For every $k \geq 1$, the estimator $\hat{\theta}^{(k)}(T)$ is asymptotically normal with rate $\sqrt{T}$:*

$$\lim_{T\to+\infty} \sqrt{T}\big(\hat{\theta}^{(k)}(T) - \theta\big) \stackrel{d}{=} \mathcal{N}\big(0, 2\theta/k^2\big). \tag{6.1.23}$$

Proof To stay in line with the notations from Theorems 6.1.4 and 6.1.7, write $\varepsilon = 1/T$ and define

$$X_\varepsilon(t) = -\frac{1}{\sqrt{T}} \int_0^{Tt} u_k(s)dw_k(s),\ \eta_\varepsilon = \frac{1}{qT}\int_0^T u_k^2(s)ds.$$

Then (6.1.15) implies

$$\sqrt{T}\big(\hat{\theta}^{(k)}(T) - \theta\big) = -\frac{T^{-1/2}\int_0^T u_k(t)dw_k(t)}{k^2(qT)^{-1}\int_0^T u_k^2(t)dt} = \frac{X_\varepsilon(1)}{k^2\eta_\varepsilon}. \tag{6.1.24}$$

Note that, for each $\varepsilon > 0$, X_ε is a square-integrable martingale and

$$\langle X_\varepsilon\rangle(1) = \eta_\varepsilon.$$

Also take a standard Brownian motion $w = w(t)$ and define the Gaussian martingale $X = X(t)$ by

$$X(t) = w(t)/\sqrt{2\theta k^2}.$$

We know from (6.1.19) that, with probability one,

$$\lim_{\varepsilon\to 0} \eta_\varepsilon = \frac{1}{2\theta k^2} = \langle X\rangle(1).$$

Therefore, by Theorem 6.1.4,

$$\lim_{\varepsilon\to 0} X_\varepsilon(1) \stackrel{d}{=} X(1) \stackrel{d}{=} \mathcal{N}\big(0, 1/(2\theta k^2)\big).$$

Now (6.1.23) follows from (6.1.24) and Theorem 6.1.7:

$$\begin{aligned}\lim_{T\to+\infty} \sqrt{T}\big(\hat{\theta}^{(k)}(T) - \theta\big) &= \lim_{\varepsilon\to 0} \frac{X_\varepsilon(1)}{k^2\eta_\varepsilon} \stackrel{d}{=} \frac{w(1)/\sqrt{2\theta k^2}}{k^2/(2\theta k^2)} \\ &= \frac{w(1)\sqrt{2\theta}}{k^2} \stackrel{d}{=} \mathcal{N}(0, 2\theta/k^2).\end{aligned}$$

This completes the proof of Proposition 6.1.8.

Let us now comment about usefulness of q. In the course of the proof of Proposition 6.1.8, there is a potential danger of manipulating various fractions the wrong way and getting, for example, $1/(2\theta k^2)$ as the limit variance in (6.1.23). It is therefore convenient to have an independent way to check whether a certain equality makes sense, and one such way is `dimensional analysis`. Whenever there is time evolution present, it is natural to measure the time variable t in the units of time $[t]$, such as seconds, minutes, etc. This forces the Brownian motion w_k to have the units of $[t]^{1/2}$: $[w_k(t)] = [t]^{1/2}$. If we want to have the process u_k dimensionless, then, to ensure consistent dimensions in (6.1.10), we need to measure θ in the units of inverse time: $[\theta] = [t]^{-1}$, and we need to introduce q with $[q] = [t]^{-1/2}$. Then

$$[\widehat{\theta}^{(k)}(T)] = [t]^{-1}, \quad [\sqrt{T}(\widehat{\theta}^{(k)}(T) - \theta] = [t]^{-1/2},$$

indicating that the limit variance in (6.1.23) should be measured in $[t]^{-1}$, which is indeed the case. Note that this limit variance does not depend on q.

Because the quantity $2\theta/k^2$ in (6.1.23) is a decreasing function of k, it is natural to try the following: instead of keeping k fixed and increasing T, keep T fixed and let k go to infinity. The result is a different kind of asymptotic normality.

Proposition 6.1.9 *For every $T > 0$,*

$$\lim_{k\to+\infty} k\big(\hat{\theta}^{(k)}(T) - \theta\big) \stackrel{d}{=} \mathcal{N}(0, 2\theta/T). \tag{6.1.25}$$

Proof To begin, note that (6.1.25) makes sense as far as dimensional analysis: the units of the limit variance should indeed be $[t]^{-2}$.

Recall that, by (6.1.15),

$$k\big(\hat{\theta}^{(k)}(T) - \theta\big) = -\frac{qk\int_0^T u_k(t)dw_k(t)}{k^2\int_0^T u_k^2(t)dt}.$$

The main objective is to understand what happens to the process $u_k(\cdot)$ as $k \to \infty$. By (6.1.17),

$$\begin{aligned} u_k(t) &= q\int_0^t e^{-\theta k^2 t + \theta k^2 s} dw_k(s) = q\int_0^{k^2 t} e^{-\theta k^2 t + \theta r} dw_k(r/k^2) \\ &\stackrel{\mathcal{L}}{=} \frac{q}{k}\int_0^{k^2 t} e^{-\theta k^2 t + \theta r} dw_1(r) \stackrel{\mathcal{L}}{=} \frac{1}{k}u_1(k^2 t), \end{aligned} \tag{6.1.26}$$

that is

$$u_k(t) = \frac{1}{k}u_1(k^2 t).$$

Then

$$\int_0^T u_k^2(t)dt \overset{d}{=} \frac{1}{k^2}\int_0^T u_1^2(k^2t)dt = \frac{1}{k^4}\int_0^{k^2T} u_1^2(s)ds,$$

and by (6.1.19),

$$\lim_{k\to+\infty} \frac{1}{k^2}\int_0^{k^2T} u_1^2(s)ds = \frac{Tq^2}{2\theta}$$

with probability one. Thus, for every $T \geq 0$,

$$\lim_{k\to+\infty} k^2\int_0^T u_k^2(t)dt = \frac{Tq^2}{2\theta} \tag{6.1.27}$$

in distribution, hence in probability, because the limit $Tq^2/(2\theta)$ is non-random. Next, define

$$X_k(t) = kq\int_0^t u_k(s)dw_k(s);$$

as in the previous proof, it is convenient to have X_k dimensionless. The process X_k is a square-integrable martingale with

$$\langle X_k\rangle(t) = q^2k^2\int_0^t u_k^2(s)ds.$$

If $w = w(t)$ is a standard Brownian motion and $X(t) = q^2w(t)/\sqrt{2\theta}$, then (6.1.27) and Theorem 6.1.4 with $\varepsilon = 1/k$ imply

$$\lim_{k\to\infty} kq\int_0^T u_k(t)dw_k(t) \overset{d}{=} \mathcal{N}\big(0, Tq^4/(2\theta)\big).$$

Since

$$k\big(\hat{\theta}^{(k)}(T) - \theta\big) = \frac{kq\int_0^T u_k(t)dw_k(t)}{k^2\int_0^T u_k^2(t)dt},$$

equality (6.1.25) follows from Theorem 6.1.7.
This completes the proof of Proposition 6.1.9.

Exercise 6.1.10 (B) Show that (6.1.25) implies

$$\lim_{k\to+\infty} \hat{\theta}^{(k)}(T) = \theta$$

in probability, for every $T > 0$.

Hint. Let $\zeta \overset{d}{=} \mathcal{N}(0, 2\theta/T)$. Given $\epsilon > 0$, find C_ϵ so that $\mathbb{P}(|\zeta| > C_\epsilon) < \epsilon$. Then, given $\delta > 0$, take k so that $k\delta > C_\epsilon$.

Propositions 6.1.8 and 6.1.9 quantify the intuitive idea: the more information we have, the closer the estimator is to the true value. In Proposition 6.1.8 extra information comes from increasing time interval; in Proposition 6.1.9 extra information comes from additional Fourier coefficients. It is also reasonable to expect that there are more efficient ways to incorporate the information provided by the Fourier coefficients $u_1, \dots u_k$ than simply using the last one.

Accordingly, let us now assume that the trajectories of $u_k(t)$ are observed for all $0 < t < T$ and all $k = 1, \dots, N$, and let us combine the estimators (6.1.14) for different k as follows:

$$\widehat{\theta}_N = -\frac{\sum_{k=1}^{N} \int_0^T k^2 u_k(t) du_k(t)}{\sum_{k=1}^{N} \int_0^T k^4 u_k^2(t) dt}. \tag{6.1.28}$$

First suggested by Huebner et al. in [83], (6.1.28) is consistent and asymptotically normal in the limit $N \to +\infty$, and is closely related to the maximum likelihood estimator of θ based on the observations $u_k(t)$, $k = 1, \dots, N$, $0 < t < T$. The actual MLE $\widehat{\boldsymbol{\theta}}_N$ of θ takes into account that θ must be positive, whereas $\widehat{\theta}_N$ can be either positive or negative:

$$\widehat{\boldsymbol{\theta}}_N = \begin{cases} \widehat{\theta}_N, & \text{if } \widehat{\theta}_N > 0, \\ 0, & \text{if } \widehat{\theta}_N \le 0. \end{cases} \tag{6.1.29}$$

If $\widehat{\theta}_N$ is consistent, then getting $\widehat{\theta}_N \le 0$ is rather unlikely for large N. We postpone derivation of (6.1.28) and (6.1.29) until Sect. 6.2.2.

Let us try to see that $\widehat{\theta}_N$ is indeed better than $\hat{\theta}^{(N)}$ (now that we fixed T, we will no longer write T as an argument in the estimators).

It follows from (6.1.10) and (6.1.28) that

$$\widehat{\theta}_N - \theta = -\frac{\sum_{k=1}^{N} \int_0^T k^2 u_k(t) dw_k(t)}{\sum_{k=1}^{N} \int_0^T k^4 u_k^2(t) dt}. \tag{6.1.30}$$

Note that both the top and the bottom of the fraction on the right-hand side of (6.1.30) are sums of independent random variables, and the analysis of the properties of the estimator $\widehat{\theta}_N$ is reduced to the study of these sums.

By (6.1.12), as $N \to \infty$,

$$\sum_{k=1}^{N} \int_0^T k^4 \mathbb{E} u_k^2(t) dt \simeq \frac{T}{2} \sum_{k=1}^{N} k^2 \simeq \frac{N^3 T}{6}, \tag{6.1.31}$$

where notation $a_N \simeq b_N$ means $\lim_{N\to\infty}(a_N/b_N) = 1$. Since

$$\mathbb{E}\int_0^T k^2 u_k(t)dw_k(t) = 0,$$

it is reasonable to conjecture that

- by the law of large numbers, $\lim_{N\to+\infty}(\widehat{\theta}_N - \theta) = 0$ with probability one;
- by the central limit theorem, the sequence of random variables $\{N^{3/2}(\widehat{\theta}_N - \theta),\ N \geq 1\}$ converges in distribution to a zero-mean Gaussian random variable.

In other words, we expect the rate of convergence for $\widehat{\theta}_N$ to be $N^{3/2}$, which is indeed better than the rate N for $\hat{\theta}^{(N)}$, and the proof is indeed a straightforward application of the strong law of large numbers and the central limit theorem for independent but not identically distributed random variables. We carry out the proof in a more general setting in Sect. 6.2.3 below.

What is much less obvious is that there are no "reasonable" ways to improve $\widehat{\theta}_N$ beside replacing it with $\widehat{\boldsymbol{\theta}}_N$ according to (6.1.29); the proof of this in a more general setting is in Sect. 6.2.4. There are certainly "unreasonable" improvements, for example, defining an estimator to be a fixed number $\theta_0 > 0$: on the one hand, it is not a very useful estimator, but, on the other hand, if indeed $\theta = \theta_0$, then nothing can beat it.

Now that we have some understanding of the estimation problem for (6.1.9), let us try to imagine what to expect from the estimation problem for (6.1.2). First of all, analysis of (6.1.9) suggests that a similar analysis of (6.1.2) will require the asymptotic formula for ν_k. This turns out to be a classical result, cf. 174:

$$\nu_k \sim k; \tag{6.1.32}$$

recall that $a_k \sim b_k$ means that a finite positive limit of the ratio a_k/b_k exists as $k \to \infty$.

Exercise 6.1.11 (A) Verify (6.1.32) (or an easier version $\nu_k \asymp k$) when G is the square $[0, \pi] \times [0, \pi]$.

Hint. The eigenvalues are of the form $k^2 + m^2$, $k, m \geq 1$. Start by enumerating them.

Next, there are two parameters to estimate rather than one. To get some idea about the resulting estimators, we pretend that estimating two parameters at once and estimating each parameter individually while treating the other one as known leads to similar asymptotic results (which turns out to be true in this case). Then estimating κ in (6.1.2) should be very similar to estimating θ in (6.1.9). The only difference is that, instead of $\nu_k = k^2$, we now have $\nu_k \sim k$, and (assuming for simplicity that $q_k = 1$), this should change (6.1.12) to

$$\int_0^T \mathbb{E}u_k^2(t)dt \sim \frac{T}{k} \tag{6.1.33}$$

and (6.1.31), to

$$\sum_{k=1}^{N} \int_0^T k^2 \mathbb{E} u_k^2(t)dt \sim N.$$

Even without looking at the estimator $\widehat{\kappa}_N$, we can conjecture that $\{N(\widehat{\kappa}_N - \kappa),\ N \geq 1\}$ converges in distribution to a zero-mean normal random variable.

Similarly, estimation of λ in (6.1.2) will lead to

$$\sum_{k=1}^{N} \int_0^T \mathbb{E} u_k^2(t)dt \sim \ln N, \tag{6.1.34}$$

and the corresponding rate of convergence $\sqrt{\ln N}$.

Exercise 6.1.12 (C) Verify (6.1.33) if we have $q_k = 1$ in (6.1.7).

Exercise 6.1.13 (B) Convince yourself that no consistent estimator of θ is possible in the model

$$u_t = u_{xx} + \theta u + \dot{W},\ x \in (0, \pi)$$

with zero boundary conditions, if $t \in [0, T]$ and T is fixed.

Hint. By analogy with (6.1.34), the rate of convergence should be

$$\left(\sum_{k=1}^{N} \int_0^T \mathbb{E} u_k^2(t)dt \right)^{1/2},$$

which, with $\int_0^T \mathbb{E} u_k^2(t)dt \sim 1/k^2$, is a convergent series. In other words, $\widehat{\theta}_N - \theta$ stays a non-degenerate random variable in the limit $N \to +\infty$.

6.1.3 Problems

All problems below are related to the estimator $\hat{\theta}^{(k)}(T)$ defined in (6.1.14) on page 384. Problem 6.1.1 offers one way to incorporate several processes u_k into a single estimator. Problem 6.1.2 suggests yet another asymptotic regime: small noise. Problem 6.1.3 investigates the influence of the initial condition $u_k(0)$ in the large space asymptotic. Problems 6.1.4 and 6.1.5 investigate $\hat{\theta}^{(k)}(T)$ in the large time asymptotic when the process u_k is not ergodic.

Problem 6.1.1 Investigate the weighted average of the estimators $\hat{\theta}^{(k)}(T)$. That is, consider expressions of the form

$$\sum_{k=1}^{N} \alpha_{k,N}(T) \hat{\theta}^{(k)}(T), \tag{6.1.35}$$

with non-random numbers (weights) $\alpha_{k,N}(T)$ satisfying

$$\alpha_{k,N}(T) \geq 0, \ \sum_{k=1}^{N} \alpha_{k,N}(T) = 1.$$

The weights can depend on the observation time interval T and on the total number N of the available Fourier coefficients. Is it possible to select the weights so that the convergence rate as $T \to +\infty$ is better than $\sqrt{T}$? Is it possible to select the weights so that the convergence rate as $N \to +\infty$ is better than N?

Problem 6.1.2 Consider (6.1.10) on page 383 with initial condition $u_k(0) \overset{d}{=} \mathcal{N}(m, \sigma^2)$ where $m \in \mathbb{R}$ and $\sigma \geq 0$ are known. With k and T fixed, investigate the corresponding estimator $\hat{\theta}^{(k)}(T)$ from (6.1.14) on page 384 in the `small noise asymptotic`, that is, in the limit $q \to 0$.

Problem 6.1.3 Consider Eq. (6.1.10) on page 383 with initial condition $u_k(0) \overset{d}{=} \mathcal{N}(m_k, \sigma_k^2)$ where $m_k \in \mathbb{R}$ and $\sigma_k \geq 0$ are known. Assume that the collection $\{u_k(0),\ k \geq 1\}$ is independent of $\{w_k,\ k \geq 1\}$. Investigate the estimator $\hat{\theta}^{(k)}(T)$ from (6.1.14) in the limit $k \to +\infty$, T fixed. Is it possible to choose m_k and/or σ_k so that the rate of convergence of $\hat{\theta}^{(k)}(T)$ is better than k?

Problem 6.1.4 Consider the collection of random variables

$$\hat{\theta}(T) = \frac{\int_0^T X(t)dX(t)}{\int_0^T X^2(t)dt}, \ T > 0,$$

where

$$dX = Xdt + \sigma dw,$$

and the initial condition $X(0)$ is a Gaussian random variable independent of the standard Brownian motion w. Determine the limit in distribution of $e^T\big(\hat{\theta}(T) - 1\big)$.

Problem 6.1.5 Consider the collection of random variables

$$\hat{\theta}(T) = \frac{\int_0^T X(t)dX(t)}{\int_0^T X^2(t)dt}, \ T > 0,$$

where

$$X(t) = X_0 + \sigma dw,$$

and the initial condition X_0 is a Gaussian random variable independent of the standard Brownian motion w. Find a positive increasing function $v = v(T)$ such that, as $T \to +\infty$, a non-degenerate limit of $v(T)\hat{\theta}(T)$ exists in distribution, and determine the limit.

6.2 Maximum Likelihood Estimator (MLE): One Unknown Parameter

6.2.1 The Forward Problem

Introduce the following objects:

1. G, a sufficiently regular bounded domain in $\mathbb{R}^{\mathrm{d}}$ or a smooth compact d-dimensional manifold;
2. $H = L_2(G)$, a separable Hilbert space with an orthonormal basis $\{\mathfrak{h}_k,\ k \geq 1\}$;
3. $W^{\mathrm{Q}} = W^{\mathrm{Q}}(t)$, a Q-cylindrical Brownian motion on H;
4. $\mathrm{A}_0, \mathrm{A}_1$, linear differential or pseudo-differential operators on H.

For $\theta \in \mathbb{R}$, consider the equation

$$\dot{u}(t) + (\mathrm{A}_0 + \theta \mathrm{A}_1)u(t) = \dot{W}^{\mathrm{Q}}(t) \tag{6.2.1}$$

with initial condition $u(0)$. Even though the solution $u = u_\theta(t, x)$ depends not only on t but also on the parameter θ and the spacial variable $x \in G$, not to mention the elementary outcome ω, most of the time we will not indicate the dependence on θ, x, and ω explicitly.

Definition 6.2.1 Equation (6.2.1) is called `diagonal`, or `diagonalizable`, if the operators A_0, A_1, and Q have pure point spectrum and a common system of eigenfunctions $\{\mathfrak{h}_k,\ k \geq 1\}$, and

$$u(0) = \sum_{k \geq 1} u_k(0)\mathfrak{h}_k, \tag{6.2.2}$$

where $u_k(0)$ are independent Gaussian random variables with mean m_k and variance σ_k^2, and the collection $\{u_k(0),\ k \geq 1\}$ is independent of W^{Q}.
We will introduce conditions on the convergence of (6.2.2) when we discuss existence and uniqueness of solution of (6.2.1).

Equation (6.2.1) is an attempt to strike a reasonable balance between the concrete examples from Sect. 6.1 and a completely abstract framework that has little or no connection with SPDEs. Although abstract, Eq. (6.2.1) is a stochastic evolution equation (in fact, an SPDE if the operators $\mathrm{A}_0, \mathrm{A}_1$ are differential). There are sufficient conditions for a (pseudo)-differential operator to have pure point spectrum and for the corresponding eigenfunctions to form an orthonormal basis in $L_2(G)$. It is even more important for our purposes that the eigenvalues of the operators in this setting have a particular growth rate in k:

$$k\text{-th eigenvalue} \sim k^{\text{order of the operator}/\mathrm{d}}; \tag{6.2.3}$$

see, for example, [201, Theorem 1.2.1]. In particular, for the Laplacian $\boldsymbol{\Delta}$, a second-order operator, we get

$$k\text{-th eigenvalue of } \boldsymbol{\Delta} \sim k^{2/\mathrm{d}};$$

in fact, we already used this result before with $\mathrm{d} = 2$: see pages 174 and 392.

We will not discuss various technical questions, such as: how regular should G be, what is the meaning of $L_2(G)$ when G is a manifold, what is the order of a pseudo-differential operator (and what is a pseudo-differential operator to begin with), etc. Nonetheless, these questions can arise naturally even when not originally present. For example, an equation with periodic boundary conditions leads to a manifold (circle if $\mathrm{d} = 1$, torus if $\mathrm{d} = 2$, etc.) Pseudo-differential operators can appear if, for example, in Eq. (6.2.1) we decide to switch from W^{Q} to W. After making this change, only two operator rather than three will have to have a common system of eigenfunctions, but in the resulting equation for $v = \mathrm{Q}^{-1/2}u$,

$$dv(t) + (\bar{\mathrm{A}}_0 + \theta\bar{\mathrm{A}}_1)v(t)dt = dW(t),$$

the operators $\bar{\mathrm{A}}_i = Q^{-1/2}\mathrm{A}_i Q^{1/2}$, $i = 0, 1$, will be pseudo-differential rather than differential. In the end, though, all we need is to know that all these technical questions have been worked out in special books such as [201] or [208].

Denote by κ_k, ν_k, q_k, and $\mu_k(\theta)$ the eigenvalues of the operators A_0, A_1, Q, and $\mathrm{A}_0 + \theta\mathrm{A}_1$:

$$\mathrm{A}_0\mathfrak{h}_k = \kappa_k\mathfrak{h}_k, \ \ \mathrm{A}_1\mathfrak{h}_k = \nu_k\mathfrak{h}_k, \ \ \mathrm{Q}\mathfrak{h}_k = q_k^2\mathfrak{h}_k, \ \ \mu_k(\theta) = \kappa_k + \theta\nu_k.$$

The reason to write the operators on the left-hand side of (6.2.1) is that it is more convenient to have $\mu_k(\theta) > 0$.

Our objective is an inverse problem for Eq. (6.2.1), that is, estimation of θ from the observations of u. Rigorous analysis of an inverse problem is impossible without first addressing the forward problem: existence, uniqueness, and regularity of the solution of (6.2.1). While we have a general result (Theorem 4.4.3 on page 199), there is some work to be done: we need to specify the corresponding normal triple and then state the corresponding conditions in the same terms as in Definition 6.2.1.

There is a more philosophical question in the background: what kinds of functions can be observed? Since we are working with Fourier series in $L_2(G)$, we will assume that every function from $L_2\big(\Omega; \mathcal{C}\big((0,T); L_2(G)\big)\big)$ is observable. In the normal triple, (V, H, V'), we therefore take $H = L_2(G)$ and then make sure that Theorem 4.4.3 applies to Eq. (6.2.1). To define V, denote by $2\mathrm{m}$ the order of the operator $\mathrm{A}_0 + \theta\mathrm{A}_1$ (which naturally leads to the assumption that the order of $\mathrm{A}_0 + \theta\mathrm{A}_1$ is the same for all $\theta \in \Theta$). Then we can take $V = H^{\mathrm{m}}(G)$, where $H^{\gamma}(G), \gamma \in \mathbb{R}$, are the Sobolev spaces in G, as constructed in Example 3.1.29 on page 87 with $\lambda_k = k^{1/\mathrm{d}}$:

$$\|f\|_{\gamma}^2 = \sum_{k\geq 1} k^{2\gamma/\mathrm{d}} f_k^2. \tag{6.2.4}$$

Note that, by (6.2.3), the eigenvalues of the operator $\sqrt{-\boldsymbol{\Delta}}$ on $L_2(G)$ have the same asymptotic as λ_k.

Definition 6.2.2 A diagonal equation (6.2.1) is called `parabolic` of order 2m if there exist numbers $c_1 > 0, c_2 \geq 0, c_3 > 0$, possibly depending on θ, such that, for all $k \geq 1$,

$$c_1 k^{2\mathrm{m}/\mathrm{d}} \leq \mu_k(\theta) + c_2 \leq c_3 k^{2\mathrm{m}/\mathrm{d}}. \tag{6.2.5}$$

If condition (6.2.5) holds for some $\theta \in \mathbb{R}$, then, by continuity, it holds in some neighborhood of θ. Accordingly, we define the set $\Theta_0 \subseteq \mathbb{R}$ as the largest open set such that (6.2.5) holds for every $\theta \in \Theta_0$ (recall that any union, countable or uncountable, of open sets is open).

Example 6.2.3 Let G be a smooth bounded domain in $\mathbb{R}^{\mathrm{d}}$ or a smooth compact d-dimensional manifold with a smooth measure, $H = L_2(G)$, and let $\boldsymbol{\Delta}$ be the Laplace operator on G (with zero boundary conditions if G is a domain). It is known (see, for example, Safarov and Vassiliev [201] or Shubin [208]) that

1. $\boldsymbol{\Delta}$ has a complete orthonormal system of eigenfunctions in H;
2. the corresponding eigenvalues ν_k are negative, can be arranged in decreasing order, and there is a positive number $c_\circ$ such that

$$|\nu_k| \simeq c_\circ k^{2/\mathrm{d}};$$

see (6.2.10) below for more details, including an explicit formula for $c_\circ$.

Then each of the following equations is diagonalizable and parabolic:

$$\begin{aligned} u_t - \theta\boldsymbol{\Delta} u &= \dot{W},\ \Theta_0 = (0, +\infty); \\ u_t - \boldsymbol{\Delta} u + \theta u &= \dot{W},\ \Theta_0 = \mathbb{R}; \\ u_t + \boldsymbol{\Delta}^2 u + \theta\boldsymbol{\Delta} u &= \dot{W},\ \Theta_0 = \mathbb{R}. \end{aligned} \tag{6.2.6}$$

We can now address the forward problem for Eq. (6.2.1).

Theorem 6.2.4 *Assume that Eq.* (6.2.1) *is diagonal and parabolic of order* 2m,

$$\sum_{k\geq 1} q_k^2 < \infty, \tag{6.2.7}$$

and the initial condition satisfies

$$\sum_{k\geq 1} (m_k^2 + \sigma_k^2) < \infty. \tag{6.2.8}$$

Then, for every $\theta \in \Theta_0$, Eq. (6.2.1) *has a unique variational solution in the sense of Definition 4.4.1 on page 199, and, with $H = L_2(G)$ and $V = H^{\mathrm{m}}(G)$,*

$$\mathbb{E} \sup_{0<t<T} \|u(t)\|_H^2 + \mathbb{E}\int_0^T \|u(t)\|_V^2 dt \leq C \sum_{k\geq 1}(m_k^2 + \sigma_k^2 + q_k^2),$$

with C depending only on T and the numbers c_1, c_2, c_3 from (6.2.5).

Proof Equation (6.2.1) is a particular case of Eq. (4.4.1) on page 198 with $\mathrm{A} = -\mathrm{A}_0 - \theta \mathrm{A}_1$, $f(t) = 0$, $\mathrm{M}_k = 0$, $g_k = q_k \mathfrak{h}_k$, $u_0 = \sum_k u_k(0)\mathfrak{h}_k$. Accordingly, the result follows from Theorem 4.4.3. Indeed,

$$\sum_{k\geq 1} \|g_k\|_H^2 = \sum_{k\geq 1} q_k^2 < \infty, \ \mathbb{E}\|u(0)\|_H^2 = \sum_{k\geq 1} \mathbb{E}u_k^2(0) = \sum_{k\geq 1}(m_k^2 + \sigma_k^2) < \infty.$$

Next, by definition of $H^{\gamma}(G)$ and assumption (6.2.5),

$$\|\mathrm{A}v\|_{V'}^2 = \sum_{k\geq 1} v_k^2 \mu_k^2(\theta) k^{-2\mathrm{m}/\mathrm{d}} \leq c_3 \sum_{k\geq 1} v_k^2 k^{2\mathrm{m}/\mathrm{d}} = c_3 \|v\|_V^2,$$

$$[\mathrm{A}v, v] = -\sum_{k\geq 1} \mu_k(\theta) v_k^2 \leq -c_2 \|v\|_H^2 - c_1 \|v\|_V^2.$$

This completes the proof of Theorem 6.2.4.

To conclude the section, here are more details about (6.2.3). Introduce the notations

$$\alpha = (\alpha_1, \ldots, \alpha_{\mathrm{d}}), \ \alpha_i \in \{0, 1, 2, \ldots\}, \ |\alpha| = \sum_{i=1}^{\mathrm{d}} \alpha_i, \ D^{\alpha} = \frac{\partial^{|\alpha|}}{\partial x_1^{\alpha_1} \cdots \partial x_{\mathrm{d}}^{\alpha_{\mathrm{d}}}}.$$

Let G be a smooth bounded domain in $\mathbb{R}^{\mathrm{d}}$ or, more generally, smooth d-dimensional manifold, with or without a boundary. A partial differential operator $\mathbf{B}$ of order m on G is written (in local coordinates, if G is a manifold) as

$$\mathbf{B} = \sum_{|\alpha|\leq \mathrm{m}} b_{\alpha}(x) D^{\alpha},$$

where the functions b_{α} are usually assumed to be infinitely differentiable. Boundary conditions, if any, can also be expressed in terms of partial differential operators of order strictly smaller than m.

For $x \in G$ and $y \in \mathbb{R}^{\mathrm{d}}$ (or more precisely, $(x, y) \in T^*(G)$, the cotangent bundle of G), define

$$B_{\mathrm{m}}(x, y) = (-\mathrm{i})^{\mathrm{m}} \sum_{|\alpha|=\mathrm{m}} b_{\alpha}(x) y^{\alpha},$$

where $y^{\alpha} = \prod_{i=1}^{\mathrm{d}} y_i^{\alpha_i}$. The standard assumptions on the operator **B** are as follows:

1. Ellipticity: $B_{\mathrm{m}}(x, y) \neq 0$ if $y \neq 0$;
2. Positivity: $(\mathbf{B}f, f)_{L_2(G)} > 0, f \not\equiv 0$;
3. Symmetry: $(\mathbf{B}f, g)_{L_2(G)} = (f, \mathbf{B}g)_{L_2(G)}$.

Positivity implies $B_{\mathrm{m}}(x, y) > 0$, $y \neq 0$, which in turn implies that the order m of the operator is even.

Finally, define

$$C_{\mathbf{B},G} = \frac{1}{(2\pi)^{\mathrm{d}}} \int_{\{(x,y):B_{\mathrm{m}}(x,y)<1\}} dx\, dy,$$

and let $\nu_1 \leq \nu_2 \leq \ldots$ be the eigenvalues of **B**, with $=$ for multiple eigenvalues.

The main result (see [201, Theorem 1.2.1]) is that

$$\lim_{k\to+\infty} \frac{\nu_k}{k^{\mathrm{m}/\mathrm{d}}} = \left(C_{\mathbf{B},G}\right)^{-\mathrm{m}/\mathrm{d}}. \tag{6.2.9}$$

Moreover,

$$\left| \nu_k - k^{\mathrm{m}/\mathrm{d}} \left(C_{\mathbf{B},G}\right)^{-m/\mathrm{d}} \right| \leq C k^{(\mathrm{m}-1)/\mathrm{d}}.$$

For example, if $\mathbf{B} = -\boldsymbol{\Delta}$ and $G \in \mathbb{R}^{\mathrm{d}}$ is a bounded domain, then

$$C_{-\boldsymbol{\Delta},G} = \frac{\omega_{\mathrm{d}}|G|}{(2\pi)^{\mathrm{d}}}, \tag{6.2.10}$$

where

$$\omega_{\mathrm{d}} = \frac{\pi^{\mathrm{d}/2}}{\Gamma((\mathrm{d}/2)+1)}$$

is the `volume of the unit ball` in $\mathbb{R}^{\mathrm{d}}$ and $|G|$ is the Lebesgue measure of G in $\mathbb{R}^{\mathrm{d}}$. When $\mathrm{d} = 1, 2, 3$, we get

$$\nu_k \simeq \left(\frac{\pi}{|G|}\right)^2 k^2, \quad \nu_k \simeq \frac{4\pi}{|G|} k, \quad \nu_k \simeq \left(\frac{6\pi^2}{|G|}\right)^{2/3} k^{2/3}, \tag{6.2.11}$$

respectively.

6.2.2 *Simplified MLE (sMLE) and MLE*

Under conditions of Theorem 6.2.4, the solution u of (6.2.1) has the Fourier series expansion

$$u(t) = \sum_{k\geq 1} u_k(t)\mathfrak{h}_k, \tag{6.2.12}$$

and

$$du_k(t) = -\mu_k(\theta)u_k(t)dt + q_k dw_k(t),\ u_k(0) \stackrel{d}{=} \mathcal{N}(m_k, \sigma_k^2), \tag{6.2.13}$$

that is,

$$u_k(t) = u_k(0)\, e^{-\mu_k(\theta)t} + q_k \int_0^t e^{-\mu_k(\theta)(t-s)} dw_k(s).$$

One can justify (6.2.13) with full rigor by using $\mathfrak{h}_k$ as a test function in the definition of the variational solution of (6.2.1).

Exercise 6.2.5 (B) Note that all the equations in (6.2.6) have $q_k = 1$, so none satisfies (6.2.7). On the other hand, it is clear that the right-hand side of (6.2.12) can be an element of $L_2(G)$ without (6.2.7). Find a more general condition, involving both q_k and $\mu_k(\theta)$ to ensure that the right-hand side of (6.2.12) is an element of $L_2(G)$, and verify that the result can be considered a `mild solution` of (6.2.1).

Hint. For example, $\sum_k q_k^2/\mu_k(\theta) < \infty$ works.

The only unknown object in Eq. (6.2.1) is the number θ; the numbers κ_k, ν_k, and q_k are assumed to be know. As far as q_k, note that (6.1.8) still holds.

The inverse problem for Eq. (6.2.1) now becomes as follows: estimate θ given the observations

$$U^{N,\theta} = \{u_k(t),\ t \in [0,T], k = 1,\ldots,N\}.$$

At this point, it becomes important to indicate the dependence of the observations on the parameter θ.

Given the problem, it is natural to assume from now on that, for all $k \geq 1$,

$$\nu_k \neq 0,\ q_k > 0. \tag{6.2.14}$$

If, in fact, $\nu_k = 0$ for some k, then the corresponding process u_k carries no information about θ and can be dropped from the observations. The sign of q_k does not matter (see (6.1.8)), and if $q_k = 0$, then there are two options. Option one is zero initial condition $u_k(0) = 0$ (that is, $m_k = \sigma_k = 0$). Then $u_k(t) = 0$ for all $t > 0$ and is dropped from observations. Option two is non-zero initial condition $u_k(0) \neq 0$. Then

$$u_k(t) = u_k(0)e^{-\mu_k(\theta)t}$$

and θ can be determined exactly from the observations of u_k at two distinct time moments. For example,

$$u_k(1) = u_k(0)e^{-\kappa_k - \theta\nu_k},\ u_k(2) = u_k(0)e^{-2\kappa_k - 2\theta\nu_k},\ \ln\frac{u_k(1)}{u_k(2)} = \kappa_k + \theta\nu_k,$$

and so

$$\theta = \frac{1}{\nu_k}\left(\ln\frac{u_k(1)}{u_k(2)} - \kappa_k\right). \tag{6.2.15}$$

The natural choice of the estimator for θ is the maximum likelihood estimator (MLE) $\widehat{\boldsymbol{\theta}}_N$, which, intuitively, is the value θ in Θ_0 for which model (6.2.13) is most likely to produce the observed realization of the vector $U^{N,\theta}$. The question is then to find the likelihood function for $U^{N,\theta}$.

If we observed each function u_k at discrete time moments $t_1, \ldots, t_K$, then the corresponding observation would be a $K \times N$-dimensional Gaussian vector with a very explicit, but very complicated, probability density function. When evaluated at the corresponding observations, this probability density function becomes the likelihood function.

With observations in continuous time, the process $U^{N,\theta}$ takes values in the space $\mathcal{C}([0,T];\mathbb{R}^N)$ of $\mathbb{R}^N$-valued continuous functions and generates the measure P_T^{θ} in that space according to the rule

$$P_T^{N,\theta}(A) = \mathbb{P}(U^{N,\theta} \in A)$$

for every Borel subset A of $\mathcal{C}([0,T];\mathbb{R}^N)$. With no Lebesque measure in infinite dimensions, there is no clear analogue of the probability density function for $U^{N,\theta}$. In particular, trying to pass to the limit of shrinking time step in the pdf corresponding to the discrete observations is not a promising idea.

It turns out that the measures $P_T^{N,\theta}$ are equivalent for different values of $\theta \in \Theta_0$ (in fact, for all $\theta \in \mathbb{R}$), and thus any one of them can serve as a reference measure. Now, the fact that $U^{N,\theta}$ is a Gaussian process is not as important as the fact that $U^{N,\theta}$ is a diffusion process. Moreover, since the processes u_k are independent for different k, $P_{T,N}^{\theta}$ is the product measure corresponding to individual u_k:

$$P_T^{N,\theta}(A_1 \times \cdots \times A_N) = \prod_{k=1}^{N} \mathbb{P}(u_k \in A_k), \tag{6.2.16}$$

where A_k is a Borel subset of $\mathcal{C}((0,T);\mathbb{R})$.

Let us start with the general result for scalar diffusion processes. Consider the processes $X = X(t)$ and $Y = Y(t)$ defined by stochastic ordinary differential equations

$$\begin{aligned} dX(t) &= A\big(t, X(t)\big)dt + \sigma\big(t, X(t)\big)dw(t),\\ dY(t) &= a\big(t, Y(t)\big)dt + \sigma\big(t, Y(t)\big)dw(t), \end{aligned} \tag{6.2.17}$$

where w is a standard Brownian motion, the functions a, A, σ satisfy the conditions to ensure existence of a unique strong solution of the corresponding equation (for example, uniformly Lipschits continuous in the second argument), the initial conditions are independent of w, and, for simplicity, the function σ is uniformly bounded away from zero: $\sigma(t,x) \geq \sigma_0 > 0$. The important assumptions are

1. The initial condition is the same in both equations: $X(0) = Y(0)$;
2. The diffusion coefficient σ is the same in both equations;
3. The functions a, A, σ are non-random.

The claim is as follows: if P_T^X and P_T^Y are the measures generates by X and Y in $\mathcal{C}((0,T);\mathbb{R})$, then

$$\begin{aligned} \frac{dP_T^X}{dP_T^Y}(Y) = \exp\Bigg(&\int_0^T \frac{A\big(t, Y(t)\big) - a\big(t, Y(t)\big)}{\sigma^2\big(t, Y(t)\big)} dY(t)\\ &- \frac{1}{2}\int_0^T \frac{A^2\big(t, Y(t)\big) - a^2\big(t, Y(t)\big)}{\sigma^2\big(t, Y(t)\big)} dt \Bigg). \end{aligned} \tag{6.2.18}$$

For the detailed statement and proof, see Liptser and Shiryaev [139, Theorem 7.19] (although a complete understanding of the proof might require a thorough review of the previous 18 theorems in Chap. 7 of [139]). Below, we will try to get some general ideas why something like (6.2.18) should be true.

We start by commenting on two visual features of (6.2.18). First, the structure of the formula: measure of the top process over the measure of the bottom process, evaluated at the bottom process, is an expression involving the drift of the top process minus the drift of the bottom process, divided by the common diffusion squared, all evaluated at the bottom process. Note also the difference of the squares rather than square of the difference. Of course, using the equation for Y, we can re-write (6.2.18) as

$$\begin{aligned} \frac{dP_T^X}{dP_T^Y}(Y) = \exp\Bigg(&\int_0^T \frac{A\big(t, Y(t)\big) - a\big(t, Y(t)\big)}{\sigma\big(t, Y(t)\big)} dw(t)\\ &- \frac{1}{2}\int_0^T \frac{\Big(A\big(t, Y(t)\big) - a\big(t, Y(t)\big)\Big)^2}{\sigma^2\big(t, Y(t)\big)} dt \Bigg), \end{aligned} \tag{6.2.19}$$

which is correct even though the right-hand side does not seem to be the function of Y. This leads to the second observation: while the density $\Phi = dP_T^X/dP_T^Y$ is a functional on $\mathcal{C}((0,T);\mathbb{R})$, there is no nice closed-form expression for $\Phi(x)$ if $x = x(t)$ is an arbitrary element from $\mathcal{C}((0,T);\mathbb{R})$. Formula (6.2.18) gives the expression for $\Phi(Y)$, and this is about as much as we can get. Of course, there is an equivalent formula (6.2.19), and, using $dP_T^Y/dP_T^X = 1/\Phi$, we can also get $\Phi(X)$.

Exercise 6.2.6 (C)

(a) Verify (6.2.19).
(b) Verify that

$$\frac{dP_T^X}{dP_T^Y}(X) = \exp\Bigg(\int_0^T \frac{A\big(t,X(t)\big) - a\big(t,X(t)\big)}{\sigma^2\big(t,X(t)\big)}dX(t) - \frac{1}{2}\int_0^T \frac{A^2\big(t,X(t)\big) - a^2\big(t,X(t)\big)}{\sigma^2\big(t,X(t)\big)}dt\Bigg), \tag{6.2.20}$$

(c) What happens if we replace dX in (6.2.20) according to (6.2.17), and how does the result compare with (6.2.19)?

To understand (6.2.18) better, we need the result known as the `Girsanov theorem`, after the Russian mathematician IGOR VLADIMIROVICH GIRSANOV (1934–1967) who established it in 1960.

Theorem 6.2.7 (Girsanov's Theorem) *Let $w = w(t)$ be a standard Brownian motion, and let $h = h(t)$ be a random process such that $\mathbb{E}\int_0^T h^2(t)dt < \infty$ and*

$$\mathbb{E}\exp\left(\int_0^T h(s)dw(s) - \frac{1}{2}\int_0^T h^2(t)dt\right) = 1. \tag{6.2.21}$$

On the original stochastic basis $(\Omega, \mathcal{F}, \{\mathcal{F}_t\}_{t\geq 0}, \mathbb{P})$, define a new probability measure $\widetilde{\mathbb{P}}$ by

$$\widetilde{\mathbb{P}}(A) = \mathbb{E}\left(\mathbf{1}_A(\omega)\exp\left(\int_0^T h^2(s)dw(s) - \frac{1}{2}\int_0^T h^2(t)dt\right)\right), \tag{6.2.22}$$

$A \in \mathcal{F}$. Then the process $X = X(t)$ defined by

$$X(t) = -\int_0^t h(s)ds + w(t),\ 0 \leq t \leq T,$$

is a standard Brownian motion on $(\Omega, \mathcal{F}, \{\mathcal{F}_t\}_{\leq t\leq T}, \widetilde{\mathbb{P}})$.

Proof Define the process

$$Z(t) = \exp\left(\int_0^T h(s)dw(s) - \frac{1}{2}\int_0^T h^2(t)dt\right).$$

By the Itô formula,

$$dZ = hZdw,$$

and by (6.2.21), Z is a martingale. Fix $\lambda \in \mathbb{R}$, $s \in (0,T)$, and, for $t > s$, define

$$g(t) = e^{\mathrm{i}\lambda(X(t)-X(s))},\ G(t) = g(t)Z(t),\ F(t) = \widetilde{\mathbb{E}}\big(g(t)\big|\mathcal{F}_s\big),$$

where $\widetilde{\mathbb{E}}$ is the expectation with respect to the measure $\widetilde{\mathbb{P}}$. Note that, similar to (4.4.87) on page 228,

$$F(t) = \frac{\mathbb{E}\big(G(t)\big|\mathcal{F}_s\big)}{Z(s)}, \tag{6.2.23}$$

which allows us to switch between $\mathbb{E}$ and $\widetilde{\mathbb{E}}$ in our computations. To complete the proof, we need to show that

$$F(t) = e^{-\lambda^2(t-s)/2}. \tag{6.2.24}$$

By the Itô formula,

$$dg = \mathrm{i}\lambda g dX - \frac{\lambda^2}{2} g dt = -\mathrm{i}\lambda g h dt + \mathrm{i}\lambda g dw - \frac{\lambda^2}{2} g dt,$$

$$dG = gdZ + Zdg + \mathrm{i}\lambda gZhdt = (h + \mathrm{i}\lambda)gZdw - \frac{\lambda^2}{2}gZdt.$$

or, since $G(r) = g(r)Z(r)$, $r \geq s$, and $G(s) = Z(s)$,

$$G(t) = Z(s) + \int_s^t (h(r) + \mathrm{i}\lambda)G(r)dw(r) - \frac{\lambda^2}{2}\int_s^t G(r)dr. \tag{6.2.25}$$

Taking $\mathbb{E}\big(\,\cdot\,\big|\mathcal{F}_s\big)$ on both sides of (6.2.25) eliminates the stochastic integral, and then (6.2.23) results in

$$F(t)Z(s) = Z(s) - \frac{\lambda^2}{2}\int_s^t F(r)Z(s)dr, \text{ or } F(t) = 1 - \frac{\lambda^2}{2}\int_s^t F(r)dr,$$

which is the same as (6.2.24).

This completes the proof of Theorem 6.2.7.

Exercise 6.2.8 (B) (a) Verify that (6.2.24) is equivalent to the statement of the theorem. **Hint.** Lévy's characterization of the Brownian motion. (b) Verify (6.2.23).

There are two places in the Girsanov theorem where the function h appears: in the process X and in the change of measure Z. In one of those places, h should have the minus sign (in our formulation, this happens in X). The Itô formula shows that the negative sign is indeed necessary to achieve the proper cancelation of terms.

Condition (6.2.21) is essential: otherwise, it will be impossible to change the measure according to (6.2.22). In general, $\mathbb{E}\int_0^T h^2(t)dt < \infty$ implies

$$\mathbb{E}\exp\left(\int_0^T h(s)dw(s) - \frac{1}{2}\int_0^T h^2(t)dt\right) \leq 1, \tag{6.2.26}$$

and equality in (6.2.26) is usually guaranteed by the `Novikov condition`

$$\mathbb{E}\exp\left(\frac{1}{2}\int_0^T h^2(t)dt\right) < \infty; \tag{6.2.27}$$

without additional information about h, one cannot do much better than (6.2.27); see [139, Sect. 6.2.4].

Let us now see how the Girsanov theorem leads to the density formula (6.2.18). Denote by $L(Y)$ the right-hand side of (6.2.18), define the function

$$B(t,x) = \frac{A(t,x) - a(t,x)}{\sigma(t,x)}$$

and then the process

$$\widetilde{w}(t) = w(t) - \int_0^t B\big(t, Y(s)\big)ds.$$

Forgetting for the moment about (6.2.21), we conclude from Theorem 6.2.7 that $\widetilde{w}$ is a standard Brownian motion under the measure $\widetilde{\mathbb{P}}$ with

$$\frac{d\widetilde{\mathbb{P}}}{d\mathbb{P}} = L(Y);$$

at this point, an alternative formula (6.2.19) for $L(Y)$ could be more convenient. After some simple algebraic manipulations,

$$\begin{aligned} Y(t) &= X(0) + \int_0^t a\big(s, Y(s)\big)ds + \int_0^t \sigma\big(s, Y(s)\big)dw(s) \\ &= X(0) + \int_0^t A\big(s, Y(s)\big)ds + \int_0^t \sigma\big(s, Y(s)\big)d\widetilde{w}(s), \end{aligned}$$

that is, the equation satisfied by Y under $\widetilde{\mathbb{P}}$ is the same as the equation satisfied by X under $\mathbb{P}$. In other words, for every Boreal set A in $\mathcal{C}((0,T))$,

$$P^X_T(A) = \mathbb{P}(X \in A) = \widetilde{\mathbb{P}}(Y \in A).$$

It follows that P^X_T is absolutely continuous with respect to P^Y_T, that is, if $P^Y_T(A) = 0$ for some Borel sub-set of $\mathcal{C}((0,T))$, then $P^X_T(A) = 0$. Indeed, since $\widetilde{\mathbb{P}}$ is absolutely continuous with respect to $\mathbb{P}$, we have

$$P^Y_T(A) = 0 \Leftrightarrow \mathbb{P}(Y \in A) = 0 \Rightarrow \widetilde{\mathbb{P}}(Y \in A) \Leftrightarrow \mathbb{P}(X \in A) = 0 \Leftrightarrow P^X_T(A) = 0.$$

Then

$$P^X_T(A) = \int_A \Phi(x)dP^Y_T(x) = \int_{\{\omega:Y\in A\}}, \Phi(Y)d\mathbb{P},$$

where the first equality follows by the Radon-Nykodim theorem, and the second, after the change of variables. On the other hand,

$$P^X_T(A) = \mathbb{P}(X \in A) = \widetilde{\mathbb{P}}(Y \in A) = \int_{\{\omega:Y\in A\}} L(Y)d\mathbb{P},$$

showing that $\Phi(Y) = L(Y)$.

As far as satisfying (6.2.21), since we are not dealing with an arbitrary process h but a rather specific process $B\big(t, Y(t)\big)$, we can get (6.2.21) without (6.2.27), via a localization argument using special stopping times. For details, see [139, Theorem 7.19].

Let us now use (6.2.18) to write the likelihood function $L_{T,N}(\theta)$ for $U^{N,\theta}$. To this end, denote by θ_0 the true value of the parameter, whatever it might be, and let the reference measure be the measure P^{N,θ_0}_T on $\mathcal{C}((0,T);\mathbb{R}^N)$ generated by the N-dimensional process U^{N,θ_0}. A particular realization of U^{N,θ_0} is the observations.

Also, denote by $P^{\theta}_{k,T}$ the measure on $\mathcal{C}((0,T);\mathbb{R})$, generated by the process $u_k = u_{k,\theta}$ from (6.2.13); through the end of this section we will indicate the dependence of u_k on the parameter θ, cumbersome notations notwithstanding. Then

$$L_{N,T}(\theta) = \frac{dP^{N,\theta}_T}{dP^{N,\theta_0}_T}\left(U^{N,\theta_0}\right), \tag{6.2.28}$$

and, because, for every $\theta \in \Theta_0$, the processes $u_{k,\theta}$ are independent for different k, we have

$$\frac{dP^{N,\theta}_T}{dP^{N,\theta_0}_T}\left(U^{N,\theta_0}\right) = \prod_{k=1}^{N} \frac{dP^{\theta}_{k,T}}{dP^{\theta_0}_{k,T}}\left(u_{k,\theta_0}\right) \tag{6.2.29}$$

Recall that

$$du_{k,\theta} = -\mu_k(\theta)u_{k,\theta}(t)dt + q_k dw_k(t). \tag{6.2.30}$$

By (6.2.18), with $X = u_{k,\theta}$, $Y = u_{k,\theta_0}$, $A(t,x) = -\mu_k(\theta)x$, $a(t,x) = -\mu_k(\theta_0)x$, $\sigma(t,x) = q_k$,

$$\frac{dP^{\theta}_{k,T}}{dP^{\theta_0}_{k,T}}\left(u_{k,\theta_0}\right) = \exp\left(-\frac{\mu_k(\theta)-\mu_k(\theta_0)}{q_k^2}\int_0^T u_{k,\theta_0}(t)du_{k,\theta_0}(t)\right.$$
$$\left. - \frac{\mu_k^2(\theta)-\mu_k^2(\theta_0)}{2q_k^2}\int_0^T u^2_{k,\theta_0}(t)dt\right). \tag{6.2.31}$$

Combining (6.2.28)–(6.2.31), we get

$$L_{N,T}(\theta) = \exp\left(-\sum_{k=1}^{N}\frac{\mu_k(\theta)-\mu_k(\theta_0)}{q_k^2}\int_0^T u_{k,\theta_0}(t)du_{k,\theta_0}(t)\right.$$
$$\left. -\sum_{k=1}^{N}\frac{\mu_k^2(\theta)-\mu_k^2(\theta_0)}{2q_k^2}\int_0^T u^2_{k,\theta_0}(t)dt\right).$$

By definition, the maximum likelihood estimator $\widehat{\boldsymbol{\theta}}_N$ of θ is

$$\widehat{\boldsymbol{\theta}}_N = \operatorname*{argmax}_{\theta\in\bar{\Theta}_0} L_{N,T}(\theta); \tag{6.2.32}$$

$\bar{\Theta}_0$ is the closure of Θ_0.

Recall that $\mu_k(\theta) = \kappa_k + \theta\nu_k$, and $\nu_k \neq 0, q_k > 0, k \geq 1$ (see the discussion on page 400). Therefore, as a function of θ, $\ln L_{N,T}(\theta)$ is a quadratic facing down, and

$$\frac{\partial \ln L_{N,T}(\theta)}{\partial\theta} = -\sum_{k=1}^{N}\frac{\nu_k}{q_k^2}\int_0^T u_{k,\theta_0}(t)du_{k,\theta_0}(t)$$
$$-\sum_{k=1}^{N}\frac{\nu_k(\kappa_k+\theta\nu_k)}{q_k^2}\int_0^T u^2_{k,\theta_0}(t)dt. \tag{6.2.33}$$

As (6.2.33) shows, there is a unique point $\widehat{\theta}_N$ where $\partial \ln L_{N,T}(\theta)/\partial\theta = 0$ and

$$\widehat{\theta}_N = -\frac{\sum_{k=1}^{N}\int_0^T \frac{\nu_k}{q_k}u_{k,\theta_0}(t)\big(du_{k,\theta_0}(t) - \kappa_k u_{k,\theta_0}(t)dt\big)}{\sum_{k=1}^{N}\int_0^T \frac{\nu_k^2}{q_k^2}u^2_{k,\theta_0}(t)dt}; \tag{6.2.34}$$

It follows from (6.2.30) that

$$\widehat{\theta}_N - \theta_0 = -\frac{\sum_{k=1}^N \int_0^T \frac{\nu_k}{q_k} u_{k,\theta_0}(t)dw_k(t)}{\sum_{k=1}^N \int_0^T \frac{\nu_k^2}{q_k^2} u_{k,\theta_0}^2(t)dt}. \tag{6.2.35}$$

Thus, the maximum likelihood estimator $\widehat{\boldsymbol{\theta}}_N$ of θ always exists and is unique. The formula for $\widehat{\boldsymbol{\theta}}_N$ depends on the underlying parameter set Θ_0. For example,

- if $\Theta_0 = \mathbb{R}$, then $\widehat{\boldsymbol{\theta}}_N = \widehat{\theta}_N$;
- if $\Theta_0 = (a, b)$, then

$$\widehat{\boldsymbol{\theta}}_N = \begin{cases} \widehat{\theta}_N, & \text{if } \widehat{\theta}_N \in (a, b), \\ b, & \text{if } \widehat{\theta}_N \geq b, \\ a, & \text{if } \widehat{\theta}_N \leq a. \end{cases}$$

Accordingly, we will refer to $\widehat{\theta}_N$ as the `simplified MLE`, or sMLE.

In some problems, there are no a priori restrictions on θ. For example, in the equation

$$u_t = \boldsymbol{\Delta} u + \theta u + \dot{W}^{\mathrm{Q}},$$

θ can be any real number, and then we will have $\widehat{\boldsymbol{\theta}}_N = \widehat{\theta}_N$. On the other hand, in the equation

$$u_t = \boldsymbol{\Delta} u + \theta u + \dot{W}^{\mathrm{Q}},$$

θ has to be positive and so $\Theta_0 = (0, +\infty)$. In this case,

$$\widehat{\boldsymbol{\theta}}_N = \begin{cases} \widehat{\theta}_N, & \text{if } \widehat{\theta}_N > 0 \\ 0, & \text{if } \widehat{\theta}_N \leq 0. \end{cases} \tag{6.2.36}$$

Because a closed-form expression for the MLE $\widehat{\boldsymbol{\theta}}_N$ might not be as convenient as the one for $\widehat{\theta}_N$, we will first establish conditions for consistency and asymptotic normality of $\widehat{\theta}_N$ as $N \to +\infty$ using the closed-form expression (6.2.34). Then we will combine these results with a general theorem from [84] to investigate $\widehat{\boldsymbol{\theta}}_N$; this investigation will not require a closed-form expression for $\widehat{\boldsymbol{\theta}}_N$.

To conclude this section, let us briefly discuss some of the changes and challenges in the non-diagonal case. An alternative way to write (6.2.34) is with the help of the operator Π^N, the orthogonal projection in H on the linear subspace generated by $\{\mathfrak{h}_1, \ldots, \mathfrak{h}_N\}$:

$$\widehat{\theta}_N = -\frac{\int_0^T \big(\Pi^N \mathrm{Q}^{-1}\mathrm{A}_1 u_{\theta_0}(t), du_{\theta_0}(t) - \mathrm{A}_0 u_{\theta_0}(t)dt\big)_H}{\int_0^T \|\Pi^N \mathrm{A}_1 \mathrm{Q}^{-1/2} u_{\theta_0}(t)\|_H^2 dt}, \tag{6.2.37}$$

where u_{θ_0} is the solution of (6.2.1) corresponding to $\theta = \theta_0$. This form allows generalizations to non-diagonal equations: given an orthonormal basis, the right-hand side of (6.2.37) is defined and can be studied both for linear non-diagonal equations (e.g. [149]) and the never diagonal non-linear equations (e.g. [27]).

The main technical difference from the diagonal case is that there is no longer a closed-form expression for the likelihood ratio, and as a result the connections between (6.2.37) and the corresponding MLE in the non-diagonal case are so far unknown.

To write the likelihood ratio in the non-diagonal case, note that if the projection operator Π^N does not commute with A_0 and A_1, there is no immediate *equation* for $\Pi^N u$, only an equality

$$d\Pi^N u = \Pi^N\big(\mathrm{A}_0 + \theta \mathrm{A}_1 u\big)dt + d\Pi^N W^{\mathrm{Q}}(t). \tag{6.2.38}$$

We can use [139, Theorem 7.12] and reduce (6.2.38) to a system of stochastic differential equations. Indeed, with the notations

$$\Pi^N u(t) = \sum_{k=1}^N v_{k,N}(t)\mathfrak{h}_k, \;\; \Pi^N \mathrm{A}_i u(t) = \sum_{k=1}^N z_{i,k,N}(t)\mathfrak{h}_k, \; i = 0, 1,$$

the process $v = (v_{1,N}, \ldots, v_{N,N})$ satisfies

$$dv_{k,N} = \big(F_{0,k,N}(t, v) + \theta F_{1,k,N}(t, v)\big)dt + \sum_{j=1}^N G_{k,j,N}\, d\bar{W}_j^N,$$

where

$$F_{i,k,N}(t, v) = \mathbb{E}\Big(z_{i,k,N}\Big|v_{k,N}(s),\; k = 1, \ldots, N,\; 0 \le s \le t\Big),\; i = 0, 1,$$

the matrix $G_N = (G_{k,j,N},\, k,j = 1, \ldots, N)$ is invertible, and $\bar{W}_k^N$, $k = 1, \ldots N$, are independent standard Brownian motions. After that, a version of formula (6.2.18) applies (although it is a much less explicit version, see [139, Theorem 7.2 or 7.17]). Note that in this construction it does not matter whether the operators A_i are linear or not, but, unless the equation is diagonal, there is usually no way to compute the functionals $F_{i,k,N}$. Accordingly, we will not discuss non-diagonal equations any further.

6.2.3 Consistency and Asymptotic Normality of sMLE

Consider the diagonal parabolic equation

$$\dot{u}(t) + (\mathrm{A}_0 + \theta \mathrm{A}_1)u(t) = \dot{W}^{\mathrm{Q}}(t). \tag{6.2.39}$$

Under conditions of Theorem 6.2.4 on page 397,

$$u \in L_2\big(\Omega; \mathcal{C}\big([0,T]; L_2(G)\big)\big)$$

and

$$u(t) = \sum_{k\geq 1} u_k(t)\mathfrak{h}_k,$$

where

$$u_k(t) = u_k(0)\, e^{-\mu_k(\theta)t} + q_k \int_0^t e^{-\mu_k(\theta)(t-s)} dw_k(s); \tag{6.2.40}$$

for now we do not indicate the dependence of u_k and u on θ. Also recall that we are assuming $q_k > 0$ and $\nu_k \neq 0$ for all k (see page 400).

In this section we investigate the simplified MLE $\widehat{\theta}_N$ from (6.2.34), or, more precisely, the estimation error (6.2.35), for one fixed value of θ such that the corresponding equation (6.2.39) is well-posed in $L_2(\Omega \times [0,T] \times G)$.

Accordingly, (6.2.35) becomes

$$\widehat{\theta}_N - \theta = -\frac{\sum_{k=1}^N \int_0^T \frac{\nu_k}{q_k} u_k(t) dw_k(t)}{\sum_{k=1}^N \int_0^T \frac{\nu_k^2}{q_k^2} u_k^2(t) dt}, \tag{6.2.41}$$

with u_k given by (6.2.40).

Denote by m_1 the order of the operator A_1; recall that 2m is order of the operator $\mathrm{A}_0 + \theta\mathrm{A}_1$ and d is the dimension of the spacial region G. The key number turns out to be

$$\varpi = \frac{2 \text{ order of } \mathrm{A}_1 - \text{ order of } (\mathrm{A}_0 + \theta\mathrm{A}_1)}{\mathrm{d}},$$

that is,

$$\varpi = \frac{2(\mathrm{m}_1 - \mathrm{m})}{\mathrm{d}}, \tag{6.2.42}$$

As (6.2.3) on page 395 suggests, it is reasonable to assume that the following limits exist:

$$\bar{\mu}(\theta) = \lim_{k\to\infty} \frac{\mu_k(\theta)}{k^{2\mathrm{m}/\mathrm{d}}} > 0, \tag{6.2.43}$$

$$\bar{\nu} = \lim_{k\to\infty} \frac{\nu_k}{k^{\mathrm{m}_1/\mathrm{d}}} \neq 0. \tag{6.2.44}$$

As we discussed earlier, this is a typical behavior of the eigenvalues of differential operators, and, for our purposes, it is more convenient to *assume* (6.2.43) and (6.2.44) rather than to write out the specific conditions on the operators A_0 and A_1. The parameter set Θ_0 is the largest open sub-set of $\mathbb{R}$ such that (6.2.43) and (6.2.44) hold for every $\theta \in \Theta_0$.

Define the number $\sigma = \sigma(\theta) > 0$ by

$$\sigma^2(\theta) = \begin{cases} \dfrac{2(\varpi+1)\bar{\mu}(\theta)}{\bar{\nu}^2}, & \text{if } \varpi > -1, \\[2ex] \dfrac{2\bar{\mu}(\theta)}{\bar{\nu}^2}, & \text{if } \varpi = -1. \end{cases} \tag{6.2.45}$$

Note that, similar to θ, $\bar{\mu}(\theta)$ is measured in $[t]^{-1}$, and therefore so is $\sigma^2(\theta)$.

The objective of the section is to prove the following result.

Theorem 6.2.9 *Assume that equation*

$$\dot{u} + (\mathrm{A}_0 + \theta\mathrm{A}_1)u = \dot{W}^Q$$

is diagonal and parabolic of order $2\mathrm{m}$ *for every* $\theta \in \Theta_0$, *and assume that the initial conditions* $u_k(0) \stackrel{d}{=} \mathcal{N}(m_k, \sigma_k^2)$, $k \geq 1$, *are such that*

$$\lim_{k\to\infty} \frac{m_k^2 + \sigma_k^2}{q_k^2} = 0, \text{ and} \tag{6.2.46}$$

$$\lim_{k\to\infty} \frac{k^{2\mathrm{m}/\mathrm{d}}\sigma_k^2(\sigma_k^2 + m_k^2)}{q_k^4} = 0. \tag{6.2.47}$$

If

$$\varpi \geq -1, \tag{6.2.48}$$

then the estimator $\widehat{\theta}_N$ *of* θ *is strongly consistent and asymptotically normal as* $N \to +\infty$. *More precisely,*

$$\lim_{N\to+\infty} \widehat{\theta}_N = \theta \tag{6.2.49}$$

with probability one, and

$$\lim_{N\to+\infty} \sqrt{I(N)}\big(\widehat{\theta}_N - \theta\big) \stackrel{d}{=} \mathcal{N}\big(0, \sigma^2(\theta)/T\big), \tag{6.2.50}$$

where

$$I(N) = \begin{cases} N^{\varpi+1}, & \text{if } \varpi > -1 \\ \ln N, & \text{if } \varpi = -1. \end{cases} \tag{6.2.51}$$

Both (6.2.46) and (6.2.47) are technical assumptions which come up naturally in the course of the proof to ensure that the initial conditions do not influence the asymptotic properties of the estimator. Without (6.2.46) and (6.2.47), the estimator can have different properties than those stated in the theorem.

Under (6.2.7), condition (6.2.46) is stronger than (6.2.8) and involves the covariance operator on the noise. On the other hand, condition (6.2.48) does not involve q_k.

Here are two situations when (6.2.46) and (6.2.47) hold:

1. $\sum_{k\geq 1}(m_k/q_k)^2 < \infty$ and $\sigma_k = 0$ (non-random initial conditions). In fact, (6.2.47) always holds if $\sigma_k = 0$.
2. $m_k = 0$ and $\sigma_k^2 = q_k^2/(2\mu_k(\theta))$ (stationary initial conditions: if $u_k(0) \stackrel{d}{=} \mathcal{N}(0, q_k^2/(2\mu_k(\theta)))$, then $u_k(t) \stackrel{d}{=} \mathcal{N}(0, q_k^2/(2\mu_k(\theta)))$ for all $t > 0$).

Conditions (6.2.46) and (6.2.47) are typically applied in the following setting, which is a version of the Toeplitz lemma: if

$$\lim_n x_n = a, \text{ and } b_n > 0, \ \sum_n b_n = +\infty, \tag{6.2.52}$$

then

$$\lim_n \frac{\sum_{k=1}^n x_k b_k}{\sum_{k=1}^n b_k} = a. \tag{6.2.53}$$

To verify (6.2.53), fix an $\epsilon > 0$, choose ℓ so that $|x_k - a| < \epsilon$ for all $k > \ell$ and then, after breaking the sum on top into $\sum_{k=1}^{\ell} + \sum_{k=\ell+1}^{n}$, pass to the limit $n \to \infty$.

Exercise 6.2.10 (C) Fill in the details in the proof of (6.2.53).

To prove (6.2.49), we need a version of the strong law of large numbers for independent but not necessarily identically distributed random variables.

Theorem 6.2.11 (The Strong Law of Large Numbers) *Let ζ_n, $n \geq 1$, be a sequence of independent random variables and $b_n, n \geq 1$, a sequence of positive numbers such that $b_{n+1} \geq b_n$, $\lim_{n\to\infty} b_n = +\infty$, and*

$$\sum_{n\geq 1} \frac{\operatorname{Var}\zeta_n}{b_n^2} < \infty. \tag{6.2.54}$$

Then

$$\lim_{n\to\infty} \frac{\sum_{k=1}^{n}(\zeta_n - \mathbb{E}\zeta_n)}{b_n} = 0$$

with probability one.

Proof See, for example, Shiryaev [207, Theorem IV.3.2]. An interested reader can recover the proof from (6.2.53).

We now use Theorem 6.2.11 to show strong consistency of $\widehat{\theta}_N$, that is, that the right-hand side of (6.2.41) converges to zero with probability one.

With no loss of generality, we will assume that $\mu_k(\theta) > 0$ for all $k \geq 1$. Indeed, (6.2.43) implies that $\mu_k(\theta) > 0$ for all sufficiently large k, so, if necessary, we can work with sums $\sum_{k=k_0}^{N}$ for a sufficiently large k_0: the very idea of taking the limit $N \to +\infty$ suggests that any finite number of observations u_k should not matter.

Proof of (6.2.49), *Step 1.* Using (6.2.40) and independence of $u_k(0)$ and w_k,

$$\begin{aligned}\mathbb{E}u_k^2(t) &= \mathbb{E}u_k^2(0)\, e^{-2\mu_k(\theta)t} + q_k^2 \int_0^t e^{-2\mu_k(\theta)(t-s)} ds \\ &= (m_k^2 + \sigma_k^2)e^{-2\mu_k(\theta)t} + \frac{q_k^2}{2\mu_k(\theta)}(1 - e^{-2\mu_k(\theta)t}).\end{aligned} \tag{6.2.55}$$

Proof of (6.2.49), *Step 2.* For fixed $T > 0$, using (6.2.55) and condition (6.2.46), together with (6.2.53),

$$\begin{aligned}\sum_{k=1}^{N} \frac{\nu_k^2}{q_k^2} \int_0^T \mathbb{E}u_k^2(t)dt &\simeq \sum_{k=1}^{N} \left(\frac{m_k^2 + \sigma_k^2}{q_k^2 T} + 1\right) \frac{\nu_k^2 T}{2\mu_k(\theta)} \simeq \sum_{k=1}^{N} \frac{\nu_k^2 T}{2\mu_k(\theta)} \\ &\simeq \frac{\bar{\nu} T}{2\bar{\mu}(\theta)} \sum_{k=1}^{N} \frac{k^{2\mathrm{m}_1/\mathrm{d}}}{k^{2\mathrm{m}/\mathrm{d}}} = \frac{\bar{\nu} T}{2\bar{\mu}(\theta)} \sum_{k=1}^{N} k^{\varpi} \simeq \frac{I(N)T}{\sigma^2(\theta)},\ N \to +\infty.\end{aligned} \tag{6.2.56}$$

Proof of (6.2.49), *Step 3.* We have

$$\lim_{N\to\infty} \frac{\sum_{k=1}^{N} \frac{\nu_k}{q_k} \int_0^T u_k(t)dw_k(t)}{\sum_{k=1}^{N} \frac{\nu_k}{q_k} \int_0^T \mathbb{E}u_k^2(t)dt} = 0 \text{ with probability one.} \tag{6.2.57}$$

Indeed, apply Theorem 6.2.11 with

$$\zeta_n = \int_0^T \frac{\nu_k}{q_k} u_k(t)dw_k(t) \text{ and } b_n = \sum_{k=1}^{N} \frac{\nu_k^2}{q_k^2} \int_0^T \mathbb{E}u_k^2(t)dt$$

so that, as we saw in (6.2.56),

$$\operatorname{Var}\zeta_n \sim n^{\varpi},\ b_n \sim I(n),$$

and thus

$$\frac{\operatorname{Var}\zeta_n}{b^2(n)} \sim \frac{\operatorname{Var}\zeta_n}{I^2(n)} \sim \begin{cases} n^{-2}, & \text{if } \varpi > -1, \\ 1/(n(\ln n)^2), & \text{if } \varpi = -1, \end{cases}$$

meaning that the series $\sum_n n^{\varpi}/I^2(n)$ converges for all $\varpi \geq -1$.

Exercise 6.2.12 (C) Verify that, in fact, the series $\sum_n \operatorname{Var}\zeta_n/I^2(n)$ converges for all ϖ, but if $\varpi < -1$, then we do not get convergence to zero in (6.2.57) because $I(n)$ stays bounded.

Proof of (6.2.49), *Step 4.* By Theorem 6.2.11,

$$\lim_{N\to\infty} \frac{\sum_{k=1}^{N} \frac{\nu_k^2}{q_k^2} \int_0^T u_k^2(t)dt}{\sum_{k=1}^{N} \frac{\nu_k^2}{q_k^2} \int_0^T \mathbb{E}u_k^2(t)dt} = 1 \text{ with probability one.} \tag{6.2.58}$$

Indeed, take $\zeta_n = \frac{\nu_k^2}{q_k^2} \int_0^T u_k^2(t)dt$ and $b_n = \sum_{k=1}^{N} \mathbb{E}\zeta_k$. Since $b_n \sim I(n)$ (see (6.2.56)), condition (6.2.54) becomes

$$\sum_n \frac{\operatorname{Var}\zeta_n}{I^2(n)} < \infty.$$

To get a bound on $\operatorname{Var}\zeta_n$, note that

$$\begin{aligned}\operatorname{Var}\int_0^T u_k^2(t)dt &= \mathbb{E}\left(\int_0^T u_k^2(t)\right)^2 - \left(\mathbb{E}\int_0^T u_k^2(t)dt\right)^2 \\ &= \mathbb{E}\left(\int_0^T \left(u_k^2(t) - \mathbb{E}u_k^2(t)\right)dt\right)^2.\end{aligned}$$

By the Cauchy-Schwarz inequality,

$$\operatorname{Var}\int_0^T u_k^2(t)dt \leq T\int_0^T \mathbb{E}\big(u_k^2(t) - \mathbb{E}u_k^2(t)\big)^2 dt = T\int_0^T \operatorname{Var}u_k^2(t)\, dt. \tag{6.2.59}$$

Using (6.2.40),

$$u_k(t) = A + B,\ A = u_k(0)e^{-\mu_k(\theta)(t-s)},\ B = q_k\int_0^t e^{-\mu_k(\theta)(t-s)}dw_k(s).$$

Note that A and B are independent Gaussian and $\mathbb{E}B = \mathbb{E}B^3 = 0$. Then

$$u_k^2(t) = A^2 + B^2 + 2AB$$

is a sum of three uncorrelated random variables and

$$\operatorname{Var} u_k^2(t) = \operatorname{Var} A^2 + \operatorname{Var} B^2 + 4\mathbb{E}A^2\,\mathbb{E}B^2. \tag{6.2.60}$$

Recall that if $\zeta \stackrel{d}{=} \mathcal{N}(0,\sigma^2)$, then $\mathbb{E}\zeta^4 = 3\sigma^4$. Then

$$\operatorname{Var} B^2 = 3\frac{q_k^4}{4\mu_k^2}(1 - e^{-2\mu_k(\theta)t})^2.$$

Since $u_k(0) - m_k = \mathcal{N}(0,\sigma_k^2)$, we similarly find

$$\mathbb{E}A^4 = \mathbb{E}((u_k(0) - m_k) + m_k)^4 e^{-4\mu_k(\theta)t} = \left(3\sigma_k^4 + m_k^4 + 6\sigma_k^2 m_k^2\right)e^{-4\mu_k(\theta)t}$$

and

$$\operatorname{Var} A^2 = 2\sigma_k^2(\sigma_k^2 + 2m_k^2)e^{-4\mu_k(\theta)t}.$$

Putting everything back into (6.2.60),

$$\operatorname{Var} u_k^2(t) = 2\sigma_k^2(\sigma_k^2 + 2m_k^2)e^{-4\mu_k(\theta)t} + 3\frac{q_k^4}{4\mu_k^2}(1 - e^{-2\mu_k(\theta)t})^2$$
$$+ \frac{4q_k^2(m_k^2 + \sigma_k^2)}{2\mu_k}(1 - e^{-2\mu_k(\theta)t})e^{-2\mu_k(\theta)t}.$$

Now integrate in time and re-arrange the terms to find

$$\frac{\nu_k^4}{q_k^4}\int_0^T \operatorname{Var} u_k^2(t)dt \sim \frac{\nu_k^4}{\mu_k^2(\theta)}\left(1 + \frac{4(m_k^2 + \sigma_k^2)}{q_k^2} + \frac{2\mu_k(\theta)\sigma_k^2(\sigma_k^2 + 2m_k^2)}{q_k^4}\right)$$

as $k \to +\infty$, or

$$\frac{\nu_k^4}{q_k^4}\int_0^T \operatorname{Var} u_k^2(t)dt \sim \frac{\nu_k^4}{\mu_k^2(\theta)} \sim k^{2\varpi},\ k \to +\infty, \tag{6.2.61}$$

where the first relation follows from (6.2.46) and (6.2.47), and the second relation follows from (6.2.43) and (6.2.44). As a result,

$$\sum_n \frac{\operatorname{Var}\zeta_n}{I^2(n)} \le C\sum_n \frac{n^{2\varpi}}{I^2(n)},$$

and, with $I(n)$ defined in (6.2.51), the series on the right-hand side converges: for every $\varpi \geq -1$, we have

$$\frac{n^{2\varpi}}{I^2(n)} \leq \frac{c}{n^2}.$$

This completes the proof of (6.2.49).

Exercise 6.2.13 (C) Compute the rate $I(N)$ for each of the equations in (6.2.6).

Let us now establish asymptotic normality (6.2.50) using a version of the martingale central limit theorem (Theorem 6.1.4 on page 386).

Thinking of $1/N$ as ε (to connect with the notations of Theorem 6.1.4), define

$$X_N(t) = \frac{\sum_{k=1}^N \frac{\nu_k}{q_k} \int_0^t u_k(s)dw_k(s)}{\sqrt{I(N)}}.$$

Then X_N is a continuous square-integrable martingale,

$$\langle X_N\rangle(t) = \frac{\sum_{k=1}^N \frac{\nu_k^2}{q_k^2} \int_0^t u_k^2(t)dt}{I(N)},$$

and

$$\sqrt{I(N)}\big(\widehat{\theta}_N - \theta\big) = \frac{I(N)}{\sum_{k=1}^N \frac{\nu_k^2}{q_k^2} \int_0^T u_k^2(t)dt} X_N(T)$$

By (6.2.56) and (6.2.58),

$$\lim_{N\to\infty} \langle X_N\rangle(t) = \frac{t}{\sigma^2(\theta)},$$

with $\sigma^2(\theta)$ from (6.2.45). Since $t/\sigma^2(\theta) = \langle X\rangle(t)$, where $X(t) = w(t)/\sigma(\theta)$ and w is a standard Brownian motion, the result follows:

$$\lim_{N\to+\infty} X_N(T) = \lim_{N\to+\infty} \frac{\sum_{k=1}^N \frac{\nu_k}{q_k} \int_0^T u_k(s)dw_k(s)}{\sqrt{I(N)}} \stackrel{d}{=} \mathcal{N}(0, T/\sigma^2(\theta)), \tag{6.2.62}$$

and

$$\lim_{N\to+\infty} \frac{I(N)}{\sum_{k=1}^N \frac{\nu_k^2}{q_k^2} \int_0^T u_k^2(t)dt} = \sigma^2(\theta)/T$$

with probability one.

This completes the proof of Theorem 6.2.9.

If condition (6.2.48) does not hold, that is, if the series $\sum_k \nu_k^2/\mu_k(\theta)$ converges, then, by (6.2.56), so do

$$\sum_k \int_0^T \frac{\nu_k}{q_k} u_k(t)dw_k(t) \text{ and } \sum_k \int_0^T \frac{\nu_k^2}{q_k^2} u_k^2(t)dt.$$

Therefore

$$\lim_{N\to+\infty} \widehat{\theta}_N = \theta - \frac{\sum_{k=1}^{\infty} \int_0^T \frac{\nu_k}{q_k} u_k(t)dw_k(t)}{\sum_{k=1}^{\infty} \int_0^T \frac{\nu_k^2}{q_k^2} u_k^2(t)dt},$$

and a consistent estimator of θ is only possible in either long time ($T \to +\infty$) or small noise $\sup_k q_k \to 0$ asymptotic. We are not considering these regimes, but here are some references: [113, 145] (long time), [81, 85–88, 190] (small noise).

6.2.4 Asymptotic Efficiency of the MLE

In this section, the dependence of various objects on θ will be of central interest and will therefore be shown explicitly everywhere. With the observation time interval $[0, T]$ fixed, dependence on T will often be omitted.

Let us recall the setting. There is a stochastic partial differential equation

$$\dot{u}_\theta(t) + (\mathrm{A}_0 + \theta \mathrm{A}_1)u_\theta(t) = \dot{W}^{\mathrm{Q}}(t) \tag{6.2.63}$$

such that its solution is written as a Fourier series

$$u(t) = \sum_{k\geq 1} u_{k,\theta}(t)\mathfrak{h}_k,$$

where

$$du_{k,\theta} = -\mu_k(\theta)u_{k,\theta}dt + q_k dw_k, \tag{6.2.64}$$

that is,

$$u_{k,\theta}(t) = u_k(0)\, e^{-\mu_k(\theta)t} + q_k \int_0^t e^{-\mu_k(\theta)(t-s)}dw_k(s),$$

$$u_k(0) \stackrel{d}{=} \mathcal{N}(m_k, \sigma_k^2).$$

The differential operators A_1 and $\mathrm{A}_0 + \theta \mathrm{A}_1$ have orders m_1 and $2\mathrm{m}_0$, respectively, and the corresponding eigenvalues ν_k, $\mu_k(\theta)$ have the asymptotic

$$\nu_k \simeq \bar{\nu} k^{\mathrm{m}_1/\mathrm{d}}, \ |\bar{\nu}| > 0, \tag{6.2.65}$$

$$\mu_k(\theta) \simeq \bar{\mu}(\theta) k^{2\mathrm{m}/\mathrm{d}}, \ \bar{\mu}_k(\theta) > 0, \tag{6.2.66}$$

where d is the dimension of the (missing) spacial variable in Eq. (6.2.63).

The objective is to estimate the parameter $\theta \in \mathbb{R}$ from the observations $U^{N,\theta} = \{u_{k,\theta}(t),\ k = 1, \ldots, N, t \in [0, T]\}$. The observations generate the measure $P_T^{N,\theta}$ on the space of continuous on $[0, T]$ functions with values in $\mathbb{R}^N$. The likelihood ratio is

$$\begin{aligned} L_{N,T}(\theta) &= \frac{dP_T^{N,\theta}}{dP_T^{N,\theta_0}} \left(U^{(N,\theta_0)}\right) \\ &= \exp\left(-\sum_{k=1}^{N} \frac{\mu_k(\theta) - \mu_k(\theta_0)}{q_k^2} \int_0^T u_{k,\theta_0}(t) du_{k,\theta_0}(t) \right. \\ &\quad \left. - \sum_{k=1}^{N} \frac{\mu_k^2(\theta) - \mu_k^2(\theta_0)}{2q_k^2} \int_0^T u_{k,\theta_0}^2(t) dt \right). \end{aligned} \tag{6.2.67}$$

While (6.2.67) does not suggest any restrictions on the possible values of θ, such restrictions can come from the underlying SPDE (6.2.63). To solve Eq. (6.2.63), it was convenient to introduce condition (6.2.5), while analysis of the estimators is more convenient under the equivalent condition (6.2.66). Recall that Θ_0 denotes the largest open set in $\mathbb{R}$ such that (6.2.66) (equivalently, (6.2.5)) holds for every $\theta \in \Theta_0$. For example, if $\mathrm{A}_0 = 0$ and $\mathrm{A}_1 = -\boldsymbol{\Delta}$, then $\Theta_0 = (0, +\infty)$.

Recall the notations

$$\begin{aligned} \varpi &= \frac{2(\mathrm{m}_1 - \mathrm{m})}{\mathrm{d}}, \text{ with the standing assumption } \varpi \geq -1, \\ \sigma^2(\theta) &= \begin{cases} \dfrac{2(\varpi + 1)\bar{\mu}(\theta)}{\bar{\nu}^2}, & \text{if } \varpi > -1, \\[2ex] \dfrac{2\bar{\mu}(\theta)}{\bar{\nu}^2}, & \text{if } \varpi = -1, \end{cases} \\ I(N) &= \begin{cases} N^{\varpi+1}, & \text{if } \varpi > -1 \\ \ln N, & \text{if } \varpi = -1. \end{cases} \end{aligned} \tag{6.2.68}$$

The key technical result of Sect. 6.2.3 is that, with probability one,

$$\sum_{k=1}^{N} \frac{\nu_k^2}{q_k^2} \int_0^T u_{k,\theta}^2(t) dt \simeq \sum_{k=1}^{N} \frac{\nu_k^2}{q_k^2} \int_0^T \mathbb{E} u_{k,\theta}^2(t) dt \simeq \frac{I(N)T}{\sigma^2(\theta)}; \tag{6.2.69}$$

assumptions (6.2.46) and (6.2.47) in Theorem 6.2.9 ensure that the initial conditions $u_k(0)$ do not affect the asymptotic relations in (6.2.69).

Theorem 6.2.9 establishes strong consistency and asymptotic normality of the simplified maximum likelihood estimator $\widehat{\theta}_N$, the un-restricted maximizer of $L_{N,T}(\theta)$. The true MLE $\widehat{\boldsymbol{\theta}}_N$ is the maximizer of $L_{N,T}(\theta)$ over the closure of Θ_0, and, while strong consistency of $\widehat{\boldsymbol{\theta}}_N$ immediately follows from that of $\widehat{\theta}_N$, the asymptotic normality does not. We will eventually establish this asymptotic normality using rather sophisticated and general results in asymptotic statistics; Problem 6.2.2 invites the reader to investigate a direct proof for a specific equation.

The question of asymptotic normality of $\widehat{\boldsymbol{\theta}}_N$ aside, there are deeper reasons why the result of Theorem 6.2.9 should not be taken as the end of the investigation of the estimation problem. The main reason is that the question about optimality of the estimator, either $\widehat{\boldsymbol{\theta}}_N$ or $\widehat{\theta}_N$ remains unanswered, and the objective of the current section is to answer this question.

It is natural to suspect that optimality of a particular estimator, once we have a suitable definition of optimality, is a property not only of the estimator but of the underlying model: there should be some intrinsic features of the model ensuring particular kind of behavior for a large class of possible estimators. Since a statistical model is usually characterized by the likelihood ratio, we can expect that suitable properties of the likelihood ratio will imply necessary properties of the model, including, with some luck, optimality of the MLE.

Given the asymptotic nature of the problem, it is natural to study asymptotic properties of the likelihood ratio. An immediate observation is that, according to (6.2.69), nothing interesting happens to the function $L_{N,T}(\theta)$ as $N \to +\infty$ for fixed θ and T. It turns out that non-trivial asymptotic behavior of $L_{N,T}(\theta)$ in the limit $N \to +\infty$ happens if, as $N \to +\infty$, θ approaches the reference value θ_0, and does so at a certain rate. These considerations lead to the `local likelihood ratio`

$$\begin{aligned} Z_{N,\theta}(x) &= \frac{d\mathbf{P}_T^{N,\theta+x\phi_{N,\theta}}}{d\mathbf{P}_T^{N,\theta}}\left(U^{(N,\theta)}\right) \\ &= \exp\left(\sum_{k=1}^{N}\left(-\frac{\mu_k(\theta+x\phi_{N,\theta})-\mu_k(\theta)}{q_k^2}\int_0^T u_{k,\theta}(t)du_{k,\theta}(t)\right.\right. \qquad (6.2.70) \\ &\left.\left.-\frac{\mu_k^2(\theta+x\phi_{N,\theta})-\mu_k^2(\theta))}{2q_k^2}\int_0^T u_{k,\theta}^2(t)dt\right)\right), \end{aligned}$$

where $x \in \mathbb{R}$ and

$$\phi_{N,\theta} = \sqrt{\frac{\sigma^2(\theta)}{I(N)T}}. \qquad (6.2.71)$$

Comparing (6.2.70) and (6.2.67), we see that the parameter θ_0 corresponding to the reference measure is no longer fixed but becomes another variable; we will put major effort later to establish various properties of $Z_{N,\theta}(x)$ uniformly in θ over compact sets. For the measure on top, the parameter is $\theta + x\phi_{N,\theta}$. By (6.2.68), $\phi_{N,\theta} \searrow 0$, $N \to +\infty$, and this turns out to be the correct rate at which the top and bottom parameters approach each other. The additional (dimensionless) variable x controls the relative distance between the parameters. Relation (6.2.50) on page 412 makes the choice of $\phi_{N,\theta}$ very natural; even the appearance of $\sigma^2(\theta)$ and T in (6.2.71) makes sense, both by looking at (6.2.50) and doing dimensional analysis.

Exercise 6.2.14 (C) Confirm that $\phi_{N,\theta}$ and θ have the same units $[t]^{-1}$.

Substituting (6.2.64) into (6.2.70) and using that $\mu_k(\theta) = \kappa_k + \theta\nu_k$, we get a more convenient formula for $Z_{N,\theta}$:

$$Z_{N,\theta}(x) = \exp\left(\sum_{k=1}^{N}\left(-\frac{x\phi_{N,\theta}}{q_k}\int_0^T \nu_k u_{k,\theta}(t)dw_k(t)\right.\right.$$
$$\left.\left. -\frac{x^2\phi_{N,\theta}^2}{2q_k^2}\int_0^T \nu_k^2 u_{k,\theta}^2(t)dt\right)\right). \tag{6.2.72}$$

We will now summarize the main results from [84, Sects. II.12 and III.1] reformulated for our setting.

To begin, here are some of the desirable properties of $Z_{N,\theta}(x)$.

[LAN] `(Local asymptotic normality)` The local likelihood ratio $Z_{N,\theta}$ is called locally asymptotically normal (LAN), as $N \to +\infty$, at a point $\theta_0 \in \Theta_0$ if there exist random variables $\zeta_N, N \geq 1$, and $\varepsilon_N(x),\ N \geq 1, x \in \mathbb{R}$, such that

a) the sequence ζ_N converges in distribution to a standard normal random variable: $\lim_{N\to+\infty} \zeta_N \stackrel{d}{=} \mathcal{N}(0,1)$.
b) for every $x \in \mathbb{R}$, the sequence $\varepsilon_N(x)$ converges to zero in probability.
c) for every $x \in \mathbb{R}$,

$$Z_{N,\theta_0}(x) = \exp\left(x\zeta_N - \frac{x^2}{2} + \varepsilon_N(x)\right).$$

[UAN] `(Uniform asymptotic normality)` The local likelihood ratio $Z_{N,\theta}$ is called uniformly asymptotically normal, as $N \to +\infty$, in a set $\Theta \subseteq \Theta_0$ if there exist random variables $\zeta_N(\theta)$ and $\varepsilon_N(x,\theta)$ such that, for every sequence $\{\theta_N,\ N \geq 1\} \subset \Theta$ and every converging sequence $\{x_N,\ N \geq 1\} \subset \mathbb{R}$, with $\lim_{N\to\infty} x_N = x$ and $\theta_N + x_N\phi_{N,\theta_N} \in \Theta$,

a) The sequence $\zeta_N(\theta_N)$ converges in distribution to a standard normal random variable: for every $y \in \mathbb{R}$,

$$\lim_{N\to+\infty} \mathbb{P}\big(\zeta_N(\theta_N) \le y\big) = \mathbb{P}(\zeta \le y),\ \zeta \overset{d}{=} \mathcal{N}(0,1),$$

even though the sequence θ_N does not need to converge;

b) The sequence $\varepsilon_N(x_N, \theta_N)$ converges to zero in probability, that is, for every $\delta > 0$,

$$\lim_{N\to+\infty} \mathbb{P}(|\varepsilon_N(x_N, \theta_N)| > \delta) = 0;$$

again, note that the sequence θ_N does not need to converge;

c) The following equality holds:

$$Z_{N,\theta_N}(x_n) = \exp\Big(x\zeta_N(\theta_N) - \frac{x^2}{2} + \varepsilon_N(x_N, \theta_N)\Big). \tag{6.2.73}$$

[HC] `(Local Hölder continuity)` Local Hölder continuity of $Z_{N,\theta}$ in an open set $\Theta \subseteq \Theta_0$ means that, for every compact set $K \subset \Theta$, there exist positive numbers a and B such that, for every $R > 0$ and $N \ge 1$,

$$\sup_{\theta\in K}\ \sup_{|x|\le R, |y|\le R}\ |x-y|^{-2}\,\mathbb{E}\Big|\sqrt{Z_{N,\theta}(x)} - \sqrt{Z_{N,\theta}(y)}\Big|^2 \le B(1+R^a).$$

[UPI] `(Uniform polynomial integrability)` Uniform polynomial integrability of $Z_{N,\theta}$ in an open set $\Theta \subseteq \Theta_0$ means that, for every compact set $K \subset \Theta$ and every $p > 0$, there exists an $N_0 = N_0(K, p)$ such that

$$\sup_{\theta\in K}\ \sup_{N>N_0}\ \sup_{x\in\mathbb{R}} |x|^p\,\mathbb{E}\sqrt{Z_{N,\theta}(x)} < \infty. \tag{6.2.74}$$

Under conditions of Theorem 6.2.9, (6.2.62) and (6.2.69) immediately imply that $Z_{N,\theta}$ is locally asymptotically normal at every point $\theta \in \Theta_0$; recall that Θ_0 is the largest open set on which the underlying SPDE is well-posed (equivalently, (6.2.66) holds). Indeed, since $\phi_{N,\theta} = \sqrt{\sigma^2(\theta)/(I(N)T)}$, (6.2.62) implies that

$$\sum_{k=1}^{N}\Big(-\frac{\phi_{N,\theta}}{q_k}\int_0^T \nu_k u_{k,\theta}(t)dw_k(t)\Big)$$

converges in distribution to a standard normal random variable, and (6.2.69) implies that

$$\frac{x^2}{2}\sum_{k=1}^{N}\frac{\phi^2_{N,\theta}}{q_k^2}\int_0^T \nu_k^2 u^2_{k,\theta}(t)dt$$

converges with probability one to $x^2/2$.

Exercise 6.2.15 (C) Confirm local asymptotic normality of $Z_{N,\theta}$.

It turns out that local asymptotic normality of $Z_{N,\theta}$ at a point $\theta = \theta_0$ implies certain property of every sequence of estimators $\tilde{\theta}_N, N \geq 1$, in a neighborhood of θ_0.

Definition 6.2.16 An `estimator` $\tilde{\theta}_N$ is a random variable of the form $\tilde{\theta}_N = F_N(U^{N,\theta})$, where $F_N : \mathcal{C}((0,T);\mathbb{R}^N) \to \mathbb{R}$ is a measurable mapping.

To characterize the behavior of an estimator, we need the notion of a loss function.

Definition 6.2.17 A function $w = w(x)$, $x \in \mathbb{R}$ is called a `loss function` if $w(0) = 0$, $w(x)$ is continuous, $w(x) = w(-x)$, $w(x)$ is non-decreasing for $x > 0$, and $w(x_0) > 0$ for at least one $x_0 \in \mathbb{R}$. A loss functions w is said to have polynomial growth if there exists a $p > 0$ such that $w(x)/(1 + |x|^p)$ is bounded for all $x \in \mathbb{R}$.

A typical loss function is $w(x) = |x|^p, p > 0$, or $w(x) = \min(|x|^p, 1)$.

Given a loss function $w = w(x)$ and an estimator $\tilde{\theta}_N$, the number

$$\mathbb{E}w(\tilde{\theta}_N - \theta)$$

characterizes the quality of the estimator at the point θ. Local asymptotic normality at a point implies a lower bound on $\mathbb{E}w(\tilde{\theta}_N - \theta)$ in a neighborhood of θ.

Theorem 6.2.18 *If $Z_{N,\theta}$ is LAN at a point $\theta = \theta_0$ and if $w = w(x)$ is a loss function such that $\lim_{|x|\to\infty} w(x)e^{-a|x|^2} = 0$ for every $a > 0$, then, for every $\delta > 0$ and every sequence of estimators $\tilde{\theta}_N$, $N \geq 1$,*

$$\liminf_{N\to+\infty} \sup_{|\theta_0-\theta|<\delta} \mathbb{E}w\big((\tilde{\theta}_N - \theta)/\phi_{N,\theta_0}\big) \geq \mathbb{E}w(\zeta),\ \zeta \overset{d}{=} \mathcal{N}(0,1); \tag{6.2.75}$$

Proof See [84, Theorem II.12.1].

The following definition is then natural.

Definition 6.2.19 A sequence of estimators $\bar{\theta}_N, N \geq 1$, is called `asymptotically efficient` at a point θ_0 with respect to a collection W of loss functions if, for every $w \in W$,

$$\lim_{\delta\to 0} \liminf_{N\to+\infty} \sup_{|\theta_0-\theta|<\delta} \mathbb{E}w\big((\bar{\theta}_N - \theta)/\phi_{N,\theta}\big) = \mathbb{E}w(\zeta),\ \zeta \overset{d}{=} \mathcal{N}(0,1).$$

Since we know that $Z_{N,\theta}$ is locally asymptotically normal at every point θ of the open set Θ_0, we conclude that every sequence of estimators $\bar{\theta}_N$ that is asymptotically normal with rate $\sqrt{I(N)}$ (for example, sMLE $\widehat{\theta}_N$ from (6.2.34) on page 407) is asymptotically efficient at every point $\theta \in \Theta_0$ with respect to bounded loss functions. This is almost obvious by the dominated convergence theorem, but a rigorous proof also relies on continuity of $\boldsymbol{\sigma}(\theta)$ as a function of θ.

Exercise 6.2.20 (B) Confirm that $\widehat{\theta}_N$ is asymptotically efficient with respect to bounded loss functions.

While the LAN property of $Z_{N,\theta}$ gives us some optimality of the simplified MLE, there are at least two reasons to proceed further: we have yet to get any information about MLE beyond consistency, and it is often desirable to have asymptotic efficiency with respect to loss functions that are not bounded. This is where we use the other three properties of $Z_{N,\theta}$. The underlying result is as follows.

Theorem 6.2.21 *If the local likelihood ratio has properties [UAN], [HC], and [UPI] in an open set Θ, then, for every $\theta \in \Theta$ the maximum likelihood estimator is consistent, asymptotically normal with rate $1/\varphi_{N,\theta}$, and asymptotically efficient with respect to rate functions of polynomial growth.*

Proof See [84, Theorem III.1.1 and Corollary II.1.1].

Accordingly, here is the main result of this section. As a brief reminder, we are estimating the number θ in

$$\dot{u} + (\mathrm{A}_0 + \theta \mathrm{A}_1)u = \dot{W}^{\mathrm{Q}}$$

where the partial differential operator A_1 and $\mathrm{A}_0 + \theta\mathrm{A}_1$ have orders m_1 and 2m, respectively. The MLE is based on the first N Fourier coefficient $u_1, \ldots, u_N$ of the solution and is defined in (6.2.32) on page 407. We already know consistency of the MLE from consistency of sMLE.

Theorem 6.2.22 *Assume that Eq. (6.2.39) is diagonal and parabolic of order* 2m *for every $\theta \in \Theta_0$, and assume that the initial conditions $u_k(0) \overset{d}{=} \mathcal{N}(m_k, \sigma_k^2)$, $k \geq 1$, are such that*

$$\lim_{k\to\infty} \frac{m_k^2 + \sigma_k^2}{q_k^2} = 0 \text{ and} \tag{6.2.76}$$

$$\lim_{k\to\infty} \frac{k^{2\mathrm{m}/\mathrm{d}}\sigma_k^2(\sigma_k^2 + m_k^2)}{q_k^4} = 0. \tag{6.2.77}$$

If

$$\frac{2\mathrm{m}_1 - 2\mathrm{m}}{\mathrm{d}} \geq -1, \tag{6.2.78}$$

then, as $N \to +\infty$, the maximum likelihood estimator $\widehat{\boldsymbol{\theta}}_N$ is asymptotically normal with rate $\sqrt{I(N)}$ and is asymptotically efficient with respect to the loss functions of polynomial growth, for every $\theta \in \Theta_0$.

To prove Theorem 6.2.22, we need to verify conditions [UAN], [HC], and [UPI] for the local likelihood ratio (6.2.72). Verification of conditions [UAN] and [HC] is relatively straightforward; verification of [UPI] is more demanding.

Uniform asymptotic normality. The objective is to show that, for every compact $K \subset \Theta_0$, every $y \in \mathbb{R}$, and every $\delta > 0$, the following two equalities hold uniformly in $\theta \in K$:

$$\lim_{N\to+\infty} \mathbb{P}\Big(\sum_{k=1}^{N} \frac{\phi_{N,\theta}}{q_k} \int_0^T \nu_k u_{k,\theta}(t)dw_k(t) \le y\Big) = \mathbb{P}(\zeta \le y),\ \zeta \stackrel{d}{=} \mathcal{N}(0,1) \tag{6.2.79}$$

and

$$\lim_{N\to+\infty} \mathbb{P}\left(\left|\frac{\phi_{N,\theta}^2}{q_k^2} \int_0^T \nu_k^2 u_{k,\theta}^2(t)dt - 1\right| > \delta\right) = 0. \tag{6.2.80}$$

Compared to local asymptotic normality [LAN], the difference is proving that the convergence is uniform in θ, which requires deriving explicit bounds on the rate of convergence that do not depend on θ.

Exercise 6.2.23 (C) Recall that condition [UAN] also involves a converging sequence x_N. Accordingly, identify the random variables $\zeta_N(\theta_N)$ and $\varepsilon_N(x_N, \theta_N)$ in (6.2.73) and confirm that (6.2.79) and (6.2.80) indeed imply uniform asymptotic normality of $Z_{N,\theta}$.

Fix a compact set $K \subset \Theta_0$. The plan is as follows:

1. First, we use (6.2.56) on page 413 to show that

$$\lim_{N\to\infty} \sum_{k=1}^{N} \frac{\phi_{N,\theta}^2}{q_k^2} \int_0^T \nu_k^2 \mathbb{E} u_{k,\theta}^2(t)dt = 1 \tag{6.2.81}$$

   uniformly in $\theta \in K$.
2. Second, we use the Chebychev inequality to show (6.2.80). Here, the key relation is (6.2.61) on page 415.
3. Finally, we use (6.2.81) and a Berry-Esseen-type result to verify (6.2.79).

Many arguments in the proofs of (6.2.81) and (6.2.80) consist in going over various computations from the proof of Theorem 6.2.9 and verifying that the corresponding asymptotic relations hold uniformly in $\theta \in K$. A key technical component in the proof is the uniform version of the Toeplitz lemma (6.2.53): if $x_n = x_n(\theta)$ and $b_n = b_n(\theta)$ are such that conditions (6.2.52) hold uniformly in $\theta \in K$, then (6.2.53) also holds uniformly in $\theta \in K$.

Exercise 6.2.24 (C) State and prove a uniform version of (6.2.53).

To proceed, note that, by Definition 6.2.2 on page 397, inequalities (6.2.5) hold uniformly in $\theta \in K$. As before, with no loss of generality, we assume that $c_2 = 0$, that is, there exist positive numbers c_1, c_3 such that, for all $k \ge 1$ and all $\theta \in K$,

$$c_1 k^{2\mathrm{m}/\mathrm{d}} \le \mu_k(\theta) \le c_3 k^{2\mathrm{m}/\mathrm{d}}. \tag{6.2.82}$$

Then (6.2.81) follows by visual inspection of (6.2.56).

Next, using notations

$$A_{N,\theta} = \sum_{k=1}^{N} \frac{\nu_k^2}{q_k^2} \int_0^T u_{k,\theta}^2(t)dt, \quad B_{N,\theta} = \mathbb{E}A_{N,\theta},$$

note that

$$\begin{aligned}\sum_{k=1}^{N} \frac{\phi_{N,\theta}^2}{q_k^2} \int_0^T \nu_k^2 u_{k,\theta}^2(t)dt - 1 &= \phi_{N,\theta}^2 A_N - 1 \\ &= \phi_{N,\theta}^2\Big(A_{N,\theta} - B_{N,\theta}\Big) + (\phi_{N,\theta}^2 B_{N,\theta} - 1);\end{aligned}$$

uniform convergence of the second (non-random) term is equivalent to (6.2.81). For the first term, we use the Chebychev inequality and relation (6.2.61) on page 415 to compute

$$\begin{aligned}\mathbb{P}\Big(\Big|\phi_{N,\theta}^2(A_{N,\theta} - B_{N,\theta})\Big| > \delta\Big) &\le \frac{C\phi_{N,\theta}^2 \sum_{k=1}^{N} \frac{\nu_k^4}{\mu_k^2(\theta)}}{\delta} \\ &\le \frac{C_1 \sum_{k=1}^{N} k^{2\varpi}}{I^2(N)\delta} \to 0,\ N \to +\infty.\end{aligned}$$

After some visual inspection of the calculations leading to (6.2.61), with (6.2.82) in mind, we conclude that the number C_1 does not depend on $\theta \in K$.

This completes the proof of (6.2.80).

Let us summarize some of our findings in a single chain of equivalence relations, all holding uniformly in θ over compact sub-sets of Θ_0 under the assumptions (6.2.65), (6.2.66), (6.2.78), (6.2.76), and (6.2.77):

$$I(N) \sim \frac{1}{\phi_{N,\theta}^2} \simeq \sum_{k=1}^{N} \frac{\nu_k^2}{q_k^2} \int_0^T \mathbb{E}u_{k,\theta}^2(t)dt \simeq \sum_{k=1}^{N} \frac{\nu_k^2 T}{2\mu_k(\theta)} \sim \sum_{k=1}^{N} k^{2\varpi}. \tag{6.2.83}$$

To establish (6.2.79), we use a `Berry-Esseen bound` on the rate of convergence in the central limit theorem:

Theorem 6.2.25 *Let $\zeta_1, \ldots, \zeta_N$ be independent random variables with zero mean and $\sum_{k=1}^{N} \mathbb{E}\zeta_k^2 = 1$, and let $\zeta \overset{d}{=} \mathcal{N}(0, 1)$. Then*

$$\sup_{y\in\mathbb{R}} \left| \mathbb{P}\left(\sum_{k=1}^{N} \zeta_k \le y\right) - \mathbb{P}(\zeta \le y) \right| \le 9.4 \sum_{k=1}^{N} \mathbb{E}|\zeta_k|^3. \tag{6.2.84}$$

Proof See [23, Theorem 3.6]. The number 9.4 on the right-hand side of (6.2.84) is certainly interesting, but not important for our purposes.

Define

$$\zeta_{k,\theta} = \frac{\frac{\nu_k}{q_k}\int_0^T u_{k,\theta}(t)dw_k(t)}{\sum_{k=1}^N \frac{\nu_k^2}{q_k^2}\int_0^T \mathbb{E}u_{k,\theta}^2(t)dt}.$$

By the BDG inequality and (6.2.81),

$$\mathbb{E}|\zeta_{k,\theta}|^3 \le \frac{C_1\frac{|\nu_k|^3}{q_k^3}\mathbb{E}\left(\int_0^T u_{k,\theta}^2(t)dt\right)^{3/2}}{I^{3/2}(N)},$$

where the number C_1 does not depend on $\theta \in K$. Writing

$$u_{k,\theta}(t) = u_k(0)e^{-\mu_k(\theta)t} + \bar{u}_k(t),\ \bar{u}_{k,\theta}(t) = q_k\int_0^t e^{-\mu_k(\theta)(t-s)}dw_k(s),$$

we find

$$\int_0^T u_{k,\theta}^2(t)dt \le \frac{u_k^2(0)}{\mu_k(\theta)} + 2\int_0^T \bar{u}_{k,\theta}^2(t)dt,$$

and then

$$\mathbb{E}\left(\int_0^T u_{k,\theta}^2(t)dt\right)^{3/2} \le C_2\left(\frac{\mathbb{E}|u_k(0)|^3}{\mu_k^{3/2}(\theta)} + \int_0^T \mathbb{E}|\bar{u}_{k,\theta}(t)|^3 dt\right).$$

Recall that if $\eta \overset{d}{=} \mathcal{N}(m,\sigma^2)$, then, for $p > 0$,

$$\begin{aligned}\mathbb{E}|\eta|^p &= \mathbb{E}|(\eta - m) + m|^p \le c(p)(\mathbb{E}|\eta - m|^p + |m|^p)\\ &\le c_1(p)(\sigma^p + |m|^p).\end{aligned}$$

Also recall that

$$u_k(0) \overset{d}{=} \mathcal{N}(m_k,\sigma_k^2),\ \bar{u}_{k,\theta}(t) \overset{d}{=} \mathcal{N}\left(0, \frac{q_k^2}{2\mu_k(\theta)}(1 - e^{-2\mu_k(\theta)t})\right).$$

Therefore,

$$\int_0^T \mathbb{E}|u_{k,\theta}(t)|^3 dt \le C_3\left(\frac{|m_k|^3 + \sigma_k^3 + q_k^3T^{3/2}}{\mu_k^{3/2}(\theta)}\right),$$

and

$$\mathbb{E}|\zeta_{k,\theta}|^3 \leq \frac{C_4}{I^{3/2}(N)} \frac{|\nu_k|^3}{\mu_k^{3/2}(\theta)} \left(T^{3/2} + \frac{|m_k|^3 + \sigma_k^2}{q_k^3} \right) \leq \frac{C_5 T^{3/2}}{I^{3/2}(N)} k^{3\varpi/2}, \tag{6.2.85}$$

where the last inequality follows from (6.2.76).

Next, using the notation

$$B_{N,\theta} = \sum_{k=1}^{N} \frac{\nu_k^2}{q_k^2} \int_0^T \mathbb{E}u_{k,\theta}^2(t)dt,$$

we get

$$\sum_{k=1}^{N} \frac{\phi_{N,\theta}}{q_k} \int_0^T \nu_k u_{k,\theta}(t)dw_k(t) = \phi_{N,\theta} B_{N,\theta} \sum_{k=1}^{N} \zeta_{k,\theta}.$$

With (6.2.81) in mind, take an $\epsilon \in (0, 1)$ and then find N_0 such that, for all $N > N_0$ and $\theta \in K$,

$$|\phi_{N,\theta} B_{N,\theta} - 1| \leq \epsilon.$$

To simplify notations even further, write

$$F(y) = \mathbb{P}(\zeta \leq y), \; \zeta \stackrel{d}{=} \mathcal{N}(0, 1).$$

If $y \geq 0$, then, by (6.2.83)–(6.2.85),

$$\begin{aligned} &\left| \mathbb{P}\Big(\sum_{k=1}^{N} \frac{\phi_{N,\theta}}{q_k} \int_0^T \nu_k u_{k,\theta}(t)dw_k(t) \leq y \Big) - F(y) \right| \\ &\leq \left| \mathbb{P}\Big(\sum_{k=1}^{N} \zeta_{k,\theta} \leq \frac{y}{1-\epsilon} \Big) - F(y/(1-\epsilon)) \right| + \big| F(y) - F(y/(1-\epsilon)) \big| \\ &\leq 9.4 \frac{C_5 T^{3/2}}{I^{3/2}(N)} \sum_{k=1}^{N} k^{3\varpi/2} + \big| F(y) - F(y/(1-\epsilon)) \big| \\ &\leq 9.4 \frac{C_5 T^{3/2}}{N^{1/2}} + \big| F(y) - F(y/(1-\epsilon)) \big|, \end{aligned} \tag{6.2.86}$$

and (6.2.79) follows because ϵ is arbitrary and the normal distribution function F is continuous.

This concludes the proof of uniform asymptotic normality of $Z_{N,\theta}$.

Exercise 6.2.26 (C) One result not mentioned in the derivation of (6.2.86) is that, if X, Y are random variables and $a > 0,\ b > 0,\ 0 < \delta < b$, then

$$\begin{aligned}\mathbb{P}(XY \le a) &= \mathbb{P}(XY \le a, |Y-b| \le \delta) + \mathbb{P}(XY \le x, |Y-b| > \delta)\\ &\le \mathbb{P}(X \le a/(b-\delta)) + \mathbb{P}(|Y-b| > \delta).\end{aligned}$$

Use the result to confirm (6.2.86), and then derive an analogue of (6.2.86) when $y \le 0$.

Local Hölder continuity. The objective is to show that, for every compact set $K \subset \Theta_0$, there exist positive numbers a and B such that, for every $R > 0$ and $N \ge 1$,

$$\sup_{\theta\in K}\ \sup_{|x|\le R, |y|\le R} |x-y|^{-2}\,\mathbb{E}\left|\sqrt{Z_{N,\theta}(x)} - \sqrt{Z_{N,\theta}(y)}\right|^2 \le B(1+R^a). \tag{6.2.87}$$

We start with the equality

$$f(x) - f(y) = (x-y)\int_0^1 f'\big(\tau x + (1-\tau)y\big)d\tau,$$

true for every continuously differentiable function $f = f(x), x \in \mathbb{R}$, and apply the equality to the function

$$f(x) = \sqrt{Z_{N,\theta}(x)} = Z_{N,\theta}^{1/2}(x),$$

with fixed N and θ. By (6.2.72) on page 420,

$$Z_{N,\theta}^{1/2}(x) = \exp\left(\frac{x}{2}A_{N,\theta} - \frac{x^2}{4}V_{N,\theta}\right),$$

where

$$A_{N,\theta} = -\sum_{k=1}^{N}\frac{\phi_{N,\theta}}{q_k}\int_0^T \nu_k u_{k,\theta}(t)dw_k(t),\ \ V_{N,\theta} = \sum_{k=1}^{N}\frac{\phi_{N,\theta}^2}{q_k^2}\int_0^T \nu_k^2 u_{k,\theta}^2(t)dt.$$

Then, with

$$F_{N,\theta}(x) = A_{N,\theta} - xV_{N,\theta},$$

we have

$$\frac{\partial Z_{N,\theta}^{1/2}(x)}{\partial x} = \frac{1}{2}F_{N,\theta}(x)\, Z_{N,\theta}^{1/2}(x).$$

Therefore,

$$\mathbb{E}\Big|Z_{N,\theta}^{1/2}(x) - Z_{N,\theta}^{1/2}(y)\Big|^2$$
$$= \frac{1}{4}|x-y|^2\,\mathbb{E}\left(\int_0^1 F_{N,\theta}\big(\tau x + (1-\tau)y\big)\,Z_{N,\theta}^{1/2}\big(\tau y + (1-\tau)x\big)d\tau\right)^2$$
$$\le \frac{1}{4}|x-y|^2 \int_0^1 \mathbb{E}\Big(F_{N,\theta}^2\big(\tau x + (1-\tau)y\big)\,Z_{N,\theta}\big(\tau y + (1-\tau)x\big)\Big)d\tau.$$

To proceed, let us recall the original definition of $Z_{N,\theta}(x)$ as a density; see (6.2.70) on page 419. Then, for every functional $g : \mathcal{C}((0,T);\mathbb{R}^N) \to \mathbb{R}$,

$$\mathbb{E}\Big(g(U^{N,\theta})Z_{N,\theta}(x)\Big) = \mathbb{E}g(U^{N,\theta+x\phi_{N,\theta}}).$$

As a result, setting

$$\theta(\tau) = \theta + (\tau x + (1-\tau)y)\,\phi_{N,\theta}$$

and using the definition of $F_{N,\theta}$, we find:

$$\mathbb{E}\Big|Z_{N,\theta}^{1/2}(x) - Z_{N,\theta}^{1/2}(y)\Big|^2$$
$$= \frac{1}{4}|x-y|^2 \int_0^1 \mathbb{E}F_{N,\theta(\tau)}^2\big(\tau x + (1-\tau)y\big)d\tau$$
$$\le \frac{1}{2}|x-y|^2\left(\int_0^1 \mathbb{E}A_{N,\theta(t)}^2 d\tau + \int_0^1 \big(\tau x + (1-\tau)y\big)^2\,\mathbb{E}V_{N,\theta(\tau)}^2 d\tau\right).$$

If x and y are in a compact set in $\mathbb{R}$ and θ is in an compact sub-set of Θ_0, then, for all sufficiently large N, $\theta(\tau)$ will also be in a compact set of Θ_0, because $\lim_{N\to+\infty}\phi_{N,\theta} = 0$ uniformly over θ in a compact set. Finally, according to (6.2.56) and (6.2.61), for every compact set $K \subset \Theta 0$, there exists a number C such that, for all $N \ge 1$ and $\theta \in K$,

$$\mathbb{E}A_{N,\theta}^2 = \mathbb{E}V_{N,\theta} \le C \text{ and } \mathbb{E}V_{N,\theta}^2 \le C.$$

Exercise 6.2.27 (C) Confirm that it is enough to establish (6.2.87) for all sufficiently large N.

This completes the proof of local Hölder continuity of $Z_{N,\theta}$.

Uniform polynomial integrability. The objective is to show (6.2.74) on page 421, that is,

$$\sup_{x\in\mathbb{R}} |x|^p\,\mathbb{E}Z_{N,\theta}^{1/2}(x) < \infty \tag{6.2.88}$$

for all sufficiently large N, uniformly in θ over compact subsets of Θ_0. With

$$A_{N,\theta} = -\sum_{k=1}^{N} \frac{\phi_{N,\theta}}{q_k} \int_0^T \nu_k u_{k,\theta}(t) dw_k(t), \; V_{N,\theta} = \sum_{k=1}^{N} \frac{\phi_{N,\theta}^2}{q_k^2} \int_0^T \nu_k^2 u_{k,\theta}^2(t) dt,$$

we have

$$Z_{N,\theta}^{1/2}(x) = \exp\left(\frac{x}{2} A_{N,\theta} - \frac{x^2}{4} V_{N,\theta}\right).$$

Note that, for every $y \in \mathbb{R}$, the Itô formula implies

$$\mathbb{E} \exp\left(y A_{N,\theta} - \frac{y^2}{2} V_{N,\theta}\right) = 1.$$

Also note that

$$\frac{1}{2}\left(\frac{x}{2}\right)^2 = \frac{x^2}{8} < \frac{x^2}{4},$$

suggesting that we can use the Hölder inequality to reduce $\mathbb{E} Z_{N,\theta}^{1/2}(x)$ to $\mathbb{E} \exp(-\varepsilon x^2 V_N)$ with $\varepsilon > 0$. For example, if $p = 3/2$ and $q = 3$, then

$$\begin{aligned}
&\mathbb{E} \exp\left(\frac{x}{2} A_{N,\theta} - \frac{x^2}{4} V_{N,\theta}\right) \\
&= \mathbb{E}\left(\exp\left(\frac{x}{2} A_{N,\theta} - \frac{3x^2}{16} V_{N,\theta}\right) \exp\left(-\frac{x^2}{16} V_{N,\theta}\right)\right) \\
&\le \left(\mathbb{E} \exp\left(\frac{3x}{4} A_{N,\theta} - \frac{9x^2}{32} V_{N,\theta}\right)\right)^{2/3} \left(\mathbb{E} \exp\left(-\frac{3x^2}{16} V_{N,\theta}\right)\right)^{1/3} \\
&= \left(\mathbb{E} \exp\left(-\frac{3x^2}{16} V_{N,\theta}\right)\right)^{1/3}.
\end{aligned}$$

Exercise 6.2.28 (C) The idea behind above computations is to write

$$-\frac{x^2}{4} = -\frac{\alpha x^2}{4} - \frac{(1-\alpha)x^2}{4}$$

and then, given $\alpha \in (1/2, 1)$, to find $p > 1$ so that

$$\frac{p^2}{8} = \frac{p\alpha}{4},$$

that is, $p = 2\alpha$. In particular, if $\alpha = 3/4$, then $p = 3/2$ and $q = p/(p-1) = 3$. Choose a different α, find the corresponding p and q, and then write the resulting inequality.

To complete the proof of (6.2.88), it is enough to show that, for every compact $K \subset \Theta_0$, there exist $c > 0$ and $N_0 \geq 1$ such that, for every $z > 0$, $\theta \in K$, and $N \geq N_0$,

$$\mathbb{E}\exp\Big(-zV_{N,\theta}\Big) \leq e^{-c\min(z,\sqrt{z})}. \tag{6.2.89}$$

Inequality (6.2.89) is not easy to guess, and its derivation requires significant computational effort. On the bright side, the computations present independent interest and can potentially be helpful in other problems, such as Problem 6.2.5 below (proving asymptotic efficiency of sMLE).

For an alternative proof of (6.2.88) without relying on (6.2.89), see [82, Lemma 3.1].

To begin, we notice that $V_{N,\theta}$ is a sum of independent random variables, and therefore

$$\mathbb{E}\exp\Big(-zV_{N,\theta}\Big) = \prod_{k=1}^{N}\mathbb{E}\exp\left(-z\phi_{N,\theta}^2\frac{v_k^2}{q_k^2}\int_0^T u_{k,\theta}^2(t)dt\right).$$

Each $u_{k,\theta}$ is an Ornstein-Uhlebeck process

$$du_{k,\theta} = -\mu_k(\theta)u_{k,\theta}(t) + q_k dw_k(t),$$

suggesting the following result as a starting point.

Proposition 6.2.29 *Let $X = X(t;a,\sigma)$ be the solution of*

$$dX = -aXdt + \sigma dw,\ 0 < t < T, \tag{6.2.90}$$

with parameters $a \geq 0$, $\sigma > 0$, a standard Brownian motion w, and initial condition $X(0;a,\sigma) \overset{d}{=} \mathcal{N}(x_0,\sigma_0^2)$ independent of w.

For $y \geq 0$, define the functions

$$\psi(y;a,\sigma) = \mathbb{E}\exp\left(-y\int_0^T X^2(t;a,\sigma)dt\right),$$

$$\varrho(y;a,\sigma) = (a^2 + 2\sigma^2 y)^{1/2},$$

$$\psi_0(y;a,\sigma) = \left(\frac{e^{aT}}{\cosh(\varrho T) + (a/\varrho)\sinh(\varrho T)}\right)^{1/2},$$

$$\Psi(y;a,\sigma) = \frac{\varrho - a}{2\sigma^2}\left(1 - \frac{e^{-\varrho T}}{\cosh(\varrho T) + (a/\varrho)\sinh(\varrho T)}\right).$$

Then

$$\psi(y;a,\sigma) = \frac{\psi_0(y;a,\sigma)}{\sqrt{1+2\sigma_0^2\Psi(y;a,\sigma)}}\exp\left(-\frac{\Psi(y;a,\sigma)x_0^2}{1+2\sigma_0^2\Psi(y;a,\sigma)}\right), \tag{6.2.91}$$

Exercise 6.2.30 (C) Use (6.2.91) to verify the following particular case of the `Cameron-Martin formula`:

$$\mathbb{E}\exp\left(-\frac{1}{2}\int_0^T w^2(t)dt\right) = \frac{1}{\sqrt{\cosh T}}.$$

The general Cameron-Martin formula in this context deals with a quadratic form of the multi-dimensional Brownian motion: see [139, Theorem 7.21].

It is the proof of Proposition 6.2.29 that takes the most effort. Accordingly, let us postpone the proof for now and instead derive (6.2.89) from (6.2.91).

First, note that $a \le \varrho$, and therefore $\psi \le \psi_0$ for all $a \ge 0$ and $y \ge 0$. In other words, it is enough to establish (6.2.89) when $u_{k,\theta}(0) = 0$. Then

$$\mathbb{E}\exp\left(-z\phi_{N,\theta}^2\frac{\nu_k^2}{q_k^2}\int_0^T u_{k,\theta}^2(t)dt\right) = \psi_0\left(\frac{z\nu_k^2\phi_{N,\theta}^2}{q_k^2};\mu_k(\theta),q_k\right),$$

and we need a suitable upper bound on the function ψ_0.

According to (6.2.91) with

$$y = \frac{z\phi_{N,\theta}^2\nu_k^2}{q_k^2},\quad a=\mu_k(\theta),\ \sigma=q_k,\ \varrho=\sqrt{\mu_k^2(\theta)+2z\phi_{N,\theta}^2\nu_k^2},$$

we have

$$\begin{aligned}\psi^2(y;a,\sigma) &= \frac{e^{aT}}{\cosh\varrho T+(a/\varrho)\sinh\varrho T}\\ &\le \frac{e^{aT}}{(a/\varrho)(\cosh\varrho T+\sinh\varrho T)} = \frac{\varrho}{a}e^{-(\varrho-a)T}.\end{aligned} \tag{6.2.92}$$

Because of (6.2.66), there is no loss of generality by assuming that

$$a = \mu_k(\theta) > 2/T$$

for all $k \ge 1$, uniformly in $\theta \in K$. By the mean value theorem applied to the function $f(x) = \ln x$, we conclude that, with $q \in [0,1]$,

$$\ln\varrho - \ln a = \frac{\varrho-a}{qa+(1-q)\varrho} \le \frac{\varrho-a}{a} \le \frac{(\varrho-a)T}{2},$$

that is,

$$\frac{\varrho}{a} \le e^{(\varrho-a)T/2}.$$

Then (6.2.92) becomes

$$\psi_0(y; a, \sigma) \le e^{-(\varrho-a)T/4}. \tag{6.2.93}$$

Next, we need a lower bound on

$$\varrho - a = \sqrt{\mu_k^2(\theta) + 2z\phi_{N,\theta}^2 \nu_k^2} - \mu_k(\theta).$$

To shorten the formulas, introduce one more notation

$$\tau = z\phi_{N,\theta}^2.$$

Because $2\mathrm{m} \ge \mathrm{m}_1$, conditions (6.2.65) and (6.2.66) imply that, with no loss of generality, we can assume existence of a positive number β such that

$$0 \le \nu_k^2/\mu_k^2(\theta) \le \beta$$

for all $k \ge 1$, uniformly in $\theta \in K$. Therefore

$$\begin{aligned}\sqrt{\mu_k^2(\theta) + 2\tau\nu_k^2} - \mu_k(\theta) &= \frac{2\tau\nu_k^2}{\sqrt{\mu_k^2(\theta) + 2\tau\nu_k^2} + \mu_k(\theta)} \\ &= \frac{2\tau(\nu_k^2/\mu_k(\theta))}{\sqrt{1 + 2\tau(\nu_k/\mu_k(\theta))^2} + 1} \ge \frac{\nu_k^2}{\mu_k(\theta)} \frac{\min(\tau, \sqrt{\tau})}{\sqrt{1+2\beta}}.\end{aligned} \tag{6.2.94}$$

Indeed, if $0 \le \tau \le 1$, then

$$\frac{2\tau}{\sqrt{1 + 2\tau(\nu_k/\mu_k(\theta))^2} + 1} \ge \frac{2\tau}{\sqrt{1+2\beta} + 1} \ge \frac{\tau}{\sqrt{1+2\beta}},$$

and if $\tau \ge 1$, then

$$\frac{2\tau}{\sqrt{1 + 2\tau(\nu_k/\mu_k(\theta))^2} + 1} \ge \frac{2\sqrt{\tau}}{\sqrt{\tau^{-1} + 2\beta} + \tau^{-1/2}} \ge \frac{2\sqrt{\tau}}{\sqrt{1+2\beta} + 1}.$$

Combining (6.2.93) and (6.2.94),

$$\psi_0(\tau\nu_k^2/q_k^2; \mu_k(\theta), q_k) \le \exp\left(-\frac{\nu_k^2}{\mu_k(\theta)} \frac{T}{4\sqrt{1+2\beta}} \min(\tau, \sqrt{\tau}) \right),$$

or, recalling that $\tau = z\phi_{N,\theta}^2$,

$$\begin{aligned}\mathbb{E}\exp\Big(-zV_{N,\theta}\Big) &\le \prod_{k=1}^{N}\psi_0\left(\frac{z\nu_k^2\phi_{N,\theta}^2}{q_k^2};\mu_k(\theta),q_k\right)\\ &\le \exp\left(-\frac{T}{4\sqrt{1+2\beta}}\min\left(z\phi_{N,\theta}^2,\sqrt{z\phi_{N,\theta}^2}\right)\sum_{k=1}^{N}\frac{\nu_k^2}{\mu_k(\theta)}\right)\\ &\le \exp\left(-\frac{T}{4\sqrt{1+2\beta}}\left(\phi_{N,\theta}^2\sum_{k=1}^{N}\frac{\nu_k^2}{\mu_k(\theta)}\right)\min\left(z,\sqrt{z/\phi_{N,\theta}^2}\right)\right).\end{aligned}$$

We now look at (6.2.83) on page 425 and make two observations: (a) there exists a positive number γ such that

$$\phi_{N,\theta}^2\sum_{k=1}^{N}\frac{\nu_k^2}{\mu_k(\theta)} \ge \gamma$$

for all $N \ge 1$ and $\theta \in K$; (b) there exists $N_0 \ge 1$ such that $1/\phi_{N,\theta}^2 > 1$ for all $N \ge N_0$ and $\theta \in K$. Then, for $N \ge N_0$,

$$\min\left(z,\sqrt{z/\phi_{N,\theta}^2}\right) \ge \min(z,\sqrt{z}),$$

and we get (6.2.89) with

$$c = \frac{\gamma T}{4\sqrt{1+2\beta}}.$$

Now that we see how Proposition 6.2.29 leads to (6.2.88), let us prove Proposition 6.2.29.

We start with a general, and well-known, property of a Gaussian random variable.

Proposition 6.2.31 *Let ζ be a Gaussian random variable with mean μ and variance κ^2 and let λ be a real number such that $2\lambda\kappa^2 < 1$. Then*

$$\mathbb{E}e^{\lambda\zeta^2} = \frac{1}{\sqrt{1-2\lambda\kappa^2}}\exp\left(\frac{\lambda\mu^2}{1-2\lambda\kappa^2}\right). \tag{6.2.95}$$

Proof The proof of (6.2.95) is by direct computation using the equality

$$\frac{1}{\sqrt{2\pi}}\int_{-\infty}^{+\infty} e^{-a(x-\mu)^2/2}dx = \frac{1}{\sqrt{a}},\ a>0,\ \mu\in\mathbb{R}.$$

Then

$$
\begin{aligned}
\mathbb{E}e^{\lambda\zeta^2} &= \frac{1}{\sqrt{2\pi\kappa^2}}\int_{-\infty}^{+\infty}\exp\left(\lambda x^2 - \frac{(x-\mu)^2}{2\kappa^2}\right)dx\\
&= \frac{1}{\sqrt{2\pi\kappa^2}}\int_{-\infty}^{+\infty}\exp\left(\lambda x^2 + \frac{2\mu x - x^2 - \mu^2}{2\kappa^2}\right)dx\\
&= \frac{1}{\sqrt{2\pi\kappa^2}}\int_{-\infty}^{+\infty}\exp\left(-\frac{1-2\lambda\kappa^2}{2\kappa^2}\left(x-\frac{\mu}{1-2\lambda\kappa^2}\right)^2 - \mu^2\left(\frac{1}{2\kappa^2}-\frac{1}{2\kappa^2(1-2\lambda\kappa^2)}\right)\right)dx\\
&= \exp\left(\frac{\lambda\mu^2}{1-2\lambda\kappa^2}\right)\frac{1}{\sqrt{2\pi\kappa^2}}\int_{-\infty}^{+\infty}\exp\left(-\frac{1-2\lambda\kappa^2}{2\kappa^2}\left(x-\frac{\mu}{1-2\lambda\kappa^2}\right)^2\right)dx\\
&= \frac{1}{\sqrt{1-2\lambda\kappa^2}}\exp\left(\frac{\lambda\mu^2}{1-2\lambda\kappa^2}\right).
\end{aligned}
$$

This completes the proof of Proposition 6.2.31.

To establish (6.2.91), take $b \in \mathbb{R}$ and consider the process $X(t; b, \sigma)$, defined by (6.2.90) with initial condition $X(0; b, \sigma) = X_0$ and with b instead of a in the drift term. By the density formula (6.2.18) on page page 402,

$$
\begin{aligned}
\psi(y; a, \sigma) &= \mathbb{E}\exp\left(-y\int_0^T X^2(t; a, \sigma)dt\right)\\
&=\mathbb{E}\exp\left(-y\int_0^T X^2(t; b, \sigma)dt - \frac{a-b}{\sigma^2}\int_0^T X(t; b, \sigma)dX(t; b, \sigma)\right.\\
&\quad\left. - \frac{a^2-b^2}{2\sigma^2}\int_0^T X^2(t; b, \sigma)dt\right).
\end{aligned}
\tag{6.2.96}
$$

If we choose $b > 0$ so that

$$
-y - \frac{a^2-b^2}{2\sigma^2} = 0 \text{ or } b^2 = 2\sigma^2 y + a^2,
$$

then

$$
b = \varrho(y; a, \sigma)
$$

and

$$
\begin{aligned}
\psi(y; a, \sigma) &= \mathbb{E}\exp\left(-\frac{a-b}{\sigma^2}\int_0^T X(t; b, \sigma)dX(t; b, \sigma)\right)\\
&= \mathbb{E}\exp\left(\frac{(b-a)\big(X^2(T; b, \sigma) - X_0^2 - \sigma^2 T\big)}{2\sigma^2}\right),
\end{aligned}
\tag{6.2.97}
$$

with the last equality following from the Itô formula, as the quadratic variation of $X(t; b, \sigma)$ is $\sigma^2 t$.

To proceed, define

$$\bar{X}(T) = \sigma \int_0^T e^{-b(t-s)} dw(s), \tag{6.2.98}$$

so that

$$X(T; b, \sigma) = \bar{X}(T) + X_0 e^{-bT}.$$

Then $\bar{X}(T)$ is a Gaussian random variable independent of X_0, with

$$\mathbb{E}\bar{X}(T) = 0, \quad \operatorname{Var} \bar{X}(T) = \sigma^2(1 - e^{-bT})/(2b).$$

To evaluate the right-hand side of (6.2.97), we use the following result: if χ and η are independent random variables and $f = f(s,t)$ is a measurable function, then, as long as the corresponding expectations are defined,

$$\mathbb{E}f(\chi, \eta) = \mathbb{E}g(\eta), \text{ where } g(x) = \mathbb{E}f(\chi, x).$$

We take $\chi = \bar{X}(T)$, $\eta = X_0$,

$$f(\chi, \eta) = \exp\left(\frac{(b-a)\big((\chi + \eta\, e^{-bT})^2 - \eta^2 - \sigma^2 T\big)}{2\sigma^2}\right).$$

Then

$$\begin{aligned} g(x) &= \mathbb{E}\exp\left(\frac{(b-a)\big((\bar{X}(T) + xe^{-bT})^2 - x^2 - \sigma^2 T\big)}{2\sigma^2}\right) \\ &= e^{-(b-a)T/2}\, e^{-(b-a)x^2/(2\sigma^2)}\mathbb{E}\exp\left(\frac{(b-a)(\bar{X}(T) + xe^{-bT})^2}{2\sigma^2}\right). \end{aligned}$$

Using (6.2.95) with

$$\zeta = \bar{X}(T) + \mu, \ \mu = xe^{-bt}, \ \lambda = (b-a)/(2\sigma^2)\ \kappa^2 = \sigma^2(1 - e^{-bT})/(2b),$$

and noticing that

$$2\lambda\kappa^2 = \frac{(b-a)(1-e^{-bT})}{2b} \le 1/2,$$

we find:

$$g(x) = \left(\frac{e^{-(b-a)T}}{1-(b-a)(1-e^{-2bT})/(2b)}\right)^{1/2} \exp\left(\frac{((b-a)/(2\sigma^2))x^2e^{-2bT}}{1-(b-a)(1-e^{-2bT})/(2b)}\right)$$

$$\times \exp\left(\frac{-(b-a)x^2}{2\sigma^2}\right) = \left(\frac{e^{-(b-a)T}}{1-(b-a)(1-e^{-2bT})/(2b)}\right)^{1/2}$$

$$\times \exp\left(-\frac{b-a}{2\sigma^2}\left(1-\frac{e^{-2bT}}{1-(b-a)(1-e^{-2bT})/(2b)}\right)x^2\right).$$

Using the definitions of the hyperbolic functions,

$$\frac{e^{-(b-a)T}}{1-(b-a)(1-e^{-2bT})/(2b)} = \frac{e^{aT}}{\cosh(bT)+(a/b)\sinh(bT)}$$

and

$$\frac{e^{-2bT}}{1-(b-a)(1-e^{-2bT})/(2b)} = \frac{e^{-bT}}{\cosh(bT)+(a/b)\sinh(bT)}.$$

Therefore, keeping in mind that $b = \varrho(y; a)$,

$$g(x) = \psi_0(y; a, \sigma)e^{-\Psi(y;a,\sigma)x^2}.$$

We now use (6.2.95) once again, this time with $\zeta = X_0$, $\mu = x_0$, $\lambda = -\Psi(y; a, \sigma)$, $\kappa^2 = \sigma_0^2$ to find

$$\psi(y; a, \sigma) = \mathbb{E}g(X_0) = \frac{\psi_0(y; a, \sigma)}{\sqrt{1+2\sigma_0^2\Psi(y; a, \sigma)}} \exp\left(-\frac{\Psi(y; a, \sigma)x_0^2}{1+2\sigma_0^2\Psi(y; a, \sigma)}\right),$$

which establishes (6.2.91).

This concludes the proof of Theorem 6.2.22.

6.2.5 Problems

The main part of Sect. 6.2 is analyzing the MLE using the local likelihood ratio. While the computations are rather long, it turns out that a straightforward analysis of the MLE, without using the local likelihood ratio, leads to essentially the same computations, and Problems 6.2.1–6.2.4 illustrate this in various settings. Problem 6.2.5 is a potentially interesting connection between SPDEs and geometry. Problem 6.2.6 is of different kind and invites the reader to compare Girsanov's theorem with a related result due to Cameron and Martin.

Problem 6.2.1 Recall the estimator $\hat{\theta}^{(k)}(T)$ from (6.1.14) on page 384. As we now know, it is not an MLE but rather an sMLE. A possible choice of the corresponding MLE is

$$\hat{\boldsymbol{\theta}}^{(k)}(T) = \begin{cases} \hat{\theta}^{(k)}(T), & \text{if } \hat{\theta}^{(k)}(T) > 0, \\ 0, & \text{if } \hat{\theta}^{(k)}(T) \le 0. \end{cases} \tag{6.2.99}$$

Investigate $\hat{\boldsymbol{\theta}}^{(k)}(T)$ in the limits (a) $T \to +\infty$, k fixed; (b) $k \to +\infty$, T fixed.

Problem 6.2.2 Prove directly (that is, without using results like Theorem 6.2.22) that the MLE (6.2.36) of $\theta > 0$ in

$$u_t = \theta u_{xx} + \dot{W}^{\mathrm{Q}},\ 0 < t < T,\ 0 < x < \pi,$$

with zero initial and boundary conditions, using the first N Fourier coefficients $u_1, \ldots, u_N$ of the solution, is asymptotically normal with rate $N^{3/2}$.

Problem 6.2.3 Consider the problem of estimating $\theta \in \mathbb{R}$ in the equation

$$u_t = u_{xx} + \theta\, u + \dot{W}^{\mathrm{Q}},\ 0 < t < T,\ 0 < x < \pi,$$

with zero initial and boundary conditions.

(a) Verify that the first N Fourier coefficients $u_1, \ldots, u_N$ of the solution do not lead to a consistent estimator of θ in the limit $N \to +\infty$.
(b) To construct a consistent estimator in the limit $T \to +\infty$, one can use the MLE based on the first Fourier coefficient or on all Fourier coefficients. Is there any difference in the asymptotic properties of the estimators?

Problem 6.2.4 (a) Verify that, for every integer $p \ge 1$,

$$\mathbb{E}\left(\sum_{k=1}^{N} \frac{\nu_k^2}{q_k^2} \int_0^T u_{k,\theta}^2(t)dt\right)^p \le C(p) I^p(N)$$

uniformly over θ in a compact set $K \subset \Theta_0$.
(b) Verify that the simplified MLE $\widehat{\theta}_N$ is asymptotically efficient with respect to loss functions of polynomial growth.

Problem 6.2.5 Suppose you observe the solution of the equation

$$u_t = \boldsymbol{\Delta} u + \dot{W}^{\mathrm{Q}},\ 0 \le t \le T,\ x \in G \subset \mathbb{R}^2,$$

with zero initial and boundary conditions. Is it possible to estimate the area of G?

Problem 6.2.6 Let P_T^w be the Wiener measure on $\mathcal{C}((0,T);\mathbb{R})$. There is a result, due to Cameron and Martin and known as the `Cameron-Martin theorem`, saying that, given a function $y \in \mathcal{C}((0,T);\mathbb{R})$, the shifted measure P_T^{w+y} is absolutely continuous with respect to P_T^w if and only if $y(t) = \int_0^t h(s)ds$ and $h \in L_2((0,T))$. The `Cameron-Martin formula` gives the formula for the corresponding density:

$$\frac{dP_T^{w+y}}{dP_T^w}(x) = \exp\left(I_y(x) - \frac{1}{2}\int_0^T h^2(t)dt\right)$$

where I_y is $\mathcal{N}(0, (1/2)\int_0^T h^2(t)dt)$ under P_T^w but in general there is no closed-form expression for $I_y(x)$, just a limiting procedure for computing it.

Compare and contrast this result with Girsanov's theorem.

6.3 Several Parameters and Further Directions

6.3.1 The Heat Balance Equation

Recall Eq. (6.1.1) on page 381 that was eventually simplified to

$$u_t = \kappa \mathrm{A} u - \lambda u + \dot{W}^Q,\ 0 < t \le T,\ x \in G;\ u(t,0) = 0, \tag{6.3.1}$$

where A is the Laplace operator in a smooth bounded domain $G \subset \mathbb{R}^2$ with zero boundary conditions. The objective is to estimate the numbers κ and λ from the observations $U^N = (u_1, \ldots, u_N)$, where

$$du_k(t) = -(\kappa\,\nu_k + \lambda)u_k(t)dt + q_k dw_k(t),\ u_k(0) = 0, \tag{6.3.2}$$

and $-\nu_k, k \ge 1$, are the eigenvalues of A. Recall that, by (6.2.3) on page 395, $\nu_k \sim k$. Also, we will choose the corresponding eigenfunctions $\mathfrak{h}_k,\ k \ge 1$, of A to form an orthonormal basis in $L_2(G)$.

Theorem 6.3.1 *Assume that* $\kappa > 0$ *and* $\sum_{k\ge1} q_k^2/k < \infty$. *Then*

$$u(t) = \sum_{k\ge1} u_k(t)\mathfrak{h}_k$$

is a mild solution of (6.3.1).

The solution is unique in the class $L_2(\Omega \times [0,T] \times G)$.

If, in addition $\sum_{k\ge1} q_k^2/k^\delta < \infty$ *for some* $\delta < 1$, *then*

$$u \in L_2(\Omega; \mathcal{C}((0,T); L_2(G))).$$

Exercise 6.3.2 (C) Prove Theorem 6.3.1.

Hint. For $T > 0$,

$$\mathbb{E}\int_0^T u_k^2(t)dt \sim \frac{q_k^2 T}{2k}. \tag{6.3.3}$$

To verify continuity, note that

$$|1 - e^{-x}| \le x^{\varepsilon}$$

for all $x > 0$ and $0 < \varepsilon \le 1$, and use the Kolmogorov criterion, keeping in mind that u is Gaussian.

Exercise 6.3.3 (B) Formulate and prove an analogue of Theorem 6.3.1 for the variational solution.

Note that the physical restriction $\lambda > 0$ (cooling as opposed to heating) is not necessary for Eq. (6.3.1) to be well posed.

We can now address the question of estimating the numbers κ and λ from the observations $U^{N,T} = \{u_k(t),\ k = 1, \ldots, N,\ t \in [0,T]\}$. By (6.2.18) on page 402, the likelihood function is

$$L_{T,N}(\kappa,\lambda) = \exp\left(\sum_{k=1}^{N}\left(-\frac{(\kappa-\kappa_0)\nu_k + (\lambda-\lambda_0)}{q_k^2}\int_0^T u_k(t)du_k(t)\right.\right.$$
$$\left.\left. - \frac{(\kappa\,\nu_k + \lambda)^2 - (\kappa_0\,\nu_k + \lambda_0)^2}{2q_k^2}\int_0^T u_k^2(t)dt\right)\right);$$

the process u_k on the right-hand side in this representation satisfies (6.3.2) with $\kappa = \kappa_0$ and $\lambda = \lambda_0$. It is convenient to introduce notations

$$A_{0,N} = -\sum_{k=1}^{N}\frac{1}{q_k^2}\int_0^T u_k(t)du_k(t),\ A_{1,N} = -\sum_{k=1}^{N}\frac{\nu_k}{q_k^2}\int_0^T u_k(t)du_k(t),$$

$$B_{0,N} = \sum_{k=1}^{N}\frac{1}{q_k^2}\int_0^T u_k^2(t)dt,\ B_{1,N} = \sum_{k=1}^{N}\frac{\nu_k}{q_k^2}\int_0^T u_k^2(t)dt,$$

$$B_{2,N} = \sum_{k=1}^{N}\frac{\nu_k^2}{q_k^2}\int_0^T u_k^2(t)dt.$$

Then

$$\ln L_{N,T}(\kappa,\lambda) = (\kappa-\kappa_0)A_{1,N} + (\lambda-\lambda_0)A_{0,N}$$
$$+\frac{\kappa_0^2-\kappa^2}{2}B_{2,N} + (\kappa_0\lambda_0 - \kappa\lambda)B_{1,N} + \frac{\lambda_0^2-\lambda^2}{2}B_{0,N}.$$

The simplified MLEs $\widehat{\kappa}_N$ of κ and $\widehat{\lambda}_N$ of λ solve the system of equations

$$\frac{\partial \ln L_{N,T}(\kappa,\lambda)}{\partial \kappa}=0,\ \frac{\partial \ln L_{N,T}(\kappa,\lambda)}{\partial \lambda}=0,$$

or

$$\begin{cases}\kappa B_{2,N}+\lambda B_{1,N}=A_{1,N}\\ \kappa B_{1,N}+\lambda B_{0,N}=A_{0,N},\end{cases}$$

that is,

$$\widehat{\kappa}_N=\frac{A_{1,N}B_{0,N}-A_{0,N}B_{1,N}}{B_{2,N}B_{0,N}-B_{1,N}^2},\ \widehat{\lambda}_N=\frac{-A_{1,N}B_{1,N}+A_{0,N}B_{2,N}}{B_{2,N}B_{0,N}-B_{1,N}^2},$$

or, in the matrix-vector form,

$$\begin{pmatrix}\widehat{\kappa}_N\\ \widehat{\lambda}_N\end{pmatrix}=\begin{pmatrix}B_{2,N} & B_{1,N}\\ B_{1,N} & B_{0,N}\end{pmatrix}^{-1}\begin{pmatrix}A_{1,N}\\ A_{0,N}\end{pmatrix} \tag{6.3.4}$$

The true MLEs are then

$$\widehat{\kappa}_N=\begin{cases}\widehat{\kappa}_N, & \text{if } \widehat{\kappa}_N>0\\ 0, & \text{if } \widehat{\kappa}_N\le 0,\end{cases}\quad \widehat{\lambda}_N=\begin{cases}\widehat{\lambda}_N, & \text{if } \widehat{\lambda}_N>0\\ 0, & \text{if } \widehat{\lambda}_N\le 0.\end{cases}$$

Theorem 6.3.4 *For every $\kappa_0>0$ and $\lambda_0\in\mathbb{R}$,*

$$\lim_{N\to\infty}\begin{pmatrix}\widehat{\kappa}_N\\ \widehat{\lambda}_N\end{pmatrix}=\begin{pmatrix}\kappa_0\\ \lambda_0\end{pmatrix} \tag{6.3.5}$$

with probability one, and

$$\lim_{N\to\infty}\begin{pmatrix}N\big(\widehat{\kappa}_N-\kappa_0\big)\\ \sqrt{\ln N}\big(\widehat{\lambda}_N-\lambda_0\big)\end{pmatrix}\stackrel{d}{=}\begin{pmatrix}\zeta_1\\ \zeta_2\end{pmatrix}, \tag{6.3.6}$$

where the zero-mean Gaussian random variables ζ_1 and ζ_2 independent.

Proof The arguments are almost identical to the single parameter case. Using the notations

$$\bar{A}_{0,N}=\sum_{k=1}^{N}\frac{1}{q_k}\int_0^T u_k(t)dw_k(t),\ \bar{A}_{1,N}=\sum_{k=1}^{N}\frac{\nu_k}{q_k}\int_0^T u_k(t)dw_k(t), \tag{6.3.7}$$

we get

$$\widehat{\kappa}_N - \kappa_0 = -\frac{\bar{A}_{1,N}B_{0,N} + \bar{A}_{0,N}B_{1,N}}{B_{2,N}B_{0,N} - B_{1,N}^2},\ \widehat{\lambda}_N - \lambda_0 = -\frac{\bar{A}_{1,N}B_{1,N} + \bar{A}_{0,N}B_{2,N}}{B_{2,N}B_{0,N} - B_{1,N}^2}. \tag{6.3.8}$$

Exercise 6.3.5 (B) Verify (6.3.8).

Hint. Use (6.3.4).

By (6.3.3) and the strong law of large numbers (Theorem 6.2.11 on page 412),

$$B_{0,N} \sim \mathbb{E}B_{0,N} \sim \ln N,\ B_{1,N} \sim \mathbb{E}B_{1,N} \sim N,\ B_{2,N} \sim \mathbb{E}B_{1,N} \sim N^2. \tag{6.3.9}$$

In particular,

$$\lim_{N\to+\infty} \frac{B_{1,N}^2}{B_{0,N}B_{2,N}} = 0 \tag{6.3.10}$$

with probability one. Then the same strong law of large numbers implies (6.3.5).

Exercise 6.3.6 (C) Verify the asymptotic of $B_{i,N}$, $i = 0, 1, 2$.

To establish (6.3.6), we need a multi-dimensional version of the martingale central limit theorem.

Theorem 6.3.7 *Let* $M = (M_1(t), \ldots, M_{\mathrm{d}}(t))$, $0 \le t \le T$, *be a* d*-dimensional continuous Gaussian martingale with* $M(0) = 0$, *and let* $M_\varepsilon = (M_{\varepsilon,1}(t), \ldots, M_{\varepsilon,\mathrm{d}}(t))$, $\varepsilon \ge 0$, $0 \le t \le T$, *be a family of continuous square-integrable* d*-dimensional martingales such that* $M_\varepsilon(0) = 0$ *for all* ε *and, for some* $t_0 \in [0, 1]$ *and all* $i, j = 1, \ldots, \mathrm{d}$,

$$\lim_{\varepsilon\to 0} \langle M_{\varepsilon,i}, M_{\varepsilon,j}\rangle(t) = \langle M_i, M_j\rangle(t)$$

in probability. Then $\lim_{\varepsilon\to\infty} M_\varepsilon(t_0) \stackrel{d}{=} M(t_0)$.

Exercise 6.3.8 (C) Derive (6.3.6) from Theorem 6.3.7. **Hint.** The key to independence of the two limits is (6.3.10).

Exercise 6.3.9 (A) Prove Theorem 6.3.7.

This concludes the proof of Theorem 6.3.4.

Next, we look at the local likelihood ratio. First, we re-rewrite $L_{N,T}$ using the notations (6.3.4) and the equality

$$du_k(t) = -(\kappa_0\,\nu_k + \lambda_0)u_k(t)dt + q_k dw_k(t),$$

as

$$L_{N,T}(\kappa,\lambda) = \exp\Bigg((\kappa_0-\kappa)\bar{A}_{1,N} + (\lambda_0-\lambda)\bar{A}_{0,N} - \frac{(\kappa-\kappa_0)^2}{2}B_{2,N} - (\kappa-\kappa_0)(\lambda-\lambda_0)B_{1,N} - \frac{(\lambda-\lambda_0)^2}{2}B_{0,N}\Bigg).$$

Then

$$Z_{N;\kappa,\lambda}(x_1,x_2) = \exp\Bigg(x_1\phi_{N,\kappa}\bar{A}_{1,N} + x_2\phi_{N,\lambda}\bar{A}_{0,N} - \frac{x_1^2\phi_{N,\kappa}^2}{2}B_{2,N} - x_1x_2\phi_{N,\kappa}\phi_{N,\lambda}B_{1,N} - \frac{x_2^2\phi_{N,\lambda}^2}{2}B_{0,N}\Bigg),$$

where $\phi_{N,\kappa} \sim N$ and $\phi_{N,\lambda} \sim \sqrt{\ln N}$, is the local likelihood ratio. It follows from (6.3.9) and Theorem 6.3.7 that $Z_{N;\kappa,\lambda}$ is LAN (locally asymptotically normal) and has the form

$$Z_{N;\kappa,\lambda}(x_1,x_2) = \exp\left(x_1\bar{\zeta}_{1,N} + x_2\bar{\zeta}_{2,N} - \frac{x_1^2+x_2^2}{2} + \varepsilon_N(x_1,x_2)\right),$$

where the sequence of vectors $(\bar{\zeta}_{1,N}, \bar{\zeta}_{2,N})$ converges in distribution to a two-dimensional standard Gaussian vector and $\lim_{N\to+\infty}\varepsilon_N(x_1,x_2) = 0$ in probability.

Exercise 6.3.10 (A) Verify the LAN property of $Z_{N;\kappa,\lambda}$.

At this point, we are not aware of an analogue of Theorem 6.2.22 for Eq. (6.3.1).

6.3.2 One Parameter: Beyond an SPDE

Let us go back to Eq. (6.2.1) on page 395 and identify the key assumptions we used in the analysis:

1. The equation is diagonal, that is, can be reduced to an uncoupled systems of stochastic ordinary differential equations

$$du_k = \mu_k(\theta)u_k(t)dt + q_k dw_k(t). \tag{6.3.11}$$

2. The drift coefficient $\mu_k(\theta)$ in (6.3.11) satisfies

$$\mu_k(\theta) \nearrow +\infty,\ k\to+\infty. \tag{6.3.12}$$

3. The drift coefficient $\mu_k(\theta)$ in (6.3.11) has a special structure:

$$\mu_k(\theta) = \kappa_k + \theta \nu_k. \tag{6.3.13}$$

Then condition

$$\sum_k \frac{\nu_k^2}{\mu_k(\theta)} = +\infty \tag{6.3.14}$$

is necessary and sufficient to get a consistent estimator of θ in the large space asymptotic $N \to +\infty$.

While conditions $\mu_k(\theta) \sim k^{2\mathrm{m}/\mathrm{d}}$ and $\nu_k^2 \sim k^{2\mathrm{m}_1/\mathrm{d}}$ were very helpful, and even natural for an SPDE, a closer look at the proofs shows that it is (6.3.14) that makes everything work.

Accordingly, let us extend an SPDE (6.2.1) to an abstract evolution equation. More precisely, Introduce the following objects:

1. H, a separable Hilbert space with an orthonormal basis $\{\mathfrak{h}_k,\ k \geq 1\}$;
2. $W^{\mathrm{Q}} = W^{\mathrm{Q}}(t)$, a Q-cylindrical Brownian motion on H;
3. A_0, A_1, linear operators on H.
4. A real number θ.

Consider the linear stochastic evolution equation

$$\dot{u} + (\mathrm{A}_0 + \theta\, \mathrm{A}_1)u = \dot{W}^{\mathrm{Q}}(t),\ 0 < t < T, \tag{6.3.15}$$

with initial condition $u(0) \in L_2(H)$ independent of W^{Q}.

Definition 6.3.11 Equation (6.3.15) is called `diagonal` if the operators A_i, $i = 0, \ldots, \ell$ and Q have a common system of eigenfunctions:

$$\mathrm{A}_0\mathfrak{h}_k = \rho_k\mathfrak{h}_k,\ \mathrm{A}_1\mathfrak{h}_k = \nu_k\mathfrak{h}_k,\ \mathrm{Q}\mathfrak{h}_k = q_k\mathfrak{h}_k \tag{6.3.16}$$

and

$$u(0) = \sum_{k\geq 1} u_k(0)\, \mathfrak{h}_k,\ u_k(0) \stackrel{d}{=} \mathcal{N}(m_k, \sigma_k^2),\ \sum_{k\geq 1}(m_k^2 + \sigma_k^2) < \infty, \tag{6.3.17}$$

where the random variables $\{u_k(0),\ k \geq 0\}$ are independent.

Define the numbers

$$\mu_k(\theta) = \rho_k + \theta \nu_k$$

and the random processes

$$du_k = -\mu_k(\theta)u_k dt + q_k dw_k,\ 0 < t < T,\ k \geq 1, \tag{6.3.18}$$

with initial condition $u_k(0)$ from (6.3.30).

Let us address the forward problem for Eq. (6.3.15).

Theorem 6.3.12 *If $\mu_k(\theta) > 0$ for all but finitely many $k \geq 1$ and*

$$\sum_k \frac{q_k^2}{\mu_k(\theta)} < \infty,$$

then

$$u(t) = \sum_{k\geq 1} u_k(t)\mathfrak{h}_k$$

is a mild solution of (6.3.28).

The solution is unique in the class $L_2(\Omega \times [0, T]; H)$.

If

$$\sum_k \frac{q_k^2}{\mu_k^{\delta}(\theta)} < \infty,$$

for some $0 < \delta < 1$, then $u \in L_2(\Omega; \mathcal{C}((0, T); H))$.

Proof We have, for all sufficiently large k,

$$\mathbb{E}u_k^2(t) \leq \frac{q_k^2}{\mu_k(\theta)}$$

and

$$\mathbb{E}|u_k(t) - u_k(s)|^2 \leq \frac{q_k^2}{\mu_k^{1-\delta}(\theta)}|t - s|^{\delta}.$$

The conclusions of the theorem now follow.

Before we proceed, note that, when it comes to parameter estimation, the story is not so much about Eq. (6.3.15) but the `multi-channel system` (6.3.18): a system of independent but maybe not identically distributed processes, with each process carrying information about the unknown number θ. Indeed, if finitely many of the processes u_k are observed, it is not really important that these processes can be put together to make a solution of (6.3.28). If we can somehow observe u_k, $k = 1, \ldots, N$, then, for estimation purposes, the interesting conditions are those that insure a consistent estimation of θ in the limit $N \to +\infty$ rather than existence of a mild solution of (6.3.15). In fact, conditions of Theorem 6.3.12 are not necessary for consistent estimation of the parameter: think of N iid observations of $dX = \theta X dt + dw(t)$. Moreover, the operators A_i are defined in terms of eigenvalues and eigenfunctions, and can have nothing to do with differential or even pseudo-differential operators: try $\rho_k = \ln k$ and $\nu_k = e^k$. In other words, we are making a somewhat unexpected conclusion that statistical inference for diagonal SPDEs is not so much about SPDEs but about multi-channel systems. There are certainly multi-channel systems that do not correspond to an SPDE, and there are plenty of SPDEs that are not diagonal and therefore do not correspond to a multi-channel system. Still, when it comes to statistical inference, it is the structure provided by the multi-channel system that makes the problem tractable. While not immediately related to the subject of this book, statistical analysis of multi-channel systems appears as an interesting research area.

Next, we look at the inverse problem for (6.3.15), or equivalently, for the multi-channel system (6.3.18). To stay within the subject of the book, we will assume that conditions of Theorem 6.3.12 hold, that is, (6.3.18) corresponds to the mild solution of (6.3.15); moreover, these assumptions do simplify the analysis. As before (see the discussion on page 400), we assume that, for all $k \geq 1$,

$$q_k > 0,\ \mu_k(\theta) > 0, \nu_k \neq 0. \tag{6.3.19}$$

By (6.2.18) on page 402, the likelihood function for the observations $U^{N,T} = \{u_1(t),\ \ldots, u_N(t),\ t \in [0,T]\}$ is

$$L_{N,T}(\theta) = \exp\left(\sum_{k=1}^{N}\left(-\frac{\mu_k(\theta)-\mu_k(\theta_0)}{q_k^2}\int_0^T u_k(t)du_k(t)\right.\right.$$
$$\left.\left.-\frac{\mu_k^2(\theta)-\mu_k^2(\theta_0)}{2q_k^2}\int_0^T u_k^2(t)dt\right)\right).$$

The simplified (unrestricted) MLE $\widehat{\theta}_N$ of θ maximizes $\ln L_{N,T}(\theta)$ over $\mathbb{R}$:

$$\widehat{\theta}_N = -\frac{\sum_{k=1}^N \frac{\nu_k}{q_k^2}\int_0^T u_k(t)du_k(t)}{\sum_{k=1}^N \frac{\nu_k^2}{q_k^2}\int_0^T u_k^2(t)dt}. \tag{6.3.20}$$

Define

$$I_{N,\theta} = \sum_{k=1}^N \frac{\nu_k^2}{\mu_k(\theta)}. \tag{6.3.21}$$

Theorem 6.3.13 *Assume that $u_k(0) = 0$ and that $\theta \in \mathbb{R}$ is such that*

$$\mu_k(\theta) \nearrow +\infty,\ k \nearrow +\infty;\ \sup_k \frac{|\nu_k|}{\mu_k(\theta)} < \infty.$$

If $\lim_{N\to+\infty} I_{N,\theta} + \infty$, *then*

$$\lim_{N\to+\infty} \widehat{\theta}_N = \theta \ \textit{(with probability one)},$$

$$\lim_{N\to+\infty} I_{N,\theta}^{1/2}(\widehat{\theta}_N - \theta) \stackrel{d}{=} \mathcal{N}(0, 2/T),$$

and $\widehat{\theta}_N$ *is asymptotically efficient at* θ *with respect to the loss functions of polynomial growth.*

If $\lim_{N\to+\infty} I_{N,\theta} < \infty$, *then*

$$\lim_{N\to+\infty} \widehat{\theta}_N = -\frac{\sum_{k=1}^{\infty} \frac{\nu_k}{q_k^2} \int_0^T u_k(t)du_k(t)}{\sum_{k=1}^{\infty} \frac{\nu_k^2}{q_k^2} \int_0^T u_k^2(t)dt}. \tag{6.3.22}$$

Proof While the arguments are very similar to the proofs of Theorem 6.2.9 (for consistency and asymptotic normality) and Theorem 6.2.22 (for asymptotic efficiency), there are also many differences. Some of the differences, such as lack of explicit asymptotic for μ_k and ν_k, make the arguments harder. Other differences, namely, the fact that we study the sMLE at a specific point θ as opposed to MLE in a set Θ_0, make the arguments easier. An insignificant technical simplification is the assumption that the initial conditions are zero.

It follows from (6.3.18) that

$$\widehat{\theta}_N - \theta = -\frac{\sum_{k=1}^{N} \frac{\nu_k}{q_k} \int_0^T u_k(t)dw_k(t)}{\sum_{k=1}^{N} \frac{\nu_k^2}{q_k^2} \int_0^T u_k^2(t)dt}$$

and

$$\sum_{k=1}^{N} \mathbb{E} \int_0^T \frac{\nu_k^2}{q_k^2} u_k^2(s)ds \simeq \frac{I_{N,\theta}T}{2}, \tag{6.3.23}$$

To proceed, we need a more delicate bound on the variance of $\int_0^T u_k^2(t)dt$ than (6.2.61). Note that

$$\mathbb{E}\left(\int_0^T u_k^2(t)dt\right)^2 = 2\int_0^T \int_0^t \mathbb{E}u_k^2(t)u_k^2(s)dsdt,$$

and since u_k is a Gaussian process, the expression can be evaluated in closed form. With details left to the reader in the form of Problem 6.3.4, here is the result:

$$\mathrm{Var} \int_0^T \frac{\nu_k^2}{q_k^2} u_k^2(t)dt \sim \frac{\nu_k^4}{\mu_k^3(\theta)}, \ k \to +\infty.$$

Next, assume that the $I_{N,\theta} \nearrow +\infty$, that is, the series $\sum_{k\geq 1} \nu_k^2/\mu_k(\theta)$ diverges. Then

$$\sum_n \frac{\nu_n^2 \mu_n^{-1}(\theta)}{\left(\sum_{k=1}^{n} \nu_k^2 \mu_k^{-1}(\theta)\right)^2} < \infty.$$

Indeed, setting $a_n = \nu_n^2 \mu_n^{-1}(\theta)$ and $A_n = \sum_{k=1}^n a_k$, we notice that

$$\sum_{n\geq 1} \frac{a_n}{A_n^2} \leq \sum_{n\geq 2} \left(\frac{1}{A_{n-1}} - \frac{1}{A_n}\right) = \frac{1}{A_1}.$$

Then the strong law of large numbers (Theorem 6.2.11 on page 412), together with the equality

$$\mathbb{E} \int_0^T u_k(s) dw_k(s) = 0, \; k \geq 1,$$

implies

$$\lim_{N\to\infty} \frac{\sum_{k=1}^N \int_0^T \frac{\nu_k}{q_k} u_k(s) dw_k(s)}{\sum_{k=1}^N \mathbb{E} \int_0^T \nu_k^2 u_k^2(s) ds} = 0 \quad \text{with probability one.}$$

Next,

$$\sum_{n\geq 1} \frac{\nu_n^4 \mu_n^{-3}(\theta)}{\left(\sum_{k=1}^{n} \nu_k^2 \mu_k^{-1}(\theta)\right)^2} < \infty, \tag{6.3.24}$$

because by assumption $|\nu_k/\mu_k(\theta)|$ stays bounded. Then another application of the strong law of large numbers shows that

$$\lim_{N\to\infty} \frac{\sum_{k=1}^N \int_0^T \frac{\nu_k^2}{q_k^2} u_k^2(s) ds}{\sum_{k=1}^N \mathbb{E} \int_0^T \frac{\nu_k^2}{q_k^2} u_k^2(s) ds} = 1 \tag{6.3.25}$$

with probability one, completing the proof of consistency. Asymptotic normality follows from (6.3.23) and Theorem 6.1.4 in the same way as in the proof of Theorem 6.2.9.

Next, define

$$\varphi_{N,\theta} = \sqrt{\frac{2}{I_{N,\theta} T}}$$

and the corresponding local likelihood ratio

$$Z_{N,\theta}(x) = \exp\left(\sum_{k=1}^{N}\left(-\frac{x\phi_{N,\theta}}{q_k}\int_0^T \nu_k u_{k,\theta}(t)dw_k(t)\right.\right.$$
$$\left.\left.-\frac{x^2\phi_{N,\theta}^2}{2q_k^2}\int_0^T \nu_k^2 u_{k,\theta}^2(t)dt\right)\right).$$

Then LAN property of $Z_{N,\theta}$ follows in the same way as asymptotic normality of $\widehat{\theta}_N$. By Theorem 6.2.18 on page 422, asymptotic efficiency of $\widehat{\theta}_N$ will follow from

$$\lim_{N\to+\infty} \mathbb{E}w\big((\widehat{\theta}_N - \theta)/\phi_{N,\theta}\big) = \mathbb{E}w(\zeta),\ \zeta \stackrel{d}{=} \mathcal{N}(0,1)$$

for every loss function of polynomial growth. Given asymptotic normality of $\widehat{\theta}_N$, all we need is

$$\sup_{N\geq 1} I_{N,\theta}^p \mathbb{E}\left|\widehat{\theta}_N - \theta\right|^{2p} < \infty$$

for every integer $p \geq 1$, which follows from the BDG inequality and (6.2.89); see Problem 6.2.4 for details.

If the series $\sum_k \nu_k^2/\mu_k(\theta)$ converges, then, by (6.3.23), so does

$$\sum_{k\geq 1}\frac{\nu_k^2}{q_k^2}\int_0^T u_k^2(t)dt,$$

which, in turn implies convergence of

$$\sum_{k\geq 1}\frac{\nu_k}{q_k}\int_0^T u_k(t)dw_k(t)$$

(by the two series theorem). Passing to the limit in (6.3.20) results in (6.3.22). *This concludes the proof of Theorem 6.3.13.*

We conclude this section with a brief discussion of a general statistical principle, namely, that possibility to identify a model is connected with singularity of the corresponding probability measure. We will see how this principle works in our setting and how the idea can be used to understand and even predict the behavior of estimators in some other models.

To begin, we recall some basic definitions and results. Let P and $\tilde{P}$ be probability measures on $(\Omega, \mathcal{F})$.

1. $\tilde{P} \ll P$, that is, $\tilde{P}$ is `absolutely continuous` with respect to P, means that, for every $A \in \mathcal{F}$, $P(A) = 0$ implies $\tilde{P}(A) = 0$;
2. $\tilde{P} \sim P$, that is, $\tilde{P}$ and P are `equivalent`, if $\tilde{P} \ll P$ and $P \ll \tilde{P}$.

3. $\tilde{P} \perp P$, that is, $\tilde{P}$ and P are `singular`, means that there exists a set $A \in \mathcal{F}$ such that $\tilde{P} = 1$ and $P(A) = 0$.
4. `Radon-Nikodym theorem`: if $\tilde{P} \ll P$, then there exists a random variable ζ, denoted by $d\tilde{P}/dP$, such that, for every $A \in \mathcal{F}$,

$$\tilde{P}(A) = \int_{\Omega} \zeta dP.$$

5. `Hájek-Feldman theorem` [51, 73, 74]: two Gaussian measures on a Banach space are either equivalent or singular. The result holds even on a locally convex space [13, Theorem 2.7.2]
6. `Kakutani theorem` (a particular case, see [13, Example 2.7.6].) If P is a product measure of $\mathcal{N}(0, \sigma_k^2)$ and $\tilde{P}$ is a product measure of $\mathcal{N}(0, \kappa_k^2)$, then $P \sim \tilde{P}$ if and only if

$$\sum_k \left(\frac{\sigma_k}{\kappa_k} - 1\right)^2 < \infty.$$

A `statistical model` (or experiment) generated by random elements $X(\theta)$ is a collection $\mathcal{P} = \{\mathcal{X}, \mathfrak{X}, \mathbf{P}^\theta, \theta \in \Theta\}$, where each $\mathbf{P}^\theta$ is a probability measures on a measurable space $(\mathcal{X}, \mathfrak{X})$ such that $\mathbf{P}^\theta(A) = \mathbb{P}(X(\theta) \in A), A \in \mathfrak{X}$. In the parametric models, Θ is a subset of a finite-dimensional Euclidean space. The conventional wisdom in statistical inference is that if $\mathbf{P}^{\theta_1} \perp \mathbf{P}^{\theta_2}$ for $\theta_1 \neq \theta_2$, then it should be possible to determine the exact value of θ from an observation of $X(\theta)$. Since singular measures are truly different, this idea certainly makes sense. Below we illustrate how the idea works in the examples we studied so far and how the same idea can be used to predict, or at least understand, the results for other models.

Let us start with the evolution equation (6.3.15). If $\sum_k q_k^2 < \infty$, then, by Theorem 6.3.12, the solution $u = u_\theta$ generates a Gaussian measure P_T^θ on $\mathcal{C}((0,T);H)$. By [113, Proposition 1], if the series $\sum_k \nu_k^2/\mu_k(\theta_0)$ converges, then $P_T^\theta \ll P^{\theta_0}$. Note that $\mu_k(\theta) - \mu_k(\theta_0) = (\theta - \theta_0)\nu_k$, so θ does not appear in the condition to have $P_T^\theta \ll P^{\theta_0}$. In the case of SPDEs, the corresponding order condition

$$\frac{2 \text{ order of } \mathrm{A}_1 - \text{ order of } (\mathrm{A}_0 + \theta \mathrm{A}_1)}{\mathrm{d}} < -1, \tag{6.3.26}$$

which is the opposite of (6.2.48), is known to be necessary and sufficient to have $P_T^\theta \ll P^{\theta_0}$ even for non-diagonal equations [115, Theorem 3].

By the Hájek-Feldman theorem, conditions of Theorem 6.3.13 ensure that the measures P_T^θ are singular for different values of θ. It is therefore not at all surprising that, under conditions of Theorem 6.3.13 we can determine the value of θ given the solution of (6.3.15):

$$\theta = -\lim_{N \to +\infty} \frac{\sum_{k=1}^N \frac{\nu_k}{q_k^2} \int_0^T u_k(t) du_k(t)}{\sum_{k=1}^N \frac{\nu_k^2}{q_k^2} \int_0^T u_k^2(t) dt}.$$

In Sect. 6.1 we encountered two similar formulas:

$$\theta = \left(\frac{1}{T} \lim_{M\to\infty} \sum_{m=1}^{M} \Big(X\big((m+1)\Delta t_M\big) - X\big(m\Delta t_M\big) \Big)^2 \right)^{1/2},$$

if $\theta > 0$ is the unknown parameter in

$$dX = \theta dw(t),$$

and

$$\theta = \lim_{T\to+\infty} \frac{\int_0^T X(t)dX(t)}{\int_0^T X^2(t)dt},$$

if $\theta \in \mathbb{R}$ is the unknown parameter in

$$dX = \theta X dt + dw(t).$$

In both cases, we are dealing with singular measures. Indeed, for every $T > 0$, the measures on $\mathcal{C}((0,T);\mathbb{R})$ generated by the solutions of $dX = \sigma dw(t)$ for different values of σ are singular (for example, by looking at (1.1.1) on page 3 and using the Kakutani theorem). Similarly, the measures on $\mathcal{C}((0,+\infty);\mathbb{R})$ generated by the solutions of $dX = \theta X dt + dw(t)$ for different values of θ are singular [235].

Let us now go back to Eq. (6.3.15) and try to predict what happens in the estimation problem if the observations are in discrete time. As before, we assume that $\sum_k q_k^2 < \infty$. Then, for fixed $t > 0$, the solution of (6.3.15) generates a Gaussian measure on H. In fact, it is a product measure of $\mathcal{N}(0, \sigma_k^2(\theta))$ and $\sigma_k^2(\theta) \simeq q_k^2/\mu_k(\theta)$. Recall that $\mu_k(\theta) = \kappa_k + \theta \nu_k$. Then, for fixed θ_0 and $\theta \neq \theta_0$,

$$\begin{aligned}\left(\frac{\sigma_k(\theta_0)}{\sigma_k(\theta)} - 1\right)^2 &\sim \left(\frac{\sqrt{\mu_k(\theta)} - \sqrt{\mu_k(\theta_0)}}{\sqrt{\mu_k(\theta_0)}}\right)^2 \\ &= \left(\frac{\mu_k(\theta) - \mu_k(\theta_0)}{\sqrt{\mu_k(\theta_0)}\big(\sqrt{\mu_k(\theta)} + \sqrt{\mu_k(\theta_0)}\big)}\right)^2 \sim \frac{\nu_k^2}{\mu_k^2(\theta_0)}.\end{aligned}$$

By Kakutani's theorem, we expect that if

$$\sum_k \frac{\nu_k^2}{\mu_k^2(\theta_0)} = +\infty,$$

the measures will be singular for different values of θ in some neighborhood of θ_0. Thus, divergence of the series $\sum_k \frac{\nu_k^2}{\mu_k^2(\theta_0)}$ should be the condition to have a consistent estimator of θ_0 from the discrete time observations of $u_1, \ldots, u_N$ in the limit $N \to +\infty$. Our guess turns out to be correct: see [187]. For partial differential operators, the divergence of the series is equivalent to the order condition

$$\frac{\text{order of A}_1 - \text{ order of } (\text{A}_0 + \theta \text{A}_1)}{\text{d}} \geq -\frac{1}{2};$$

this is indeed the opposite of the condition

$$\frac{\text{order of A}_1 - \text{ order of } (\text{A}_0 + \theta \text{A}_1)}{\text{d}} < -\frac{1}{2} \tag{6.3.27}$$

from [115, Theorem 4] for absolute continuity of the measures. Comparing (6.3.26) and (6.3.27), we come to a very reasonable conclusion that it is easier for the measures to be equivalent at a fixed time than on the whole time interval.

While the above discussion suggests that it is singularity of measures that makes consistent estimation possible, an estimation problem requires absolutely continuous measures to write the likelihood ratio. To reconcile these two seemingly contradictory requirements, note that a typical estimator (including all the above examples) gives the true value of the parameter as a limit, and the setting before the limit corresponds to absolutely continuous measures. For example,

1. Time discretizations of $\sigma w(t)$ produce finite-dimensional Gaussian vectors with equivalent distributions;
2. For every finite T, the measures generated by the solution of $dX = \theta X dt + dw(t)$ on $\mathcal{C}((0, T); \mathbb{R})$ are equivalent for different values of θ;
3. For every finite N, the measures generated by the first N Fourier coefficients of the solution of (6.3.15) are always equivalent for different values of θ, whether in discrete or continuous time.

In other words, the underlying singular model is approximated by a collection of absolutely continuous models, making it possible to write the corresponding likelihood ratio and the MLE; consistency of the MLE is ensured by singularities of the measures in the limit.

Infinite-dimensional setting can lead to models so singular that an exact value of the unknown parameter is computable without passing to the limit, leading to a `closed-form exact estimator`. For example, consider the stochastic heat equation

$$du = \theta u_{xx} dt + u dw(t), \ \theta > 0,$$

on the interval $(0, \pi)$ with zero boundary conditions; the initial conditions must be non-zero to have a non-trivial solution. In the equation, $w = w(t)$ is the standard Brownian motion: the only way to get a diagonal equation with multiplicative noise is to have noise independent of space. Then $u(t, x) = \sum_k u_k(t) \mathfrak{h}_k(x)$, and

$$du_k(t) = -\theta k^2 u_k dt + u_k dw(t), \ u_k(t) = u_k(0) e^{-(\theta k^2 + (1/2))t + w(t)}.$$

If $u_1(0) \neq 0$ and $u_2(0) \neq 0$, then

$$\frac{u_1(1)}{u_2(1)} = \frac{u_1(0)}{u_2(0)} e^{3\theta},$$

that is,

$$\theta = \frac{1}{3} \ln \frac{u_1(1)u_2(0)}{u_2(1)u_1(0)}.$$

The key observation for this example is that the measures in $\mathcal{C}((0,T);\mathbb{R}^2)$ generated by the vector (u_1, u_2) are singular for different values of θ. Note that even though the process u_k is not Gaussian, $\ln u_k$ is. For more on this and related examples see [26, 28].

Exercise 6.3.14 (A+) Is there a closed-form exact estimator of θ in

$$du = \theta u_{xx} dt + dw(t)$$

on $(0, \pi)$ with zero initial and boundary conditions? What if, instead of Dirichlet zero boundary conditions, we consider Neumann zero boundary conditions?

6.3.3 Several Parameters

Let us start with the multi-parameter, and a more abstract, version of equation (6.3.15) on page 444. Introduce the following objects:

1. H, a separable Hilbert space with an orthonormal basis $\{\mathfrak{h}_k,\ k \geq 1\}$;
2. $W^{\mathrm{Q}} = W^{\mathrm{Q}}(t)$, a Q-cylindrical Brownian motion on H;
3. $\mathrm{A}_0, \mathrm{A}_1, \ldots, \mathrm{A}_\ell$, linear operators on H.
4. Real numbers $\theta_1, \ldots, \theta_\ell$.

Consider the linear stochastic evolution equation

$$\dot{u} + (\mathrm{A}_0 + \sum_{i=1}^{\ell} \theta_i \mathrm{A}_i)u = \dot{W}^{\mathrm{Q}}(t),\ 0 < t < T, \tag{6.3.28}$$

with initial condition $u(0) \in L_2(H)$ independent of W^{Q}.

Definition 6.3.15 Equation (6.3.28) is called `diagonal` if the operators A_i, $i = 0, \ldots, \ell$ and Q have a common system of eigenfunctions:

$$\mathrm{A}_0 \mathfrak{h}_k = \rho_k \mathfrak{h}_k,\ \mathrm{A}_i \mathfrak{h}_k = \nu_{i,k} \mathfrak{h}_k,\ \mathrm{Q}\mathfrak{h}_k = q_k \mathfrak{h}_k \tag{6.3.29}$$

and

$$u(0) = \sum_{k\geq 1} u_k(0)\, \mathfrak{h}_k,\ u_k(0) \stackrel{d}{=} \mathcal{N}(m_k, \sigma_k^2),\ \sum_{k\geq 1}(m_k^2 + \sigma_k^2) < \infty, \tag{6.3.30}$$

where the random variables $\{u_k(0),\ k \geq 0\}$ are independent.

Define the vectors

$$\theta = \begin{pmatrix} \theta_1 \\ \theta_2 \\ \vdots \\ \theta_\ell \end{pmatrix},\quad \nu_k = \begin{pmatrix} \nu_{1,k} \\ \nu_{2,k} \\ \vdots \\ \nu_{\ell,k} \end{pmatrix},$$

the numbers

$$\mu_k(\theta) = \rho_k + \sum_{i=1}^{\ell} \theta_i \nu_{i,k} = \rho_k + \theta^\top \nu_k,$$

and the random processes

$$du_k = -\mu_k(\theta) u_k dt + q_k dw_k,\ 0 < t < T,\ k \geq 1, \tag{6.3.31}$$

with initial condition $u_k(0)$ from (6.3.30).

Let us address the forward problem for Eq. (6.3.28).

Theorem 6.3.16 *If $\mu_k(\theta) > 0$ for all but finitely many $k \geq 1$ and*

$$\sum_k \frac{q_k^2}{\mu_k(\theta)} < \infty,$$

then

$$u(t) = \sum_{k\geq 1} u_k(t) \mathfrak{h}_k$$

is a mild solution of (6.3.28).

The solution is unique in the class $L_2(\Omega \times [0,T]; H)$.

If

$$\sum_k \frac{q_k^2}{\mu_k^\delta(\theta)} < \infty,$$

for some $0 < \delta < 1$, then $u \in L_2(\Omega; \mathcal{C}((0,T); H))$.

The proof is identical to that of Theorem 6.3.12 with one parameter.

Let us now discuss the problem of estimating θ from the observations $U^{N,T} = \{u_1(t), \ldots, u_N(t),\ t \in [0,T]\}$. Assume that $q_k > 0$ for all $k \geq 1$ and the observations correspond to the value θ_0 of the parameter so that $\mu_k(\theta_0) > 0$ for all $k \geq 1$. By (6.2.18) on page 402, the likelihood function is

$$L_{N,T}(\theta) = \exp\left(\sum_{k=1}^{N}\left(-\frac{\mu_k(\theta)-\mu_k(\theta_0)}{q_k^2}\int_0^T u_k(t)du_k(t)\right.\right.$$
$$\left.\left.-\frac{\mu_k^2(\theta)-\mu_k^2(\theta_0)}{2q_k^2}\int_0^T u_k^2(t)dt\right)\right).$$

We will now compute the simplified (unrestricted) MLE of θ by maximizing $\ln L_{N,T}(\theta)$ over $\mathbb{R}^\ell$.

Recall that

$$\mu_k(\theta) = \rho_k + \theta^\top \nu_k$$

Therefore the corresponding gradients are

$$\nabla_\theta \mu_k(\theta) = \nu_k,\ \nabla_\theta \mu_k^2(\theta) = 2\rho_k \nu_k + (\nu_k \nu_k^\top)\theta.$$

Similar to our analysis of the heat balance equation with two unknown parameters, it is convenient to introduce the matrix $B_N \in \mathbb{R}^{\ell\times\ell}$ and vectors $A_N \in \mathbb{E}^\ell$ and $\bar{A}_N \in \mathbb{R}^\ell$ with components

$$B_{ij,N} = \sum_{k=1}^{N}\frac{\nu_{i,k}\nu_{j,k}}{q_k^2}\int_0^T u_k^2(t)dt,\ \bar{A}_{i,N} = -\sum_{k=1}^{N}\frac{\nu_{i,k}}{q_k}\int_0^T u_k(t)dw_k(t) \qquad (6.3.32)$$

$$A_{i,N} = -\sum_{k=1}^{N}\frac{\nu_{i,k}}{q_k^2}\int_0^T u_k(t)\big(du_k(t) - \rho_k u_k(t)dt\big).$$

Then

$$\nabla_\theta \ln L_{N,T}(\theta) = A_N - B_N\theta, \qquad (6.3.33)$$

and to proceed, we need to invert the matrix B_N.

Proposition 6.3.17 *The matrix B_N is invertible with probability one if and only if the vectors $\nu_1, \ldots, \nu_N$ span $\mathbb{R}^\ell$.*

Proof The result is immediate from the equalities

$$B_N = \sum_{k=1}^{N} \frac{v_k v_k^\top}{q_k^2} \int_0^T u_k^2(t)dt$$

and

$$y^\top B_N y = \sum_{k=1}^{N} \frac{(y^\top v_k)^2}{q_k^2} \int_0^T u_k^2(t)dt, \ y \in \mathbb{R}^\ell.$$

Since we expect N to be much larger than ℓ, the condition that $v_1, \ldots, v_N$ span $\mathbb{R}^\ell$ is equivalent to identifiability of the original model (6.3.28) in the sense that the operators $\mathrm{A}_1, \ldots, \mathrm{A}_\ell$ are all different.

It is therefore natural to assume that the vectors $v_1, \ldots, v_N$ form a basis in $\mathbb{R}^\ell$ for all sufficiently large N. Then (6.3.33) implies

$$\widehat{\theta}_N = B_N^{-1} A_N.$$

Exercise 6.3.18 (C) Verify that

$$\widehat{\theta}_N - \theta_0 = B_N^{-1} \bar{A}_N. \tag{6.3.34}$$

The matrix-vector stricture of the problem makes the analysis more complicated compared to the one-parameter case. Accordingly, before stating and proving the corresponding theorem we will try to come up with the suitable assumptions. As much as possible, we will try to keep (6.3.28) an abstract equation. For a more concrete SPDE setting, see [80].

Define the diagonal matrix $S_N = \mathrm{diag}(S_{1,N}, \ldots, S_{\ell,N})$ with

$$S_{i,N} = \left(\sum_{k=1}^{N} \frac{v_{i,k}^2}{\mu_k(\theta_0)} \right)^{1/2}. \tag{6.3.35}$$

By the Cauchy-Schwartz inequality,

$$\frac{\sum_{k=1}^{N} |v_{i,k} v_{j,k}| / \mu_k(\theta_0)}{S_{i,N} S_{j,N}} \leq 1 \tag{6.3.36}$$

for all i and j. Therefore, to simplify the analysis, assume that, for every $i,j = 1, \ldots, \ell$, the limits

$$G_{ij} = \lim_{N \to +\infty} \frac{\sum_{k=1}^{N} v_{i,k} v_{j,k} / \mu_k(\theta_0)}{S_{i,N} S_{j,N}} \tag{6.3.37}$$

exist, and define the matrix $G = (G_{ij}, i,j = 1, \ldots, N)$.

By (6.3.36), there is always a possibility to define G along a converging subsequence; our simplifying assumption simply ensures uniqueness of G.

For example, the matrix G is well-defined if all the eigenvalues have the algebraic asymptotic

$$\rho_k \simeq c_0 k^{\alpha_0}, \ \nu_{i,k} \simeq c_i k^{\alpha_i}. \tag{6.3.38}$$

Exercise 6.3.19 (A)

(a) Compute G_{ij} under the assumption (6.3.38).
(b) Give an example when the limit in (6.3.37) does not exist.

Exercise 6.3.20 (C) Let $\bar{A}$ be the vector defined in (6.3.32). Using Theorem 6.3.7, verify that if

$$\lim_{N\to+\infty} S_{i,N} = +\infty, \ i = 1, \ldots, \ell, \tag{6.3.39}$$

and $\sup_k |\nu_{i,k}|/\mu_k(\theta_0) < \infty$ for all i, then

$$\lim_{N\to+\infty} S_N^{-1} B_N S_N^{-1} = TG/2 \quad \text{with probability one} \tag{6.3.40}$$

and

$$\lim_{N\to+\infty} S_N^{-1}\bar{A} \stackrel{d}{=} \zeta, \tag{6.3.41}$$

where ζ is a Gaussian vector with mean zero and covariance matrix $TG/2$.

Hint. See the proof of Theorem 6.3.13.

As (6.3.40) suggests $B_N \sim S_N G S_N$, the key question becomes non-degeneracy of the matrix G. The question is non-trivial indeed, because G can be singular even if every B_N is not. As an example, consider $\ell = 2$, $\mathrm{A}_0 = 0$, $\nu_{1,k} = k$ and $\nu_{2,k} = k + 1$, $\theta_1 > 0$, $\theta_2 > 0$. Then direct computations show that any collection of the vectors ν_k, $k \geq 1$, is linearly independent, but

$$G = \frac{1}{2(\theta_1 + \theta_2)} \begin{pmatrix} 1 & 1 \\ 1 & 1 \end{pmatrix}.$$

If operators A_i are (pseudo)-differential, then, assuming the power asymptotic (6.3.38) for the corresponding eigenvalues, one can show that G is non-degenerate if no two of the operators A_i, A_j have the same order.

Exercise 6.3.21 (A) Assume that (6.3.38) holds and $\alpha_i \neq \alpha_j$ for $i \neq j$. Show that G is invertible.

Theorem 6.3.22 *Assume that the vectors $\nu_1, \ldots, \nu_N$ are linearly independent in $\mathbb{R}^\ell$ for all sufficiently large N, and assume that the vector $\theta_0 \in \mathbb{R}^\ell$ is such that*

1. $\mu_k(\theta_0) > 0$ *for all* $k \geq 1$ *and* $\lim_{k\to+\infty} \mu_k(\theta_0) = +\infty$;
2. $\sup_k |\nu_{i,k}|/\mu_k(\theta_0) < \infty$, $i = 1, \ldots, \ell$;
3. *the series* $\sum_k \nu_{i,k}^2/\mu_k(\theta_0)$ *diverges for every* $i = 1, \ldots, \ell$;
4. *the matrix G is well-defined according to* (6.3.37) *and is invertible.*

Then

$$\lim_{N\to+\infty} |\widehat{\theta}_N - \theta_0| = 0 \quad \textit{with probability one}$$

and

$$\lim_{N\to+\infty} S_N(\widehat{\theta}_N - \theta_0) \stackrel{d}{=} \zeta,$$

where the matrix S_N is from (6.3.35) *and ζ is a Gaussian random vector with mean zero and covariance matrix* $(2/T)G^{-1}$.

Proof Since

$$\widehat{\theta}_N - \theta_0 = B_N^{-1}\bar{A}_N,$$

we have, with a suitable matrix norm $\|\cdot\|$,

$$|\widehat{\theta}_N - \theta_0| \leq \|S_N B_N^{-1} S_N)\| \frac{|\bar{A}_N|}{\sum_{i=1}^N \sum_{k=1}^N \nu_{i,k}^2/\mu_k(\theta_0)}.$$

By assumption and using the strong law of large numbers, the matrix norm on the right-hand side stays bounded, and the rest converges to zero with probability one.

Similarly,

$$S_N(\widehat{\theta}_N - \theta_0) = S_N B_N^{-1} S_N S_N^{-1} \bar{A}_N$$

and the result follows from (6.3.41).

6.3.4 Problems

The objective of the problems below is to bring the presentation in this section closer to a more detailed analysis of the SPDE with one parameter in Sect. 6.2. Problem 6.3.1 is about extending Theorem 6.3.4 to non-zero initial conditions. Problem 6.3.2 is about alternative ways to define the solution of the abstract evolution equation (6.3.15). Problem 6.3.3 is an easy example of extending Theorem 6.3.13 beyond an evolution equation. Problem 6.3.4 is a technical result about the asymptotic behavior of the integral of the square of an Ornstein-Uhlenbeck process; this result is the key for the proof of consistency and asymptotic normality in Theorem 6.3.13. Problem 6.3.5 is an invitation to extend the results of Theorem 6.2.22 to some of the models in this section.

Problem 6.3.1 Suppose that the initial condition in (6.3.2) is Gaussian with mean m_k and variance σ_k^2. Find sufficient conditions on m_k and σ_k for (6.3.9) to hold.

Problem 6.3.2

(a) Discuss the possibility of defining the variational solution for the abstract evolution equation (6.3.15).
(b) Discuss the possibility of defining a mild solution for the abstract evolution equation (6.3.15) when $q_k = 1$ for all k.

Problem 6.3.3 State and prove an analogue of Theorem 6.3.13 when the processes (6.3.18) are iid.

Problem 6.3.4 Let $X = X(t; a, \sigma)$ be the solution of

$$dX = -aXdt + \sigma dw(t),\ t > 0,\ a > 0,$$

with initial condition $X(0; a, \sigma) \overset{d}{=} \mathcal{N}(x_0, \sigma_0^2)$ independent of the standard Brownian motion w. Verify that

$$\lim_{a\to+\infty} a\,\mathbb{E}\int_0^T X^2(t; a, \sigma)dt = \frac{\sigma^2 T + \sigma_0^2 + x_0^2}{2},$$

$$\mathbb{E}\left(\int_0^T X^2(t; a, \sigma)dt\right)^n \le C(n)\frac{x_0^{2n} + \sigma_0^{2n} + T^n\sigma^{2n}}{a^n},\ n = 1, 2, \ldots,\ a > 0,$$

$$\operatorname{Var}\int_0^T X^2(t; a, \sigma)dt = \frac{\operatorname{Var} X_0^2}{4a^2} + \frac{\sigma^4 T + 4(x_0^2 + \sigma_0^2)\sigma^2}{2a^3} + o(a^{-3}),\ a \to +\infty.$$

Problem 6.3.5 Discuss potential difficulties in extending Theorem 6.2.22 to (a) two-parameter SPDE model (6.3.1); (b) one-parameter diagonal abstract evolution equation (6.3.15) under condition of Theorem 6.3.13 (that is, without assuming any particular asymptotic for μ_k and ν_k).

Problems: Answers, Hints, Further Discussions

Problems of Chap. 2

2.1.1 (page 35)

For every $t > 0$, the function $B^H(t,\cdot)$ is Hölder continuous of any order less than $1/2$; for every $x \in (0,\pi)$, the function $B^H(\cdot,x)$ is Hölder continuous of any order less than H.

Indeed, B^H is Gaussian;

$$\begin{aligned}\mathbb{E}|B^H(t,x) - B^H(t,y)|^2 &= t^{2H}\sum_{k\geq 1}\frac{|\cos(kx)-\cos(ky)|^2}{k^2}\\ &\leq t^{2H}|x-y|^{1-\delta}\sum_{k\geq 1}k^{-1-\delta},\ \delta\in(0,1),\end{aligned}$$

because, by direct computation, $\cos(a) - \cos(b) = 2\sin((a+b)/2)\sin((b-a)/2)$ and $|\sin x| \leq |x|^{\varepsilon}$ for every $\varepsilon \in (0,1]$;

$$\mathbb{E}|B^H(t,x) - B^H(s,x)|^2 = |t-s|^{2H}\sum_{k\geq 1}\frac{(1-\cos(kx))^2}{k^2}.$$

2.1.2 (page 35)

If $\inf_k H_k = H_0 > 0$, then the analysis of the previous solution shows that the upper bound on the order of the Hölder continuity in time will be H_0. The case $H_0 = 0$ is left as a challenge to the reader.

2.1.3 (page 35)

The covariance function has to be a positive-definite kernel (see page 94). An easy way to prove that a function is a positive-definite kernel is to interpret the

S.V. Lototsky, B.L. Rozovsky, *Stochastic Partial Differential Equations*,
Universitext, DOI 10.1007/978-3-319-58647-2

function as the covariance of some Gaussian field, but this is unlikely to work in this problem. The details are left to the reader.

2.2.1 (page 50)

(a) Direct computation.
(b) Note that the Fourier transform $\widehat{v} = \widehat{v}(t, y)$ of the homogeneous equation satisfies

$$\widehat{v}_t = -a(t)y^2\widehat{v}(t, y).$$

We need the solution for $t \geq s$ satisfying $\widehat{v}(s, y) = 1$. If $A(t) = \int_0^t a(s)ds$, then we conclude that $\Phi(t, s)$ is the integral operator

$$\Big(\Phi(t, s)h\Big)(x) = \frac{1}{\sqrt{4\pi\big(A(t) - A(s)\big)}} \int_{-\infty}^{\infty} \exp\left(-\frac{(x - y)^2}{4\big(A(t) - A(s)\big)}\right) h(y)dy.$$

2.2.2 (page 50)

(a), (b) Direct computations.
(c) There is a difficulty interpreting the integral

$$\int_0^t \frac{X(t)}{X(s)} f(s)dw(s),$$

because $X(t)$ is not $\mathcal{F}_s^w$-measurable for $s < t$. Note that the integral is NOT the same as

$$X(t) \int_0^t \frac{f(s)}{X(s)} dw(s).$$

While the difficulty can be resolved using anticipating stochastic calculus (see Nualart [175, Sect. 3.2]), it is much easier to study the equation without using the closed-form solution.
(d) An $\mathcal{F}_t^w$-adapted locally square-integrable a will work. In this case, Itô formula shows that

$$\Phi(t, s) = \exp\left(\int_s^t a(r)dw(r) - \frac{1}{2}\int_s^t a^2(r)dr\right).$$

2.2.3 (page 51)

(a) Here is one possible definition. Assume that A is a linear, and possibly unbounded, operator on a Banach space V. Then u is a classical solution of $u(t) = u_0 + \int_0^t \mathrm{A}u(s)ds$ if $u : [0, T] \to V$ is a continuous function for every $T > 0$, $u(t)$ is in the domain of A for all $t \geq 0$, $u(0) = u_0$, and $u(t) = u_0 + \int_0^t \mathrm{A}u(s)ds$ in V for all $t \geq 0$. For the same equation in the differential form, the mapping $u : [0, T] \to V$ would have to be continuously differentiable and $\mathrm{A}u : [0, T] \to V$, continuous.

(b) The operator A should be acting on functions from $L_2(G)$, where $G \subset \mathbb{R}^{\mathrm{d}}$, and the formal adjoint A^* of A must be defined using integration by parts so that

$$(\mathrm{A}f, g)_{L_2(G)} = (f, \mathrm{A}^* g)_{L_2(G)}$$

for all smooth compactly supported f, g. Then the equality in [W5] becomes

$$\mu_t[\varphi] = \mu_0[\varphi] + \int_0^t \mu_s[\mathrm{A}^*\varphi] ds.$$

For $\mathbf{A}_3$, the function a can be just bounded and measurable; for $\mathbf{A}_2$ we would need one bounded generalized derivative in x, and for $\mathbf{A}_1$, two. Note that positivity of a is not necessary to define the solution, but is usually required to find one.

2.2.4 (page 51)

Substitution of $\bar{v}$ produces the term $\mathrm{A}N(t)$ and changes $F(t, u)$ to $F\big(t, u - N(t)\big)$. When the operator A is unbounded, the process N_A typically has better regularity properties than either $N(t)$ or $\mathrm{A}N(t)$ (think of $\mathrm{A} = \boldsymbol{\Delta}$ and $\Phi(t)$, the heat semi-group), meaning that $v(t) = u(t) - N_A(t)$ is usually the better choice. A trivial example when $\bar{v}$ is easier is a linear equation such that $\mathrm{A}N(t) = 0$.

2.2.5 (page 51)

Note that, by the Kolmogorov criterion, $W(t, x)$ as a function of x is Hölder continuous of every order less than $1/2$. Thus, the equation cannot have a classical solution or solutions [W1], [W2]. The class of the test functions should reflect the boundary conditions (for example, for [W3], the test functions can be infinitely differentiable on [0, 1] with derivatives equal to zero at 0 and 1).

Solutions [W3] and [W4] are equivalent for this equation and the resulting function u is actually continuous in t and x (see Walsh [223, Theorem 3.2]). A measure-valued solution also exists and has a density for $t > 0$ [120, Sect. 8.3].

2.2.6 (page 51)

This is an open-ended question. Note that, in the case of stochastic equations, one can take random test functions $\varphi(x)$ or $\varphi(t, x)$, and then take expectation along with integration in space and/or time. To take the expectation, one would have to impose a priori integrability conditions on the solution.

2.3.1 (page 68)

(a,b) $X_i(t) = \sum_{k=1}^{\infty} \sigma_{ik} w_k(t)$;

$$v_t = \sum_{i,j=1}^{\mathrm{d}} \left(a_{ij} - (1/2) \sum_{k=1}^{\infty} \sigma_{ik}\sigma_{jk} \right) D_i D_j v;$$

the matrix $\left(a_{ij} - (1/2) \sum_{k=1}^{\infty} \sigma_{ik}\sigma_{jk} \right)$ must be non-negative definite.

(c) Same equation for v, but the equation is now considered in the domain $\{(t, x+X(t)) : t > 0, x \in G\}$ with the boundary condition $v(t, x + X(t)) = 0$ for x on the boundary of G.

2.3.2 (page 68)

(a) Use (2.2.7) on page 43. The integral in space is, up to a constant, a convolution of two normal densities and can be evaluated. The result is

$$u(t,x) = \int_0^t \left(\frac{f(s)}{(t-s+f(s))}\right)^{d/2} \exp\left(-\frac{x^2}{2(t-s+f(s))}\right) dw(s).$$

Then

$$\mathbb{E}u^2(t,x) = \int_0^t \left(\frac{f(s)}{(t-s+f(s))}\right)^{d} \exp\left(-\frac{x^2}{t-s+f(s)}\right) ds.$$

Limiting distribution can exist (for example, if $f(t) = 1$ and $d > 1$).

(b)

$$u(t,x) = \frac{8}{\pi}\sum_{k=0}^{\infty} \frac{\sin((2k+1)x)}{(2k+1)^3} \exp\left(-\left(k^2+\frac{1}{2}\right)t + w(t)\right);$$

$$\mathbb{E}u^2(t,x) = e^t\left(\frac{8}{\pi}\sum_{k=0}^{\infty} \frac{\sin((2k+1)x)}{(2k+1)^3}\, e^{-k^2 t}\right)^2;$$

no limiting distribution.

(c)

$$u(t,x) = \sqrt{\frac{2}{\pi}}\sum_{k=1}^{\infty}\int_0^t y_k(t-s)f(s)dw_k(s)\ \sin(kx),$$

where $y_k''(t) + by_k'(t) + k^2y_k(t) = 0$, $y_k(0) = 0$, $y_k'(0) = 1$ (see (2.3.26) on page 58);

$$\mathbb{E}u^2(t,x) = \frac{2}{\pi}\sum_{k=1}^{\infty}\frac{1}{k^2}\int_0^t y_k^2(t-s)f^2(s)ds\ \sin^2(kx).$$

Limiting distribution exists, for example, if $b > 0$ and $f = 1$.

2.3.3 (page 69) Note that, for each $k \geq 1$, $u(t,x) = k^{-2}e^{k^2t+2\mathrm{i}kw(t)+\mathrm{i}kx}$ is a solution.

2.3.4 (page 69)
Similar to (2.3.18) on page 56, we have

$$\mathbb{E}\|u(t,\cdot)\|^2_{L_2(G)} = \sum_{k\geq 1} \frac{e^{2\lambda_k t}-1}{2\lambda_k},$$

where $\{\lambda_k,\ k \geq 1\}$ are the eigenvalues of the Laplace operator in G with zero boundary conditions. The key observation is that $\lim_{k\to\infty} \frac{\lambda_k}{k} = -c$ for some positive number c, which is true for every smooth bounded domain $G \subset \mathbb{R}^2$, with c depending only on the domain: see, for example, Safarov and Vassiliev [201, Sect. 1.2].

2.3.5 (page 69)
Taking u to be the closed-form solution

$$u(t,x) = \int_0^t \int_{\mathbb{R}^{\mathrm{d}}} K(t-s, x-y))dW(s,y),\ K(t,x) = \frac{1}{(4\pi t)^{\mathrm{d}/2}} e^{-|x|^2/(4\pi t)},$$

use (1.1.9) on page 4 to conclude that

$$\mathbb{E}u^2(t,x) < \infty \Leftrightarrow \int_0^t \frac{ds}{(t-s)^{\mathrm{d}/2}} < \infty \Leftrightarrow \mathrm{d} = 1.$$

2.3.6 (page 69)

(a)

$$u(x,y) = \frac{2}{\pi} \sum_{n,k=1}^{\infty} \frac{\xi_k}{k^2+n^2} \sin(nx)\ \sin(ky);$$

$$\mathbb{E} \iint\limits_G u^2(x,y)\,dxdy = \sum_{n,k=1}^{\infty} \frac{1}{(n^2+k^2)^2} < \frac{\pi}{4} \sum_{k=1}^{\infty} k^{-3} < \frac{3\pi}{8}.$$

(b) If $\mathrm{d} = 1, 2, 3$.

2.3.7 (page 70)
We have

$$u(x) = \frac{1}{2}\int_{-\infty}^{+\infty} e^{-|x-y|} dW(y) = \frac{1}{2}\int_x^{+\infty} e^{x-y} W(y)dy - \frac{1}{2}\int_{-\infty}^{x} e^{y-x} W(y)dy,$$

where the first equality follows by the Fourier transform and the second, after integration by parts. Uniqueness follows because the general solution of $u'' - u = 0$

is $c_1e^x + c_2e^{-x}$; it is unbounded unless $c_1 = c_2 = 0$. Next,

$$\mathbb{E}u^2(x) = \frac{1}{4}\int_{-\infty}^{+\infty} e^{-2|x-y|}dy = \frac{1}{4}.$$

As a result, $\mathbb{E}\|u\|^2_{H_2^\gamma(\mathbb{R})} = +\infty$ for all $\gamma \in \mathbb{R}$, just like the non-zero constant function is not in any Sobolev space $H^\gamma\mathbb{R}^{\mathrm{d}}$). If you need to be more rigorous, see Problem 4.2.3 on page 181.

2.3.8 (page 70)

(a) The solution is

$$u(t,x) = \frac{\xi_0}{\sqrt{\pi}} + \sqrt{\frac{2}{\pi}}\sum_{k\geq 1} e^{-k^2t}\xi_k\cos(kx),$$

where ξ_k are iid standard normal random variables. This function is infinitely differentiable in the region $\{(t,x) : t > 0, x \in (0,\pi)\}$.

(b) The solution is

$$u(t,x) = \sqrt{\frac{2}{\pi}}\sum_{k\geq 1}\frac{\xi_k}{k}\sin(kx)\sin(kt).$$

By the Kolmogorov criterion, this function is Hölder continuous of any order less than $1/2$ in the region $\{(t,x) : 0 \leq t \leq T, x \in [0,\pi]\}$ for every $T > 0$.

(c) Let $x = r\cos\theta,\ y = r\sin\theta$. Note that the functions $r^k\cos k\theta,\ r^k\sin k\theta,\ k = 0,1,2,\ldots$ are harmonic. Therefore

$$u(r,\theta) = \frac{\xi_0}{\sqrt{2\pi}} + \sum_{k=1}^{\infty}\left(\frac{\xi_k}{\sqrt{\pi}}\cos(k\theta) + \frac{\eta_k}{\sqrt{\pi}}\sin(k\theta)\right)r^k, \tag{S.1}$$

where $\xi_0, \xi_1, \eta_1, \xi_2, \eta_2, \ldots$ are iid standard Gaussian random variables; recall that an orthonormal basis in $L_2((0,2\pi))$ is $\{\frac{1}{\sqrt{2\pi}}, \frac{\cos k\theta}{\sqrt{\pi}}, \frac{\sin k\theta}{\sqrt{\pi}}, k \geq 1\}$. This function is analytic in the region $\{(x,y) : x^2 + y^2 < 1\}$: the series in (S.1) converges uniformly in (x,y,ω) in the region $\{(x,y) : x^2 + y^2 < R\}$ for every $R < 1$.

2.3.9 (page 70)

(a) Note that $\mathbb{E}B^2(0) = \mathbb{E}B^2(T) = 0$.

(b) For the first three, verify (2.3.70). For the last, show that the limit is that same as the third representation (see also page 14 and [71, Theorem 3.3]). As far as simulations, note that the first representation is not adapted and the second has a (removable) singularity at the end point.

2.3.10 (page 71)

(b) Note that

$$\frac{\partial^n}{\partial t^n} e^{xt-(t^2/2)}\Big|_{t=0} = e^{x^2/2} \frac{\partial^n}{\partial t^n} e^{-(x-t)^2/2}\Big|_{t=0}.$$

(c) Compute

$$\mathbb{E}\Big(e^{s\xi-(s^2/2)} e^{t\eta-(t^2/2)}\Big)$$

directly and using (2.3.37), and compare the results.

2.3.13 (page 72)
Since

$$\lim_{\varepsilon\to 0} \int_{\varepsilon<|y|<c} \frac{dy}{y} = 0$$

for every $c > 0$, we have

$$\langle f, \varphi\rangle = \lim_{c\to\infty} \int_{|y|<c} \frac{\varphi(1-y)-\varphi(1)}{y}\, dy = \lim_{\varepsilon\to 0} \int_{|y|>\varepsilon} \frac{\varphi(1-y)}{y}\, dy$$
$$= \int_{\mathbb{R}} \varphi'(1-y)\, \ln|y|\, dy$$

which is finite for all $\varphi \in \mathcal{S}(\mathbb{R})$ and is not zero for $\varphi(x) = e^{-x^2/2}$.

2.3.14 (page 72)
Note that

$$\mathcal{E}_x = \sum_{k\geq 0} \frac{x^k}{k!} \mathrm{H}_k(\xi). \tag{S.2}$$

(a) By (2.3.39) on page 61, if $u(x) = \sum_{k\geq 0} u_k(s)\mathrm{H}_k(\xi)$ and $\breve{u}(y;x) = \mathbb{E}\big(u(y)\mathcal{E}_x\big)$, then

$$\breve{u}(y;x) = \sum_{k\geq 0} u_k(y) x^k. \tag{S.3}$$

[Keep in mind that, if u were square-integrable, then we would have $u_k = (k!)^{-1}\, \mathbb{E}\big(u\mathrm{H}_k(\xi)\big)$.] Since $\xi \diamond \mathrm{H}_k(\xi) = \mathrm{H}_1(\xi) \diamond \mathrm{H}_k(\xi) = \mathrm{H}_{k+1}(\xi)$, we also have

$$u(y) \diamond \xi = \sum_{k\geq 1} u_{k-1}(y)\mathrm{H}_k(\xi),$$

and the result follows.

(b) Using (S.3), we can define $\breve{u}(\cdot; x)$ as long as the power series in x on the right-hand side has a non-zero radius of convergence, and then (2.3.73) holds. Since we know that $u_k(y) = \sin y$, it follows that, for every $y \in (0, 1\pi)$, the function $\breve{u}(y; x)$ is defined and is an analytic function of x for $x \in (-1, 1)$.
Note that Eq. (2.3.71) makes sense without any integrability assumptions on u. If the solution is analytic for $h = 0$, then we can use (2.3.73) to define u. This is the underlying idea of the `Wiener Chaos solution`.

2.3.15 (page 73)
Use the Hopf-Cole transformation to reduce (2.3.74) to a bi-linear heat equation.

2.3.16 (page 73)
The result is $U = -2a\nabla \ln v$, where $v_t = a\boldsymbol{\Delta} v - \frac{1}{2a} vG$.

Problems of Chap. 3

3.1.1 (page 96)
Let $\{\mathfrak{h}_k,\ k \geq 1\}$ be an orthonormal basis in H and let ℓ_2 be the space of square-integrable real sequences. Show that $x \mapsto \{(x, \mathfrak{h}_k)_H,\ k \geq 1\}$ defines an isomorphism between H and ℓ_2.

3.1.2 (page 96)
If $\mathfrak{h}_k, k \geq 1$, is an orthonormal basis in H, then $\tilde{\mathfrak{h}}_k = \mathfrak{h}_k/q_k$ is an orthonormal basis in $\tilde{H}$. We have $\mathfrak{j}(\mathfrak{h}_k) = \mathfrak{h}_k$ and

$$\big(\mathfrak{j}(\mathfrak{h}_k), \tilde{\mathfrak{h}}_k\big)_{\tilde{H}} = q_k = \big(\mathfrak{h}_k, \mathfrak{j}^*(\tilde{\mathfrak{h}}_k)\big)_H,$$

and so $\mathfrak{j}^*(\tilde{\mathfrak{h}}_k) = q_k\mathfrak{h}_k$. Then $\big(\mathfrak{j}\mathfrak{j}^*(\tilde{\mathfrak{h}}_k), \tilde{\mathfrak{h}}_k\big)_{\tilde{H}} = \|\mathfrak{j}^*(\tilde{\mathfrak{h}}_k)\|_H^2 = q_k^2$ and the result follows.
Computing $\mathrm{tr}(\mathfrak{j}^*\mathfrak{j})$ is easier. Indeed,

$$\mathrm{tr}(\mathfrak{j}^*\mathfrak{j}) = \sum_{k\geq 1} \big(\mathfrak{j}^*\mathfrak{j}(\mathfrak{h}_k), \mathfrak{h}_k\big)_H = \sum_{k\geq 1} \|\mathfrak{j}(\mathfrak{h}_k)\|_{\tilde{H}}^2 = \sum_{k\geq 1} q_k^2.$$

3.1.3 (page 96)

(a) The "if" part is the one we really need to prove. If h is an element of the dense subset of H, then write $\|h - h_n\|_H^2 = \|h\|_H^2 - 2(h_n, h)_H + \|h_n\|_H^2$ and pass to the limit $n \to \infty$. If not, let $\|h - x_m\|_H \to 0$, $m \to \infty$, write $\|h - h_n\|_H^2 = \|(h - x_m) + (x_m - h_n)\|_H^2$, expand and pass to the limit, first in n, then in m.
(b) The idea is as follows. Write $\dot{u}(t) = f(t)$ and conclude that $\|u(t)\|_H^2 = \|u_0\|_H^2 + \int_0^t [f(s), u(s)]ds$. This gives continuity of the norm and the required bound. Weak continuity on a dense subset of H is clear, because, for $v \in V$, $(u(t), v)_H = (u_0, v) + \int_0^t [f(s), v]ds$.

For a more detailed proof in a less abstract setting, see Evans [50, Theorem 5.9.3].

3.1.4 (page 97)

We only need to prove completeness, and this follows from $L_2((0,T);X)$ being a Hilbert space if X is a Hilbert space.

3.1.5 (page 97)

These are all standard results and can be found, for example, in [61] or [171].

3.1.6 (page 98)

(a) Note that

$$\|(\mathrm{A}\otimes\mathrm{B})(x\otimes y)\|_{X\otimes Y}=\|\mathrm{A}x\|_X\,\|\mathrm{B}y\|_Y,$$

and

$$(\mathrm{A}\otimes\mathrm{B})\left(\sum_{k=1}^{N}c_kx_k\otimes y_k\right)=\sum_{k=1}^{N}c_k(\mathrm{A}x_k)\otimes(\mathrm{B}y_k).$$

For more details, see Murphy [171, Lemma 6.3.2].

(b) (i) $\mathrm{A}\otimes\mathrm{B}$ is compact: if A and B have finite-dimensional ranges, so does $\mathrm{A}\otimes\mathrm{B}$, then pass to the limit in the operator norm $\|\cdot\|_0$.

(ii) $\mathrm{A}\otimes\mathrm{B}$ is Hilbert-Schmidt: given orthonormal bases in X and Y, use (3.1.14) on page 82 to conclude that

$$\sum_{k,n\geq 1}\|(\mathrm{A}\otimes\mathrm{B})(\mathfrak{m}_k\otimes\mathfrak{h}_n)\|_{X\otimes Y}^2=\left(\sum_{k\geq 1}\|\mathrm{A}\mathfrak{m}_k\|_X^2\right)\left(\sum_{n\geq 1}\|\mathrm{B}\mathfrak{h}_n\|_Y^2\right).$$

(iii) The operator $\mathrm{A}\otimes\mathrm{B}$ is nuclear: use the same argument as in (ii) and Theorem 3.1.41(b) on page 91.

3.1.7 (page 98)

By definition, $K_H(g,h)=(g,h)_H$. There is no contradiction if $H=L_2(\mathbb{R})$, because we consider different sets S: in Exercise 3.1.47(a), $S=\mathbb{R}$, and in this problem, $S=H$.

3.1.8 (page 98)

(a) Denote by $\mathrm{j}:H\to V'$ the inclusion operator of H into V'. Then the reproducing kernel K_V of V is $K_V(u,v)=(\mathrm{j}^*u,\mathrm{j}^*v)_H$, $u,v\in V'$.

(b) There is no contradiction, as we use different representations of the elements of $L_2(\mathbb{R})$: in Exercise 3.1.47(a), they are functions on $\mathbb{R}$, while in this problem, they are functionals on $H^{-2}(\mathbb{R})$. In other words, in Exercise 3.1.47(a), $S=\mathbb{R}$, and in this problem, $S=H^{-2}(\mathbb{R})$.

3.1.9 (page 99)

Verify that, by definition, the reproducing kernel Hilbert space corresponding to K is the closure of the range of the operator A under the norm corresponding to the inner product $(x, y)_H = (\mathrm{A}x, y)_X = (\sqrt{\mathrm{A}}x, \sqrt{\mathrm{A}}y)_X$ for x, y in the range of A.

3.1.10 (page 99)

Note that (i) $K(t, s) = \mathbb{E}\big(w(t)w(s)\big)$, where w is a standard Brownian motion; (ii) for fixed t, $K(t, \cdot)$ is the anti-derivative of the indicator function $\mathbf{1}_{[0,t]}$ of the interval $[0, t]$; (iii) $K(t, s) = \big(\mathbf{1}_{[0,t]}, \mathbf{1}_{[0,s]}\big)_{L_2((0,1))}$. Conclude that the corresponding reproducing kernel Hilbert space H is the `Cameron-Martin space` of the Wiener measure:

$$H = \left\{ f = f(t) : f(t) = \int_0^t h(s)ds,\ \|f\|_H^2 = \int_0^1 h^2(t)dt \right\}.$$

3.1.11 (page 99)

Fix an orthonormal basis $\{\mathfrak{m}_i,\ i \geq 1\}$ in X and an orthonormal basis $\{\mathfrak{u}_j,\ j \geq 1\}$ in Y. Given an element

$$z = \sum_{i,j} a_{ij}\mathfrak{m}_i \otimes \mathfrak{u}_j \in X \otimes Y, \tag{S.4}$$

define the operator $\mathrm{A}_z : X \to Y$ by

$$\mathrm{A}_z x = \sum_{i,j} a_{ij}(\mathfrak{m}_i, x)_X \mathfrak{u}_j$$

and note that

$$\|\mathrm{A}_z\|_2^2 = \sum_{k \geq 1} \|\mathrm{A}_z \mathfrak{m}_k\|_Y^2 = \sum_{k,j} a_{kj}^2 = \|z\|_{X\otimes Y}^2.$$

Conversely, given an operator $\mathrm{A} \in L_2(X, Y)$, define $a_{ij} = (\mathrm{A}\mathfrak{m}_i, \mathfrak{u}_j)_Y$, and then define the corresponding z by (S.4). Note the difference between this result and Problem 3.1.1: all separable Hilbert spaces are isomorphic, but some are isomorphic in a special way.

3.2.1 (page 125)

(a) A quick solution is to switch the order of integration in

$$\dot{W}_{[f]}(g) = \int_{\mathbb{R}^d} \left(\int_{\mathbb{R}^d} f(x - y)dW(y) \right) g(x)dx.$$

A more careful solution is to work with the sum

$$\dot{W}_{[f]}(g) = \sum_{k\geq 1}\left(\int_{\mathbb{R}^{\mathrm{d}}} f_k(x)g(x)dx\right)\xi_k,$$

where $\{h_k,\ k \geq 1\}$ is an orthonormal basis in $L_2(\mathbb{R}^{\mathrm{d}})$ such that $h_k \in \mathcal{S}(\mathbb{R}^{\mathrm{d}})$ for all k, and

$$f_k(x) = \int_{\mathbb{R}^{\mathrm{d}}} f(x-y)h_k(y)dy = (f * h_k)(x).$$

(b) This follows from part (a) because condition $\varphi(x) = \varphi(-x)$ implies $(f \bullet \varphi_\varepsilon)(y) = (f * \varphi_\varepsilon)(y)$, the convolution, so that

$$\mathbb{E}\Big(\dot{W}_{[\varphi_\varepsilon]}(f) - \dot{W}(f)\Big)^2 = \|f - f * \varphi_\varepsilon\|^2_{L_2(\mathbb{R}^{\mathrm{d}})}.$$

Other conditions on φ imply that φ_ε is a smooth approximation of the delta-function.

(c) A slight modification of the argument in part (b) is necessary because $W_r(x) = \dot{W}_{[\psi]}(x)$ and ψ is a normalized indicator function of the ball with radius r and center at the origin. That is, ψ is not smooth, but otherwise, it is the same idea as in part (b).

(d) This part is left to the reader.

3.2.2 (page 125) Let Ω be the set of real-valued functions on $[0,1]^{\mathrm{d}}$, $\mathcal{F}$, the cylindrical sigma-algebra on Ω, and $\mathbb{P}$, a probability measure on $(\Omega, \mathcal{F})$ such that all finite-dimensional projections of $\mathbb{P}$ are centered Gaussian with covariance K. Consider the canonical process $X(f) = f$ on the probability space $(\Omega, \mathcal{F}, \mathbb{P})$ and note that

$$\mathbb{E}|f(x) - f(y)|^2 = K(x,x) + K(y,y) - 2K(x,y).$$

Conclude that $\mathbb{P}(V) = 1$. The result follows.

3.2.3 If the vector X is n-dimensional and has a density in $\mathbb{R}^n$, then $H_X = \mathbb{R}^n$.

3.2.4 (page 126)

(a) A minimal-effort solution would consist in verifying that the functions $\mathfrak{m}_k(t) = \sin((k-1/2)\pi t)$ are eigenfunctions of the integral operator $\mathrm{K} : f(t) \mapsto \int_0^1 f(s)\min(t,s)\,ds$. A more elaborate solution would consist in computing the eigenfunctions of this operator.

Representations (3.2.49) and (3.2.24) are expansions of $W = W(t)$ in different bases in $L_2((0,1))$; note that neither basis is orthonormal in $L_2((0,1))$.

(b) Note that the functions $\mathfrak{m}_k$ are eigenfunctions of the Laplace operator on the interval $(0, 1)$ with a Dirichlet boundary condition at the left end-point and the Neumann boundary condition at the right end-point.

3.2.5 (page 126)

Let ν be a regular measure on $[0, 1]$. Define $\varphi_\nu(x) = \nu([0, 1]) - \nu([0, x])$. Then

$$C(\nu, \sigma) = \int_0^1 \varphi_\nu(t)\varphi_\sigma(t)dt.$$

Indeed, keeping in mind that the canonical process on V under μ is the standard Wiener process w, we find

$$\begin{aligned} C(\nu, \sigma) &= \int_V \nu(f)\sigma(f)\mu(df) \\ &= \mathbb{E}\big(\nu(w)\sigma(w)\big) = \mathbb{E}\left(\int_0^1 w(t)\nu(dt) \int_0^1 w(t)\sigma(dt)\right) \\ &= \mathbb{E}\left(\int_0^1 \varphi_\nu(t)dw(t) \int_0^1 \varphi_\sigma(t)dw(t)\right) \\ &= \int_0^1 \varphi_\nu(t)\varphi_\sigma(t)dt, \end{aligned}$$

where the last equality is Itô isometry, and the previous equality follows after integration by parts. Taking $\nu = \delta_s$ and $\sigma = \delta_t$, we recover (3.2.3) on page 101.

3.2.6 (page 126)

Just verify the corresponding definitions; there are several of them, in different places.

DISCUSSION. The result raises the following question. Let $X = X(t)$, $t \in [0, 1]$ be a zero-mean Gaussian process. Its covariance function $K(s, t) = \mathbb{E}\big(X(s)X(t)\big)$ is a positive-definite kernel on $[0, 1]$. Let H_K be the corresponding reproducing kernel Hilbert space.

On the other hand, the probability distribution of X is a Gaussian measure μ on a suitable space of functions on $[0, 1]$. This measure has a `Cameron-Martin space` H_μ, which is the collection of functions f such that the measure μ_f, defined by $\mu_f(A) = \mu(f + a, a \in A)$, is absolutely continuous with respect to μ.

Under what additional conditions will we have $H_K = H_\mu$?

3.2.7 (page 126)

(a) See Gelfand and Vilenkin [61, Theorem 6, page 169].
(b) Direct computations.
(c) If $\mathfrak{B}$ is regular and $q(x, y) = \mathbb{E}\big(B(x)B(y)\big)$, then $q(x, y) = r(|x - y|)$ for some function $r = r(t)$, $t \geq 0$, so that, up to a constant, $\mu(\mathbb{R}^\mathrm{d})$ is $r(0)$. In particular, $\int_{\mathbb{R}^\mathrm{d}} q(x, x)dx$ cannot be finite, and so B is not an element of $L_2(\Omega; L_2(\mathbb{R}^\mathrm{d}))$.

If the measure μ is finite, define $B(x)$ as the limit of $\mathfrak{B}(\varphi_n^x)$, where φ_n^x is a smooth function approximating the delta-function at x; for more details, see Peszat and Zabczyk [184, page 190].

3.2.8 (page 127)

(a) For d = 1 only. The explicit expression for the regular representation is left to the reader.
(b) Let $\dot{W}$ be a Gaussian white noise on $H^0 = L_2(\mathbb{R}^{\mathrm{d}})$ and let $\Lambda = \sqrt{1-\boldsymbol{\Delta}}$. Then $\mathfrak{X} = \Lambda^{-1}\dot{W}$, and it remains to repeat the arguments from Example 3.2.21 on page 109, with Fourier transform instead of Fourier series. In this case, Definition 3.2.19 implies that $\mathfrak{X}$ is a Gaussian white noise on $H^1(\mathbb{R}^{\mathrm{d}})$. For more detailed arguments, see Example 4.2.3.

3.2.9 (page 127)
The arguments are the same as in the finite-dimensional case.

3.2.10 (page 128)
Repeat, with obvious modifications, the arguments in the case of real-valued martingales; see, for example, Karatzas and Shreve [103, Proposition 1.5.23] or Kunita [125, Theorem 2.1.2]. Note that, by the BDG inequality,

$$\mathbb{E} \sup_{0<t<T} \|M_n(t) - M_m(t)\|_H^2 \leq C\mathbb{E}\langle M_n - M_m\rangle(T).$$

Keep in mind that the space $\mathcal{C}([0,T];H)$ of deterministic continuous mappings $f : [0,T] \mapsto H$ with norm $\sup_{0<t<T} \|f(t)\|_H$ is complete.

3.3.1 and **3.3.2** (page 141)
Even though W^{Q} is not necessarily an H-valued martingale, all the steps in the construction of the integral $\int_0^t (f(s), dM(s))_H$ can be repeated (see page 131). In particular, writing

$$f(t) = \sum_{k\geq 1} f_k(t)\mathfrak{m}_k$$

for some orthonormal basis $\{\mathfrak{m}_k,\ k \geq 1\}$ in H, we set

$$W_f^{\mathrm{Q}}(t) = \sum_{k\geq 1} \int_0^t f_k(s)dM_k(s),$$

where $M_k(t) = W_{\mathfrak{m}_k}^{\mathrm{Q}}(t)$ is a standard Brownian motion and

$$\langle M_k, M_\ell\rangle(t) = t\left(\mathrm{Q}\mathfrak{m}_k, \mathfrak{m}_\ell\right)_H.$$

Then show that $W_f^{\mathrm{Q}}(t)$ does not depend on the choice of the basis in H. Once we have $W_f^{\mathrm{Q}}(t)$ for $f \in H$, we define $W_{\mathrm{B}}^{\mathrm{Q}}(t)$ for an operator B (see Definition 3.3.7 on page 134).

3.3.3 (page 142)
See Chow [24, Lemma 6.4.3].

3.3.4 (page 142)
This is an essay question.

Problems of Chap. 4

4.1.1 (page 163)

(a) Local Lipschitz continuity of the coefficients implies uniqueness and local existence. The function F ensures non-explosion (global existence). The case of linear growth is included, with $F(x) = 1 + |x|^2$.
(b) This is a particular case of the result proved by Krylov and Rozovskii [122, Theorem 3.14]. Roughly speaking, condition (4.1.44) ensures uniqueness of the solution, and (4.1.45) ensures global existence (non-explosion).

4.1.2 (page 165)

(a) A straightforward approach is to enclose the domain in a box (hyper-cube or parallelepiped) and apply repeatedly the arguments from (4.1.11) (page 149) along each side of the box. Zeidler [233, Proposition 18.19] presents all the details for d $= 2$.
An alternative approach is to write $u = \sum_k u_k\varphi_k$, where $-\boldsymbol{\Delta}\varphi_k = \lambda_k\varphi_k$, $\|\varphi_k\|_{L_2(G)} = 1$, $(\varphi_k, \varphi_m)_{L_2(G)} = 0$ if $k \neq m$, and then note that
$$\|\nabla u\|_{L_2(G)}^2 = -\int_G u\boldsymbol{\Delta}u\,dx = \sum_k \lambda_k u_k^2 \geq \lambda_1 \sum_k u_k^2.$$
This also gives $C_\pi \leq 1/\lambda_1$.
More refined proofs seek the best value of C_π (both analytically and numerically), as well as other conditions on the boundary behavior of u for the inequality still to hold.
(b) Integrate by parts and use (4.1.46) to conclude that (4.1.10) holds with $c_A = 1/(1 + C_\pi)$. Note that in Example 4.1.4 we have $C_\pi = 1/2$ and $c_A = 2/3$.
(c) Note that if $\mathrm{A}\varphi = \lambda_k\varphi$, then $[\mathrm{A}\varphi, \varphi] = \lambda_k\|\varphi\|_H^2$. Compared with part (b), we have the same spaces, but with different, and equivalent, norms. As a result, we get the same inequalities, possibly with different constants.
(d) The most straightforward way is to use integration by parts.

As an illustration, consider the operator $\mathbf{A}_2$. For brevity, we write $\boldsymbol{a} = \big(a_{ij},\ i,j = 1,\dots,\mathrm{d}\big)$, $\boldsymbol{b} = (b_i,\ i = 1,\dots,\mathrm{d})$, $\tilde{\boldsymbol{b}} = (\tilde{b}_i,\ i = 1,\dots,\mathrm{d})$, and use $\cdot$ to denote the inner product in $\mathbb{R}^{\mathrm{d}}$. Then, for $u \in \mathcal{C}_0^{\infty}(G)$,

$$[\mathbf{A}_2 u, u] = (\mathbf{A}_2 u, u)_H = \int_G (\boldsymbol{a}\nabla u \cdot \nabla u)(x)dx + \big((\boldsymbol{b} - \tilde{\boldsymbol{b}}) \cdot \nabla u, u\big)_H + (cu, u)_H.$$

By Green'f formula, for every smooth vector function $\boldsymbol{h}$, $\int_G \nabla \cdot (\boldsymbol{h}u^2)(x)dx = 0$ (because $u|_{\partial G} = 0$), and therefore, since $\nabla \cdot (\boldsymbol{h}u^2) = 2u(\boldsymbol{h} \cdot \nabla u) + (\nabla \cdot \boldsymbol{h})u^2$, we find (assuming both b_i and $\tilde{b}_i$ are continuously differentiable)

$$\big((\boldsymbol{b} - \tilde{\boldsymbol{b}}) \cdot \nabla u, u\big)_H = \frac{1}{2}\big(u\nabla \cdot (\tilde{\boldsymbol{b}} - \boldsymbol{b}), u\big)_H.$$

Define

$$\bar{c} = \inf_{x \in G}\left(\frac{1}{2}\nabla \cdot (\tilde{\boldsymbol{b}} - \boldsymbol{b})(x) + c(x)\right), \tag{S.5}$$

and assume that there exists a positive real number c_a such that, for all $x \in G$ and all $y \in \mathbb{R}^{\mathrm{d}}$,

$$\sum_{i,j=1}^{\mathrm{d}} a_{ij}(x) y_i y_j \geq c_a \sum_{i=1}^{\mathrm{d}} y_i^2. \tag{S.6}$$

If C_π is the Poincaré constant for the domain G (from part (a) of the problem) and

$$\frac{c_a}{2C_\pi} + \min(\bar{c}, 0) > 0, \tag{S.7}$$

then the operator $\mathbf{A}_2$ satisfies (4.1.10) with c_A given by the left-hand side of (S.7). For the above computations to make sense, we need (a) bounded $a_{ij}, b_i, \tilde{b}_i, c$; (b) continuously differentiable $b_i, \tilde{b}_i$; (c) conditions (S.6) and (S.7).

Requiring b_i and/or $\tilde{b}_i$ to be differentiable can be avoided with the help of the inequality

$$|(\boldsymbol{b} \cdot \nabla u, u)_H| \leq \sup_{x \in G} |\boldsymbol{b}(x)| \left(\varepsilon \|\nabla u\|_H^2 + \varepsilon^{-1}\|u\|_H^2\right);$$

c_a will dictate the choice of ε, and this ε will appear in an analogue of (S.7).

The overall conclusions are as follows:

1. For the operators to act in the normal triple $(H_0^1(G), L_2(G), H^{-1}(G))$, all the coefficients must be bounded, and sometimes, existence of bounded derivatives of some of the coefficients must be assumed as well;
2. To have (4.1.10), it is always necessary to assume (S.6);

3. With bounded coefficients and (S.6) in place, it is always possible to choose c large enough so that (4.1.10) will hold.

4.1.3 (page 166)

(a) Let $F(t,x) = e^{tx-t^2/2}$ and $\bar{F}(t,x) = e^{2tx-t^2}$. Note that $F(t,x) = \bar{F}(t/\sqrt{2}, x/\sqrt{2})$ and compare the coefficients in the corresponding power series expansions.
(b) Using (a), reduce the question to the corresponding relation for H_n, verified in Problem 2.3.10.
(c) Orthonormality is part (b). Completeness follows from the completeness of the set of polynomials [215, Theorem 5.7.1]. For $H^r(\mathbb{R})$, $r \neq 0$, the functions are clearly not orthogonal, but still form a dense subset. Indeed, suppose that $f \in H^r(\mathbb{R})$ and $(f, \mathfrak{w}_n)_r = 0$ for all $n \geq 1$, that is,

$$\int_{\mathbb{R}} (1+y^2)^r \widehat{f}(y) \overline{\widehat{\mathfrak{w}}_n(y)}\, dy = 0. \tag{S.8}$$

We need to conclude that $f = 0$. For that, we use part (e) and notation $h_r(y) = (1+y^2)^r$ to re-write (S.8) as

$$(h_r\widehat{f}, \mathfrak{w}_n)_{L_2(\mathbb{R})} = 0.$$

By completeness of the system in $L_2(\mathbb{R})$, $h_r\widehat{f} = 0$ or, since h_r is never zero, $\widehat{f} = 0$, that is, $f = 0$.
(d) First verify that $\bar{\mathrm{H}}_n'' - 2x\bar{\mathrm{H}}_n' + 2n\bar{\mathrm{H}}_n = 0$. For an efficient way to do this, see [2, Sect. 6.1].
(e) Given (4.1.51), it is easy to see that $\mathcal{F}\mathfrak{w}_n = \lambda_n \mathfrak{w}_n$, with $|\lambda_n| = 1$, because the operator Λ is unchanged in the Fourier domain and the Fourier transform does not change the L_2 norm. Unfortunately, this argument does not provide the value of λ_n.

The complete solution comes from [2, Exercise 11 for Chap. 6] and is the following chain of equalities:

$$\begin{aligned}
\int \mathfrak{w}_{n+1}(x) e^{-\mathrm{i}xy} dx &= \int e^{-x^2} \frac{d^n}{dx^n} e^{-\mathrm{i}xy + x^2/2} dx \\
&= \mathrm{i}^n e^{y^2/2} \int e^{-x^2} \frac{d^n}{dy^n} e^{(x-\mathrm{i}y)^2/2} dx \\
&= (-\mathrm{i})^n e^{y^2/2} \frac{d^n}{dy^n} \left(e^{-y^2/2} \int e^{(-x^2/2)-\mathrm{i}xy} dx \right) \\
&= (-\mathrm{i})^n \sqrt{2\pi}\, \mathfrak{w}_{n+1}(y),
\end{aligned}$$

where the first equality follows after n integration by parts, with $(-1)^n$ in the definition of $\bar{\mathrm{H}}_n$ canceling the $(-1)^n$ incurred by the integration by parts formula;

the second equality is a two step process: first, complete the square and take $e^{y^2/2}$ out of the integral, then change the derivative from d/dx to d/dy using the relation $\partial f(x-ay)/\partial x = \partial f(x-ay)/(-a\partial y) = f'(x-ay)$; the third equality is expanding the square and rearranging the terms; the key to the last equality is that the normal density is unchanged by the Fourier transform (see (1.1.12)).

4.1.4 (page 168)

(a) Note that operator $(N+1)\mathrm{I} - \mathrm{A}$ is elliptic.
(b) These are the main steps.
[Step 1] Define $v(t) = \dot{u}(t)$ so that the equation becomes $\dot{v} = \mathrm{A}u + f$.
[Step 2] Take inner product with v in H on both sides:

$$\frac{1}{2}\frac{d}{dt}\|v\|_H^2 = [\mathrm{A}u, v] + (f, v)_H. \tag{S.9}$$

[Step 3] Note that, since the operator A does not depend on time and $[\mathrm{A}u, v] = [u, \mathrm{A}v]$, we get

$$\frac{d}{dt}[\mathrm{A}u, u] = [\mathrm{A}v, u] + [\mathrm{A}u, v] = 2[\mathrm{A}u, v]. \tag{S.10}$$

[Step 4] Integrate both sides of (S.9) with respect to time, and use (S.10):

$$\|v(t)\|_H^2 - \|v_0\|_H^2 = [\mathrm{A}u(t), u(t)] - [\mathrm{A}u_0, u_0] + 2\int_0^t \big(f(s), v(s)\big)_H ds. \tag{S.11}$$

[Step 5] Use the Cauchy-Schwarz and the epsilon inequalities to write

$$\begin{aligned} 2\left|\int_0^t \big(f(s), v(s)\big)_H ds\right| &\le \sup_{0<t<T} \|v(t)\|_H \int_0^T \|f(s)\|_H ds \\ &\le \frac{1}{2}\sup_{0<t<T} \|v(t)\|_H^2 + C\left(\int_0^T \|f(s)\|_H ds\right)^2. \end{aligned} \tag{S.12}$$

[Step 6] Use (S.12) and assumption (ii) about the operator A, as well as continuity of A from V to V', to deduce from (S.11) that, with some other number C,

$$\begin{aligned} \frac{1}{2}\sup_{0<t<T} \|v(t)\|_H^2 + c_A \sup_{0<t<T} \|u(t)\|_V^2 &\le C\|u_0\|_V^2 + \|v_0\|_H^2 \\ &+ C\left(\int_0^T \|f(t)\|_H dt\right)^2 + M \sup_{0\le t\le T} \|u(t)\|_H^2. \end{aligned} \tag{S.13}$$

[Step 7] Write

$$u(t) = u_0 + \int_0^t v(s)ds$$

and take the $\|\cdot\|_H$ norm on both sides. Note that

$$\left\| \int_0^t v(s)ds \right\|_H^2 \leq \left(\int_0^t \|v(s)\|_H ds \right)^2 \leq C(T) \int_0^T \|v(t)\|_H^2 dt.$$

[Step 8] Go back to (S.13) and eliminate the term $\int_0^T \|v(t)\|_H^2 dt$ on the right using Gronwall's inequality. The result follows.

(c) The arguments are the same as in the parabolic case.

(d) We need to verify conditions of Theorem 4.1.15 on page 156. We start by verifying (4.1.23), which will imply (4.1.20).

Note that the operator $\bar{\mathrm{A}}$ reduces the second-order in time equation $\ddot{u} = \mathrm{A}u + f$ to the system of two first-order equations

$$(\dot{u} \quad \dot{v}) = \bar{\mathrm{A}}(u \quad v) + (0 \quad f).$$

By assumption, $[\mathrm{A}u, v] = [u, \mathrm{A}v]$ and $[-\mathrm{A}u, u] \geq c_A \|u\|_V^2$. Define the equivalent norm on $X = V \times H$ by

$$\|w\|_X^2 = [-\mathrm{A}u, u] + \|v\|_H^2, \; w = (u \quad v).$$

The corresponding inner product on X is

$$(w, w_1)_X = [-\mathrm{A}u, u_1] + (v, v_1)_H$$

and so, for w in the domain of $\bar{\mathrm{A}}$,

$$(\bar{\mathrm{A}}w, w)_X = [-\mathrm{A}v, u] + (\mathrm{A}u, v)_H = 0; \tag{S.14}$$

note that, if $w = (u \quad v)$ is in the domain of $\bar{\mathrm{A}}$, then $v \in V$ and $\mathrm{A}u \in H$, so that $(\mathrm{A}u, v)_H = [\mathrm{A}u, v]$. As a result,

$$\|(\lambda \mathrm{I} - \bar{\mathrm{A}})w\|_X^2 = \big((\lambda \mathrm{I} - \bar{\mathrm{A}})w, (\lambda \mathrm{I} - \bar{\mathrm{A}})w\big)_X = \lambda^2 \|w\|_X^2 + \|\bar{\mathrm{A}}w\|_X^2 \geq \lambda^2 \|w\|_X^2$$

which is (4.1.23) with $\lambda_0 = 0$.

Verification that $\bar{\mathrm{A}}$ is densely defined and closed requires additional constructions: the normal triple (V, H, V') is not enough to describe the domain of $\bar{\mathrm{A}}$. Accordingly, we use the elliptic operator $M\mathrm{I} - \mathrm{A}$ and the material from Kreĭn et al. [117, Sect. IV.1.10] (see also Theorem 3.1.12 on page 81 and the comments after it) to construct a Hilbert scale $\{H^r, \; r \in \mathbb{R}\}$ such that $V = H^1$, $H = H^0$,

$V' = H^{-1}$. By this construction, the space $Y = H^2 \times H^1$ is the domain of $\bar{\mathrm{A}}$ and is a dense subset of X. Direct computations then show that conditions of Theorem 3.1.18 hold:

$$\|w\|_Y^2 \sim \|u\|_2^2 + \|v\|_1^2 \sim \|(M\mathrm{I} - \mathrm{A})u\|_0^2 + \|v\|_1^2 \sim \|\bar{\mathrm{A}}w\|_X^2 + \|w\|_X^2,$$

where $\sim$ means equivalence of norms. By Proposition 4.1.14, the resulting mild solution is also a variational solution in the sense of Definition 4.1.20.

(e) While the semigroup method works under more restrictive conditions, the strong continuity of the semigroup implies that $u \in \mathcal{C}\big((0,T); V\big)$, $\dot{u} \in \mathcal{C}\big((0,T), H\big)$. For the solution constructed using the Galerkin approximation, we only get $u \in L_\infty\big((0,T); V\big)$, $\dot{u} \in L_\infty\big((0,T); H\big)$. Extra work is required to prove that the solution is in fact continuous. For details, see Lions and Magenes [133, Sect. 3.8.4]).

4.1.5 (page 168)

(a) The arguments are the same as for the equation $\ddot{u} = \mathrm{A}u + f$ with time-independent A. Now (S.10) becomes

$$\frac{d}{dt}[\mathrm{A}u, u] = 2[\mathrm{A}u, v] + [\dot{\mathrm{A}}u, u];$$

the addition term $\int_0^t \|u(t)\|_V^2 ds$ is eliminated using Gronwall's inequality. The terms $\int_0^t (\mathrm{A}_1 u, v)_H^2 ds$ and $\int_0^t (\mathrm{B}v, v)_H^2 ds$ are also processed using the Cauchy-Schwarz and Growall inequalities.

(b) Writing $f(t) = f_0 + \int_0^t \dot{f}(s)ds$, and assuming at first that $f_0, \dot{f} \in V$, we integrate by parts to find

$$\int_0^t (f, v)_H ds = (f(t), u(t))_H - [f_0, u_0] - \int_0^t (\dot{f}, u)_H ds.$$

Then

$$\begin{aligned} |(f(t), u(t))_H| &\le \varepsilon \|u(t)\|_V^2 + C(\varepsilon)\|f(t)\|_{V')}^2 \\ &\le \varepsilon \|u(t)\|_V^2 + C(\varepsilon, T)\|f\|_{H^1((0,T);V')}^2 \\ \left|\int_0^t (\dot{f}, u)_H ds\right| &\le \|f\|_{H^1((0,T);V')}^2 + \int_0^t \|u\|_V^2 ds; \end{aligned}$$

Gronwall's inequality takes care of $\int_0^t \|u\|_V^2 ds$.

(c) Note that the time-dependent setting is pretty hard to handle by the semigroup method, but when it works, the semigroup method gives the statement of the Theorem 4.1.21 right away. The Galerkin approximation works in the most general setting, but, for the resulting solution, we only get $u \in L_\infty((0,T); V)$,

$\dot{u} = v \in L_\infty((0,T);H)$, and, unlike the parabolic case, we do not have general result (like the fundamental theorem of calculus) to claim continuity. Thus, some extra work is required to prove that the solution constructed by the Galerkin method is indeed continuous. For details, see Lions and Magenes [133, Sect. 3.8.4]. We repeat the arguments of Lions and Magenes for the stochastic version of the equation in the proof of Theorem 4.3.3. (see page 189).

4.1.6 (page 169)

(a) To derive the estimate, we have to assume that $v = \dot{u} \in V$, which is not the case in general. The estimate does hold for the solution constructed from the Galerkin approximation (because it holds for the approximate solutions and is preserved in the limit), but a priori it does not have to hold for other possible solutions.
(b) The idea is to repeat the derivation of the a priori bound (with $u_0 = v_0 = f = 0$) using $\psi(t) = \int_s^t u(r)dr$ (for fixed t) instead of $\dot{u}$:

$$[\dot{v}, \psi] = [\mathrm{A}u, \psi] + \dots.$$

For details, see Evans [50, Theorem 7.2.4] or Lions and Magenes [133, Sect. 3.8.2].

4.1.7 (page 169)

(a) Taking the inner product with u in H,

$$\begin{aligned}\|u(t)\|_H^2 &= \|u_0\|_H^2 + 2\int_0^t [\mathrm{A}u, u]ds + 2\int_0^t [f(s), u(s)]ds \\ &\le -2c_A\|u\|^2_{L_2((0,T);V)} + 2M\int_0^t \|u(s)\|_H^2 ds + 2\left|\int_0^t [f(s), u(s)]ds\right|.\end{aligned}$$

The term $2M\int_0^t \|u(s)\|_H^2 ds$ is eliminated by the Gronwall inequality. To process $\int_0^t [f(s), u(s)]ds$, recall that $[u, v] \le \|u\|_{V'}\|v\|_V$. Then, by the epsilon inequality,

$$\left|\int_0^t [f(s), u(s)]ds\right| \le \frac{c_A}{2}\|u\|^2_{L_2((0,T);V)} + C\|f\|^2_{L_2((0,T);V')}.$$

This establishes the a priori bound. Galerkin approximation then leads to the solution in the usual way. For alternative ideas, see Lions and Magenes [133, Sect. 3.4].

(b) To get a semigroup on H, we cannot take V as the domain of A. The domain $D(A)$ must be such that the mapping $\mathrm{A} : D(A) \to H$ is bounded. Accordingly, we use the (elliptic) operator $M\mathrm{I} - \mathrm{A}$ and the material from Kreĭn et al. [117, Sect. IV.1.10] (see also Theorem 3.1.12 on page 81 and the comments after it) to construct a Hilbert scale $\{H^r,\ r \in \mathbb{R}\}$ such that $V = H^1, H = H^0, V' = H^{-1}$. As usual, $\|\cdot\|_r$ is the norm in H^r. By this construction, the space H^2 is the domain of A and is a dense subset of H. Conditions of Theorem 3.1.18 then hold automatically: the norm $\|u\|_2^2$ is equivalent to $\|(M\mathrm{I}-\mathrm{A})u\|_0^2$. Finally,

$$\begin{aligned}\|(\lambda\mathrm{I}-\mathrm{A})u\|_H^2 &= \lambda^2\|u\|_H^2 - 2\lambda[\mathrm{A}u,u] + \|\mathrm{A}u\|_H^2\\ &\geq (\lambda^2-2M\lambda)\|u\|_H^2 = \lambda(\lambda-2M)\|u\|_H^2.\end{aligned}$$

which is (4.1.23) with $\lambda_0 = \max(2M,0)$.

4.1.8 (page 169)

Theorem 4.1.23 establishes integrability of the solution in space and time, which is different from properties of the solution in space for fixed time. For the equation in question, $\widehat{u}_0(y)$ looks like $\sin(y)/y$, and, with $u_0 \in H^r(\mathbb{R})$ for $r < 1/2$, $u \in L_2\big((0,1); H^{r+1}(\mathbb{R})\big)$ if $r < 1/2$, just as promised by the theorem. Moreover,

$$\int_0^1 \|u(t,\cdot)\|_{3/2}^2\,dt = +\infty,$$

that is, the conclusions of Theorem 4.1.23 are sharp. A repeated application of Theorem 4.1.23 for $t > 0$ shows that the solution belongs to every $H^r(\mathbb{R})$.

4.1.9 (page 169)

Let us start with equations $u_t = \big(a(t,x)u_x\big)_x$ (parabolic) and $u_{tt} = \big(a(t,x)u_x\big)_x$ (hyperbolic), both for $x \in \mathbb{R}$. To simplify comparison, we assume zero initial speed for the hyperbolic equation: $u_t(0,x) = 0$.

For the parabolic equation, we get $u \in \mathcal{C}((0,T); L_2(\mathbb{R}))$ if $u(0,\cdot) \in L_2(\mathbb{R})$, and a is bounded measurable and $a(t,x) \geq a_0 > 0$ for all t, x. For the hyperbolic equation, we get a better solution $u \in \mathcal{C}((0,T); H^1(\mathbb{R}))$ under stronger assumption about the initial condition: $u(0,\cdot) \in H^1(\mathbb{R})$, and an additional assumption about the coefficient a: $\partial a(t,x)/\partial t$ exists and is bounded and measurable. The reader can also compare the properties of u_t in both cases.

To get the same regularity of the solution for equations $u_t = a(t,x)u_{xx}$ and $u_{tt} = a(t,x)u_{xx}$, beside the usual $a(t,x) \geq a_0 > 0$, we need extra regularity of a in x to multiply objects from $H^{-1}(\mathbb{R})$; for example, $a(t,\cdot) \in \mathcal{C}^1(\mathbb{R})$ is enough. In the hyperbolic case, we also have to write

$$a(t,x)u_{xx} = \big(a(t,x)u_x\big)_x - a_x(t,x)u_x,$$

because the leading operator has to be symmetric. In particular, $\mathrm{A}_1 u = -a_x u_x$. As before, we also need $a_t(t,x)$ to exist. Note that the symmetry of the leading operator in the hyperbolic equation is not a very restrictive condition as long as the coefficients are sufficiently smooth.

An interesting question is how the condition for a function $f = f(x)$ to be a point-wise multiplier in $H^{-1}(\mathbb{R})$ (that is, for the operator $u \mapsto fu$ to be bounded on $H^{-1}(\mathbb{R})$) compares with the condition for the operator $u \mapsto f(x)u_{xx} - \big(f(x)u_x\big)_x$ to be bounded from $H^1(\mathbb{R})$ to $L_2(\mathbb{R})$.

The solution of a parabolic equation has better regularity than the initial condition: if $u(0) \in H$, then $u(t) \in V$ for almost all t. For the heat equation (and more generally, for equations with smooth coefficients), a repeated application of the theorem in suitable spaces will show that the solution is smooth for $t > 0$. The solution of a hyperbolic equation has exactly the same regularity as the initial conditions: $u(0) \in V$ implies $u(t) \in V$; $\dot{u}(0) \in H$ implies $\dot{u}(t) \in H$. No gain in spacial regularity is consistent with the explicit formulas for the wave equation.

4.1.10 (page 170)

In (a), (b) the main conditions are always boundedness of all the coefficients and (S.6). Assuming all the coefficients are infinitely differentiable, with all the derivatives bounded, is always sufficient. Minimal smoothness depends on the on the particular problem and is still subject of research.

4.2.1 (page 180)

Since we can only guarantee that $u \in L_2(G)$, the definition should go along the lines of the weak variational solution [W3]; see (2.2.13) on page 47. The technical issue is to impose the right conditions on the coefficients so that, in each case, the formal adjoint $\mathbf{A}^\top$ is defined. The details are left to the reader.

4.2.2 (page 180)

Problem 4.1.3 on page 166 provides all the necessary background. In particular,

(a) follows from $f = \sum_k f_k \mathfrak{w}_k$ and $\Lambda \mathfrak{w}_k = 2k$;
(b) follows from $\mathbb{E}\|\dot{W}\|_{\widetilde{r}}^2 = \sum_k (2k)^r$.
(c) In $\mathbb{R}^{\mathrm{d}}$, $\dot{W} \in L_2(\Omega; \widetilde{H}^{-r})$ for $r > \mathrm{d}$, because

$$\sum_{n_1+\ldots+n_{\mathrm{d}}=k} \frac{1}{(n_1+\ldots+n_{\mathrm{d}})^r} \asymp \frac{1}{k^{r+1-\mathrm{d}}}.$$

(d) Spaces $\widetilde{H}^r$ constructed using the operator Λ control both the number of the derivatives of the function and the rate of decay of the function at infinity; the Sobolev spaces only control the number of derivatives through the rate of decay of the Fourier transform. Also note that Λ and $\boldsymbol{\Delta}$ do not commute and the the function $|x|^2$ is unbounded, making the spaces $\widetilde{H}^r$ not especially convenient for the study of the heat and similar equations. On the other hand, for the usual Sobolev spaces, $\bigcap_r H^r(\mathbb{R}^{\mathrm{d}}) \neq \mathcal{S}(\mathbb{R}^{\mathrm{d}})$ (consider $f(x) = 1/(1+x^2)$ which is in every $H^r(\mathbb{R}^{\mathrm{d}})$ but whose Fourier transform is $\widehat{f}(y) \asymp e^{-|y|}$ and is not in $\mathcal{S}(\mathbb{R}^{\mathrm{d}})$), while for the spaces $\widetilde{H}^r$ constructed using the operator Λ, we have $\bigcap_r \widetilde{H}^r = \mathcal{S}(\mathbb{R}^{\mathrm{d}})$.

4.2.3 (page 181)

(a) Denote by $\mathfrak{X}$ the Fourier transform of $\dot{W}$. Since $\dot{W}$ can be considered an element of the space $\mathcal{S}'(\mathbb{R}^{\mathrm{d}})$, the Fourier transform of $\dot{W}$ exists. Linearity of the Fourier transform implies that, just as $\dot{W}$, $\mathfrak{X}$ is a zero-mean Gaussian field.

Let $\widehat{f}$ denote the Fourier transform of $f \in \mathcal{S}(\mathbb{R}^{\mathrm{d}})$. By definition,

$$\mathfrak{X}(f) = \dot{W}(\widehat{f}),$$

and therefore, for $f, g \in \mathcal{S}(\mathbb{R}^{\mathrm{d}})$,

$$\mathbb{E}\Big(\mathfrak{X}(f)\mathfrak{X}(g)\Big) = \mathbb{E}\Big(\dot{W}(\widehat{f})\dot{W}(\widehat{g})\Big) = (\widehat{f}, \widehat{g})_{L_2(\mathbb{R}^{\mathrm{d}})} = (f, g)_{L_2(\mathbb{R}^{\mathrm{d}})},$$

where the last equality is the well-known L_2-isometry of the Fourier transform. In other words, $\mathbb{E}\Big(\mathfrak{X}(f)\mathfrak{X}(g)\Big) = (f, g)_{L_2(\mathbb{R}^{\mathrm{d}})}$, that is, $\mathfrak{X}$ is indeed white noise over $L_2(\mathbb{R}^{\mathrm{d}})$.

(b) First show that the Fourier transform of $\varphi\dot{W}$ is a regular field

$$V(y) = \frac{1}{(2\pi)^{\mathrm{d}/2}} \int_{\mathbb{R}^{\mathrm{d}}} e^{\mathrm{i}xy}\varphi(x)dW(y).$$

Then, keeping in mind that the Fourier transform of a function from $L_1(\mathbb{R}^{\mathrm{d}})$ is a bounded continuous function, show that

$$\begin{aligned}\int_{\mathbb{R}^{\mathrm{d}}} (1+|y|^2)^{\gamma}\, \mathbb{E}|V(y)|^2 dy &= \int_{\mathbb{R}^{\mathrm{d}}} (1+|y|^2)^{\gamma} |\widehat{\varphi}(y)|^2 dy \\ &\le \left(\max_{y\in\mathbb{R}^{\mathrm{d}}} |\widehat{\varphi}(y)|^2\right) \int_{\mathbb{R}^{\mathrm{d}}} (1+|y|^2)^{\gamma} dy, \text{ if } \gamma < -\mathrm{d}/2,\end{aligned}$$

which completes the proof. For details, see Walsh [223, Chap. 9].

4.2.4 (page 181)

(b) See Rozanov [198, pp. 96–98]. The idea is to use the Fourier transform and get a closed-form solution:

$$u(x) = \frac{1}{(2\pi)^{3/2}} \int_{\mathbb{R}^3} \frac{e^{\mathrm{i}xy} - 1}{|y|^2}\, dW(y);$$

the zero boundary condition $u(0) = 0$ leads to the term $e^{\mathrm{i}xy} - 1$ instead of just $e^{\mathrm{i}xy}$.

(a), (c) One can try the same approach as in (b).

4.2.5 (page 181)

The easiest way to get an expression for $\mathbb{E}\big(u(f)u(g)\big)$ is via the Fourier transform (see Problem 4.2.3).

For equation $\sqrt{m^2 - \boldsymbol{\Delta}}\, u = \dot{W}$, we get

$$\mathbb{E}\big(u(f)u(g)\big) = \int_{\mathbb{R}^d} \frac{\widehat{f}(y)\overline{\widehat{g}(y)}}{m^2 + y^2} dy,$$

where $\overline{\widehat{g}(y)}$ is the complex conjugation. Note that if $\mathrm{d} \geq 3$, the integral converges for all $f, g \in L_2(\mathbb{R}^d)$ even if $m = 0$; more details are in Nelson [172].

For equation $\boldsymbol{\Delta} u = \nabla \cdot \dot{\boldsymbol{W}}$,

$$\mathbb{E}\big(u(f)u(g)\big) = \int_{\mathbb{R}^d} \frac{\widehat{f}(y)\overline{\widehat{g}(y)}}{y^2} dy,$$

which is the same as for the previous equation with $m = 0$, suggesting that the equation is natural to consider for $\mathrm{d} \geq 3$. More details for the case $\mathrm{d} \geq 3$ are in Walsh [223, Chap. IX].

For equation $(1 - \boldsymbol{\Delta})u = \mathfrak{B}$, with $\mathfrak{B}$ a Gaussian white noise on $H^1(\mathbb{R}^d)$, we write $\mathfrak{B} = \sqrt{1 - \boldsymbol{\Delta}}\, \dot{W}$ and reduce the problem to the one we just solved. For an alternative approach, see Rozanov [198, page 102].

In each case

- the solution is known as the `Euclidean free field`.
- the solution can be interpreted as a field either on $H^{-1}(\mathbb{R}^d)$ (using the Fourier transform) or on $H^1(\mathbb{R}^d)$ (using Theorem 4.2.2 on page 170).

4.2.6 (page 182)

Solve the equation $\sqrt{1 - \boldsymbol{\Delta}}\, u = \dot{W}$ using the Fourier transform and the result of Problem 4.2.3 to conclude that

$$\widehat{u}(y) = \frac{\dot{W}(y)}{\sqrt{1 + |y|^2}}$$

Using the properties of the Fourier transform and the notation $\check{f}(y) = \hat{f}(-y)$,

$$\mathbb{E}\big(u(f)u(g)\big) = \mathbb{E}\big(\widehat{u}(\check{f})\widehat{u}(\check{g})\big) = \int_{\mathbb{R}^d} \frac{\widehat{f}(y)\overline{\widehat{g}(y)}}{1 + |y|^2} dy = (f, g)_{-1};$$

this is also consistent with the interpretation of u as a homogeneous field (see Problem 3.2.8 on page 127).

4.2.7 (page 182)

This is an essay question.

4.2.8 (page 182)

The basic argument is identical to the proof of Corollary 4.2.7 (page 178). Indeed, Green's function of the equation

$$\mathbf{A}v - c\,v = f, \text{ in } G$$

with zero boundary conditions can be written as

$$K(x,y) = K_A(x,y) - \bar{K}_A(x,y),$$

where, for every $y \in G$,

$$\mathbf{A}\bar{K}_A(x,y) - c(x)\bar{K}_A(x,y) = 0, \;\; x \in G, \; \bar{K}_A(x,y)|_{x\in\partial G} = K_A(x,y).$$

It is also known that

$$\bar{K}_A(x,y) = \mathbb{E}\left(K_A\big(X^x(\tau_x), y\big) e^{-\int_0^{\tau_x} c(X^x(t))dt} \right);$$

see, for example, Freidlin [55, Theorem 2.2.1]. The technical details, such as the precise conditions on the operator **A**, function c and the domain G, are left to the reader.

A Comment Under suitable conditions on the operator **A**, the function c, and the domain G, one can show that $\bar{K}_A$ is a smooth function and the singularity of K_A for $x = y$ will be the same as the singularity at zero of the function K_{d} from (4.2.12) on page 175. The reader can use this observation to produce a proof of Theorem 4.2.6 when $\mathbf{A} \neq \boldsymbol{\Delta}$.

4.2.9 (page 183)

Here are some general ideas. The solution of $(1 - \boldsymbol{\Delta})^{\gamma/2} u = \dot{W}$ is never an element of any Sobolev space $H^r(\mathbb{R}^{\mathrm{d}})$, because otherwise $\dot{W}$ would be an element of some Sobolev space, which it is not. The Hölder regularity of the solution can still be studied if one can show that there is a locally integrable function $u = u(x)$ such that, for all $f \in \mathcal{S}(\mathbb{R}^{\mathrm{d}})$, $u(f) = \int_{\mathbb{R}^{\mathrm{d}}} u(x)f(x)dx$. On the other hand, in the notations of Problem 4.2.2, the solution of $\Lambda^{\gamma} u = \dot{W}$ is an element of suitable $\widetilde{H}^r$ and can be a regular field over $L_2(\mathbb{R}^{\mathrm{d}})$.

4.3.1 (page 197)

(a) If $f \in L_2\big(\Omega; H^1((0,T); V')\big)$, then $f(t) = f_0 + \int_0^t \dot{f}(s)ds$, with $f_0, \dot{f} \in V'$. Integrating by parts,

$$\int_0^t (f, v)_H ds = [f(t), u(t)] - [f_0, u_0] - \int_0^t [\dot{f}, u]ds,$$

and then, using epsilon inequality, $|\,[f(t), u(t)]| \leq \varepsilon \|u(t)\|_V^2 + C\|f(t)\|_{V'}^2$. Note that $\|f(t)\|_{V'}^2 \leq \|f_0\|_{V'}^2 + C(T)\|\dot{f}\|^2_{L_2((0,T);V')} \leq C(T)\|f\|^2_{H^1((0,T);V')}$.

(b) Derivation of the a priori bound goes through without changes.

4.3.2 (page 197)
Direct computations using formula (2.3.27) on page 59. For simplicity, assume $a_k = 1$.

4.3.3 (page 198)
The idea is to re-write each equation in the abstract form and to verify the conditions of Theorem 4.3.3. The most convenient boundary conditions, when necessary, should be derived using integration by parts. When coefficients are constant and noise is additive, one can look for closed-form solutions.

4.4.1 (page 230)
Let $G(x)$ be the indicator function of $(-1, 1)$ and let $g(x) = G'(x)$ as a generalized derivative. The Fourier transform of g satisfies $\widehat{g}(y) \asymp \sin(cy)$ for some c. Compute the Fourier transform of u.

4.4.2 (page 231)
Theorem 4.4.3 provides most of the results. The definition of $\boldsymbol{H}$ ensures that the solution operator is onto, that is, every element of $\boldsymbol{H}$ is a solution of some equation. Indeed, if $u \in \boldsymbol{H}$, then $u \in L_2\big(\Omega \times (0,T); V\big)$ and

$$u(t) = u_0 + \int_0^t F(s)ds + \sum_k \int_0^t G_k(s)dw(s),$$

which means that $\dot{u} = \mathrm{A}u + f + \sum_k(\mathrm{M}_k u + g_k)dw_k$, where $f = F - \mathrm{A}u$ and $g_k = G_k - \mathrm{M}_k u$.

4.4.3 (page 231)
To get an idea, derive a priori bounds and find a solution for the following equations:

$$u_t = u_{xx}/\sqrt{t},\ u_t = \xi\, u_{xx},\ 0 < t \leq 1,\ x \in \mathbb{R},$$

where ξ is a normal random variable with mean and variance both equal to one.

4.4.4 (page 231)
This is left to the reader.

4.4.5 (page 231)
Direct computations using the closed-form expression for u_k. In fact, one can allow b and σ to be functions of t. For simplicity, assume $a_k = 1$.

4.4.6 (page 232)
The idea is to re-write each equation in the abstract form and to verify the conditions of Theorem 4.4.3. The most convenient boundary conditions, when

necessary, should be derived using integration by parts. When coefficients are constant and noise is additive, one can look for closed-form solutions.

Problems of Chap. 5

5.1.1 (page 249)

(a) In one dimension, $\eta = \xi$ and $\zeta = \sum_k \frac{\mathrm{H}_k(\xi)}{k\sqrt{k!}}$ should work as a counterexample: $\eta \diamond \zeta \notin L_2(\xi)$.

As far as the rest of the problem, you are on your own...

Problems of Chap. 6

6.1.1 (page 393) For the estimator $\hat{\theta}^{(N)}(T)$, the rate in time $\sqrt{T}$ is known to be optimal and therefore cannot be improved; the rate in N is known to be not optimal and can be improved to $N^{3/2}$ by considering weights $\alpha_{k,N}(T) \sim k^2/N^3$. Since our computations suggest that $\mathrm{Var}(\hat{\theta}^{(k)}(T) - \theta) \sim 1/k^2$, the optimal weights can be determined by minimizing $\sum_{k=1}^{N} k^{-2}x_k^2$ under the constraint $\sum_{k=1}^{N} x_k = 1, x_k \geq 0$.

6.1.2 (page 393)

We have

$$u_k(t,q) = u_k(0)e^{-\theta k^2 t} + q\int_0^t e^{-\theta k^2(t-s)}dw_k(s),$$

so that

$$v_k(t) = \lim_{q\to 0} u_k(t,q) = u_k(0)e^{-\theta k^2 t};$$

the details about the convergence are up to the reader. Then

$$q^{-1}(\hat{\theta}^{(k)}(T) - \theta) = \frac{\int_0^T u_k(t,q)dw_k}{\int_0^T u_k^2(t,q)dt}$$

and the limit in distribution, as $q \to 0$, is

$$\frac{\int_0^T v_k(t)dw_k(t)}{\int_0^T v_k^2(t)dt} = \frac{2\theta k^2 \int_0^T e^{-\theta k^2 t}dw_k(t)}{u_k(0)(1-e^{-2\theta k^2 T})}$$

6.1.3 (page 393)
Similar to (6.1.26) on page 389, we have

$$u_k(t) \overset{\mathcal{L}}{=} u_k(0)e^{-\theta k^2 t} + \frac{1}{k}\bar{u}(k^2 t),$$

and

$$\bar{u}(t) = q \int_0^t e^{-\theta(t-s)} dw_1(t)$$

is ergodic. Taking $u_k(0) = k$ will result in

$$\int_0^T \mathbb{E}u_k^2(t)dt \sim 1$$

and the rate of convergence k^2 rather than k.

6.1.4 (page 394)
We have $X(t) = X(0)e^t + \sigma \int_0^T e^{t-s} dw(s)$ and

$$\hat{\theta}(T) - 1 = \frac{\sigma \int_0^T X(t)dw(t)}{\int_0^T X^2(t)dt}.$$

Define

$$\zeta = \sigma \int_0^{+\infty} e^{-s} dw(s),\ \eta \overset{d}{=} \lim_{T\to+\infty} \sigma \int_0^T e^{-(T-s)} dw(s)$$

and note that ζ and η are Gaussian and independent:

$$\mathbb{E}\eta\zeta = \lim_{T\to+\infty} \sigma^2 e^{-T} \int_0^T dt = 0.$$

Writing

$$X(t) \approx (X(0) + \zeta)e^t,$$

we find

$$e^T(\hat{\theta}(T) - 1) \approx \frac{(X(0) + \zeta)\eta\, e^{2T}}{(X(0) + \zeta)^2 \int_0^T e^{2t} dt};$$

suggesting that the limit in distribution is

$$\frac{2\eta}{X(0)+\zeta};$$

if $\mathbb{E}X(0) = 0$, then the limit distribution is Cauchy. Justifying the approximations $\approx$ is up to the reader.

6.1.5 (page 394) Using self-similarity of the Brownian motion (6.1.17) (with $c = T$), we find

$$\lim_{T\to+\infty} \hat{\theta}(T) \stackrel{d}{=} \frac{\int_0^1 \bar{w}(s)d\bar{w}(s)}{\int_0^1 \bar{w}^2(s)ds},$$

independent of the initial condition. If the initial condition is zero, then we have equality for all $T > 0$; $\bar{w}(s) = T^{-1/2}w(sT)$ is a standard Brownian motion. Note that dimensional analysis implies that both $\bar{w}$ and s are dimensionless.

6.2.1 (page 438) If $\hat{r} = \hat{\theta}^{(k)}(T) - \theta$, and $\hat{\mathbf{r}} = \hat{\boldsymbol{\theta}}^{(k)}(T) - \theta$, then

$$\hat{\mathbf{r}} = \begin{cases} \hat{r}, & \hat{\theta}^{(k)}(T) \geq 0, \\ -\theta, & \hat{\theta}^{(k)}(T) < 0. \end{cases}$$

Then, with $v = \sqrt{T}$ or $v = k$ denoting the corresponding rate,

$$\mathbb{P}(v\,\hat{\mathbf{r}} < y) = \mathbb{P}\Big(v\,\hat{r} < y, \hat{\theta}^{(k)}(T) \geq 0\Big) + v\mathbb{P}(\hat{\theta}^{(k)}(T) < 0).$$

The first term on the right will converge to the normal cdf, and the goal is to show that the second term goes to zero. Since

$$\mathbb{P}(\hat{\theta}^{(k)}(T) < 0) = \mathbb{P}(\hat{\theta}^{(k)}(T) - \theta < -\theta) \leq \mathbb{P}(|\hat{r}| > \theta) \leq \frac{\mathbb{E}|\hat{r}|^p}{\theta^p}$$

and we expect that

$$\mathbb{E}|\hat{r}|^p \sim v^{-p}, \tag{S.15}$$

we take p large enough and complete the proof. The problem is to show (S.15), which, because of the need to investigate the denominator in the fraction (6.1.15), is essentially equivalent to establishing uniform polynomial integrability of the local likelihood ratio. In other words, to complete the proof, it remains to use (6.2.91).

For the large-time asymptotic, see [140, Sect. 17.3].

6.2.2 (page 438) The arguments are the same as in the solution of Problem 6.2.1

6.2.3 (page 438) Since

$$du_1 = -(1-\theta)u_1dt + q_1dw_1(t),$$

there are three cases to consider: $\theta < 1$, $\theta = 1$, $\theta > 1$. For more ideas, see Problems 6.1.4 and 6.1.5.

6.2.4 (page 438)

(a) Use the multinomial formula and the fact that $u_k(t)$ is Gaussian.
(b) Since we have asymptotic normality, all we need is convergence of moment, which, by a version of a dominated convergence theorem, will follow from uniform integrability of all the moment of $\sqrt{I(N)}(\widehat{\theta}_N - \theta)$, which, in turn, follows after combining part (a) with (6.2.89). More precisely, $\sqrt{I(N)}(\widehat{\theta}_N - \theta)$ is a fraction; part (a) takes care of the numerator (with the help of the BDG inequality) and (6.2.89) takes care of the denominator using the equality

$$\mathbb{E}\eta^{-p} = c(p)\int_0^{+\infty} z^{p-1}\mathbb{E}e^{-\eta z}dz,\ \eta > 0.$$

6.2.5 (page 438) Pretend that you observe

$$du_k = \theta k u_k dt + q_k dw_k,$$

where $\theta = \lim_{k\to+\infty} \nu_k/k$ and d ν_k, $k \geq 1$, are the eigenvalues of the Laplacian in G with zero boundary conditions. Use (6.2.9) and (6.2.10).

6.2.6 (page 438) This is up to the reader. One clear difference is that Girsanov's theorem allows extra randomness beside the underlying Brownian motion. One similarity is that, in Girsanov's theorem,

$$X(t) = -\int_0^T h(s)ds + w(t)$$

is shifted Brownian motion, and the shift is of the same type as the one in the Cameron-Martin theorem.

6.3.1 (page 459) The same arguments as in the case of one parameter lead to

$$\lim_{k\to+\infty} \frac{m_k^2+\sigma_k^2}{q_k^2} = 0,\ \lim_{k\to+\infty} \frac{k\sigma_k^2(m_k^2+\sigma_k^2)}{q_k^4} = 0.$$

6.3.2 (page 459)

(a) If indeed $\mu_k(\theta) = e^k$, then there is no natural scale of Hilbert spaces in which to construct the variational solution.

(b) Take a separable Hilbert space X such that the embedding H to X is Hilbert-Schmidt. Then the mild solution will be an element of $L_2(\Omega; C((0,T); X))$.

6.3.3 (page 459) If $q_k = 1$ and $\mu_k(\theta) = \rho + \theta\nu = \mu(\theta)$, with known $\rho \geq 0$ and $\nu > 0$ and unknown $\theta > 0$, the conclusions of the theorem hold with

$$I_{N,\theta} = N\left(\frac{\nu^2}{\mu(\theta)} - \frac{\nu^2(1 - e^{-\mu(\theta)T})}{2\mu^2(\theta)}\right),$$

because the LLN and CLT continue to hold. the iid case hold under more general conditions. The case $\mu(\theta) \leq 0$ is left to the reader.

6.3.4 (page 459) The proof is by direct computation using the following results:

$$X(t; a, \sigma) = X(0; a, \sigma)e^{-at} + \bar{X}(t; a, \sigma),\ \bar{X}(t; a, \sigma) = \sigma^2 \int_0^t e^{-a(t-s)} dw(s),$$

and $\bar{X}$ is a zero-mean Gaussian process independent of $X(0; a, \sigma)$;

If (ξ, η) is a two-dimensional (bivariate) Gaussian vector such that $\mathbb{E}\xi = \mathbb{E}\eta = 0$, $\mathbb{E}\xi^2 = \sigma_\xi^2$, $\mathbb{E}\eta^2 = \sigma_\eta^2$, $\mathbb{E}\xi\eta = \sigma_{\xi\eta}$, then

$$\mathbb{E}\xi\eta^2 = 0,$$
$$\mathbb{E}\xi^2\eta^2 = \sigma_\xi^2\sigma_\eta^2 + 2\sigma_{\xi\eta}^2;$$

If $Y = Y(t)$, $t \geq 0$, is a Gaussian process with $\mathbb{E}Y(t) = 0$, $\mathbb{E}Y(t)Y(s) = R(t, s)$, then

$$\mathrm{Var} \int_0^T Y^2(t)dt = 4\int_0^T \int_0^t R^2(t, s)dsdt.$$

More familiar relations are also necessary, such as

$$(a + b)^n \leq 2^n(a^n + b^n),\ \mathbb{E}|\mathcal{N}(0, \sigma^2)|^p = C(p)\sigma^p.$$

6.3.5 (page 459) In both cases, one difficulty is proving uniform asymptotic normality of the local likelihood ratio. For (6.3.1), we need a multi-dimensional analogue of the Berry-Esseen inequality (6.2.84) (see, for example, [64]). For (6.3.15), the difficulty is that, without explicit asymptotic for μ_k and ν_k, the classical Berry-Esseen inequality does not provide rate of convergence in terms of $I_{N,\theta}$. This difficulty can handled as follows. Assuming zero initial conditions, we have

$$\int_0^T u_k^2(t)dt = \int_0^T \int_0^t \int_0^s e^{-\mu_k(\theta)|t-s|} dw_k(s)dw_k(t),$$

which allows us to use a result about normal approximation in chaos spaces (also known as the chaotic CLT, see, for example, [23, Theorem 14.6]), leading to an explicit bound in terms of $I_{n,\theta}$. Note that, in the SPDE case, the bound is better than $C/\sqrt{N}$.

There is a multi-dimensional version of the chaotic CLT (see [173]) suggesting that uniform asymptotic normality also holds for the abstract evolution equation (6.3.28).

References

1. N.I. Akhiezer, *The Classical Moment Problem and Some Related Questions in Analysis*. Translated by N. Kemmer (Hafner Publishing, New York, 1965)
2. G.E. Andrews, R. Askey, R. Roy, *Special Functions*. Encyclopedia of Mathematics and Its Applications, vol. 71 (Cambridge University Press, Cambridge, 1999)
3. D.G. Aronson, The porous medium equation, in *Nonlinear Diffusion Problems (Montecatini Terme, 1985)*. Lecture Notes in Mathematics, vol. 1224 (Springer, Berlin, 1986), pp. 1–46
4. N. Aronszajn, Theory of reproducing kernels. Trans. Am. Math. Soc. **68**, 337–404 (1950)
5. A. Bain, D. Crisan, *Fundamentals of Stochastic Filtering*. Stochastic Modelling and Applied Probability, vol. 60 (Springer, New York, 2009)
6. R.M. Balan, B.G. Ivanoff, A Markov property for set-indexed processes. J. Theor. Probab. **15**(3), 553–588 (2002)
7. G.I. Barenblatt, On some unsteady motions of a liquid and gas in a porous medium. Akad. Nauk SSSR. Prikl. Mat. Meh. **16**, 67–78 (1952)
8. P. Baxendale, T.E. Harris, Isotropic stochastic flows. Ann. Probab. **14**(4), 1155–1179 (1986)
9. R. Bellman, The stability of solutions of linear differential equations. Duke Math. J. **10**, 643–647 (1943)
10. L. Bertini, G. Giacomin, Stochastic Burgers and KPZ equations from particle systems. Commun. Math. Phys. **183**(3), 571–607 (1997)
11. D. Blömker, Nonhomogeneous noise and Q-Wiener processes on bounded domains. Stoch. Anal. Appl. **23**(2), 255–273 (2005)
12. D. Blömker, *Amplitude Equations for Stochastic Partial Differential Equations*. Interdisciplinary Mathematical Sciences, vol. 3 (World Scientific Publishing, Hackensack, 2007)
13. V.I. Bogachev, *Gaussian Measures*. Mathematical Surveys and Monographs, vol. 62 (American Mathematical Society, Providence, 1998)
14. B. Boufoussi, A support theorem for hyperbolic SPDEs in anisotropic Besov-Orlicz space. Random Oper. Stoch. Equ. **10**(1), 59–88 (2002)
15. B. Boufoussi, M. Eddahbi, M. N'zi, Freidlin-Wentzell type estimates for solutions of hyperbolic SPDEs in Besov-Orlicz spaces and applications. Stoch. Anal. Appl. **18**(5), 697–722 (2000)
16. R. Buckdahn, É. Pardoux, *Monotonicity Methods for White Noise Driven Quasi-Linear SPDEs*. Diffusion Processes and Related Problems in Analysis, Vol. I (Evanston, IL, 1989). Progress in Probability, vol. 22 (Birkhäuser Boston, Boston, 1990), pp. 219–233
17. A. Budhiraja, G. Kallianpur, Approximations to the solution of the Zakai equations using multiple Wiener and Stratonovich integral expansions. Stoch. Stoch. Rep. **56**(3–4), 271–315 (1996)

S.V. Lototsky, B.L. Rozovsky, *Stochastic Partial Differential Equations*,
Universitext, DOI 10.1007/978-3-319-58647-2

18. D.L. Burkholder, R.F. Gundy, Extrapolation and interpolation of quasi-linear operators on martingales. Acta Math. **124**, 249–304 (1970)
19. D.L. Burkholder, B.J. Davis, R.F. Gundy, Integral inequalities for convex functions of operators on martingales, in *Proceedings of the Sixth Berkeley Symposium on Mathematical Statistics and Probability* (University of California Press, Berkeley, CA, 1970/1971), Vol. II: Probability Theory (University of California Press, Berkeley, CA, 1972), pp. 223–240
20. R.H. Cameron, W.T. Martin, The orthogonal development of nonlinear functionals in a series of Fourier-Hermite functions. Ann. Math. **48**(2), 385–392 (1947)
21. R.A. Carmona, B. Rozovskii (eds.), *Stochastic Partial Differential Equations: Six Perspectives*. Mathematical Surveys and Monographs, vol. 64 (American Mathematical Society, Providence, 1999)
22. R.A. Carmona, M.R. Tehranchi, *Interest Rate Models: An Infinite Dimensional Stochastic Analysis Perspective*. Springer Finance (Springer, Berlin, 2006)
23. L.H.Y. Chen, L. Goldstein, Q.-M. Shao, *Normal Approximation by Stein's Method*. Probability and Its Applications (Springer, Heidelberg, 2011)
24. P.-L. Chow, *Stochastic Partial Differential Equations*. Chapman & Hall/CRC Applied Mathematics and Nonlinear Science Series (Chapman & Hall/CRC, Boca Raton, 2007)
25. P.-L. Chow, Y. Huang, Semilinear stochastic hyperbolic systems in one dimension. Stoch. Anal. Appl. **22**(1), 43–65 (2004)
26. Ig. Cialenco, Parameter estimation for SPDEs with multiplicative fractional noise. Stoch. Dyn. **10**(4), 561–576 (2010)
27. Ig. Cialenco, N. Glatt-Holtz, Parameter estimation for the stochastically perturbed Navier-Stokes equations. Stoch. Process. Appl. **121**(4), 701–724 (2011)
28. Ig. Cialenco, S.V. Lototsky, Parameter estimation in diagonalizable bilinear stochastic parabolic equations. Stat. Inference Stoch. Process. **12**(3), 203–219 (2009)
29. J.D. Cole, On a quasi-linear parabolic equation occurring in aerodynamics. Q. Appl. Math. **9**, 225–236 (1951)
30. R. Cont, Modeling term structure dynamics: an infinite dimensional approach. Int. J. Theor. Appl. Finance **8**(3), 357–380 (2005)
31. G. Da Prato, J. Zabczyk, *Stochastic Equations in Infinite Dimensions*. Encyclopedia of Mathematics and Its Applications, vol. 44 (Cambridge University Press, Cambridge, 1992)
32. G. Da Prato, J. Zabczyk, *Ergodicity for Infinite-Dimensional Systems*. London Mathematical Society Lecture Note Series, vol. 229 (Cambridge University Press, Cambridge, 1996)
33. R. Dalang, E. Nualart, Potential theory for hyperbolic SPDEs. Ann. Probab. **32**(3A), 2099–2148 (2004)
34. R.C. Dalang, J.B. Walsh, The sharp Markov property of Lévy sheets. Ann. Probab. **20**(2), 591–626 (1992)
35. R.C. Dalang, J.B. Walsh, The sharp Markov property of the Brownian sheet and related processes. Acta Math. **168**(3–4), 153–218 (1992)
36. R. Dalang, D. Khoshnevisan, C. Mueller, D. Nualart, Y. Xiao, *A Minicourse on Stochastic Partial Differential Equations*. Lecture Notes in Mathematics, vol. 1962 (Springer, Berlin, 2009). MR 1500166 (2009k:60009)
37. T. Deck, J. Potthoff, On a class of stochastic partial differential equations related to turbulent transport. Probab. Theory Relat. Fields **111**, 101–122 (1998)
38. G. Di Nunno, B. Øksendal, F. Proske, *Malliavin Calculus for Lévy Processes with Applications to Finance*. Universitext (Springer, Berlin, 2009)
39. C. Donati-Martin, Quasi-linear elliptic stochastic partial differential equation: Markov property. Stoch. Stoch. Rep. **41**(4), 219–240 (1992)
40. C. Donati-Martin, D. Nualart, Markov property for elliptic stochastic partial differential equations. Stoch. Stoch. Rep. **46**(1–2), 107–115 (1994)
41. T.E. Duncan, B. Maslowski, B. Pasik-Duncan, Stochastic equations in Hilbert space with a multiplicative fractional Gaussian noise. Stoch. Process. Appl. **115**(8), 1357–1383 (2005)
42. N. Dunford, J.T. Schwartz, *Linear Operators, I: General Theory* (Wiley-Interscience, Chichester, 1988)

43. N. Dunford, J.T. Schwartz, *Linear Operators, II: Spectral Theory* (Wiley-Interscience, Chichester, 1988)
44. W.E, E. Vanden Eijnden, Generalized flows, intrinsic stochasticity, and turbulent transport. Proc. Natl. Acad. Sci. **97**(15), 8200–8205 (2000)
45. M. Eddahbi, Large deviations for solutions of hyperbolic SPDE's in the Hölder norm. Potential Anal. **7**(2), 517–537 (1997)
46. S.D. Eidelman, *Parabolic Systems* (Wolters-Noordhoff, Groningen, 1969)
47. L. Erdős, B. Schlein, H.-T. Yau, Derivation of the cubic non-linear Schrödinger equation from quantum dynamics of many-body systems. Invent. Math. **167**(3), 515–614 (2007)
48. L. Erdős, B. Schlein, H.-T. Yau, Rigorous derivation of the Gross-Pitaevskii equation with a large interaction potential. J. Am. Math. Soc. **22**(4), 1099–1156 (2009)
49. S.N. Ethier, S.M. Krone, Comparing Fleming-Viot and Dawson-Watanabe processes. Stoch. Process. Appl. **60**(2), 171–190 (1995)
50. L.C. Evans, *Partial Differential Equations*. Graduate Studies in Mathematics, vol. 19 (American Mathematical Society, Providence, 1998)
51. J. Feldman, Equivalence and perpendicularity of Gaussian processes. Pac. J. Math. **8**, 699–708 (1958)
52. F. Flandoli, *Regularity Theory and Stochastic Flows for Parabolic SPDEs*. Stochastics Monographs, vol. 9 (Gordon and Breach Science, Yverdon, 1995)
53. C. Frankignoul, SST anomalies, planetary waves and RC in the middle rectitudes. Rev. Geophys. **23**(4), 357–390 (1985)
54. M. Freidlin, *Functional Integration and Partial Differential Equations* (Princeton University Press, Princeton, 1985)
55. M. Freidlin, *Functional Integration and Partial Differential Equations*. Annals of Mathematics Studies, vol. 109 (Princeton University Press, Princeton, 1985)
56. M.I. Freidlin, A.D. Wentzell, *Random Perturbations of Dynamical Systems*, 2nd edn. Grundlehren der Mathematischen Wissenschaften [Fundamental Principles of Mathematical Sciences], vol. 260 (Springer, New York, 1998)
57. A. Friedman, *Stochastic Differential Equations and Applications* (Dover Publications, Mineola, 2006)
58. A. Friedman, *Partial Differential Equations* (Dover Publications, Mineola, 2008)
59. L. Gawarecki, V. Mandrekar, *Stochastic Differential Equations in Infinite Dimensions with Applications to Stochastic Partial Differential Equations*. Probability and Its Applications (New York) (Springer, Heidelberg, 2011). MR 2560625 (2012a:60179)
60. K. Gawędzki, M. Vergassola, Phase transition in the passive scalar advection. Physica D **138**, 63–90 (2000)
61. I.M. Gelfand, N.Ya. Vilenkin, *Generalized Functions, vol. 4: Applications of Harmonic Analysis* (Academic, New York, 1964)
62. I.I. Gikhman, T.M. Mestechkina, The Cauchy problem for stochastic first-order partial differential equations. Theory Random Process. **11**, 25–28 (1983). "Naukova Dumka", Kiev; in Russian
63. J. Glimm, A. Jaffe, *Quantum Physics*, 2nd edn. (Springer, New York, 1987)
64. F. Götze, On the rate of convergence in the multivariate CLT. Ann. Probab. **19**(2), 724–739 (1991)
65. W. Grecksch, C. Tudor, *Stochastic Evolution Equations*. Mathematical Research, vol. 85 (Akademie-Verlag, Berlin, 1995)
66. D. Grieser, Uniform bounds for eigenfunctions of the Laplacian on manifolds with boundary. Commun. Partial Differ. Equ. **27**(7–8), 1283–1299 (2002)
67. T.H. Gronwall, Note on the derivatives with respect to a parameter of the solutions of a system of differential equations. Ann. Math. (2) **20**(4), 292–296 (1919)
68. M.E. Gurtin, R.C. MacCamy, On the diffusion of biological populations. Math. Biosci. **33**(1–2), 35–49 (1977)
69. I. Gyöngy, T. Martínez, On numerical solution of stochastic partial differential equations of elliptic type. Stochastics **78**(4), 213–231 (2006)

70. M. Hairer, *An Introduction to Stochastic PDEs*, http://www.hairer.org/notes/SPDEs.pdf (2009)
71. M. Hairer, A.M. Stuart, J. Voß, P. Wiberg, Analysis of SPDEs arising in path sampling, part I: the Gaussian case. Commun. Math. Sci. **3**(4), 587–603 (2005)
72. M. Hairer, A.M. Stuart, J. Voß, Analysis of SPDEs arising in path sampling, part II: the nonlinear case. Ann. Appl. Probab. **17**(5/6), 1657–1706 (2007)
73. J. Hájek, On a property of normal distribution of any stochastic process. Czech. Math. J. **8**(83), 610–618 (1958)
74. J. Hájek, A property of J-divergences of marginal probability distributions. Czech. Math. J. **8**(83), 460–463 (1958)
75. T. Hida, N. Ikeda, Analysis on Hilbert space with reproducing kernel arising from multiple Wiener integral, in *Proceedings of the Fifth Berkeley Symposium Mathematical Statistics and Probability* (University of California Press, Berkeley, CA, 1965/66). Vol. II: Contributions to Probability Theory, Part 1 (University of California Press, Berkeley, CA, 1967), pp. 117–143
76. T. Hida, H.-H. Kuo, J. Potthoff, L. Streit, *White Noise*. Mathematics and Its Applications, vol. 253 (Kluwer Academic, Dordrecht, 1993)
77. E. Hille, R.S. Phillips, *Functional Analysis and Semi-groups* (American Mathematical Society, Providence, 1974)
78. H. Holden, B. Øksendal, J. Ubøe, T. Zhang, *Stochastic Partial Differential Equations*, 2nd edn. Universitext (Springer, New York, 2010)
79. E. Hopf, The partial differential equation $u_t + uu_x = \mu u_{xx}$. Commun. Pure Appl. Math. **3**, 201–230 (1950)
80. M. Huebner, A characterization of asymptotic behaviour of maximum likelihood estimators for stochastic PDE's. Math. Methods Stat. **6**(4), 395–415 (1997)
81. M. Huebner, Asymptotic properties of the maximum likelihood estimator for stochastic PDEs disturbed by small noise. Stat. Inference Stoch. Process. **2**(1), 57–68 (1999). https://link.springer.com/article/10.1023%2FA%3A1009990504925
82. M. Huebner, B. Rozovskii, On asymptotic properties of maximum likelihood estimators for parabolic stochastic PDE's. Probab. Theory Relat. Fields **103**, 143–163 (1995)
83. M. Huebner, R.Z. Khasćminskiĭ, B.L. Rozovskii, *Two Examples of Parameter Estimation*, ed. by S. Cambanis, J.K. Ghosh, R.L. Karandikar, P.K. Sen. Stochastic Processes: A Volume in Honor of G. Kallianpur (Springer, New York, 1992), pp. 149–160
84. I.A. Ibragimov, R.Z. Khas'minskiĭ, *Statistical Estimation: Asymptotic Theory*. Applications of Mathematics, vol. 16 (Springer, New York, 1981)
85. I.A. Ibragimov, R.Z. Khas'minskiĭ, *Some Estimation Problems in Infinite Dimensional White Noise*, ed. by D. Pollard, E. Torgersen, G. Yang. Festschrift for Lucien LeCam (Springer, New York, 1997), pp. 259–274
86. I.A. Ibragimov, R.Z. Khas'minskiĭ, Problems of estimating the coefficients of stochastic partial differential equations. I. Teor. Veroyatnost. i Primenen. **43**(3), 417–438 (1998). Translation in Theory Probab. Appl. **43**(3), 370–387 (1999)
87. I.A. Ibragimov, R.Z. Khas'minskiĭ, Problems of estimating the coefficients of stochastic partial differential equations. II. Teor. Veroyatnost. i Primenen. **44**(3), 526–554 (1999). Translation in Theory Probab. Appl. **44**(3), 469–494 (2000)
88. I.A. Ibragimov, R.Z. Khas'minskiĭ, Problems of estimating the coefficients of stochastic partial differential equations. III. Teor. Veroyatnost. i Primenen. **45**(2), 209–235 (2000). Translation in Theory Probab. Appl. **45**(2), 210–232 (2001)
89. K. Itô, Stochastic integral. Proc. Imp. Acad. Tokyo **20**(8), 519–524 (1944)
90. K. Ito, Multiple Wiener integral. J. Math. Soc. Jpn. **3**, 157–169 (1951)
91. K. Itô, On a formula concerning stochastic differentials. Nagoya Math. J. **3**, 55–65 (1951)
92. K. Iwata, The inverse of a local operator preserves the Markov property. Ann. Scuola Norm. Sup. Pisa Cl. Sci. (4) **19**(2), 223–253 (1992)
93. J. Jacod, A.N. Shiryaev, *Limit Theorems for Stochastic Processes*, 2nd edn. Grundlehren der Mathematischen Wissenschaften, vol. 288 (Springer, New York, 2003)

94. R. Jarrow, P. Protter, *A Short History of Stochastic Integration and Mathematical Finance: The Early Years, 1880–1970*. A Festschrift for Herman Rubin. IMS Lecture Notes Monograph Series, vol. 45 (Institute of Mathematical Statistics, Beachwood, 2004), pp. 75–91
95. J.L.W.V. Jensen, Sur les fonctions convexes et les inégalités entre les valeurs moyennes. Acta Math. **30**(1), 175–193 (1906)
96. A. Jentzen, P.E. Kloeden, *Taylor Approximations for Stochastic Partial Differential Equations*. CBMS-NSF Regional Conference Series in Applied Mathematics, vol. 83 (Society for Industrial and Applied Mathematics, Philadelphia, 2011)
97. F. John, *Partial Differential Equations*, 4th edn. Applied Mathematical Sciences, vol. 1 (Springer, New York, 1991)
98. O. Juan, R. Kerivena, G. Postelnicu, Stochastic motion and the level set method in computer vision: Stochastic active contours. Int. J. Comput. Vis. **69**(1), 7–25 (2006)
99. O. Kallenberg, R. Sztencel, Some dimension-free features of vector-valued martingales. Probab. Theory Relat. Fields **88**, 215–247 (1991)
100. G. Kallianpur, *Stochastic Filtering Theory*. Applications of Mathematics, vol. 13 (Springer, Berlin, 1980)
101. G. Kallianpur, *Stochastic Filtering Theory* (Springer, New York, 1980)
102. G. Kallianpur, J. Xiong, *Stochastic Differential Equations in Infinite-Dimensional Spaces*. Institute of Mathematical Statistics Lecture Notes—Monograph Series, vol. 26 (Institute of Mathematical Statistics, Hayward, 1995)
103. I. Karatzas, S.E. Shreve, *Brownian Motion and Stochastic Calculus*, 2nd edn. Graduate Texts in Mathematics, vol. 113 (Springer, New York, 1991)
104. M. Kardar, G. Parisi, Y.-C. Zhang, Dynamic scaling of growing interfaces. Phys. Rev. Lett. **56**(9), 889–892 (1986)
105. K. Karhunen, Zur Spektraltheorie stochastischer Prozesse. Ann. Acad. Sci. Fennicae. Ser. A. I. Math.-Phys. **34**, 1–7 (1946)
106. K. Karhunen, Über lineare Methoden in der Wahrscheinlichkeitsrechnung. Ann. Acad. Sci. Fennicae. Ser. A. I. Math.-Phys. **37**, 3–79 (1947)
107. T. Kato, G. Ponce, On nonstationary flows of viscous and ideal fluids in $L^p_s(R^2)$. Duke Math. J. **55**, 487–489 (1987)
108. D. Khoshnevisan, *Multiparameter Processes: An Introduction to Random Fields*. Springer Monographs in Mathematics (Springer, New York, 2002)
109. D. Khoshnevisan, *Analysis of Stochastic Partial Differential Equations*. CBMS Regional Conference Series in Mathematics, vol. 119 (American Mathematical Society, Providence, 2014)
110. D. Khoshnevisan, E. Nualart, Level sets of the stochastic wave equation driven by a symmetric Lévy noise. Bernoulli **14**(4), 899–925 (2008)
111. J.U. Kim, On a stochastic hyperbolic integro-differential equation. J. Differ. Equ. **201**(2), 201–233 (2004)
112. T.W. Körner, *Fourier Analysis*, 2nd edn. (Cambridge University Press, Cambridge, 1989)
113. T. Koski, W. Loges, Asymptotic statistical inference for a stochastic heat flow problem. Stat. Probab. Lett. **3**(4), 185–189 (1985)
114. P. Kotelenez, *Stochastic Ordinary and Stochastic Partial Differential Equations: Transition from Microscopic to Macroscopic Equations*. Stochastic Modelling and Applied Probability, vol. 58 (Springer, New York, 2007)
115. S.M. Kozlov, Some questions of stochastic partial differential equations. Trudy Sem. Petrovsk. **4**, 147–172 (1978)
116. R.H. Kraichnan, Small-scale structure of a scalar field convected by turbulence. Phys. Fluids **11**, 945–963 (1968)
117. S.G. Kreĭn, Yu.Ī. Petunīn, E.M. Semënov, *Interpolation of Linear Operators*. Translations of Mathematical Monographs, vol. 54 (American Mathematical Society, Providence, 1982)
118. N.V. Krylov, *Introduction to the Theory of Diffusion Processes*. Translations of Mathematical Monographs, vol. 142 (American Mathematical Society, Providence, 1995)

119. N.V. Krylov, *Lectures on Elliptic and Parabolic Equations in Hölder Spaces*. Graduate Studies in Mathematics, vol. 12 (American Mathematical Society, Providence, 1996)
120. N.V. Krylov, *An Analytic Approach to SPDEs, Stochastic Partial Differential Equations*, ed. by B.L. Rozovskii, R. Carmona. Six Perspectives, Mathematical Surveys and Monographs (American Mathematical Society, Providence, 1999), pp. 185–242
121. N.V. Krylov, *Lectures on Elliptic and Parabolic Equations in Sobolev Spaces*. Graduate Studies in Mathematics, vol. 96 (American Mathematical Society, Providence, 2008)
122. N.V. Krylov, B.L. Rozovskii, Stochastic evolution equations. J. Sov. Math. **14**(4), 1233–1277 (1981). Reprinted in *Stochastic Differential Equations: Theory and Applications*, ed. by S.V. Lototsky, P.H. Baxendale. *Interdisciplinary Mathematical Sciences*, vol. 2 (World Scientific, 2007), pp. 1–70
123. N.V. Krylov, A.J. Veretennikov, On explicit formula for solutions of stochastic equations. Math. USSR Sbornik **29**(2), 239–256 (1976)
124. H. Kunita, Cauchy problem for stochastic partial differential equations arising in nonlinear filtering theory. Syst. Cont. Lett. **1**(1), 37–41 (1981)
125. H. Kunita, *Stochastic Flows and Stochastic Differential Equations*. Cambridge Studies in Advanced Mathematics, vol. 24 (Cambridge University Press, Cambridge, 1997)
126. H. Künsch, Gaussian Markov random fields. J. Fac. Sci. Univ. Tokyo Sect. IA Math. **26**(1), 53–73 (1979)
127. T.G. Kurtz, J. Xiong, Particle representations for a class of nonlinear SPDEs. Stoch. Process. Appl. **83**(1), 103–126 (1999)
128. O.A. Ladyženskaja, V.A. Solonnikov, N.N. Ural'ceva, *Linear and Quasilinear Equations of Parabolic Type*. Translated from the Russian by S. Smith. Translations of Mathematical Monographs, vol. 23 (American Mathematical Society, Providence, 1968)
129. P.D. Lax, A.N. Milgram, Parabolic equations, in *Contributions to the Theory of Partial Differential Equations*. Annals of Mathematics Studies, vol. 33 (Princeton University Press, Princeton, 1954), pp. 167–190
130. Y. LeJan, O. Raimond, Integration of Brownian vector fields. Ann. Probab. **30**(2), 826–873 (2002)
131. P. Lévy, Sur le mouvement brownien dépendant de plusieurs paramètres. C. R. Acad. Sci. Paris **220**, 420–422 (1945)
132. P. Lévy, *Processus Stochastiques et Mouvement Brownien. Suivi d'une note de M. Loève* (Gauthier-Villars, Paris, 1948)
133. J.-L. Lions, E. Magenes, *Non-homogeneous Boundary Value Problems and Applications. Vol. I* (Springer, New York, 1972)
134. P.-L. Lions, P.E. Souganidis, Fully nonlinear stochastic partial differential equations. C. R. Acad. Sci. Paris Ser. I **326**, 1085–1092 (1998)
135. P.-L. Lions, P.E. Souganidis, Fully nonlinear stochastic partial differential equations: non-smooth equations and applications. C. R. Acad. Sci. Paris Ser. I **327**, 735–741 (1998)
136. P.-L. Lions, P.E. Souganidis, Fully nonlinear stochastic partial differential equations with semilinear stochastic dependence. C. R. Acad. Sci. Paris Ser. I **331**, 617–624 (2000)
137. P.-L. Lions, P.E. Souganidis, Uniqueness of weak solutions of fully nonlinear stochastic partial differential equations. C. R. Acad. Sci. Paris Sér. I Math. **331**(10), 783–790 (2000)
138. R.Sh. Liptser, A.N. Shiryaev, *Theory of Martingales*. Mathematics and Its Applications (Soviet Series), vol. 49 (Kluwer Academic, Dordrecht, 1989)
139. R.Sh. Liptser, A.N. Shiryaev, *Statistics of Random Processes, I: General Theory*, 2nd edn. Applications of Mathematics, vol. 5 (Springer, New York, 2001)
140. R.Sh. Liptser, A.N. Shiryaev, *Statistics of Random Processes, II: Applications*, 2nd edn. Applications of Mathematics, vol. 6 (Springer, New York, 2001)
141. K. Liu, *Stability of Infinite Dimensional Stochastic Differential Equations with Applications*. Chapman & Hall/CRC Monographs and Surveys in Pure and Applied Mathematics, vol. 135 (Chapman & Hall/CRC, Boca Raton, 2006)
142. W. Liu, M. Röckner, *Stochastic Partial Differential Equations: An Introduction*. Universitext (Springer, New York, 2015)

143. M. Loève, Sur les fonctions aléatoires stationnaires de second ordre. Revue Sci. **83**, 297–303 (1945)
144. M. Loève, Quelques propriétés des fonctions aléatoires de second ordre. C. R. Acad. Sci. Paris **222**, 469–470 (1946)
145. W. Loges, Girsanov's theorem in Hilbert space and an application to statistics of Hilbert space valued stochastic differential equations. Stoch. Proc. Appl. **17**, 243–263 (1984)
146. S.V. Lototsky, A random change of variables and applications to the stochastic porous medium equation with multiplicative time noise. Commun. Stoch. Anal. **1**(3), 343–355 (2007)
147. S.V. Lototsky, B.L. Rozovskii, *Recursive Multiple Wiener Integral Expansion for Nonlinear Filtering of Diffusion Processes*, ed. by J.A. Goldstein, N.E. Gretsky, J.J. Uhl. Stochastic Processes and Functional Analysis. Lecture Notes in Pure and Applied Mathematics, vol. 186 (Marcel Dekker, New York, 1997), pp. 199–208
148. S.V. Lototsky, B.L. Rozovskii, Recursive nonlinear filter for a continuous - discrete time model: separation of parameters and observations. IEEE Trans. Aut. Contr. **43**(8), 1154–1158 (1998)
149. S.V. Lototsky, B.L. Rozovskii, Spectral asymptotics of some functionals arising in statistical inference for SPDEs. Stoch. Process. Appl. **79**(1), 69–94 (1999)
150. S.V. Lototsky, B.L. Rozovskii, Passive scalar equation in a turbulent incompressible Gaussian velocity field. Rus. Math. Surv. **59**(2), 297–312 (2004)
151. S.V. Lototsky, B.L. Rozovskii, *Stochastic Differential Equations: A Wiener Chaos Approach*, ed. by Yu. Kabanov, R. Liptser, J. Stoyanov. From Stochastic Calculus to Mathematical Finance: The Shiryaev Festschrift (Springer, New York, 2006), pp. 433–507
152. S.V. Lototsky, B.L. Rozovskii, Wiener chaos solutions of linear stochastic evolution equations. Ann. Probab. **34**(2), 638–662 (2006)
153. S.V. Lototsky, R. Mikulevicius, B.L. Rozovskii, Nonlinear filtering revisited: a spectral approach. SIAM J. Contr. Optim. **35**(2), 435–461 (1997)
154. M. Marcus, V.J. Mizel, Stochastic hyperbolic systems and the wave equation. Stoch. Stoch. Rep. **36**(3–4), 225–244 (1991)
155. T. Martínez, M. Sanz-Solé, A lattice scheme for stochastic partial differential equations of elliptic type in dimension $d \geq 4$. Appl. Math. Optim. **54**(3), 343–368 (2006)
156. B. Maslowski, D. Nualart, Evolution equations driven by a fractional Brownian motion. J. Funct. Anal. **202**(1), 277–305 (2003)
157. H.P. McKean, Brownian motion with a several-dimensional time. Teor. Verojatnost. i Primenen. **8**, 357–378 (1963)
158. M. Métivier, *Stochastic Partial Differential Equations in Infinite-Dimensional Spaces*. Scuola Normale Superiore di Pisa. Quaderni. [Publications of the Scuola Normale Superiore of Pisa] (Scuola Normale Superiore, Pisa, 1988)
159. M. Métivier, J. Pellaumail, *Stochastic Integration* (Academic, New York, 1980)
160. P.-A. Meyer, *Quantum Probability for Probabilists*. Lecture Notes in Mathematics, vol. 1538 (Springer, Berlin, 1993)
161. R. Mikulevicius, On the Cauchy problem for parabolic SPDEs in Hölder classes. Ann. Probab. **28**(1), 74–103 (2000)
162. R. Mikulevicius, B.L. Rozovskii, Linear parabolic stochastic PDE's and Wiener chaos. SIAM J. Math. Anal. **29**(2), 452–480 (1998)
163. R. Mikulevicius, B.L. Rozovskii, Stochastic Navier-Stokes equations for turbulent flows. SIAM J. Math. Anal. **35**(5), 1250–1310 (2004) (electronic)
164. R. Mikulevicius, B.L. Rozovskii, Stochastic Navier-Stokes equations for turbulent flows. SIAM J. Math. Anal. **35**(5), 1250–1310 (2004)
165. R. Mikulevicius, B.L. Rozovskii, Global L_2-solutions of stochastic Navier-Stokes equations. Ann. Probab. **33**(1), 137–176 (2005)
166. R. Mikulevicius, B.L. Rozovskii, On unbiased stochastic Navier-Stokes equations. Probab. Theory Relat. Fields **154**(3–4), 787–834 (2012)
167. A. Millet, M. Sanz-Solé, The support of the solution to a hyperbolic SPDE. Probab. Theory Relat. Fields **98**(3), 361–387 (1994)

168. A. Millet, M. Sanz-Solé, Points of positive density for the solution to a hyperbolic SPDE. Potential Anal. **7**(3), 623–659 (1997)
169. A.S. Monin, A.M. Yaglom, *Statistical Fluid Mechanics: Mechanics of Turbulence*, vol. 1 (MIT Press, Cambridge, 1971)
170. C. Mueller, R. Tribe, Stochastic P.D.E.'s arising from the long range contact and long range voter processes. Probab. Theory Relat. Fields **102**(4), 519–545 (1995)
171. G.J. Murphy, *C^*-Algebras and Operator Theory* (Academic, New York, 1990)
172. E. Nelson, The free Markoff field. J. Funct. Anal. **12**(2), 211–227 (1973)
173. I. Nourdin, G. Peccati, A. Réveillac, Multivariate normal approximation using Stein's method and Malliavin calculus. Ann. Inst. Henri Poincaré Probab. Stat. **46**(1), 45–58 (2010)
174. D. Nualart, Propriedad de markov para functiones aleatorias Gaussianas. Cuadern. Estadistica Mat. Univ. Granada Ser. A Prob. **5**, 30–43 (1980)
175. D. Nualart, *The Malliavin Calculus and Related Topics*, 2nd edn. (Springer, New York, 2006)
176. D. Nualart, *Malliavin Calculus and Its Applications*. CBMS Regional Conference Series in Mathematics, vol. 110 (American Mathematical Society, Providence, 2009)
177. D. Nualart, B.L. Rozovskii, Weighted stochastic Sobolev spaces and bilinear SPDE's driven by space-time white noise. J. Funct. Anal. **149**(1), 200–225 (1997)
178. D. Nualart, S. Tindel, Quasilinear stochastic elliptic equations with reflection. Stoch. Process. Appl. **57**(1), 73–82 (1995)
179. E. Nualart, F. Viens, The fractional stochastic heat equation on the circle: time regularity and potential theory. Stoch. Process. Appl. **119**(5), 1505–1540 (2009)
180. D. Ocone, Multiple integral expansions for nonlinear filtering. Stochastics **10**(1), 1–30 (1983)
181. B.K. Øksendal, *Stochastic Differential Equations: An Introduction with Applications*, 5th edn. (Springer, New York, 1998)
182. S. Omatu, J.H. Seinfeld, *Distributed Parameter Systems: Theory and Applications* (Clarendon Press, Oxford, 1989)
183. M. Ondreját, Existence of global martingale solutions to stochastic hyperbolic equations driven by a spatially homogeneous Wiener process. Stoch. Dyn. **6**(1), 23–52 (2006)
184. S. Peszat, J. Zabczyk, Stochastic evolution equations with a spatially homogeneous Wiener process. Stoch. Process. Appl. **72**(2), 187–204 (1997)
185. S. Peszat, J. Zabczyk, *Stochastic Partial Differential Equations with Lévy Noise*. Encyclopedia of Mathematics and Its Applications, vol. 113 (Cambridge University Press, Cambridge, 2007)
186. L. Piterbarg, B. Rozovskii, Maximum likelihood estimators in the equations of physical oceanography, in *Stochastic Modelling in Physical Oceanography*, ed. by R.J. Adler, P. Müller, B.L. Rozovskii. Progress in Probability, vol. 39 (Birkhäuser Boston, Boston, 1996), pp. 397–421
187. L. Piterbarg, B. Rozovskii, On asymptotic problems of parameter estimation in stochastic PDE's: discrete time sampling. Math. Methods Stat. **6**(2), 200–223 (1997)
188. L.D. Pitt, A Markov property for Gaussian processes with a multidimensional parameter. Arch. Rational Mech. Anal. **43**, 367–391 (1971)
189. J. Potthoff, G. Våge, H. Watanabe, Generalized solutions of linear parabolic stochastic partial differential equations. Appl. Math. Optim. **38**, 95–107 (1998)
190. B.L.S. Prakasa Rao, Nonparametric inference for a class of stochastic partial differential equations. II. Stat. Inference Stoch. Process. **4**(1), 41–52 (2001)
191. C. Prévôt, M. Röckner, *A Concise Course on Stochastic Partial Differential Equations*. Lecture Notes in Mathematics, vol. 1905 (Springer, Berlin, 2007)
192. P. Protter, *Stochastic Integration and Differential Equations*, 2nd edn. (Springer, New York, 2004)
193. J. Rauch, *Partial Differential Equations*. Graduate Texts in Mathematics, vol. 128 (Springer, New York, 1991)
194. M. Reimers, One-dimensional stochastic partial differential equations and the branching measure diffusion. Probab. Theory Relat. Fields **81**(3), 319–340 (1989)
195. C.P. Robert, G. Casella, *Monte Carlo Statistical Methods* (Springer, New York, 2004)

196. C. Rovira, M. Sanz-Solé, A nonlinear hyperbolic SPDE: approximations and support, in *Stochastic Partial Differential Equations* (London Mathematical Society, Edinburgh, 1994). Lecture Note Series, vol. 216 (Cambridge University Press, Cambridge, 1995), pp. 241–261
197. C. Rovira, M. Sanz-Solé, The law of the solution to a nonlinear hyperbolic SPDE. J. Theor. Probab. **9**(4), 863–901 (1996)
198. Yu.A. Rozanov, *Random Fields and Stochastic Partial Differential Equations*. Mathematics and Its Applications, vol. 438 (Kluwer Academic, Dordrecht, 1998)
199. B.L. Rozovskii, *Stochastic Evolution Systems*. Mathematics and Its Applications (Soviet Series), vol. 35 (Kluwer Academic, Dordrecht, 1990)
200. F. Russo, Étude de la propriété de Markov étroite en relation avec les processus planaires à accroissements indépendants, in *Seminar on Probability, XVIII*. Lecture Notes in Mathematics, vol. 1059 (Springer, Berlin, 1984), pp. 353–378
201. Yu. Safarov, D. Vassiliev, *The Asymptotic Distribution of Eigenvalues of Partial Differential Operators*. Translations of Mathematical Monographs, vol. 155 (American Mathematical Society, Providence, 1997)
202. S. Sandow, S. Trimper, Gaussian white noise as a many-particle process: the Kardar-Parisi-Zhang equation. J. Phys. A **26**(13), 3079–3084 (1993)
203. P. Santa-Clara, D. Sornette, The dynamics of the forward interest rate curve with stochastic string shocks. Rev. Financ. Stud. **14**(1), 149–185 (2001)
204. M. Sanz-Solé, *A Course on Malliavin Calculus with Applications to Stochastic Partial Differential Equations*. Fundamental Sciences (EPFL Press, Lausanne, 2005)
205. M. Sanz-Solé, I. Torrecilla, A fractional Poisson equation: existence, regularity and approximations of the solution. Stoch. Dyn. **9**(4), 519–548 (2009)
206. M. Sanz-Solé, I. Torrecilla-Tarantino, Probability density for a hyperbolic SPDE with time dependent coefficients. ESAIM Probab. Stat. **11**, 365–380 (2007) (electronic)
207. A.N. Shiryaev, *Probability*, 2nd edn. Graduate Texts in Mathematics, vol. 95 (Springer, New York, 1996)
208. M.A. Shubin, *Pseudo-Differential Operators and Spectral Theory*, 2nd edn. (Springer, New York, 2001)
209. B. Simon, *The* $P(\phi)_2$ *Euclidean (Quantum) Field Theory*. Princeton Series in Physics (Princeton University Press, Princeton, 1974)
210. W. Stannat, Stochastic partial differential equations: Kolmogorov operators and invariant measures. Jahresber. Dtsch. Math.-Ver. **113**(2), 81–109 (2011)
211. R.L. Stratonovich, A new representation for stochastic integrals and equations. Vestn. Moskov. Univer., Ser. Mat. Mekhan. **1**, 3–12 (1964)
212. W.A. Strauss, *Partial Differential Equations: An Introduction*, 2nd edn. (Wiley, Chichester, 2008)
213. D.W. Stroock, S.R.S. Varadhan, On the support of diffusion processes with applications to the strong maximum principle, in *Proceedings of the Sixth Berkeley Symposium on Mathematical Statistics and Probability* (University of California Press, Berkeley, CA, 1970/1971), Vol. III: Probability Theory (University of California Press, Berkeley, CA, 1972), pp. 333–359
214. D.W. Stroock, S.R.S. Varadhan, *Multidimensional Diffusion Processes* (Springer, Berlin, 1979)
215. G. Szegö, *Orthogonal Polynomials*. Colloquium Publications, vol. 23 (American Mathematical Society, Providence, 2003)
216. M. Tehranchi, A note on invariant measures for HJM models. Finance Stochast. **9**(3), 389–398 (2005)
217. S. Tindel, C.A. Tudor, F. Viens, Stochastic evolution equations with fractional Brownian motion. Probab. Theory Relat. Fields **127**(2), 186–204 (2003)
218. H. Triebel, *Interpolation Theory, Function Spaces, Differential Operators*, 2nd edn. (Johann Ambrosius Barth, Heidelberg, 1995)
219. J.L. Vázquez, *The Porous Medium Equation*. Oxford Mathematical Monographs (The Clarendon Press/Oxford University Press, Oxford, 2007)

220. A.D. Ventzel, On equations of the theory of conditional Markov processes. Theory Probab. Appl. **10**(2), 357–361 (1965)
221. M.I. Vishik, A.V. Fursikov, *Mathematical Problems of Statistical Hydromechanics* (Kluwer Academic, Dordrecht, 1979)
222. J.B. Walsh, A stochastic model of neural response. Adv. Appl. Probab. **13**(2), 231–281 (1981)
223. J.B. Walsh, An introduction to stochastic partial differential equations, in *'Ecole d'été de probabilités de Saint-Flour, XIV—1984*. Lecture Notes in Mathematics, vol. 1180 (Springer, Berlin, 1986), pp. 265–439
224. J. Weidmann, *Linear Operators in Hilbert Spaces*. Graduate Texts in Mathematics, vol. 68 (Springer, New York, 1980)
225. G.C. Wick, The evaluation of the collision matrix. Phys. Rev. (2) **80**, 268–272 (1950)
226. D.V. Widder, *The Laplace Transform*. Princeton Mathematical Series, vol. 6 (Princeton University Press, Princeton, 1941)
227. E. Wong, Homogeneous Gauss-Markov random fields. Ann. Math. Stat. **40**, 1625–1634 (1969)
228. E. Wong, Explicit solutions to a class of nonlinear filtering problems. Stochastics **16**(5), 311–321 (1981)
229. E. Wong, M. Zakai, Riemann-Stieltjes approximation of stochastic integrals. Z. Wahr. verw. Geb. **120**, 87–97 (1969)
230. J. Xiong, *An Introduction to Stochastic Filtering Theory*. Oxford Graduate Texts in Mathematics, vol. 18 (Oxford University Press, Oxford, 2008)
231. K. Yosida, *Functional Analysis*, 6th edn. (Springer, Berlin, 1980)
232. E. Zeidler, *Nonlinear Functional Analysis and Its Applications, I: Fixed-Point Theorems* (Springer, New York, 1986)
233. E. Zeidler, *Nonlinear Functional Analysis and Its Applications, II/A: Linear Monotone Operators* (Springer, New York, 1990)
234. T.S. Zhang, Characterizations of the white noise test functionals and Hida distributions. Stoch. Stoch. Rep. **41**(1–2), 71–87 (1992)
235. S. Zhang, Singularity of two diffusions on C_∞. Stat. Probab. Lett. **19**(2), 143–145 (1994)

List of Notations

S.V. Lototsky, B.L. Rozovsky, *Stochastic Partial Differential Equations*, Universitext, DOI 10.1007/978-3-319-58647-2

Index

S.V. Lototsky, B.L. Rozovsky, *Stochastic Partial Differential Equations*, Universitext, DOI 10.1007/978-3-319-58647-2